Basic chemistry of life

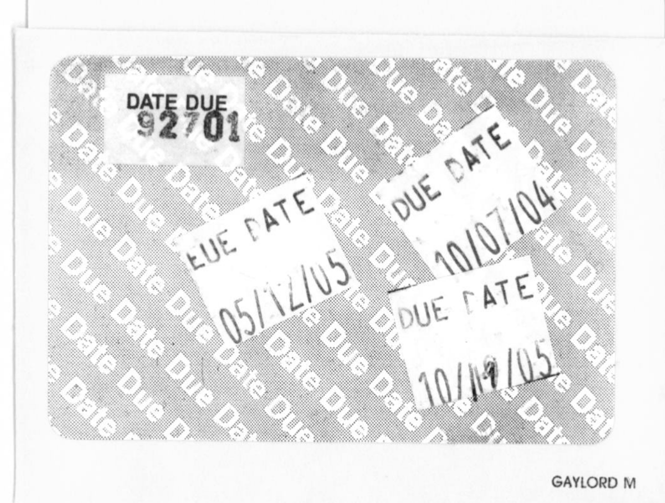

BASIC CHEMISTRY OF LIFE

MILTON TOPOREK, Ph.D.

Professor of Biochemistry,
The Jefferson Medical College of Thomas Jefferson University,
Philadelphia, Pennsylvania

with 155 illustrations

The C. V. Mosby Company

ST. LOUIS • TORONTO • LONDON 1981

Printed in the United States of America

The C. V. Mosby Company
11830 Westline Industrial Drive, St. Louis, Missouri 63141

Library of Congress Cataloging in Publication Data

Toporek, Milton, 1920-
Basic chemistry of life.

Includes bibliographies and index.
1. Biological chemistry. 2. Allied health personnel. I. Title. [DNLM: 1. Biochemistry. QU4 T675b]
QP514.2.T66 1980 612'.015 80-11980
ISBN 0-8016-5002-X

GW/M/M 9 8 7 6 5 4 3 2 02/D/225

To

Our Parents

Preface

The demands of society for more and better health care based on the best and latest information available continue to escalate. An important aspect in attempts to meet these demands is the ongoing effort in biochemical research. As a natural consequence of this continuous growth of knowledge, it becomes necessary, at appropriate intervals, to include the newly accepted information in a biochemistry textbook for students in the medical and allied health care professions: therefore, this edition of *Basic Chemistry of Life*.

Rapid and significant additions have been made to information about the prostaglandins and their effects on the atherosclerotic and clotting mechanisms and cellular responses to hormonal influences. These additions have been included in the chapters on the chemistry of lipids, lipid metabolism, blood, and hormones.

Because of the continuing increase in interest in nutrition, the chapter on nutrition has been expanded to emphasize the need for nutritional intervention and the formulation of intervention plans for various patients. When the close interdependency of information about nutrition and metabolic biochemistry is recognized, much of the misinformation about nutrition will hopefully be corrected.

Details about the enzymatic processes involved in the biosynthesis of fatty acids and of hemoglobin have been included in the chapters on lipid metabolism and blood. This information should provide students with a more complete knowledge base on which to build an understanding of the reasons for carrying out related procedures in the health care environment.

Scattered throughout the text are many other minor changes. As always, responsibility for the choice of what to include and what not to include is my own. Time and reasoned criticism will determine the validity of these choices.

Milton Toporek

Preface to the first edition

Progress in all phases of the life, or biomedical, sciences is being made at a relatively explosive rate. Chemistry has been, and continues to be, in the forefront of such efforts. This is so because, even though a living human being may be worth only a dollar or two as a collection of chemicals, many secrets of health are locked up in the intricate organization, functioning, and control mechanisms which nature has achieved with these chemicals.

Many of these secrets have yielded to researchers in recent years, leading to a better understanding of certain phases of the life process. At the same time many secrets remain to be solved. As a result, a continuously accelerated demand has developed for better medical and health services based on this new understanding. This in turn is leading to increased requirements for personnel properly educated in the biomedical sciences. It is difficult to see how anyone can function adequately in this field without a proper understanding of basic chemical principles. It is the purpose of this textbook to provide such a base in chemistry on which the student may build an understanding of the complex processes which go to make up the phenomenon of life as we know it, particularly in the human being. It is believed that this can be accomplished without forcing the student to become a chemist or biochemist, while exploring basic concepts to the fullest extent possible for nonspecialists in this field.

Major emphasis is placed on the dynamic nature of biochemistry and the interrelationships of the various metabolic pathways. At the same time, points along these pathways at which derangements may occur and the manner in which they occur are discussed in such a way as to stress the biochemistry involved in the underlying clinical situations. At the completion of the biochemistry chapters of this textbook, the student should see the living human being as the result of a miraculous coordination of thousands of chemical events occurring simultaneously and in sequence, with all sorts of provisions in readiness for any possible emergency which may arise to threaten the status quo. Included in this section is the almost unbelievable story of the molecule of life, DNA (deoxyribonucleic acid), and its partner, RNA (ribonucleic acid), and the roles they play in protein synthesis and reproduction.

Without energy, life as we know it could not exist. It is necessary for a proper understanding of the functioning of a living organism to learn how chemicals trap energy and how the energy, or at least a part of it, is recovered from the chemicals and converted to the forms required by the living organism. In addition, some knowledge of basic chemical reactions, structure, and physical properties of substances is also necessary. These topics are studied in the chapters on inorganic chemistry.

Many of the key compounds associated with life are from the world of organic chemistry. Some knowledge of the structure, properties, and reactions of these fascinating and extraordinary compounds also is necessary for a proper understanding of the functioning of a living organism. Care has been taken to include in the chapters on organic chemistry only as much as necessary to provide an adequate base for the subsequent discussions in the biochemistry chapters, omitting unnecessary ramblings in the realm of industrial and pharmaceutical applications, where organic chemicals undoubtedly play a very important role. Structures are used mainly to give the student some feeling for the significance of such considerations in the actions and functions of the various biochemical compounds.

Some historical material is included where appropriate to give the student an indication of the manner in which the superstructure of science is built up on the efforts of many investigators as they lay down a brick at a time on the established base. Occasionally, an entire structure goes up almost instantaneously, also on a preexisting base. Conceptually, the thread of development is more important than the specific dates, or historical detail, which are included for reference purposes.

Among the important problems encountered in courses for which this text is designed are those of disparity in scientific background of the students and wide variations in the number of hours assigned to such courses. Therefore, this text starts at a relatively elementary level in order to supply a good base for all students, and moves rapidly to a level in biochemistry which should provide a challenge to the best students. As an aid to all students, especially those in the shorter courses, an outline is provided with each chapter. This outline has been used successfully for some years by the author's students, as indicated by objective tests and unsolicited comments of the students. Their reactions have affirmed the author's opinion that such an outline would provide a more universally useful teaching tool than the conventional study questions at the end of each chapter. Such questions, therefore, have been omitted from this text.

Summary flow charts are used wherever possible to help the student appreciate the overall picture of the situation in the presence of the wealth of specific detail often required to describe a particular subject area. Although the trend in the health sciences is, hopefully, in the direction of upgrading courses in content and time, in the shorter courses these flow charts may contain the largest portion of the subject matter which can be successfully taught in the time allotted.

The list of suggested additional readings to be found in the Appendix is meant to serve merely as an introduction to the vast literature in the biomedical and related sciences for those classes or individual students with the time or inclination to track down an interesting subject. The list of references is obviously not exhaustive and is also not completely representative. However, an exploration starting with any single reference will usually illustrate the well-known cascade effect encountered in such literature, i.e., each reference has its own list of references, which in turn lead to many other lists of references.

A work of this type inevitably involves cooperation from many sources. It is a pleasure to acknowledge the help of all whose efforts and encouragement were of aid in completion of this venture. In particular, the many discussions with my colleague, Dr. Bernard Schepartz, Professor of Biochemistry at The Jefferson Medical College, have been of great value. Thanks are due to many individual scientists and commercial and nonprofit organizations for a large number of the illustrations used in the text, a practical indication of the interdependence of various research groups. An unexpected source of encouragement were my daughter and two sons, who followed the progress in writing with great interest in spite of their loss of my services for a significant period of time. The tedious task of transcribing the writing into a near-perfect manuscript was skillfully accomplished by my wife. Another important ingredient was supplied by the publisher in the form of the proper mixture of prodding, encouragement, and cooperation. To all, my sincere appreciation.

Contents

UNIT FOUR
BIOCHEMISTRY

UNIT ONE

INTRODUCTION

Overview

1

One of the most important tasks humans set for themselves was the seeking for any and all means of prolonging life and making things more comfortable during that lifetime. The search for the knowledge necessary to accomplish this task probably dates back many thousands of years to the time humans first found it possible to influence their environment. It is the purpose of this book to emphasize the central role modern biochemistry fills in this ever-expanding undertaking of determining what is life, defining the "normal" state of health, and discovering means of coping with situations of all kinds that pose a threat to this "normal" state.

BIOCHEMISTRY AND THE HEALTH SCIENCES

In recent years, the field of medicine and its allied health sciences have benefited greatly from the many advances being made as biochemists and other researchers continue attempts to explain the workings of the body on a molecular basis. These activities have led to the establishment and verification of the concept of *molecular diseases,* which, in turn, has led to more rational approaches to the investigation and treatment of such diseases.

Sickle cell anemia, a disease still very much under investigation, was the first disease proved to result from the presence of an abnormal molecule (hemoglobin S) replacing the normal type (hemoglobin A). This achievement began to unfold in 1945 when Dr. Linus Pauling, a two-time Nobel prize winner, proposed that red cells in an individual with sickle cell anemia assumed sickle shapes (from which the disease obtained its name) because a change in the hemoglobin molecule changed the physical characteristics of the cells under certain conditions so that they could not retain their normal, round shape. Within a few years, Dr. Pauling and his students showed that there was indeed a hemoglobin S that was distinctly different from hemoglobin A and that they differed from each other in only two out of a total of 574 amino acids in the hemoglobin molecule. As small as such a difference may appear to be on a quantitative basis, the consequences can be very serious or fatal. Further details about sickle cell anemia will be discussed later (p. 381).

Other important health problems related to biochemical molecules involve the production of abnormally high or low amounts of enzymes, the body's chemical catalysts. Strategic biochemical reactions may thus be forced to produce too great or too small amounts of vital substances required for the body to maintain its normal state of health. Since, as will be shown later, the information required for the synthesis of enzymes comes from genetic material, deoxyribonucleic acids (DNA) and ribonucleic acids (RNA), diseases or conditions resulting from deranged enzyme production are known as *inborn errors of metabolism.*

Another series of difficult health problems to contend with arises from derangements of a part or parts of the delicate mechanism that controls the net flow of the hundreds of biochemical reactions that are occurring in such a way as to respond properly to the ever-changing requirements of the body. As an example, changes in the acidity or alkalinity of body tissues must be controlled to a relatively small range. This involves many hormones and the processes they affect. A breakdown in any part of the control process can become a very serious threat to the health status of the individual.

It should be abundantly clear, even from this very brief listing of some of the major classes of difficulties to which the body may be subjected, that pinpointing the biochemical problem involved in a particular situation is the starting point for devising and carrying out a rational approach to correcting the problem. It therefore becomes imperative for anyone contemplating a career in the health sciences to achieve a basic understanding of the chemistry of life. For these and others, a corollary dividend would be the intellectual satisfaction of being able to better comprehend the magnitude of the miracle known as life.

THE STUDY OF CHEMISTRY

In the early history of chemistry, inorganic and organic chemistry were divided on the basis of the presence or absence of a relationship to life. It was believed that organic chemicals could be produced only by a living organism. The first indication that life was not absolutely necessary for the production of an organic chemical came in 1828 when a German chemist, Wöhler, prepared urea, a common product of animal metabolism, from two inorganic substances: ammonia and cyanic acid. Since then, many other organic chemicals have been produced in laboratories and factories until now several million such substances are known, with the element carbon as their common feature.

The ingenuity and ability of humans to produce organic compounds for specific purposes almost at will are well known and appreciated. Examples include synthetic fibers such as nylon for clothing, the wonder drugs such as the sulfonamides for curing diseases in the manner of Ehrlich's "magic bullet,"* and many items being used in space vehicles and in support of the systems that have sent these space vehicles to the moon and the planets. Yet many organic compounds found in nature have not been synthesized because of their exceedingly complex structure.

It is quite clear, then, that a mysterious "vital force" is not required for the preparation of organic chemicals. Why, therefore, are the subjects of inorganic and organic chemistry still studied separately? The reason is mainly one of convenience. Since carbon is only one of the 105 elements now recognized, general principles involving all of the elements are studied under inorganic chemistry. These principles are then applied to the study of organic chemistry, or compounds of carbon. Also included in the study of organic chemistry are the properties peculiar to carbon that make it possible to synthesize more compounds of carbon than has been possible to produce from the other 104 elements combined.

THE STUDY OF BIOCHEMISTRY

The importance of the study of inorganic and organic chemistry to the understanding of biochemistry should be quite obvious. From the chemical viewpoint, the living organism, with particular emphasis on the human being, is simply an organized, operating collection of organic and inorganic chemicals (Fig. 1-1).

It is the ultimate purpose of biochemistry (here broadly defined to include biologic chemistry, physiologic chemistry, and molecular biology) to define the processes occurring in the living organism on the basis of the chemical principles governing the activity of the various compounds found in the organism.

Among the most important and interesting processes to be studied are those concerned with the manner in which an organism manages to reproduce itself in the image of its parents. This involves the story of DNA, deoxyribonucleic acid, or the "molecule of life" as it sometimes called. Related to this process is the one that specifies and controls the synthesis of the important class of compounds known as proteins. Among the proteins are the enzymes, the organic catalysts that allow chemical reactions to occur in the body under much milder conditions than would be possible in a test tube or a factory. The problem of the relationship between cancer and viruses, nucleic acid–containing particles, has been receiving increasing attention from researchers. Continuing intense activity in these areas has brought an understanding, in broad outline, of many of these processes and has begun to fill in the specific details of some. Enough is now known

*Paul Ehrlich, a German chemist (1854-1915), believed in the possibility of curing diseases by destroying the invading enemy, infectious bacteria, by some chemical means without affecting the normal body tissues. Hence the concept of the "magic bullet." (de Kruif, P.: Microbe hunters, New York, 1926, Harcourt Brace Jovanovich, Inc.)

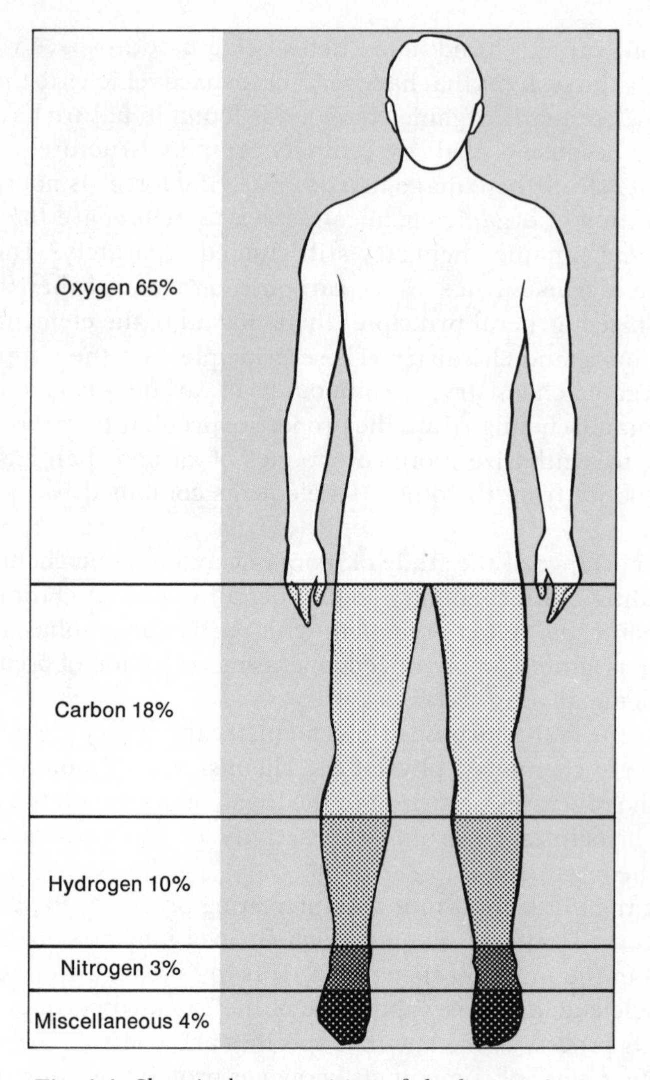

Fig. 1-1. Chemical composition of the human body.

to indicate the probability that many diseases and abnormal conditions are directly related to abnormalities found in DNA or in the processes controlled by DNA.

In fact, many studies are now under way investigating the possibilities of devising corrections for some of nature's errors and also of harnessing nature's own designs to produce large quantities of strategic substances, such as hormones, with much less effort than before or to make substances that have not yet been produced by humans. These subject areas are known popularly as genetic engineering and DNA recombinant procedures. Because of their exceedingly important social, ethical, and religious implications, the discussions of these areas of scientific innovation have been very intense among both scientists and nonscientists and will probably continue to be so in the future.

A very interesting recent example of how biochemistry progresses is found in the development of information about the prostaglandins, essential fatty acids, and aspirin. Not too many years ago it would have been very unusual to mention these substances together. Today it is known that an essential fatty acid is essential because it is used in the body to synthesize prostaglandins. One property of certain prostaglandins is that of promoting the clumping of platelets, leading to the formation of blood clots. In case of bleeding, this is a desirable property. In the case of individuals subject to heart attacks, this is a property to avoid. Recent reports have indicated that aspirin has the property of inhibiting the production of a substance, by way of prostaglandins, that causes the clumping of platelets. As a result, aspirin is now being investigated as a means of lessening the possibility of heart attacks.

An adequate background in biochemistry is required not only for an understanding of the workings of the human body, both in health and in disease states, but also for a complete realization of the potential benefits to be obtained from applications of this knowledge.

THE INFORMATION EXPLOSION

It has been estimated that the world literature in chemistry and the life sciences is doubling approximately every 8 years, and this rate has not yet leveled off. This information explosion can be explained on the bases that according to recent estimates, 90% of the scientists known to have lived are living now and that people are better educated in and more aware of scientific matters, generating ever-increasing pressure for more research and achievements in the life sciences.

Such a situation means that anyone contemplating a career in the life sciences must be prepared to discard old or superseded ideas and learn new ones at an alarmingly rapid rate. This is quite a challenge, but it can be a very exciting one, particularly if the base on which the student is to build is a sound one.

In addressing his medical students, Joseph Lister, an English surgeon (1827-1912) and the originator of antiseptic surgery, said:

> You must always be students, learning and unlearning till your life's end, and if, gentlemen, you are not prepared to follow your profession in this spirit, I implore you to leave its ranks and betake yourself to some third-class trade.*

With the inclusion of other areas of the life sciences and the inclusion of women also pursuing such careers, Lister's quotation is just as appropriate today as it was many years ago.

---

*Leeson, J. R.: Lister as I knew him, London, 1927, Baillière, Tindall, and Cassell, p. 102.

2 Energy

Energy may be defined in very simple terms as the *ability to do work.* The tremendous importance of energy considerations for living organisms cannot be overemphasized and is the reason for starting this text with a discussion of basic information about energy. Thus in the unit on inorganic chemistry, the manner in which chemicals can capture and store energy as *potential energy* will be reviewed, as will the manner by which the energy in chemicals can be extracted and utilized, or withdrawn, in the form of *kinetic energy*.

As an analogy with finances, potential energy, i.e., energy that is stored at the moment but that can be utilized as required, is like "money in the bank." Just as people must be able to keep putting money in the bank in order to pay for their needs, the supply of potential energy must be replenished in order to continue functioning properly. For the human body replenishing the supply of potential energy is accomplished by eating food. The chemical energy found in the food molecules can later be used as required for heat, motion, or metabolic reactions that require the input of energy. Thus, energy can be thought of as the "currency of life." Most of the facts about these aspects of energy will be developed in the unit on biochemistry.

The concept of constant replenishment of sources of potential energy brings up a number of questions, foremost among which is the question of the ultimate renewable source of energy that makes life possible. The answer, of course, is the sun. Green plants, through the process of photosynthesis (discussed later), capture some of the sun's energy in the form of chemical energy.

As the text develops, attention will be repeatedly focused on the energy aspects of the subject under discussion. Ultimately, although few of us would settle for the idea that we are nothing more than energy-converting machines while we live, functionally the biochemical and physiologic facts are that in many aspects this is true. In other words, the driving force behind the many reactions that make up the life process is energy.

ENERGY FORMS

Life as we know it depends on many types of energy, such as mechanical, heat, light, chemical, electrical, sound, and now, nuclear energy. One of the most important properties of these forms of energy is their interconvertibility. The different forms of energy are all easily converted to heat but some of the other interconversions still cannot be done on a practical basis.

Because the different forms of energy are all convertible to heat, all forms of energy are usually measured as heat. In these measurements, the unit of heat energy is known as the *calorie* (cal), which is defined as the heat required to raise the temperature of 1 g of water 1 Celsius (or centigrade) degree. The amount of heat energy found in foods is usually expressed in terms of the large calorie (Cal, or kilocalorie, kcal), which is equal to 1000 small calories.

PHOTOSYNTHESIS AND THE ENERGY CYCLE

For us, the sun plays a central role in providing the energy required for life. Some of the light energy reaching the earth is converted by green plants into chemical energy in the form of sugars (e.g., $C_6H_{12}O_6$), which are carbohydrates, by the combination of carbon dioxide (CO_2) and water (H_2O) in the proper proportions and in the presence of the green plant pigment, chlorophyll:

$$6CO_2 + 6H_2O + \text{light energy} \xrightarrow[\text{photosynthesis}]{\text{chlorophyll}} C_6H_{12}O_6 + 6O_2$$

This process is known as *photosynthesis* (synthesis under the influence of light) and produces oxygen at the same time that it converts light energy to chemical energy. This equation is a simple summary of the complicated events that take place. Only a very small fraction of the available light energy is converted to chemical energy by this process.

Animals cannot produce chemical energy in the form of sugars in this manner and therefore must depend on eating plants that produce sugars or eating other animals that have eaten such plants. The energy trapped in the sugars by photosynthesis can then be released by a series of chemical reactions, or *oxidations*, in the animal body and converted to the form of energy required, such as heat, mechanical, or sound energy:

$$C_6H_{12}O_6 + 6O_2 \xrightarrow[\text{by body}]{\text{oxidation}} 6CO_2 + 6H_2O + \text{energy}$$

It should be noted that this process of oxidation of the sugar by the animal body is the *reverse* of the process of photosynthesis. It requires oxygen that was produced during photosynthesis and, in turn, supplies carbon dioxide that is required for photosynthesis, completing the energy cycle diagrammed in Fig. 2-1.

The strategic position of green plants in this energy cycle should be quite clear. These plants act as the intermediary between the almost unlimited energy of the sun and the animal world, which is completely dependent on the sun as its ultimate source of energy. The appropriate management and conservation of land for planting and farming purposes are not only desirable for esthetic and economic reasons but are also necessary for survival.

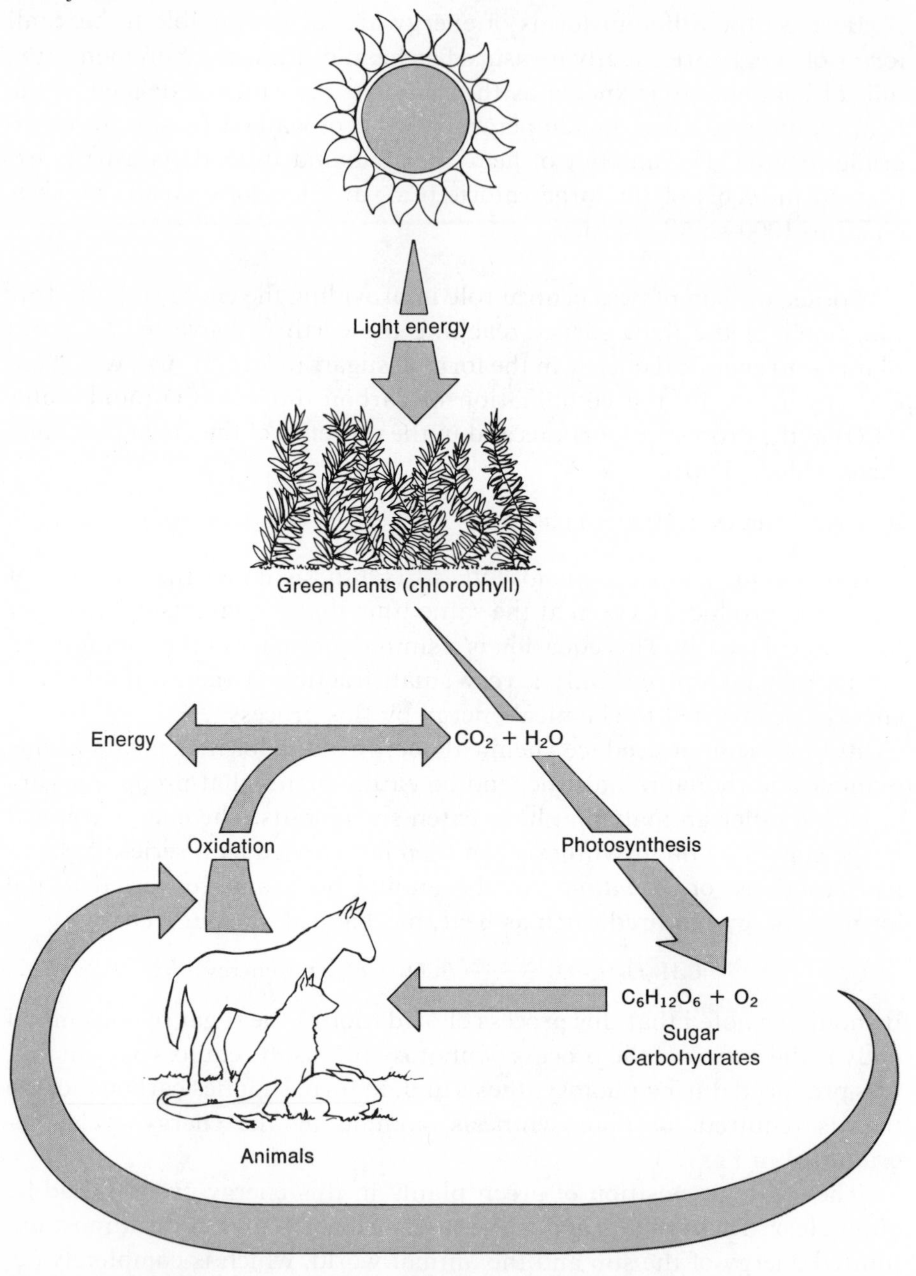

Fig. 2-1. The energy cycle that is the basis for life on earth. The sun is the ultimate source of energy input, with green plants acting as the intermediary.

LAW OF CONSERVATION OF ENERGY OR MASS

It has been found, on the basis of repeated observations and experiments, that *energy or mass can neither be created nor destroyed.* This phenomenon has long been recognized as the *law of conservation of energy or mass.* In the early 1900s Albert Einstein (1879-1955), a German-Swiss-American mathematician and physicist, added a new factor to consider in the relationship between energy and mass. He proposed that matter and energy are convertible into each other according to the equation:

$$E = mc^2$$

where E indicates energy in ergs (a unit of energy), m indicates mass in grams (a unit of weight), and c indicates the speed of light, which is 186,000 miles per second, or approximately 30 billion centimeters (a unit of length) per second.

To achieve some idea of what this means, a pound of matter (approximately 454 g) converted completely to energy would produce 11 billion kilowatt-hours of electricity, or enough to supply the need for electric power in the entire United States for several days. The actual numbers are not important in themselves, but the magnitude of the numbers is an indication of the tremendous amount of energy locked up in matter. The atomic and hydrogen bombs and nuclear power stations have supplied ample proof of Einstein's hypothesis. It therefore becomes necessary to qualify the law of conservation of mass or energy by stating that, although energy or mass can neither be created nor destroyed, *energy and mass are interconvertible.*

The practical result of the law of conservation of energy or mass is that an animal that requires energy for life must constantly replenish its supply of energy in order to function. This is quite similar to the need for repeatedly refilling the gasoline tank of an automobile in order to keep it functioning. To carry the analogy further, both the sugar for the animal body and the gasoline for the automobile are produced as the result of conversion of light energy from the sun into chemical energy on earth. Subsequently the energy in both cases is released by chemical oxidation reactions. In the following chapters, the properties of matter and the means by which it can trap energy will be described, as will the means by which this energy can be released for use in the living animal.

SUMMARY

DEFINITION

Energy is the ability to do work.

FORMS

The forms of energy are mechanical, heat, light, chemical, electrical, sound, and nuclear. All forms are convertible to heat, but the reverse is not always practical. All forms are measured as heat, the unit of measurement being the calorie (cal), which is defined as the heat required to raise the temperature of 1 g of water 1 Celsius (centigrade) degree. For referring to the energy in foods, the large calorie (Cal = 1000 small cal = 1 kcal) is used.

PHOTOSYNTHESIS Photosynthesis is the trapping of light energy by green plants (chlorophyll) in the form of chemical energy in sugars (carbohydrates):

$$6CO_2 + 6H_2O + \text{light energy} \xrightarrow[\text{photosynthesis}]{\text{chlorophyll}} C_6H_{12}O_6 + 6O_2$$

Animals cannot convert light energy to chemical energy and therefore depend on green plants as a source of energy.

OXIDATION Oxidation is the conversion, in the animal body, of the chemical energy in sugars into the required form of energy, such as mechanical or heat energy, by a series of chemical reactions that oxidize the sugar.

$$C_6H_{12}O_6 + 6O_2 \xrightarrow[\text{by body}]{\text{oxidation}} 6CO_2 + 6H_2O + \text{energy}$$

This is the reverse of photosynthesis.

LAW OF CONSERVATION OF ENERGY OR MASS Energy or mass can neither be created nor destroyed, but energy and mass are interconvertible.

REVIEW QUESTIONS

1. What is the ultimate source of energy for living organisms? Explain.
2. What is the strategic intermediary in the photosynthetic process? How is the human limited in this respect?
3. How is the energy captured in the photosynthetic process released in the body? How are these processes related to the law of conservation of energy or mass?
4. Why are ali forms of energy usually measured as heat?
5. Why are considerations of energy of importance for living things, including humans?

REFERENCES

Asimov. I.: Life and energy, Garden City, New York, 1962, Doubleday & Co., Inc.

Barber, J., editor: Primary processes of photosynthesis, New York, 1977, Elsevier International Projects Ltd.

Bassham, J. A.: Increasing crop production through more controlled photosynthesis, Science **197:**630, 1977.

Energy and power, Scientific American, San Francisco, 1971, W. H. Freeman and Co., Publishers.

Pauling, L.: General chemistry, ed. 3, San Francisco, 1970, W. H. Freeman and Co., Publishers.

Science **184:**245-389, 1974.

Science **199:**605-664, 1978.

Wilson, M.: Energy, New York, 1963, Time-Life Books.

UNIT TWO

INORGANIC CHEMISTRY

Units of measurement

3

A precise system of measurements of natural phenomena is an important requirement in science. Certain standards form the base of any system of measures, and the importance of these standards cannot be overrated. In the United States it is the function of the Institute for Basic Standards, one of four institutes of the National Bureau of Standards, to maintain the national standards.

At first the various national systems of measurements were not necessarily compatible, but in 1875 the United States and sixteen other nations created an International Bureau of Weights and Measures, which is situated at Sèvres, France, and provided for an International Conference of Weights and Measures to meet at appropriate intervals. This has led, in most instances, to a uniform international system of measurements. Something as basic as a standard of measurement might, at first thought, appear to be relatively unchangeable. However, as scientific investigations have become more and more sophisticated, the need for more accurate measurements has increased. As the means for more accurate measurements are devised, it may become necessary to redefine a standard in more precise terms or replace it with a new standard. The pressing need for more frequent reviews of such matters was recognized by the International Conference of Weights and Measures when it met in 1964 instead of the regularly scheduled meeting in 1966, an interval of 4 instead of the usual 6 years between meetings. This trend toward shorter intervals between meetings appears to be continuing. Thus, the process of developing and maintaining a system of standards is a constantly evolving one.

As a result of international cooperation, it is possible for scientists all over the world to make measurements with similar precision and to check and confirm one another's results with a great degree of confidence.

THE METRIC SYSTEM

Scientific measurements of length, mass, and volume are usually made on the basis of the metric system. The United States is still in the process of replacing the avoirdupois system of measurements in daily life with the metric system.

Fig. 3-1. United States National Bureau of Standards scientist adjusts a krypton 86 lamp in its liquid nitrogen bath (−345° F). The wavelength of the orange-red light coming from the lamp is used as the basis for the International Standard of length, the meter (m). (From Huntoon, R. D.: U.S. National Bureau of Standards, Science **150:**172, 1965. Copyright 1965 by the American Association for the Advancement of Science.)

Length

The unit of length is the *meter* (m), and the latest international standard of length is defined as 1,650,763.73 wavelengths, in vacuum, of the orange-red radiation from krypton 86 (Fig. 3-1). If a distance of 2,500 miles could be measured with the same accuracy as the standard wavelength of krypton 86, the true distance would be known to within less than ½ inch! The standard is then used to make rulers and all sorts of other length-measuring devices with an accuracy that can be no greater than that mentioned above and in most instances, much less.

As accurate as the krypton standard is (1 part in 10^8), investigations of methods for increasing the accuracy of length are continuing. If the development of an announced laser technique is successful, the accuracy of measurement of length may increase by another factor of 10^3.

Listed here are the more common multiples and subunits of the meter and some common equivalents.

Kilometer (km) = 1,000 meters (m)
Centimeter (cm) = 0.01 m
Millimeter (mm) = 0.001 m = 0.1 cm

1 inch = 2.54 cm
1 m ≅ 39.5 in. ≅ 1.1 yards

As an example of how a standard may be changed in order to provide greater accuracy of measurement, the history of the standard meter may be cited. At first the meter was defined as one ten-millionth of the north polar quadrant of the earth on the meridian through Paris. This was not a convenient standard, and the meter was next defined in terms of the length of a standard bar of platinum-iridium kept in the Archives of Paris. The latest redefinition in terms of the orange-red radiation from krypton 86 came in 1960.

Mass

The unit of mass is the *kilogram* (kg), and the international standard is the mass of a cylinder of platinum-iridium metal that is kept at the International Bureau of Weights and Measures in France. An accurate copy of this cylinder at the Institute for Basic Standards is the United States national standard of mass (Fig. 3-2).

Listed here are the more common subunits of the kilogram and some useful equivalents.

Gram (g) = 0.001 kilogram (kg)
Milligram (mg) = 0.001 g = 0.000001 kg

1 kg = 2.2 pounds (approximately)
1 ounce = 30 g (approximately)
1 g = 15 grains (gr)

The abbreviation for gram is g in the chemical literature and sometimes gm in the medical literature.

Volume

The unit of volume is the *liter* (l), and the international standard was recently redefined as the volume of a cubic decimeter. A decimeter (dm) is equal to 10 cm, therefore a cubic decimeter contains 1000 cubic centimeters (cc) (Fig. 3-3).

Fig. 3-2. The national standard of mass, Prototype Kilogram No. 20, is a platinum-iridium cylinder 39 mm in diameter and 39 mm in height. It is an accurate copy of the international standard kept at the International Bureau of Weights and Measures at Sèvres, France, and was furnished to the United States by the international bureau in accordance with the metric treaty of 1875. Recomparison with the international standard in 1948 showed that the United States standard has remained constant within one part in 50 million over a period of 60 years. (From Huntoon, R. D.: U.S. National Bureau of Standards, Science **150:**170, 1965. Copyright 1965 by the American Association for the Advancement of Science.)

Listed here are the more common subunits of the liter and some equivalents.

Cubic centimeter (cc) = 0.001 liter (l)
Milliliter (ml) = 0.001 l
Therefore, 1 cc = 1 ml

1 quart = 950 ml
1 fluid ounce = 30 ml
1 teaspoon = 4 to 5 ml
1 tablespoon = 15 ml = ½ fluid ounce

Temperature

The unit of temperature is the *degree,* in some cases indicated by a small superscript zero. The standards for temperature measurements in the more common range are usually related to the freezing point and boiling point of water. The two temperature scales used most frequently are the Fahrenheit (F) and the centigrade or Celsius (C) scales. On the Fahrenheit scale, the freezing point of water is 32° F, normal body temperature is 98.6° F, and the boiling point of water is 212° F. The comparable points on the centigrade scale are 0° C, 37° C, and 100° C (Fig. 3-4).

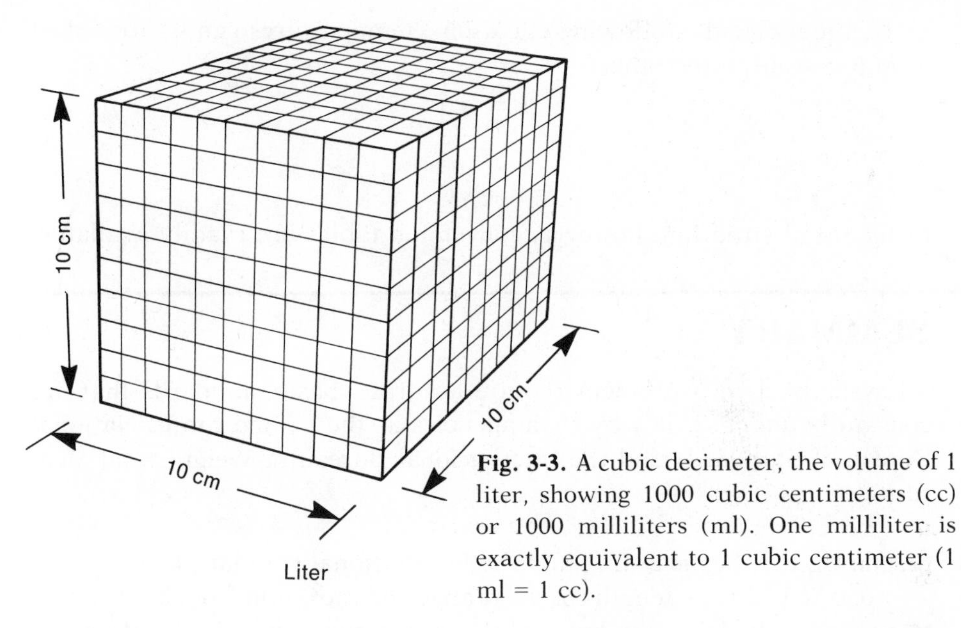

Fig. 3-3. A cubic decimeter, the volume of 1 liter, showing 1000 cubic centimeters (cc) or 1000 milliliters (ml). One milliliter is exactly equivalent to 1 cubic centimeter (1 ml = 1 cc).

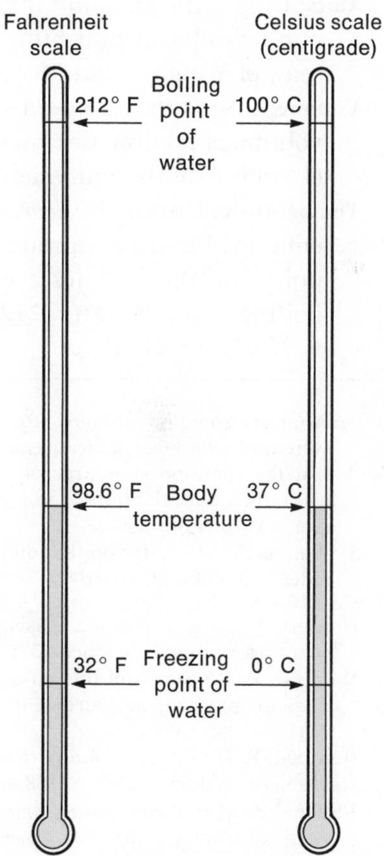

Fig. 3-4. A comparison of the Fahrenheit (F) and Celsius (C) or centigrade scales showing the corresponding values of some commonly used temperature points.

By the use of the following equations, temperatures can be converted from one scale to the other:

$$C = \frac{5}{9}(F - 32)$$

or

$$F = \frac{9}{5}C + 32$$

In the usual situation, however, conversion tables are readily available.

SUMMARY

SYSTEM OF MEASUREMENTS

All systems of measurement require accurate, basic standards that are constantly under review by such agencies as the United States National Bureau of Standards and the International Bureau of Weights and Measures.

METRIC SYSTEM

Length: Unit is the *meter* (m), and the international standard is defined as 1,650,763.73 wavelengths of the orange-red radiation from krypton 86.

Mass: Unit is the *kilogram* (kg), and the international standard is the mass of a cylinder of platinum-iridium metal kept at the International Bureau of Weights and Measures in France.

Volume: Unit is the *liter* (l), and the international standard is defined as the volume of a cubic decimeter. Under this definition, 1 cubic centimeter (1 cc) is exactly equivalent to 1 milliliter (1 ml).

Temperature: Unit is the *degree.* The two most common scales are the Fahrenheit (F) and centigrade or Celsius (C). The more common standard points on these scales are: freezing point of water, 32° F or 0° C, and boiling point of water, 212° F or 100° C. Also normal body temperature is 98.6° F or 37° C.

REVIEW QUESTIONS

1. What are some advantages of a uniform system of standards and measurements?
2. List the common standards of measurement, their abbreviations, and some commonly used subunits.
3. What is the key criterion that dictates the redefinition of a standard?
4. Which United States government agency is responsible for matters related to standards of measurements? Is there an international agency concerned with such matters?

REFERENCES

Howlett, L. E.: International basis for uniform measurement, Science **158:**72, 1967.

Huntoon, R. D.: Status of the national standards for physical measurement, Science **150:**169, 1965.

Huntoon, R. D.: Concept of a national measurement system, Science **158:**67, 1967.

National Bureau of Standards: The international system of units (SI), special pub. no. 330, Washington, D.C., 1977, Superintendent of Documents, U.S. Government Printing Office.

National Bureau of Standards: Weights and measures standards of the United States: a brief history, special pub. no. 447, Washington, D.C., 1976 Superintendent of Documents, U.S. Government Printing Office.

Matter and atomic structure; radioactivity

4

Matter is generally defined in simple terms as *anything that possesses weight and occupies space.* Matter may exist in a solid, liquid, or gaseous state. It is readily observed that solids and liquids qualify as matter from the point of view of weight and taking up space, but many gases are invisible and it usually requires more than sight and feeling to determine that gases do have weight and occupy space.

PHYSICAL AND CHEMICAL PROPERTIES OF MATTER

Characteristics by which matter or substances are described for purposes of recognition and differentiation are known as properties of these substances. Two classes of properties cover the range of characteristics that can be attributed to any substance. These are physical and chemical properties. *Physical properties* are those characteristics of substances as they exist. These properties include, among others, the state of the substance (solid, liquid, or gaseous), color, odor, taste, crystalline form, density, solubility, and melting and boiling points. *Chemical properties* include the manner and speed with which substances react with one another to form new substances.

In the health sciences it is usually very important to be able to identify particular substances quite accurately, to make certain they are pure, and to know what may and does happen to these substances in or on the human body. This is accomplished, in part, by thorough study of the physical and chemical properties of such substances so that they may be recognized without question.

PHYSICAL AND CHEMICAL CHANGES OF MATTER

When a substance undergoes a physical change, the important feature is that although some of the properties of the substance may change temporarily, *no new substances are formed.* In a very common example, when ice melts to water, a solid substance becomes a liquid, and when water is heated to boiling and changed to steam, a liquid becomes a gas:

Ice	→	Water	→	Steam
(Solid water)		(Liquid water)		(Gaseous water)

Only the state of the water changes but not its composition. Another everyday example of a physical change is the mixing of salt and water to form a solution. Only the state of the salt is changed from solid to liquid, and the salt can be recovered unchanged by evaporation of the water. Taking a big rock and breaking it down to smaller rocks and pebbles is also an example of physical change.

A *chemical change*, on the other hand, results in the formation of one or more new substances as a result of the change. Burning wood usually produces some carbon (the soot that collects on the cooler areas of a fireplace), carbon dioxide (a gas), and water vapor, and leaves behind a mixture of substances as the ash. The composition and properties of the wood are changed completely by the process of burning. Another very familiar chemical change is the rusting of a piece of iron. Whereas the original iron contains only iron, has great strength, a definite shape, and magnetic properties, the rust is composed of iron chemically combined with oxygen, forms a powder, and has no magnetic properties. Neither the unburned wood nor the original iron can be recovered from these chemical changes by any simple processes such as evaporation.

The two examples of chemical changes cited are both oxidation reactions—reactions in which a substance combines with oxygen. However, in the case of the burning wood, the reaction takes place relatively rapidly with the instantaneous production of large amounts of heat, a form of energy, whereas the iron rusts slowly over a long period of time, and detection of any heat produced under these circumstances is very difficult. Oxidation reactions represent nature's most common method for the production of energy. As will be seen later, this same type of oxidation reaction is used to supply the human body with its energy requirements.

KINDS AND COMPOSITION OF MATTER

Matter, as is found in nature and in synthetic products, can be classified into three major types: elements, compounds, and mixtures.

Elements

Elements are considered to be basic units of matter, meaning that they cannot be decomposed into simpler units by ordinary means, nor can they be formed by a combination of other simpler units. The other types of matter are made up of various combinations of the basic elements. At the present time 105 elements are generally recognized, and elements beyond 105 have been predicted. Pure oxygen, gold, and iron are examples of elements.

Atoms. The basic unit of an element is known as an atom. If it were possible to take a bar of the element iron and keep subdividing or cutting it into smaller and smaller pieces until the smallest portion was reached that still had the properties of the original iron bar, this would be an atom of iron. Atoms are exceedingly small in size and weight, a factor that was a major obstacle in the way of progress in the study of matter and its properties until 1802-1808 when John Dalton, an English scientist, proposed and developed his *atomic theory of matter*. His major points were as follows:

1. Every substance is composed of ultimate particles called atoms (Gr., something that cannot be divided) that cannot be divided, created, or destroyed. (Law of conservation of mass.)
2. All the atoms of a specific element have identical properties, including the same weight; atoms of different elements have different properties and weights.
3. Chemical compounds are formed by the union of atoms of different elements in simple numerical proportions, such as 1:1, 1:2, 2:3. (Law of definite proportions.)
4. The atoms of two or more elements may combine in more than one ratio to form different chemical compounds. (Law of multiple proportions.)

Some modification of Dalton's original ideas became necessary as time went on, and some were found to be in error, but his atomic theory as a scientific milestone remains as a remarkable achievement.

Compounds

Substances that can be made by combinations of elements or that can be decomposed into the original elements are known as compounds. The processes by which compounds are formed and decomposed are two examples of chemical change. There are hundreds of thousands of compounds known to science at the present time. One of the simplest but very important compounds necessary for life, water, can be formed by the combination of the elements of hydrogen and oxygen. Under the proper conditions, such as are found in the electrolysis of water, the water can be decomposed to form the elements hydrogen and oxygen, which are given off as gases.

Several important laws about the formation of chemical compounds and chemical changes are intimately related to the atomic theory. If atoms cannot be divided, created, or destroyed (statement 1, before), this means that in a chemical change there is a rearrangement of the atoms involved and the weight of the starting materials must equal the weight of the final products. With precise measurements this was found to be true and is a statement of the *law of conservation of mass.*

In general, when a compound is formed from two or more elements under the same conditions, it is found that it has the same properties and composition each time it is made. On analysis, the atoms of the different elements are always found to be present in simple numerical ratios. In other words, it is always a whole number of atoms of each element that reacts, and since each atom of each element has the same weight and the same properties, the resulting compound also has the same weight and properties each time it is formed. This is the basis of the *law of definite proportions* (statement 3, before). It should take little imagination to visualize the chaotic situation that would exist if this law were not true.

A little complication does creep in at times when it is found that under different sets of conditions the same elements can combine in different proportions to form different compounds, as specified in the *law of multiple proportions* (statement 4, before). In the case of each different compound formed, however, the numerical ratio of the combining atoms is specific, simple, and the same every time the compound is made, e.g., in water there are always two atoms of hydrogen to one of oxygen, i.e., the ratio is 2:1, whereas in hydrogen peroxide there are two atoms of hydrogen to two of oxygen, a ratio of 2:2.

Molecules. If a grain of ordinary sugar, more formally known as the compound sucrose, could be physically subdivided into the smallest units that still retain the properties of the compound, these units would be *molecules.* As is the case with atoms of elements, single molecules of compounds are also very small so that under ordinary circumstances very large numbers of molecules are being handled.

Mixtures

In terms of what has already been discussed, mixtures are simply physical conglomerations of the different types of matter. For example, if salt, which is a compound, water, also a compound, and sand, a mixture of very finely powdered rocks, are placed together in the right way by nature, the resultant mixture produces a seashore. If a portion of this seashore is placed in a vessel of some kind, the sand will settle to the bottom, and the liquid can be poured off. If the water is evaporated, salt will remain, and thus all the ingredients can be separated from the mixture. If the same procedure is followed at a number of seashores, the amounts of each ingredient in the mixtures will probably differ because mixtures do not have any definite composition. Also, the ingredients can often be separated from each other by simple physical means. Most of the materials commonly encountered in daily life are mixtures.

Atomic weights

As already mentioned, single atoms are much too small to be observed or weighed. For many purposes, however, it is of great advantage to describe things at least on the basis of comparative weights. Thus, in the compound water, which is made up of two atoms of hydrogen and one atom of oxygen, for every 16 parts by weight of oxygen, there are found

Table 4-1. Partial list of elements, their symbols, and atomic weights

Name	Symbol	Atomic Weight	Name	Symbol	Atomic Weight
Hydrogen	H	1.0079	Sulfur	S	32.06
Helium	He	4.00260	Chlorine	Cl	35.453
Carbon	C	12.011	Potassium	K	39.0983
Nitrogen	N	14.0067	Calcium	Ca	40.08
Oxygen	O	15.9994	Iron	Fe	55.847
Sodium	Na	22.98977	Copper	Cu	63.546
Phosphorus	P	30.97376	Iodine	I	126.9045

two parts by weight of hydrogen. On a comparative basis, then, an atom of oxygen has a weight that is 16 times greater than that of an atom of hydrogen. Therefore, if the number 16 is taken as the standard for oxygen, the lightest element, hydrogen, would have a comparative weight of 1, and the heaviest known elements would have comparative weights in the middle 200s. On such a scale, moreover, many of the elements would have comparative weights that are whole numbers or close to whole numbers. Although such a scale of atomic weights was used for many years, for purposes of very high precision work it was found desirable in 1961 to change the standard from oxygen as 16.000 to carbon as 12.011. On this new scale, oxygen is 15.9994, a difference which for most ordinary chemical purposes is negligible.

Using such a scale, if the atomic weight of carbon, rounded off to 12, is expressed in grams, this quantity is known as the *gram atomic weight*. Also, an element with an atomic weight of 24 would be twice as heavy as carbon; one with an atomic weight of 36 would be three times as heavy as carbon. These relationships expressed in grams, or gram atomic weights or multiples thereof, represent the usual way of describing the amounts of materials going into and being produced by chemical reactions.

Atomic symbols. For convenience in writing about chemicals and their reactions, symbols are used rather than complete names. The symbol is taken as the first letter, or the first letter and one other letter, of the official name of the element. If a single letter is used, it is always capitalized; if two letters are used, the first is capitalized and the second is a small letter. Many of the original names of the elements are from Latin and are different from the English equivalents. That is why the symbols for some of the elements are not identical to the first one or two letters of their English names.

In Table 4-1 are found some of the more common elements, their symbols, and atomic weights. For a complete list, see the Appendix.

Molecular weights

The *molecular weight* of a compound is the sum of the atomic weights of the atoms found in the compound, e.g., a molecule of water, H_2O, would have a molecular weight of 18, to the nearest whole number:

$$\begin{aligned} 2\,H &= 2 \\ O &= \underline{16} \\ H_2O &= 18 \end{aligned}$$

If the molecular weight of a compound is expressed in grams, this is called the *gram molecular weight.* For water this would be 18 g.

Avogadro's principle. A very important hypothesis concerning molecular weights was first proposed in 1811 by Avogadro, an Italian scientist. Avogadro's principle states: *Equal volumes of all gases, under the same conditions of temperature and pressure, contain the same number of molecules.* On confirming this principle, it is found that in a gram molecular weight of any compound there are the same number of molecules as in a gram molecular weight of any other compound. When applied to solutions, this forms the basis of a very useful system for designating the concentration of solutions. For those who may be curious, the number of molecules in a gram molecular weight of any compound is approximately 6×10^{23} (6 followed by 23 zeros). This is known as Avogadro's number. Methods for determining this number may be found in more advanced texts.

THE ATOM

Although in Dalton's atomic theory the atom was stated to be the ultimate particle of matter, and is still defined as being indivisible by ordinary means, the atom can and has been subdivided into smaller subatomic particles. There are many subatomic particles now generally recognized, and the number is constantly increasing as the size of atom smashers (cyclotrons, betatrons) continues to increase. It is not possible for a text of this scope to discuss in detail this large number of subatomic particles, nor is it necessary to do so in order to obtain a working knowledge of how atoms are put together and what features contribute to their properties. For this purpose, a discussion of the particles known as *electrons, protons,* and *neutrons* will suffice.

Major subatomic particles

Electron. This is a particle with a unit negative charge and is $^1/_{1837}$ the weight of a hydrogen atom. For many purposes this is a negligible mass.

Proton. This particle has a unit positive charge and weighs as much as a hydrogen atom, which gives it an atomic weight of 1.

Neutron. This is a particle with no charge and can be considered as made up of a proton-electron combination that has an atomic weight of 1 (proton = 1, electron mass is negligible).

Structure of atoms*

In simple terms, the structure of an atom can best be described as a miniature solar system in which the *nucleus* at the center is analogous to the sun, and the electrons revolve around the nucleus in relatively distant orbits, just as the planets are quite a distance from the sun. The nucleus, then, is the central, dense portion of an atom consisting of protons and neutrons, the heavy components of the atom. Since protons carry positive charges and neutrons have no charge, the net charge on any nucleus must be positive. Furthermore, since all matter is normally electrically neutral, it must follow that the positive charge on the nucleus must be neutralized by the electrons revolving around the nucleus. These electrons, often referred to as *planetary electrons,* must be equal in number to the protons in the nucleus. This number is known as the *atomic number* of an element.

Atomic weight. As a result of the fact that electrons have a negligible weight and protons and neutrons each have an atomic weight of 1, the total atomic weight of an atom is equal to the sum of the protons and neutrons in its structure.

The simplest atom, hydrogen, has one proton in its nucleus. Applying the principles already discussed, this means that it must have one planetary electron, and therefore its atomic number must be 1. Its atomic weight must also be 1 because it has only one proton in its nucleus. The next more complicated atom is helium, which has two protons and two neutrons in its nucleus. It must have two planetary electrons to neutralize the charge on the nucleus resulting from the two protons, which means that its atomic number is 2. The two protons and two neutrons give helium an atomic weight of 4. These relationships are depicted in Fig. 4-1.

Some orbital properties. The orbits in which planetary electrons revolve around the nucleus are not randomly chosen by the electrons but are found to be quite rigidly specified. The first orbit around the nucleus can have no more than two electrons. This orbit, given the letter designation K, is completely filled in the helium atom. The next orbit further from the nucleus, L, can have as many as eight electrons in it although the next orbits can have more than eight electrons (M can have 18 and N can have

*The Bohr representation of atomic structure is used in this text because, although it may not be as precise physically as the more modern electron cloud or orbital representation, the major descriptive considerations relating atomic structure to chemical properties are quite similar in both representations. In other words, the Bohr picture, while largely correct on a qualitative basis, is not as accurate on a quantitative basis as the more modern models of atomic structure. For the purposes of a course for which this text was prepared, it has been found that the simpler Bohr representation has led to a greater understanding of the important relations between atomic structure and chemical properties than is generally possible with a more complicated approach, especially within the constraints of time limitations and background of students.

Fig. 4-1. Structures of the hydrogen and helium atoms. Hydrogen has an atomic number of 1 and an atomic weight of 1; helium has an atomic number of 2 and an atomic weight of 4.

Table 4-2. Number of electrons in successive orbital levels

Element	Atomic number	K	L	M	N	O	P
Helium	2	2					
Neon	10	2	8				
Argon	18	2	8	8			
Krypton	36	2	8	18	8		
Xenon	54	2	8	18	18	8	
Radon	86	2	8	18	32	18	8

32). The outer orbit is always found to have a maximum of eight electrons, as illustrated in Table 4-2. This led to interesting and very important observations of the periodicity in properties of the different elements and eventually to the statement of the *periodic law* by Mendeleev, a Russian chemist, in 1869, which in modern form would be: *The physical and chemical properties of the elements are periodic functions of the atomic numbers.*

This meant that the properties of an undiscovered element could be predicted on the basis of its atomic number and the properties of the elements above and below it in the periodic table (see Appendix). As a result, the search for undiscovered elements was greatly simplified, and the properties of elements could be classified in groups. This is most obvious with elements of lower atomic numbers. For example, sodium, atomic number 11, and potassium, atomic number 19, have very similar properties and are classed as alkali metals. Note in Fig. 4-2 that each element has only one electron in its outer orbit. Fluorine, atomic number 9, and chlorine, atomic number 17, also have very similar properties and are classed as halogen gases that are quite different from alkali metals. Note that fluorine and chlorine have seven electrons in their outer orbits in contrast to sodium and potassium, which have only one. It is these electrons in the outer orbit that give the elements their chemical properties. As shall be

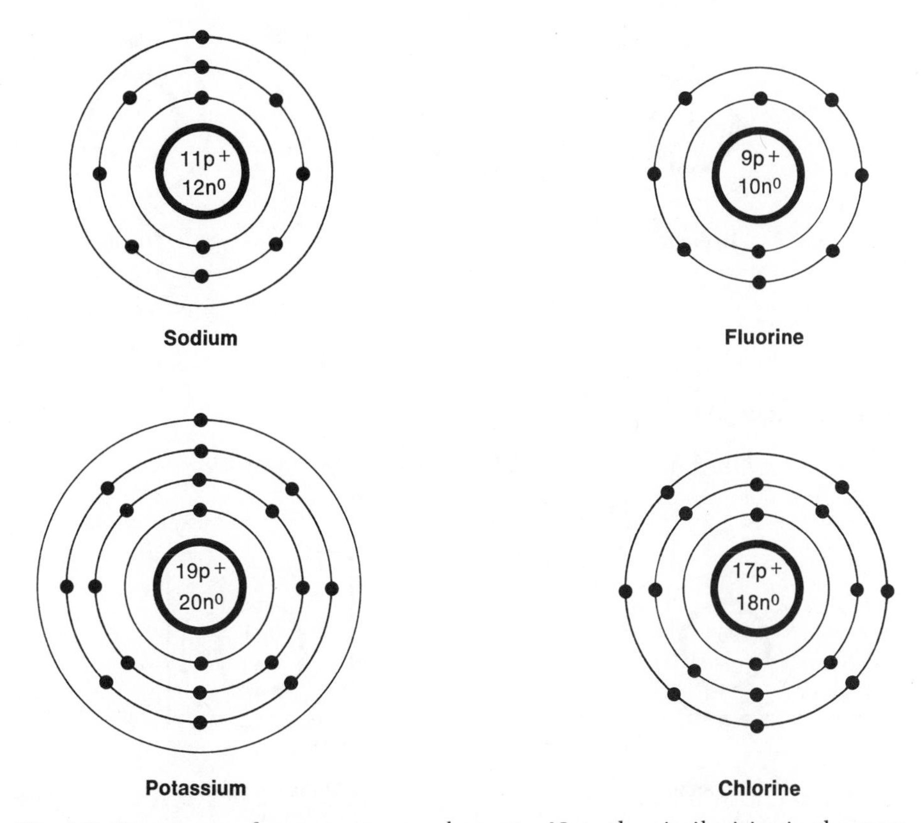

Fig. 4-2. Structures of some common elements. Note the similarities in the numbers of electrons in the outer orbits of sodium and potassium, and of fluorine and chlorine.

discussed subsequently, these outer orbit electrons are known as the *valence electrons.*

Isotopes. According to Dalton's atomic theory all atoms of the same element have the same weight. This is one of the features of the theory that has required modification as newer knowledge became available. Adding or removing a proton from an atomic nucleus would change the *weight* of the atom and at the same time *would change the atomic number.* By definition these are atoms of different elements. However, adding or removing a neutron from the nucleus would change the *weight* of the atom but *would not change the atomic number.* There would thus be atoms with *different atomic weights but the same atomic number,* or atoms of the same element but with different weights. These varieties of the same element are known as *isotopes.* The three isotopes of hydrogen are illustrated in Fig. 4-3.

Such isotopes are often found in nature and more recently have been made in the laboratory by exposing elements to bombardment with high-

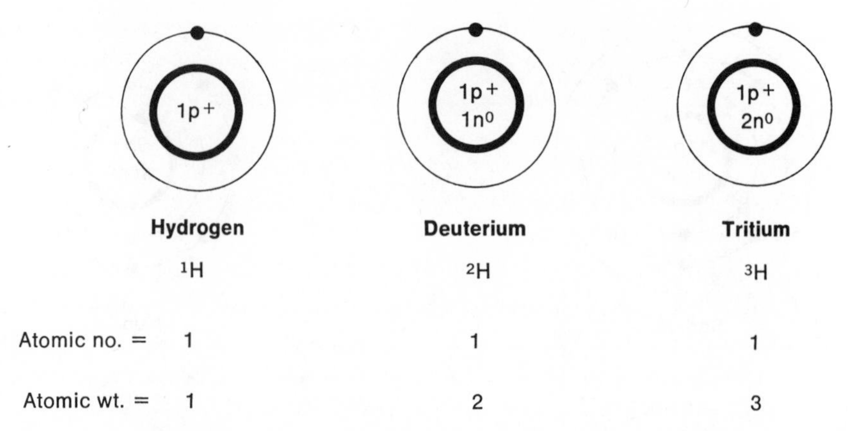

Fig. 4-3. Structures of the three isotopes of hydrogen.

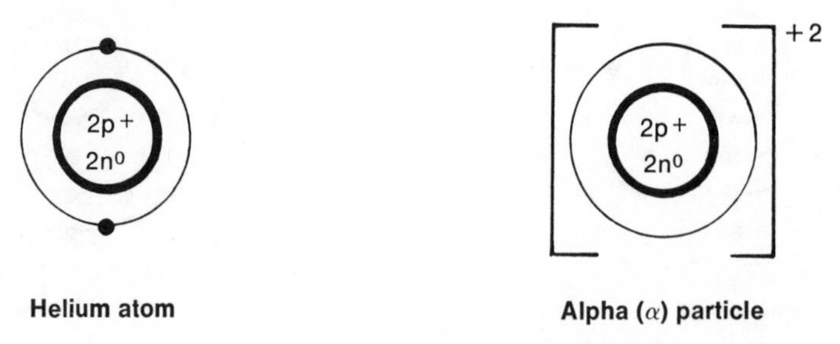

Fig. 4-4. Structures of the helium atom and the alpha (α) particle.

speed neutrons or other subatomic particles. Should a neutron strike a nucleus with sufficient force to push its way into the nucleus and remain there, an isotope of the particular atom is formed because the atomic number of the atom remains the same (no change in number of protons or planetary electrons), but the atomic weight has increased one unit (an additional neutron is now found in the nucleus). Some other types of nuclear reactions are also possible.

RADIOACTIVITY

Accidental observations at the turn of this century by one scientist, Henri Becquerel, followed by painstaking work for a number of years by Marie Curie and her husband Pièrre Curie, established the fact that there were some naturally occurring substances that spontaneously decomposed by giving off certain radiations. These radiations are known as radioactivity and the materials showing this phenomenon are called radioactive.

Radioactive substances give off three types of radiation:

1. *Alpha rays (α)* consist of alpha particles that are helium atoms with-

out the two planetary electrons, thus giving the particle a double positive charge (Fig. 4-4). Because of their weight, these particles travel relatively slowly and do not penetrate matter to any large extent.

2. *Beta rays (β)* consist of beta particles that are electrons carrying a single negative charge. Beta rays travel faster and penetrate matter further than alpha rays.
3. *Gamma rays (γ)* consist of electromagnetic waves, similar to x-rays, that carry no electric charge and penetrate much further than either alpha or beta rays. Because of their different electrical nature, alpha and beta rays subjected to a magnetic field will be deflected in opposite directions while gamma rays will be unaffected.

These radiations occur because in the nucleus there is an apparently unstable arrangement of protons and neutrons that tends toward a more stable arrangement by throwing off certain particles. When an alpha particle leaves a nucleus, the atomic weight decreases by 4 (two neutrons plus two protons), and the atomic number decreases by 2 (two protons). When a beta particle is given off by a nucleus, it can be considered as coming from the electron-proton combination of a neutron, thus leaving a new proton behind. The atomic number of the resulting element increases by 1, but the atomic weight remains the same. Adjustments in planetary electrons during such changes are made by reaction with the environment in order to maintain electrical neutrality.

The rate at which a specific radioactive element decomposes is quite constant. There are methods of measuring this rate and a particularly useful application of such measurements is the determination of the *half-life* of radioactive elements. The half-life is the time it takes for half the number of atoms of such an element to decompose, e.g., if the half-life of a radioactive element is 1 year, at the end of of 1 year only half the original material would remain unchanged. At the end of another year one-fourth ($1/2 \times 1/2$) would remain; at the end of the third year one-eighth ($1/2 \times 1/4$) would remain, etc. It should be clear then, that theoretically some portion of the original radioactive material will always remain. However, on a practical basis, after the radioactivity passes below a certain level, it has no further significance. Half-lives vary quite widely. Table 4-3 lists some of the more important radioactive elements being used in research in the life sciences.

Although some radioactive isotopes occur in nature, those used in research are generally man-made and are a direct result of the research that led to the atomic bomb and nuclear power plants. The greater availability of these radioactive isotopes has made possible many of the great advances that have occurred in the health sciences during the last 30 years or so.

Table 4-3. Isotopic designation and half-life of some radioactive elements

Name	Isotopic designation*	Half-life
Tritium	^{3}H	12.5 years
Carbon 14	^{14}C	5730 years
Phosphorus 32	^{32}P	14.3 days
Iron 59	^{59}Fe	45.1 days
Cobalt 60	^{60}Co	5.3 years
Iodine 131	^{131}I	8.0 days

*The superscript refers to the atomic weight of the isotope.

Radioisotopes in research and medicine

Because of the characteristics of radioactivity, exceedingly small amounts of radioactive substances can be measured and accounted for very accurately. This makes it possible to mix a radioactive chemical with its nonradioactive counterpart and trace its path through the body to determine what happens to the particular chemical without disturbing the normal processes. Radioactive chemicals used in this way are known as *tracers*. Some of the more commonly used tracers are iodine 131 (^{131}I), used to study thyroid function; cobalt 60 (^{60}Co), used to trace vitamin B_{12} through the body (cobalt is an atom in the structure of vitamin B_{12}); phosphorus 32 (^{32}P), used to study phosphorus metabolism in the body; and iron 59 (^{59}Fe), used to study the production of hemoglobin and red cells in various anemia and other disease states. Further details on how such studies are conducted will be presented in the biochemistry chapters.

Radioactivity in large amounts can be destructive to living tissue, and in certain instances advantage can be taken of this fact in attempts to selectively destroy tissues with radioactive substances. In order for this to be successful the target tissue must be able to concentrate the radioactive chemical in its cells. In this way, the tissue can be destroyed without surgical intervention and without serious consequences to other tissues of the body. An example of this type of procedure is the use of iodine 131 (^{131}I) in certain thyroid cases. Iodine 131 is largely concentrated in the thyroid gland, and when it is desired to decrease thyroid activity without a surgical procedure, a large dose of ^{131}I can be administered to the patient. The isotope goes mainly to the thyroid gland, where it can destroy some of the tissue and, because of its short half-life of 8 days, in a few weeks there is no longer any serious danger to the thyroid or any other body tissues. In addition, the size and condition of the thyroid gland can be visualized by a procedure known as scanning. After a dose of ^{131}I, the radioactivity in the patient's thyroid gland is plotted either by a machine that moves up and back above the position of the thyroid gland or by a

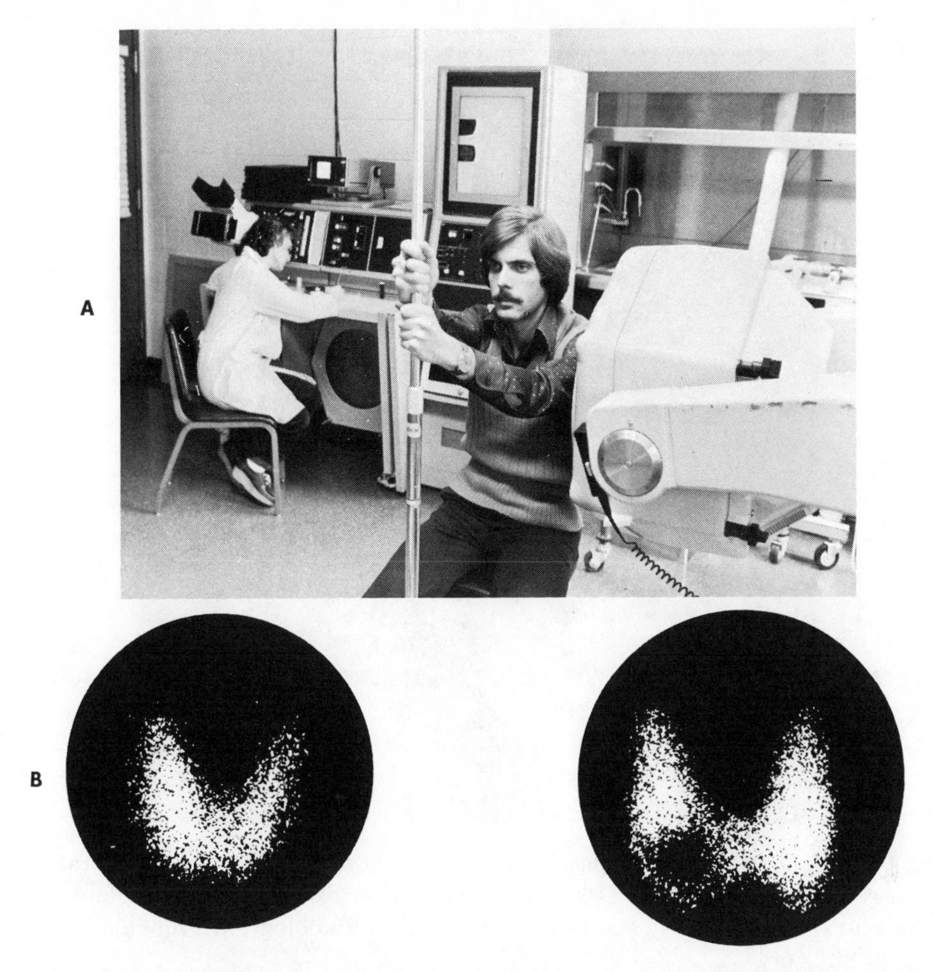

Fig. 4-5. A, Apparatus (scintillation camera) used to detect and photograph the pattern of concentration of radioactivity in an organ. **B,** Scintiphotos taken with camera. (**A,** Courtesy Mallinckrodt Institute of Radiology, Washington University, St. Louis. **B,** Courtesy of Nuclear-Chicago Corporation, Des Plaines, Illinois.)

stationary device that scans the entire organ (Fig. 4-5). Since the thyroid gland concentrates the ^{131}I, the shape of the thyroid gland will show up on the plot of the radioactivity measured. Further, if there are areas of increased metabolic activity in the thyroid gland, more radioactivity will be noted and plotted by the machine (Fig. 4-5, *B*). The reverse would be true in areas of decreased activity.

X-rays, tomography in medicine

For diagnostic purposes x-rays are often very valuable. However, most of the tissues are penetrated too easily by the rays and nothing of value shows up on the photographic plate. In some instances, particularly in the intestinal tract, administration of a suspension of barium sulfate coats the

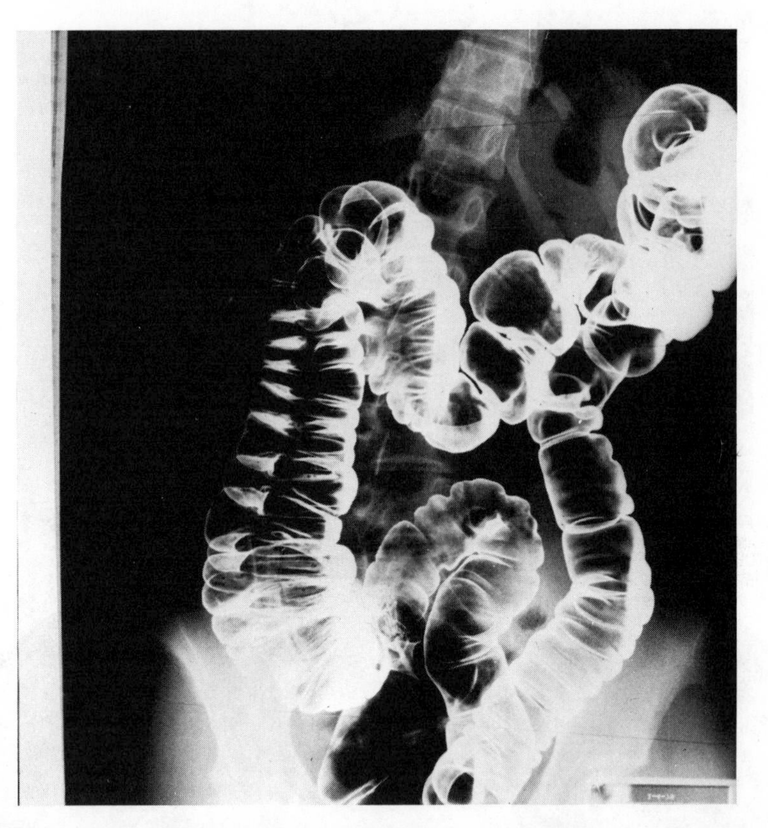

Fig. 4-6. Use of barium sulfate ($BaSO_4$) enema to visualize the lower intestinal tract in x-ray studies. The intestines are well outlined because the barium sulfate is opaque to x-rays. (Courtesy Mallinckrodt Institute of Radiology, Washington University, St. Louis.)

lining of the intestines and, as a result, the outline of the intestine shows up with good definition because the barium sulfate is opaque to the x-rays. A comparison with normal intestines will sometimes show up areas of ulceration, cancer, or other abnormalities (Fig. 4-6).

Recent advances in tomography, an application of x-rays known as computed tomography (CT) or computerized axial tomography (CAT), have capitalized on the ten to twenty-five times greater sensitivity of tomographic methods compared to ordinary x-ray methods in order to increase the diagnostic value of such studies by detecting minor differences of density, particularly between soft tissue structures. In conventional x-ray procedures, the result is registered on a photographic plate after the x-rays have penetrated the subject. In CT, the strength of the x-rays is first registered by x-ray detectors before reaching the subject and again by another set of detectors after passing through the subject. A computer de-

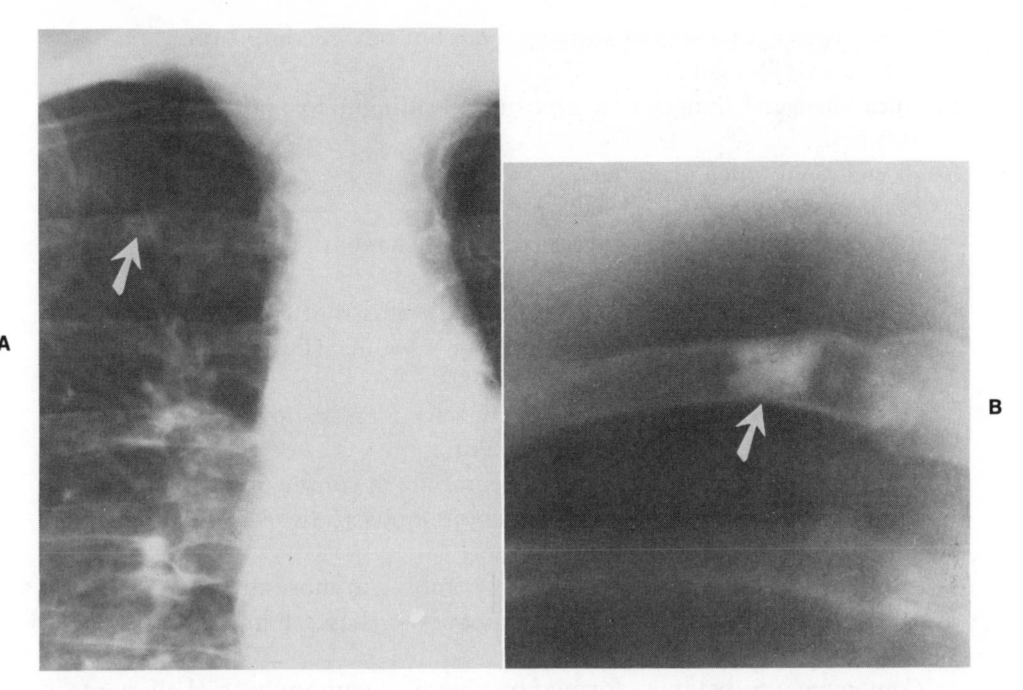

Fig. 4-7. A, Plain film, lordotic projection, showing questionable small nodular density in right apex. **B,** Tomogram confirmed the presence of the density and showed it to be a clinically nonsignificant dense bone island near the costovertebral junction of the fourth rib. (From Littleton, J. T.: Tomography: physical principles and clinical applications, Baltimore, 1976, The Williams & Wilkins Co.)

termines the differences in the resulting strength of the x-rays and stores this information until a complete series of exposures is made. The computer then assembles all the information and produces a digital readout or a television-type picture display, which can be photographed. The more x-ray dense an area is, the greater will be the difference between the strength of the x-ray entering and that of the x-ray leaving the subject. Fig. 4-7 illustrates dramatically the increased sensitivity of CT over conventional x-ray films in certain situations.

SUMMARY

MATTER

Anything that possesses weight and occupies space.

Physical properties: Characteristics of substances as they exist, such as state (solid, liquid, or gaseous), color, odor, taste.

Chemical properties: The manner in which substances react with one another to form new substances.

Physical changes: Changes of state or form but not in composition, no new substances formed.

Chemical changes: Changes in composition leading to formation of new substances.

Kinds and composition of matter:

Elements: Basic units of matter.

Atoms: Smallest recognizable unit of an element.

Atomic theory of matter:

1. Every substance is composed of atoms that cannot be divided, created, or destroyed by ordinary means. (Law of conservation of mass.)
2. All atoms of a specific element have identical properties and are different from all other elements.
3. Atoms of different elements combine in simple proportions (1 : 1, 1 : 2, 2 : 3, and so on) to form compounds. (Law of definite proportions.)
4. Atoms of some elements may combine in more than one ratio to form different chemical compounds. (Law of multiple proportions.)

Compounds: Substances formed by chemical combination of elements.

Molecules: Smallest recognizable unit of a compound.

Mixtures: Physical conglomerations of different substances with no definite composition and usually separable by simple physical means, e.g., sand and water.

Atomic weights: System of comparative weights of the elements based on carbon, atomic weight 12.011. In this system hydrogen is approximately 1, and oxygen is approximately 16.

Gram atomic weight: Atomic weight of an element expressed in grams.

Atomic symbols: Shorthand system of designating the elements. Consists of the first letter or the first and second (or other) letters of the official name of each element.

Molecular weights: Sum of the atomic weights of all the atoms in a molecule.

Gram molecular weight: Molecular weight expressed in grams.

Avogadro's Principle: Equal volumes of all gases, under the same conditions of temperature and pressure, contain the same number of molecules.

THE ATOM

Subatomic structure: The major subatomic particles are electrons, protons, and neutrons.

Electron: Particle with unit negative charge and $^1/_{1837}$ the weight of a hydrogen atom.

Proton: Particle with unit positive charge and atomic weight of 1.

Neutron: Can be considered a combination of electron and proton, with no charge and an atomic weight of 1.

Structure of atoms: Miniature solar system with the heavy nucleus (sun) in the center and electrons (planets) revolving around the nucleus in relatively distant orbits.

Nucleus: Central, dense portion of atoms consisting of protons and neutrons and carrying a net positive charge equal to the number of protons.

Planetary electrons: Electrons revolving around nucleus, equal to the number of protons in the nucleus, thus providing electrical neutrality.

Atomic number: Number of protons in the nucleus or the number of planetary electrons.

Atomic weight: Total of protons and neutrons (weight of electrons is negligible).

Orbital properties: Electrons revolve in definite orbits. First orbit, K, can have a maximum of two electrons, and the next three orbits in order are L, with a maximum of 8; M, 18; and N, 32. The outer orbit in any atom prefers a maximum of eight electrons.

Periodic law: Stated by Mendeléev in 1869: The physical and chemical properties of the elements are periodic functions of the atomic numbers.

Isotopes: Varieties of the same element with different atomic weights but the same atomic number.

RADIOACTIVITY

Spontaneous, regular decomposition of atoms with the emission of one or more of the following:

Alpha rays (α): Consist of alpha particles that are helium atoms without the two planetary electrons and therefore have a charge of +2. Low penetrating power.

Beta rays (β): Consist of beta particles that are electrons carrying a single negative charge. Travel faster and penetrate further through matter than α-rays.

Gamma rays (γ): Consist of electromagnetic waves, like x-rays; carry no charge; and penetrate matter much further than α- or β-rays.

Half-life: Time it takes for half the number of radioactive atoms of an element to decompose.

Radioisotopes in research and medicine: Exceedingly small amounts of radioactive substances can be measured and accounted for very accurately. Radioactive compounds can thus be traced through biologic processes for research purposes. Radioactive compounds in large amounts can be used to destroy tissue, such as thyroid, without surgical intervention, if the tissue concentrates the radioactive compound.

X-rays, tomography in medicine: Conventional x-rays have been used successfully for diagnostic purposes for many years. Recent advances in computed tomography (CT) have significantly extended the sensitivity of these methods for certain purposes.

REVIEW QUESTIONS

1. What is the key difference between physical changes and chemical changes?
2. What are the important postulates of the atomic theory of matter? What important laws are related to these postulates? How?
3. Is the system of atomic weights inherently related to a unit of weight? If a particular atomic weight is expressed in grams, what is it called? How is the molecular weight of a compound calculated?
4. Name the major subatomic particles and their properties with respect to electrical charge, atomic weight, and location in the atomic structure. What is the significance of the periodic law?
5. What is radioactivity? Name the particles that may arise from radioactive substances and list their properties. Are radioactive substances dangerous to health? Is it possible to use such substances to good advantage in the health sciences? How?

REFERENCES

Lagowski, J. J.: Modern inorganic chemistry, New York, 1973, Marcel Dekker, Inc.

Masterton, W. L., and Slowinski, E. J.: Chemical principles, ed. 4, Philadelphia, 1977, W. B. Saunders Co.

Pauling, L.: General chemistry, ed. 3, San Francisco, 1970, W. H. Freeman and Co., Publishers.

Petrucci, R. H.: General chemistry: principles and modern applications, ed. 2, New York, 1977, Macmillan Publishing Co., Inc.

Seaborg, G. T.: Man-made transuranium elements, Englewood Cliffs, N.J., 1963, Prentice-Hall International, Inc.

Formulas, valence, chemical equations, gas laws

5

FORMULAS

In order to facilitate discussion of the properties of various substances and the changes they may undergo, *formulas* are generally used. Such formulas, or chemical shorthand methods of representing the composition of molecules, are made up of the appropriate combination of atomic symbols (Chapter 4). A molecule of the simple compound sodium chloride, or table salt, is known to be composed of one atom of sodium (Na) and one atom of chlorine (Cl). Therefore its formula is written as NaCl, which represents not only the atomic composition of the molecule, but also stands for the molecular weight (sum of the atomic weights) of sodium chloride. When expressed in grams, this weight is known as the *gram molecular weight.*

With more complicated compounds containing more than one atom of an element in the molecule, subscripts after the atomic symbol are used. The formula for the molecule of water, which is composed of two atoms of hydrogen (H) and one atom of oxygen (O), is written as H_2O. It should be clear from the examples used that if only one atom of an element appears in a molecule, the atomic symbol with no subscript represents that single atom. It is also true that a single molecule of a substance is represented by the molecular formula without using the number 1. If two molecules of water are to be indicated, a 2 appears in front of the molecular formula: $2H_2O$ means two molecules or two molecular weights of water. Other factors involved in correct formula writing will be brought out in the discussions that follow.

VALENCE

Molecules are generally formed by the combination of two or more elements. In order for elements to combine there must be some attraction or affinity for each other. A measure of such affinity or ability of an element to combine with other elements is designated as its *valence.*

Some basic principles of atomic structure were discussed in the preceding chapter, including the observation that, except for the first planetary electron orbit, which has a maximum of two electrons, the other orbits prefer a maximum of eight electrons in the outermost orbit, the configuration found in the rare or noble gases (helium, neon, argon, etc.) that generally are very stable substances undergoing very few reactions. Also noted was the fact that elements with the same number of electrons in the outer orbit had similar properties. These observations led to the conclusion that the valence of elements was related to the number of electrons in the outer orbit. This conclusion will be borne out by the discussions that follow.

The structures of the atoms are represented by the atomic symbols standing for the nucleus and all the electrons except those in the outer orbit, with surrounding dots indicating the number of electrons in the outer orbit, e.g., hydrogen and sodium, both of which have one electron in the outer orbit are written as follows:

$$\mathrm{H}\cdot \qquad \mathrm{Na}\cdot$$

Atoms such as oxygen with six electrons and chlorine with seven electrons in their outer orbits are written as follows:

$$:\ddot{\underset{\cdot\cdot}{\mathrm{O}}} \qquad \cdot\ddot{\underset{\cdot\cdot}{\mathrm{Cl}}}:$$

The particular position of the dots around the atomic symbols is not usually of any significance but may be of convenience in illustrating a certain reaction.

Electrovalence

An electrovalent bond is the type of bond formed between atoms by the *complete loss or gain of one or more electrons* by the atoms concerned. As an example, the sodium atom, with one electron in its outer orbit, loses its electron to a suitable acceptor atom such as chlorine, with seven electrons in its outer orbit, which accepts an electron from a suitable donor atom. In giving up its single outer orbit electron, the sodium atom becomes positively charged, while, at the same time, in accepting the single electron into its outer orbit, the chlorine atom becomes negatively charged. The reaction is written as follows:

$$\mathrm{Na}\cdot + \cdot\ddot{\underset{\cdot\cdot}{\mathrm{Cl}}}: \rightarrow \mathrm{Na}^{+}:\ddot{\underset{\cdot\cdot}{\mathrm{Cl}}}:^{-}$$

The charged atoms produced in this reaction are known as *ions* and, because of their opposite electrical charges, attract each other strongly. The term *electrovalence* is derived from the fact that in the formation of such bonds electrically charged atoms are formed, with valences that are equal to the units of charge on the respective ions. Thus, in the case of sodium chloride (NaCl), the sodium ion has a valence of +1 and the chloride ion has a valence of −1, corresponding to the number of electrons lost or gained. It should also be noted, as seen in Fig. 5-1, that in giving up its electron the new outer orbit of the sodium ion contains eight electrons, and in accepting the electron from sodium, the outer orbit of the chloride ion also contains eight electrons.

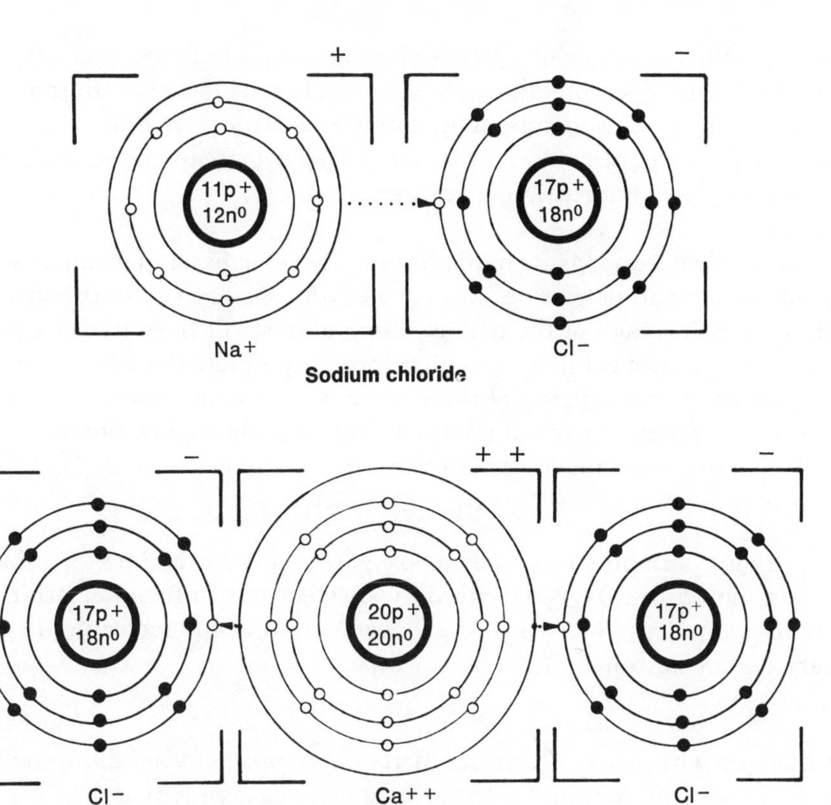

Fig. 5-1. Structures of sodium chloride (NaCl) and calcium chloride ($CaCl_2$) molecules, indicating electrovalent bonds.

Many reactions occur that involve the complete loss or gain of more than one electron. In the formation of calcium chloride ($CaCl_2$), the calcium atom gives up the two electrons in its outer orbit to become a calcium ion with an electric charge of +2, and at the same time the two chlorine atoms gain one electron each and become chloride ions, each with a charge of −1, as shown in Fig. 5-1. In simple form, the reaction may be written as follows:

$$\cdot\mathrm{Ca}\cdot + 2\cdot\underset{\cdot\cdot}{\overset{\cdot\cdot}{\mathrm{Cl}}}: \rightarrow {}^{-}:\underset{\cdot\cdot}{\overset{\cdot\cdot}{\mathrm{Cl}}}:{}^{+}\mathrm{Ca}^{+}:\underset{\cdot\cdot}{\overset{\cdot\cdot}{\mathrm{Cl}}}:{}^{-}$$

The calcium ion and the two chloride ions all contain eight electrons in their outer orbits after the molecule of calcium chloride is formed.

From these examples it can be seen that the sum of the positive charges or valences in a compound is always balanced by the sum of the negative charges. This is necessary to maintain the electrical neutrality of compounds. Another generalization that can be made from reactions of this type is that metallic elements, such as sodium and calcium, with few electrons in their outer orbits, tend to lose electrons in forming electrovalent bonds, whereas nonmetallic elements, such as chlorine, with many electrons in their outer orbits, tend to gain electrons. As will be explained later, electrovalent compounds in water solution can conduct electricity because of the charged ions that are formed.

Covalence

A covalent bond, as its name implies, is a partnership formed between atoms as a result of *sharing one or more pairs of electrons.* In the sharing of electron pairs, each of the participating atoms will have two electrons to complete the first orbit or eight electrons to complete the other outer orbit. For instance, molecules of hydrogen are formed from two atoms of hydrogen sharing their single planetary electrons so that each atom in the molecule has two electrons in its outer orbit (Fig. 5-2).

$$H\cdot + \cdot H \rightarrow H{:}H$$

Other common gases such as oxygen and chlorine form similar diatomic molecules. Oxygen, with only six electrons in its outer orbit, must share two pairs of electrons to give each atom in the molecule the necessary eight electrons in the outer orbits:

$$\ddot{\underset{\cdot\cdot}{O}}{:} + {:}\ddot{\underset{\cdot\cdot}{O}} \rightarrow \ddot{\underset{\cdot\cdot}{O}}{::}\ddot{\underset{\cdot\cdot}{O}}$$

Chlorine, with seven electrons in its outer orbit, forms diatomic molecules by sharing one pair of electrons, as is the case with hydrogen (Fig. 5-2):

$${:}\ddot{\underset{\cdot\cdot}{Cl}}\cdot + \cdot\ddot{\underset{\cdot\cdot}{Cl}}{:} \rightarrow {:}\ddot{\underset{\cdot\cdot}{Cl}}{:}\ddot{\underset{\cdot\cdot}{Cl}}{:}$$

With covalent bonds, the valence is determined by the number of *pairs of electrons shared* by the atoms involved. Thus, in the case of hydro-

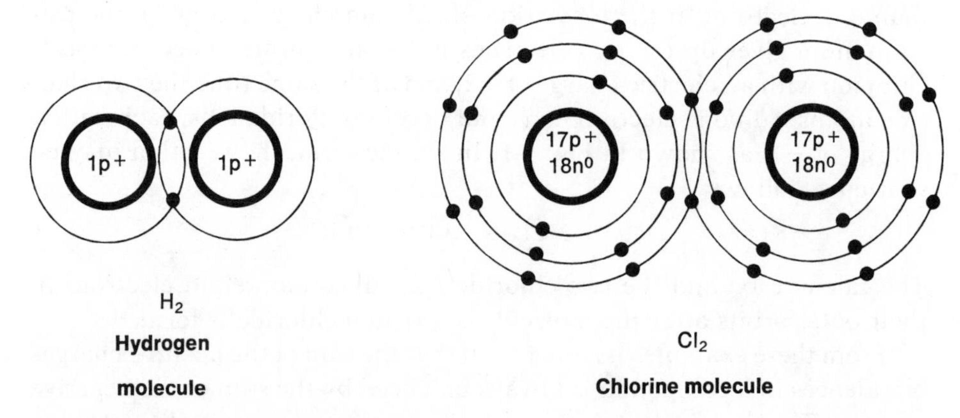

Fig. 5-2. Structures of the diatomic molecules of hydrogen (H_2) and chlorine (Cl_2) gases, showing covalent bonds.

gen and chlorine gas the valence is 1, and with oxygen gas the valence is 2. Since no electric charge of the type found in electrovalent bonds is found with covalent bonds, covalent compounds do not conduct electricity. Covalent bonds are also much more stable than electrovalent bonds as deduced from the fact that covalent compounds do not enter into chemical reactions as readily as do electrovalent compounds.

Other types of compounds formed with covalent bonds, such as water and organic compounds (compounds of carbon, hydrogen, and some other elements) that are so vital to life, are illustrated in Fig. 5-3. The molecule of methane is used to illustrate the covalent nature of the carbon-hydrogen bond so prevalent in organic compounds. In this compound it can be seen that the valence of carbon is 4, a very important feature in the structure of organic compounds.

Polar bonds. Thus far in discussing electrovalent and covalent bonds, the definitions are most suitable for compounds at one extreme or the other. Most compounds, however, are neither completely electrovalent nor completely covalent. Water is one of the best known examples of compounds found somewhere between the two extremes. It is found that the two pairs of electrons shared by the oxygen and the two hydrogen atoms are not held as tightly by the hydrogen as by the oxygen. These pairs of electrons are therefore somewhat closer to the oxygen atom, making it slightly more negative than the hydrogen atoms that therefore become slightly positive with respect to the oxygen atom, although the molecule as

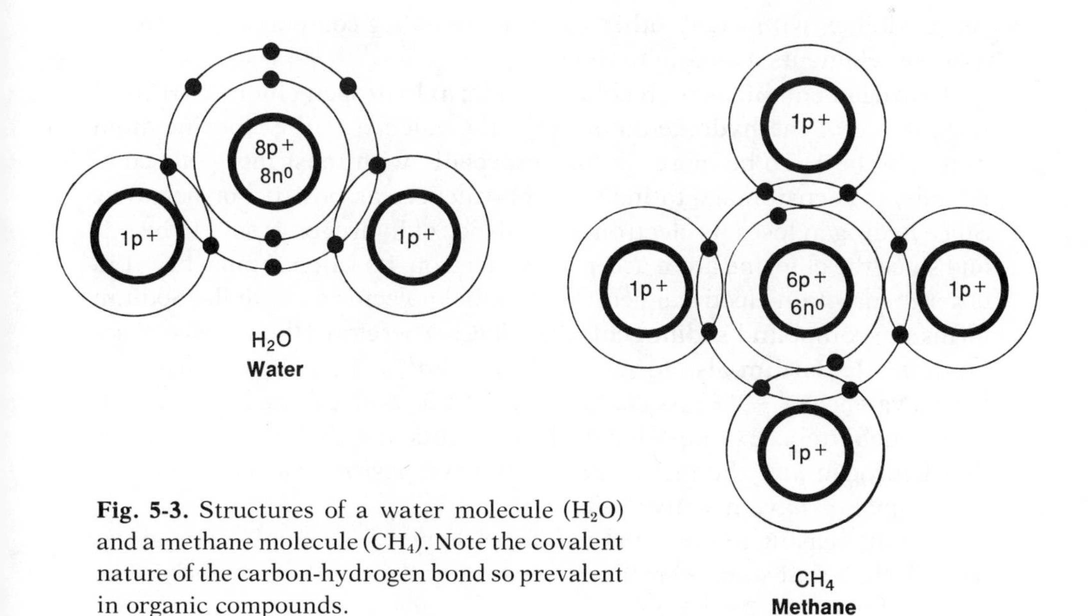

Fig. 5-3. Structures of a water molecule (H_2O) and a methane molecule (CH_4). Note the covalent nature of the carbon-hydrogen bond so prevalent in organic compounds.

a whole remains electrically neutral. These relationships can be shown as follows:

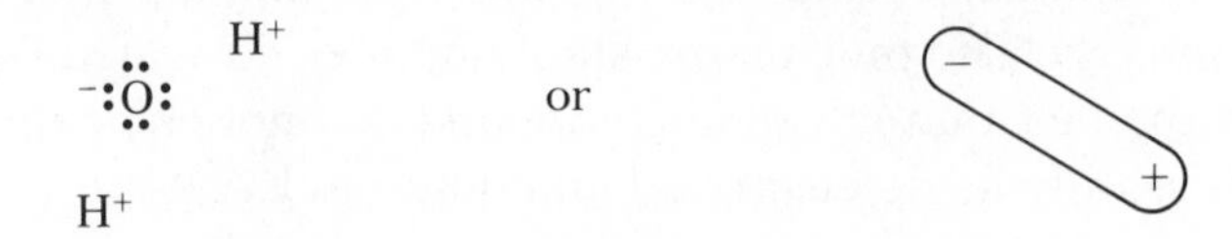

Because of the unequal distribution of the small charge, such molecules are called *dipoles,* and the type of bond is referred to as a *polar bond.* Such compounds are known as polar compounds. In contrast, molecules such as H_2 and O_2, in which there is no uneven distribution of charges, are known as nonpolar compounds.

The polar nature of the water molecule is the basis for its ability to dissolve so many other polar substances, the positive end of the water molecule attracting the negative end of the molecule being dissolved and the negative end of the water molecule attracting the positive end of the other molecule (p. 68).

Valence system

A good understanding of valence provides much information necessary for writing correct chemical formulas and determining the general chemical properties to be expected of the elements. It has been found convenient to assign hydrogen a valence number of 1 (it has only one electron that can participate in a chemical reaction and therefore can combine with only one other atom) as the basis of the valence system. Also, hydrogen combines with many other elements, making comparisons between different elements a simple matter.

Hydrogen combines with chlorine to form hydrogen chloride (HCl). If the valence of the hydrogen atom is 1, the valence of the chlorine atom must also be 1. To be more specific, especially with most inorganic compounds, it is customary to indicate the valences as positive or negative. Since hydrogen loses its electron, the valence of hydrogen is said to be $+1$, and since the chlorine atom accepts the electron, its valence must be -1 in order to maintain electrical neutrality of the molecule as a whole. Sodium forms the compound sodium chloride, NaCl. Therefore, the valence of sodium is $+1$. Barium also forms a chloride, $BaCl_2$. This shows barium to have a valence of $+2$. Cross comparisons of this type can be continued almost indefinitely. Examination of the elements and their valences shows that hydrogen and the metals generally have positive valences whereas the nonmetals have negative valences.

With increasing atomic number of the elements, a complication arises from the fact that some elements can exhibit more than one valence in the different compounds they may form. For instance, iron can form two chlorides: $FeCl_2$ in which the valence of iron is +2, and $FeCl_3$ in which the valence of iron is +3. In order to differentiate between these two compounds, the one having the lower valence, $FeCl_2$, is known as ferrous chloride, whereas the one with the higher valence, $FeCl_3$, is named ferric chloride. Other common metals having more than one valence are copper and mercury, both of which show valences of +1 or +2. The +1 compounds are called cuprous and mercurous compounds, and the +2 compounds are called cupric and mercuric compounds, respectively.

Radicals. Sodium also forms the compound sodium hydroxide, NaOH. Since sodium has a valence of +1, the combination of atoms OH must have a valence of −1. Barium also forms a hydroxide, $Ba(OH)_2$. Since barium has a valence of +2 and two combinations of the atoms OH are required to form the compound, this confirms the fact that the OH combination has a valence of −1. Such combinations of atoms that do not change throughout a series of chemical changes are known as *radicals* and can be considered similar to a single atom. When more than one radical appears in a compound, as in barium hydroxide above, the radical is enclosed in parentheses and the proper subscript is used as with single atoms. The OH combination is known as the hydroxyl radical. Other commonly encountered radicals are listed here:

Radical	*Name*
NH_4^+	ammonium
CO_3^{-2}	carbonate
HCO_3^-	bicarbonate
NO_3^-	nitrate
SO_4^{-2}	sulfate

Writing formulas. It is customary in writing formulas to put the positive valence element or radical first, followed by the negative valence element or radical. If the valence of the positive and negative components of a substance are known, it then becomes necessary only to balance the total positive and negative charges in order to be able to write the formula of the substance. As an example, ferric iron, Fe^{+3}, and the sulfate radical, SO_4^{-2}, form a compound called ferric sulfate. Since the +3 charge does not balance the −2 charge, more than one unit of one or both components may be required. As a first approximation, take two Fe^{+3} units with a total charge of +6. This can be balanced by three SO_4^{-2} units with a total charge of −6. Therefore the correct formula for ferric sulfate must show two Fe units and three SO_4 units taking into account the principles and customs already discussed. When this is done, the formula is written as $Fe_2(SO_4)_3$.

CHEMICAL EQUATIONS

One of the most important purposes for learning to write formulas correctly is to be able to understand the significance of a chemical equation, which is a shorthand representation of a chemical reaction. For every chemical reaction in which the formulas of the reactants and the products are all known a relatively simple chemical equation can be written, which, when interpreted properly, provides a great deal of information about the reaction with a minimum of writing. As an example, in the reaction between hydrogen gas (H_2) and oxygen gas (O_2) to produce water (H_2O), a first approach to writing the correct equation would be to show in a qualitative sense what is occurring, as shown:

$$H_2 + O_2 \rightarrow H_2O$$

This equation states that hydrogen reacts with oxygen to form water and is correct as far as it goes.

In order to obtain the full significance from a chemical equation it must also be *balanced* in a quantitative sense. From the law of conservation of mass it is known that under ordinary circumstances matter is neither created nor destroyed. This means that *the same number of atoms of each element that appear on one side of the equation must appear on the other side.* A quick glance at the qualitative equation just given shows that two hydrogen atoms appear on the left side and two on the right. This is balanced for hydrogen atoms thus far. However, two oxygen atoms appear on the left but only one on the right. Since the smallest unit of oxygen gas that can react is a diatomic molecule, O_2, a 2 can be placed in front of the H_2O to balance the oxygen atoms:

$$H_2 + O_2 \rightarrow 2H_2O$$

At this point the O atoms are balanced, but the hydrogen atoms are now unbalanced with two H atoms on the left and four on the right. This is easily remedied by adding a 2 in front of the H_2 on the left, giving a completely balanced equation:

$$2H_2 + O_2 \rightarrow 2H_2O$$

$$\begin{bmatrix} \text{4 H atoms} & \rightarrow & \text{4 H atoms} \\ \text{2 O atoms} & & \text{2 O atoms} \end{bmatrix}$$

This equation now states, in addition, that two molecules of hydrogen, or two gram molecular weights of hydrogen, react with one molecule of oxygen, or one gram molecular weight of oxygen, to produce two molecules of water, or two gram molecular weights of water. Since the gram molecular weights can be obtained by adding the appropriate gram atomic weights from the atomic weight table, the exact amounts of reactants required to produce a definite amount of product can easily be calculated.

Similar considerations and manipulations as just described are all that is required for the balancing of almost any other chemical equation.

Classification of chemical reactions

There are five major types of chemical reactions, some of which have already been used as illustrations in previous discussions.

1. Synthesis: Reactions in which two or more elements combine to form compounds. The reaction between hydrogen and oxygen to form water is a good example of this type.
2. Decomposition: As the name implies, this type of reaction results in the breakdown of a substance into its component parts. If water is subjected to an electric current under the proper conditions, *electrolysis* (*lysis* = breaking down; *electro* = under influence of electric current) occurs in which the water is broken down to hydrogen and oxygen gases:

$$2H_2O \xrightarrow{\text{electrolysis}} 2H_2\uparrow + O_2\uparrow$$

   The ↑ indicates a gaseous substance.
3. Metathesis: Reactions between two compounds in which two new compounds are formed as a result of an exchange of components. One of the new compounds formed is often insoluble, as in the reaction between the two salts, barium chloride and sodium sulfate:

$$BaCl_2 + Na_2SO_4 \rightarrow BaSO_4\downarrow + 2NaCl$$

   The ↓ indicates an insoluble substance that appears as a precipitate.
4. Replacement: As suggested by the name, a component of an existing compound is replaced by a new component in this type of reaction. A common example is the replacement of hydrogen from a strong acid, such as sulfuric acid, by a metal, such as zinc, to produce gaseous hydrogen and a salt, zinc sulfate:

$$Zn + H_2SO_4 \rightarrow H_2\uparrow + ZnSO_4$$

5. Oxidation-reduction: Reaction in which one substance is oxidized while another is being reduced. This type of reaction, with examples, will be discussed in the next chapter.

Factors affecting rate of chemical reactions

It is well known from the kitchen to the laboratory that the application of heat will cause most reactions to occur at a greater rate than otherwise. The effect of light on photosynthesis in the production of carbohydrates by plants and the effect of electricity on the decomposition of water have already been discussed. Other factors, such as solution and ionization of reactants, will be discussed in subsequent chapters.

Another factor of importance in controlling or affecting chemical reactions is the phenomenon of *catalysis.* In catalysis, a substance (known as a catalyst) is added, in small quantity, to a reaction that can then occur at a faster rate and under much less drastic conditions than possible without the catalyst. In the process, the catalyst itself remains unchanged. A catalyst can operate in this way only in reactions that can occur whether or not the catalyst is present. A catalyst cannot make a reaction occur that does not occur because of the chemical properties of the substances involved. Catalysts appear to act as expediters of chemical reactions where they do have an effect.

Catalysts are of great importance in industry in decreasing the costs of the required energy and in increasing the yield of products. More important for the purposes of this text is the fact that certain catalysts in the body, the enzymes, allow necessary reactions to take place in the body under such mild conditions that would be impossible to duplicate any other way in the laboratory or factory.

GAS LAWS

A short discussion of the kinetic molecular theory and the gas laws at this time will illustrate some important relationships between energy and the behavior of matter, particularly the gaseous state of matter.

Kinetic molecular theory

Among the postulates of the kinetic molecular theory are the following.

1. *Gases are composed of molecules separated by great expanses of empty space.* In fact, the volume occupied by a gas is largely empty space.
2. *Gas molecules are in constant motion and their velocities are related to the temperature. The pressure exerted by a gas results from the constant bombardment of the walls of the container by the gas molecules.*
3. *Gas molecules are elastic* (i.e., they bounce away from each other after collisions without loss of energy) *and have very little attraction for each other.* Gas molecules, therefore, do not settle into more compact masses such as liquids or solids.
4. *The average kinetic energy of all different gas molecules is the same at the same temperature:*

$$\text{Kinetic energy (KE)} = \tfrac{1}{2}\,mv^2, \text{ where m = mass, and v = velocity}$$

It should be clear from this equation that the molecules of a lighter gas must travel faster than the molecules of a heavier gas. Note the following example:

$$\text{Mass of gas A} = 2$$
$$\text{Mass of gas B} = 8$$

$$\text{Therefore, } KE_A = \tfrac{1}{2} \times 2v_A^2 = v_A^2$$
$$KE_B = \tfrac{1}{2} \times 8v_B^2 = 4v_B^2$$

$$\text{Since } KE_A = KE_B$$
$$v_A^2 = 4v_B^2$$
$$\text{or, } v_A = 2v_B$$

Therefore, v_A, the velocity of the lighter gas, must be greater than v_B, the velocity of the heavier gas.

Boyle's law

At constant temperature, the volume of a gas is inversely proportional to the pressure, i.e., if the volume is halved, the pressure is doubled:

$$V \propto 1/P, \text{ where T is constant}$$

This law is illustrated in Fig. 5-4. Since the temperature is constant, the kinetic energy is the same for the molecules of gas in both the large and small volumes, and the molecules will be striking the walls of their containers at the same rate. With the same rate of bombardment of the walls but half the wall space available, the gas in the smaller container will have double the pressure.

Charles' law

At constant pressure, the volume of a gas varies directly with the temperature (absolute), i.e., the higher the temperature, the greater the volume must be to keep the pressure constant:

$$V \propto T, \text{ where P is constant}$$

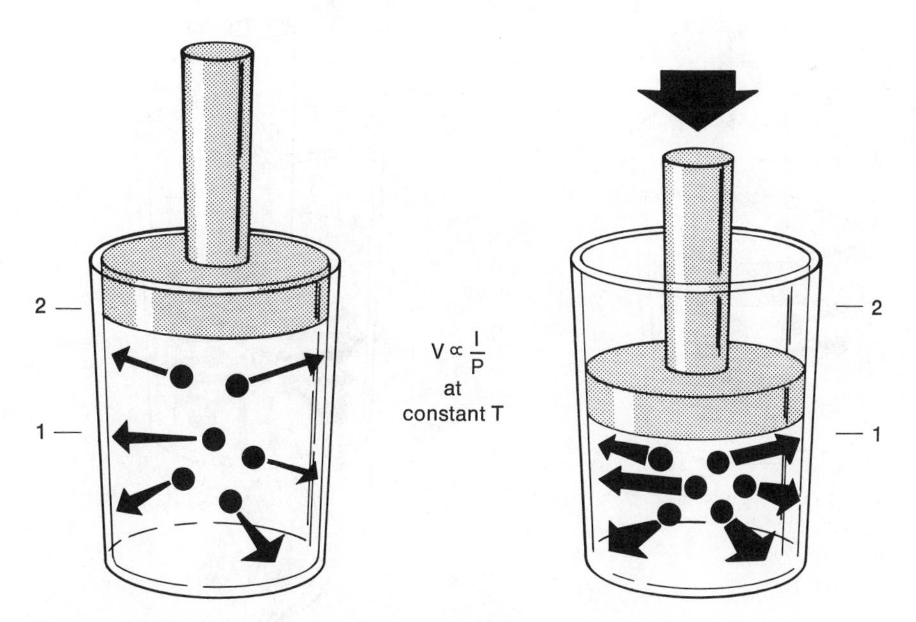

Fig. 5-4. Boyle's law. Average kinetic energy of the gas molecules is the same in both vessels at constant temperature (T).

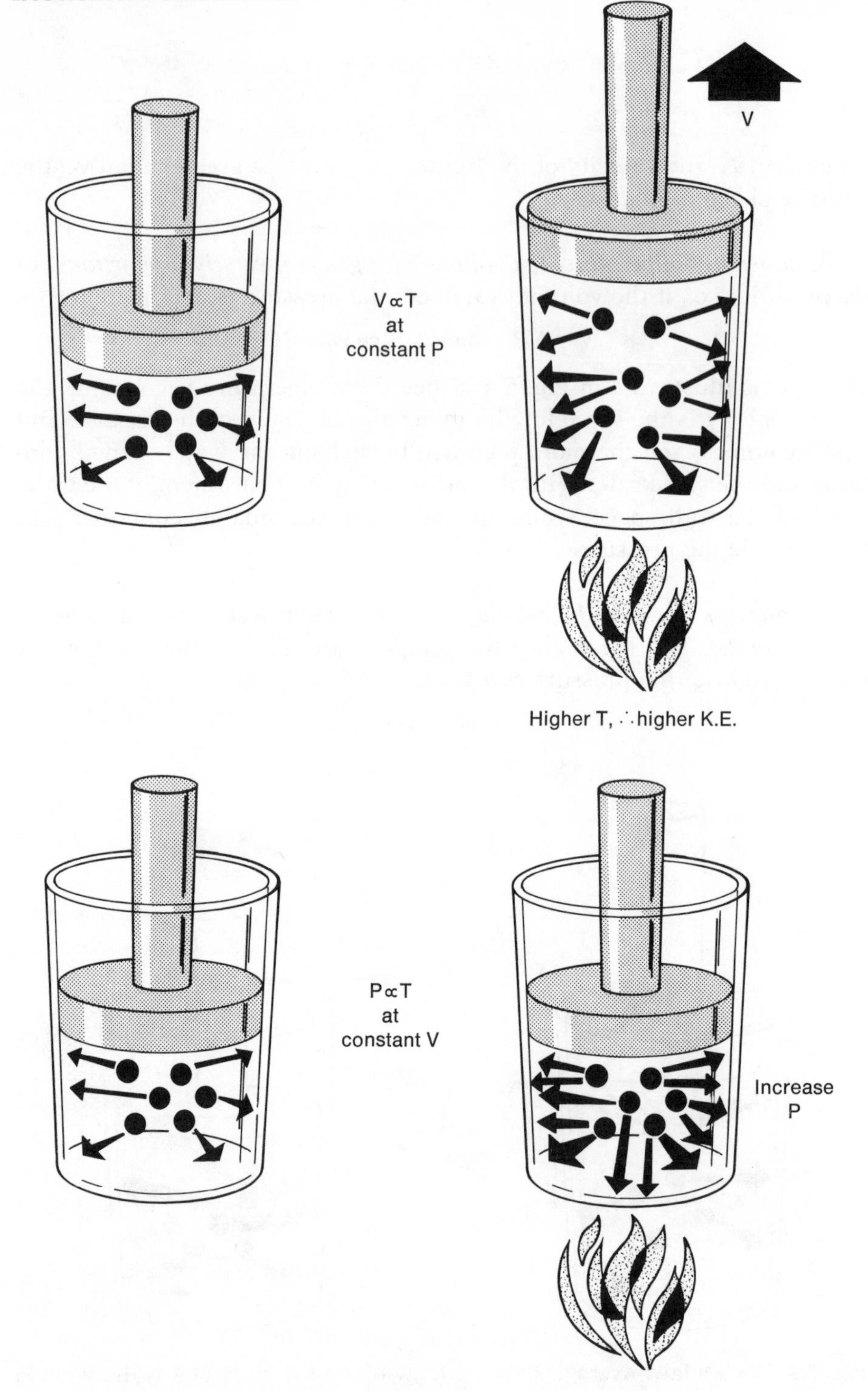

Fig. 5-5. Charles' law. At a higher temperature (T), the average kinetic energy of the gas molecules in the vessels on the right is greater than in the vessels on the left.

This law is illustrated in Fig. 5-5. As T increases, the kinetic energy of the gas molecules increases, meaning that they will strike the walls at a faster rate, unless the volume increases. Thus, the volume must increase to keep the pressure constant as the temperature increases. As an extension of Charles' law, it is also found that *at constant volume, the pressure of a gas varies directly with the temperature,* i.e., the higher the temperature, the higher the pressure becomes at constant volume:

$$P \propto T, \text{ where V is constant}$$

This relationship is also illustrated in Fig. 5-5. Since the kinetic energy increases with temperature and the volume remains constant, the gas molecules will be striking the walls at a greater rate (than at the lower temperature), and consequently the pressure will increase.

SUMMARY

FORMULAS

Chemical shorthand representation of the composition of molecules made up of the proper combination of atomic symbols. More than one atom of an element is indicated by a subscript following the symbol, more than one molecule by an appropriate number in front of the formula. Formula also stands for molecular weight and gram molecular weight of the compound.

VALENCE

Measure of ability of an element to combine with other elements. Valence and chemical properties of elements are related to the number of electrons in the outer orbit.

Electrovalence: Bond formed between atoms by complete loss or gain of one or more electrons. Each atom usually ends up with its outer orbit containing zero or two electrons if the outer orbit is the first orbit, or eight electrons if the outer orbit is other than the first orbit, the configuration found in the very stable rare gases. The atom that loses the electron(s) becomes positively charged, has positive valence, and the atom that gains the electron(s) becomes negatively charged, has negative valence. The sum of the positive valences or charges must equal the sum of the negative valences or charges to maintain electrical neutrality of the molecule. Electrovalent compounds conduct electricity when dissolved in water.

Covalence: Bond formed between atoms as a result of sharing one or more pairs of electrons. Many of the common gases, such as hydrogen and oxygen, form diatomic molecules, H_2 and O_2, by sharing the appropriate number of pairs of electrons to give each atom the configuration found in the rare gases. Covalent compounds do not conduct electricity and are more stable (less reactive) than electrovalent compounds.

Polar bonds. Bonds in between the extremes of complete electrovalent or covalent bonds in which the shared electrons are closer to one atom than the other. This results in a dipole, or polar compound, in which one atom, or end of the molecule, is slightly more negative or positive than the other end. Compounds with an even distribution of charges are nonpolar.

Valence system: Basis for the system is that the valence of hydrogen = 1, since hydrogen is the smallest element and can lose, gain, or share only one electron and combines with many elements. Valences of other elements can be determined from examination of compounds of hydrogen and other compounds formed by elements that combine with hydrogen. With increasing atomic number of the elements, some elements can have more than one valence in the different compounds formed. The lower valence form name ends in *-ous,* and the higher valence form in *-ic,* e.g., Fe^{+2} = ferrous, and Fe^{+3} = ferric iron.

Radicals: Combinations of atoms, such as OH and SO_4, that act as single atoms in chemical reactions.

Writing formulas: Positive valence element or radical is written first, then the negative valence element or radical. Positive and negative valences must balance.

CHEMICAL EQUATIONS

Shorthand representation of chemical reactions. An equation can be written for every reaction in which the reactants and the products are all known. Qualitatively, a chemical equation states that certain reactants will produce certain products. Quantitatively, the exact amounts of reactants required to produce a definite amount of product can be calculated on the basis of the balanced equation. An equation is balanced when the same number of atoms of each element found on one side of the equation appear on the other side (law of conservation of mass).

Classification of chemical reactions:

Synthesis: Combination of two or more elements to form compounds:

$$2H_2 + O_2 \rightarrow 2H_2O$$

Decomposition: Breakdown of compounds into component elements:

$$2H_2O \xrightarrow{\text{electrolysis}} 2H_2\uparrow + O_2\uparrow$$

Metathesis: Reaction in which partners are exchanged between compounds to form new compounds:

$$BaCl_2 + Na_2SO_4 \rightarrow BaSO_4\downarrow + 2NaCl$$

Replacement: Reaction in which a component of compound is replaced by a new component:

$$Zn + H_2SO_4 \rightarrow ZnSO_4 + H_2\uparrow$$

Oxidation-reduction: Reaction in which one substance is oxidized while another is reduced (to be discussed in the next chapter).

Factors affecting rate of chemical reactions: Heat, light, electricity, solution, catalysis, etc. In catalysis, a substance in small quantity (catalyst) can cause a reaction to occur at a faster rate and under much less drastic conditions than otherwise. The catalyst itself remains unchanged and cannot cause a reaction to occur where it is otherwise impossible. It merely expedites reactions that are possible even without the catalyst. Enzymes of the body are very important catalysts in processes of life.

GAS LAWS

Kinetic molecular theory:

1. Gases are composed of molecules separated by great expanses of empty space.
2. Gas molecules are in constant motion, and their velocities are related to the temperature. The pressure exerted by a gas results from the constant bombardment of the walls of the container by the gas molecules.
3. Gas molecules are elastic and have very little attraction for each other.
4. The average kinetic energy of all different gas molecules is the same at the same temperature:

$$\text{Kinetic energy} = \tfrac{1}{2}\,mv^2$$

$$\begin{pmatrix} m = \text{mass} \\ v = \text{velocity} \end{pmatrix}$$

Boyle's law: At constant temperature, the volume of a gas is inversely proportional to the pressure:

$$V \propto 1/P, \text{ where T is constant}$$

Charles' law: At constant pressure, the volume of a gas varies directly with the temperature (absolute):

$$V \propto T, \text{ where P is constant}$$

It is also found that, at constant volume, the pressure of a gas varies directly with the temperature:

$$P \propto T, \text{ where V is constant}$$

REVIEW QUESTIONS

1. What information is required in order to write a correct formula for a specific chemical compound? What are some of the things such a formula may represent?
2. Name the major types of bonds formed between atoms and list their important characteristics.
3. Define valence and list important aspects of the valence system. What are the relationships between valence, chemical properties, and atomic structure?
4. What must hold true of a balanced equation? What law is this related to? What information is provided by such an equa-

tion? What are the major types of chemical reactions? Can you write balanced equations as examples of such reactions?

5. How is the pressure of a gas related to the postulates of the kinetic molecular theory? Explain the major gas laws in these terms.

REFERENCES

Lagowski, J. J.: Modern inorganic chemistry, New York, 1973, Marcel Dekker, Inc.

Masterton, W. L., and Slowinski, E. J.: Chemical principles, ed. 4, Philadelphia, 1977, W. B. Saunders Co.

Pauling, L.: General chemistry, ed. 3, San Francisco, 1970, W. H. Freeman and Co., Publishers.

Petrucci, R. H.: General chemistry: principles and modern applications, ed. 2, New York, 1977, Macmillan Publishing Co., Inc.

Oxygen, hydrogen, oxidation-reduction

6

OXYGEN

Although oxygen is the most abundant of all the elements and is intimately involved in the functioning of living organisms and the changes that occur in their environment, it was only discovered long after many more obscure elements were known. Scheele, a Swedish pharmacist, was the first to prepare oxygen in 1771, but he did not publish his results until 1777. In 1775, Priestley, an English chemist, prepared oxygen and published his observations immediately. Two years later, Lavoisier, a French chemist, showed that air was a mixture, one part of which supported life and combustion whereas the other part was inactive in these respects. The active component, which he thought was present in all acids, he named oxygen ("acid former"), and the inactive component he named azote, now called nitrogen.

Animals require oxygen for reactions in which food chemicals release their energy for use by the animal as required, and plants need oxygen for the production of the major food classes: carbohydrates, proteins, and fats, which eventually are eaten by animals. Oxygen also comprises eight-ninths the weight of water, which is a compound making up about two-thirds the weight of the animal body. Oxygen causes iron to rust and many natural substances to decay, and is required for the combustion of fuels for the production of heat and other forms of energy for industrial purposes. Oxygen is a very important element in life processes.

Occurrence

As the most abundant element, it is estimated that the weight of all the oxygen in and on the earth's crust, contained in soil, rocks, water, air, plants, and animals, is approximately equal to the combined weight of all the other elements. As Lavoisier noted, air is composed of approximately 21% oxygen, with the remainder now known to be about 78% nitrogen, leaving a residue of 1% composed mainly of argon and trace amounts of other gases, such as carbon dioxide (0.04%). In air, oxygen occurs free in the form of diatomic molecules, O_2. It also combines with many other elements and is found in the human body, composing 60% of its total weight.

Preparation

Because of oxygen's relative abundance in the free state in air, air would seem to be an economic source for the preparation of oxygen, as indeed it turns out to be. In order to obtain oxygen from air, it must be separated from nitrogen and small amounts of other gases. To do this, air is first liquefied by being subjected to an appropriate combination of high pressure and low temperature. The liquid air is then allowed to warm up slowly. The more volatile nitrogen evaporates and the oxygen remains behind. The oxygen, as well as the nitrogen, can then be stored in steel cylinders under high pressure. This product, although quite useful, does contain as impurities some of the other gases found in air.

For a purer commercial product, the electrolysis of water is usually the method used. In this process, water with a trace of a substance such as salt (required for the conduction of electricity) is subjected to the action of an electric current. It is found that relatively pure oxygen forms at the positive pole, the anode, and hydrogen forms at the negative pole, the cathode

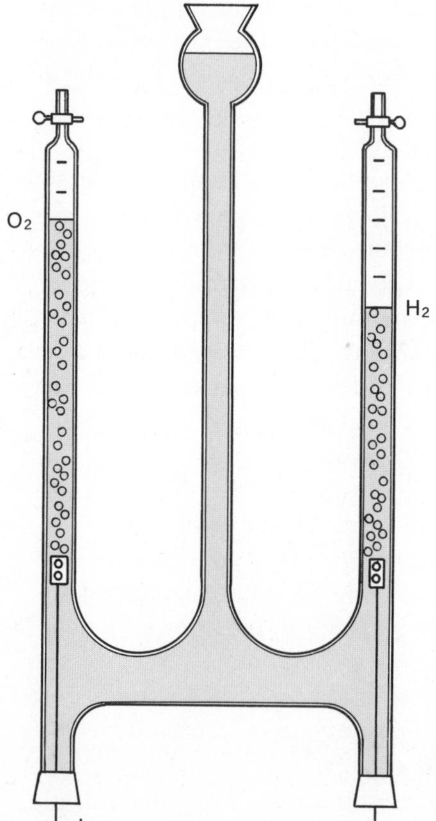

Fig. 6-1. Electrolysis of water. Note that the volume of hydrogen formed is twice as much as that of oxygen, as required by the equation: $2H_2O \rightarrow 2H_2\uparrow + O_2\uparrow$.

(Fig. 6-1). It should be noted here that twice as much hydrogen forms as oxygen, as required by the equation for the reaction:

$$2H_2O \xrightarrow{\text{electrolysis}} 2H_2\uparrow + O_2\uparrow$$

This process is also used for the preparation of relatively pure hydrogen.

On a small scale in the laboratory, oxygen is conveniently prepared by heating potassium chlorate:

$$2KClO_3 \underset{\Delta}{\rightarrow} 2KCl + 3O_2\uparrow$$

$$(\Delta = \text{symbol for heat})$$

The oxygen is usually collected by the displacement of water, as is usually the case with gases that are relatively insoluble in water (Fig. 6-2). An interesting phenomenon occurs with this reaction. Although the reaction will occur as described, it does so rather sluggishly. However, if a small amount of manganese dioxide (MnO_2) is added, the reaction proceeds at a much more rapid rate and the manganese dioxide can be recovered unchanged at the end of the reaction. This is a good example of the process of catalysis (p. 48).

Physical properties

Oxygen is a colorless, odorless, tasteless gas that can be liquefied and solidified under appropriate conditions (blue color as liquid and solid). It is slightly heavier than air (molecular weight of air is about 29, that of oxygen is 32) and is soluble in water (about 3 to 4 ml in 100 ml of water at 20° to 0° C), a fact that is very important to all forms of aquatic life.

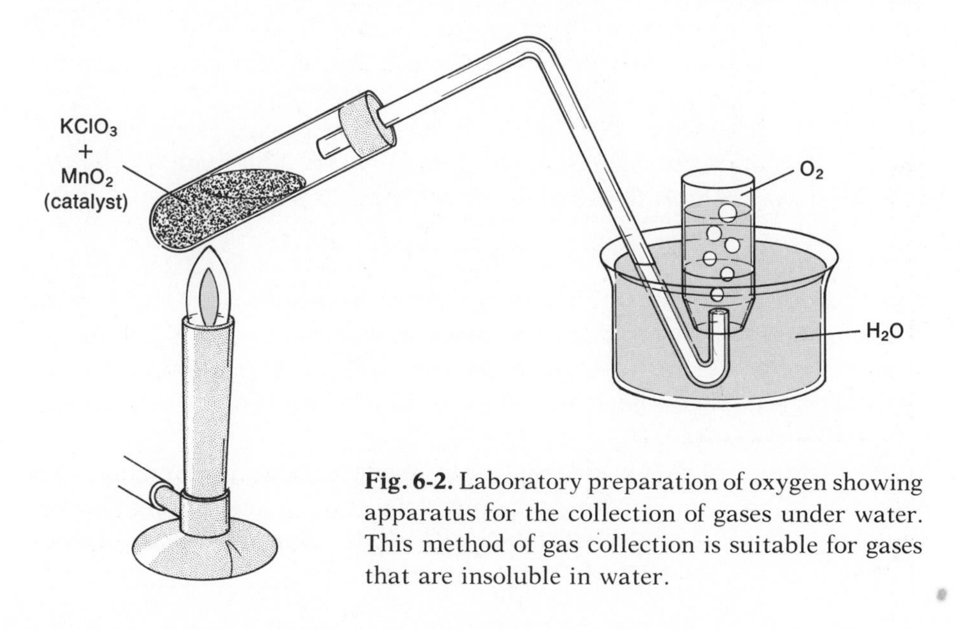

Fig. 6-2. Laboratory preparation of oxygen showing apparatus for the collection of gases under water. This method of gas collection is suitable for gases that are insoluble in water.

Chemical properties

Oxygen, with six electrons in its outer orbit, tends to accept two electrons from most other elements, metal and nonmetal, generally at higher temperatures, although moderate activity does occur at ordinary and low temperatures.

With metals oxygen reacts to form oxides that, when dissolved in water, form bases, as will be discussed subsequently; e.g., calcium reacts to form calcium oxide:

$$2Ca + O_2 \rightarrow 2CaO$$

With nonmetals oxygen forms oxides that produce acids when dissolved in water; e.g., sulfur burns to give sulfur dioxide (coal gas):

$$S + O_2 \rightarrow SO_2\uparrow$$

These and other reactions in which oxygen combines with other elements are known as *oxidation* reactions (p. 63).

Combustion. Another very important property of oxygen is that it supports combustion, a type of vigorous and rapid oxidation accompanied by the release of energy in the form of heat and light. This is an example of the conversion of chemical energy to heat and light energy.

Substances that burn readily are variously called *combustible, flammable,* or *inflammable,* whereas those that do not burn are designated as *incombustible, nonflammable,* or *noninflammable.* The terms *combustible* and *incombustible,* while the least ambiguous, are not universally used. Since oxygen is usually the limiting factor in combustion reactions, some substances that burn reluctantly or not at all in air, which is 21% oxygen, will burn much more vigorously in pure oxygen. For example, iron, which may rust or oxidize slowly in the atmosphere, burns brightly in pure oxygen.

There are reactions other than those involving oxygen that are also classified as combustions in which heat and light are liberated; e.g., hydrogen can burn in chlorine to produce hydrogen chloride:

$$H_2\uparrow + Cl_2\uparrow \rightarrow 2HCl\uparrow$$

Under ordinary circumstances, combustions involve oxygen; and, most commonly, molecular oxygen is the reactant. However, some elements can react vigorously enough to obtain the oxygen from compounds; e.g., the metal magnesium can burn almost as well in carbon dioxide as it does in air.

One important factor to emphasize about oxidation or combustion reactions is that usually a large amount of energy is released by such reactions. As will be seen in subsequent discussions, the human body obtains its energy requirements from similar oxidation reactions in which the hydrogen in food compounds is oxidized to water:

$$2H_2 + O_2 \rightarrow 2H_2O + \text{energy}$$

When this reaction is carried out in the laboratory it is usually explosive. In the body, with the aid of enzymes (catalysts), the same type of reaction is carried out at body temperature, 37° C, and at such a rate that a good percentage of the energy produced can be put to good use, rather than being dissipated in one great burst (p. 285).

Kindling temperature. Not all substances will burn as readily as others; some require much more heat from an outside source before they ignite and continue to burn. The point at which ignition and self-sustained combustion of a substance occurs is known as the *kindling temperature.* To cause an oak log to burn, the kindling temperature of the log is usually approached in stepwise fashion. A match, with a low kindling temperature, is used to light some paper, which produces enough heat to ignite some small twigs, "kindling wood," which in turn then produce enough heat to ignite the log.

Spontaneous combustion. Whether a substance oxidizes or burns slowly or rapidly, the same amount of heat is produced. If a substance oxidizes slowly, the heat is produced over a longer period of time and is generally dissipated into the environment without any significant temperature rise. However, under circumstances that do not allow for such continuous dissipation of the heat, the temperature of the substance being oxidized continues to rise and the rate of oxidation increases until the kindling temperature is reached. At this point the substance burns. This phenomenon is known as *spontaneous combustion* and may occur with something like a large, crumpled, oily rag, where the oxidation occurring on the inside of the rag produces heat that cannot be dissipated to the surroundings as rapidly as it is produced.

Extinguishing fires. The two cardinal rules to remember when trying to extinguish a fire are: (1) a fire will go out when deprived of its oxygen source; and (2) a substance will stop burning when the temperature is lowered below its kindling point.

The best types of fire extinguishers will accomplish both things at the same time; e.g., the pressurized carbon dioxide, or "dry ice," fire extinguisher aims a blast of cold carbon dioxide at the base of the flame and lowers the temperature of the burning substance as it surrounds it with a gas that does not support combustion. In recent years, extinguishers of dry powder under pressure have been found to be effective. When the powder is heated by the fire, carbon dioxide is produced, displacing the oxygen from around the fire. These types of extinguishers are particularly useful in electrical fires because no water is involved.

Water is still generally relied on for most large wood or paper fires and operates by lowering the temperature of the burning substances. Water absorbs substantial amounts of heat in the process of being vaporized, and as a result the temperature drops.

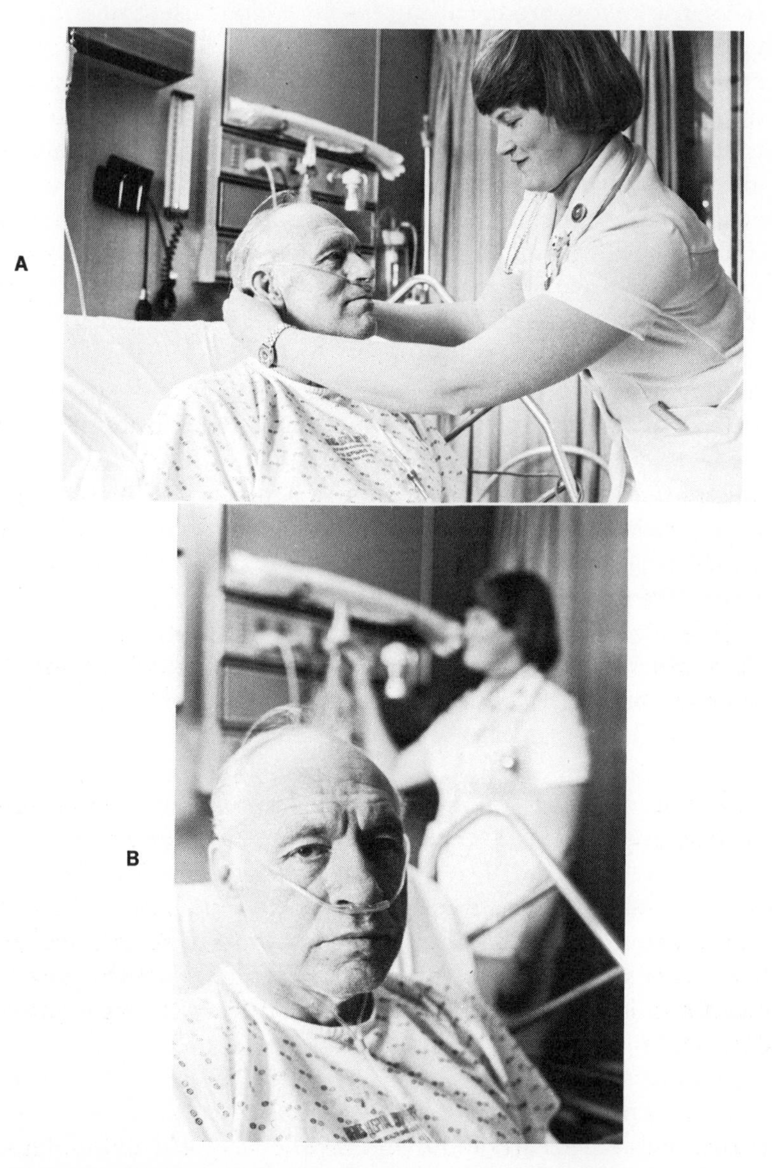

Fig. 6-3. Medical use of oxygen. (Courtesy Barnes Hospital, St. Louis.)

Oil fires, which can float and spread on top of water, are successfully extinguished by an extinguisher that produces a foam of carbon dioxide bubbles in an appropriate carrier, which can float on the burning oil and displace the oxygen around the fire.

Medical uses

Under normal, healthy circumstances, an adequate supply of oxygen is distributed to the tissues in the body by way of hemoglobin in the blood. After disposing of the oxygen in the tissues, the blood carries carbon dioxide away from the tissues to the lungs, where the carbon dioxide is exchanged for oxygen. The mechanics of this process will be discussed later. In abnormal conditions that decrease the opportunity for exchange of carbon dioxide for oxygen in the lungs, such as lung congestion, decreased rate of breathing for any of a number of reasons, exposure to decreased oxygen concentrations in the air, or decreased hemoglobin concentration in the blood, it may become necessary to use pure oxygen for breathing purposes in order to provide an adequate supply for distribution by hemoglobin (Fig. 6-3). At times oxygen may be used in combination with a small amount of carbon dioxide, which stimulates breathing, or it may be used in combination with helium, which is much lighter than the nitrogen in air and therefore requires less effort for breathing. When gaseous anesthetics are administered, oxygen must also be administered at the same time.

OZONE

Certain substances that can exhibit different forms in the same physical state are known as *allotropic modifications*. Although oxygen, as a gas, normally appears as O_2, an allotropic modification of oxygen with a formula of O_3, ozone, also exists as a gas.

Preparation

Ozone is usually prepared by producing an electric discharge, like lightning, in the presence of oxygen:

$$\text{Energy} + 3O_2\uparrow \rightarrow 2O_3\uparrow$$

In the formation of ozone, large amounts of energy must be absorbed. Therefore, ozone will contain more chemical energy than oxygen.

Properties

Ozone is obviously heavier than oxygen, is pale blue, has a distinctive odor, and is much more soluble than oxygen in water. The chemical properties of ozone are similar to those of oxygen, but ozone is much more reactive. When ozone decomposes, it releases the additional energy required to produce it:

$$2O_3 \rightarrow 3O_2 + \text{energy}$$

This explains its greater activity as compared to oxygen, and, because of its strong oxidizing activity, it is used for air and water purification, sterilization, disinfection, and bleaching. Rubber items exposed to ozone lose their elasticity and become brittle, requiring frequent replacement.

HYDROGEN

Hydrogen was recognized as an element in 1776 by the English scientist, Cavendish. He called it "inflammable air," but it was renamed hydrogen ("water former") by Lavoisier, who showed that it was a constituent of water.

Occurrence

Hydrogen is not found free to any appreciable extent on the earth or in the atmosphere, except at very high altitudes. It does appear in large quantities in the sun and stars. In the combined state, hydrogen is quite abundant in water, acids, bases, and many biologically important organic compounds such as carbohydrates, proteins, lipids, and other body components.

Preparation

As noted in the discussion of the preparation of oxygen, the electrolysis of water is a good method for producing relatively pure hydrogen. Another simple method convenient for laboratory use is the displacement of hydrogen from acids by certain metals:

$$Zn + 2HCl \rightarrow H_2\uparrow + ZnCl_2$$

In addition to these methods, there are many others that are used in industry.

Properties

Hydrogen is a very light, colorless, odorless, tasteless, combustible gas, which forms explosive mixtures with the oxygen in the air. At higher temperatures hydrogen is quite reactive, combining usually with many nonmetallic elements, although some compounds of hydrogen and metals (hydrides) can also be formed. The rate of many of the reactions of hydrogen can be speeded up at lower temperatures by using platinum catalysts. This is also true of the enzyme catalysts in the body. In its chemical reactions, hydrogen is known as a *reducing agent.*

Uses

Hydrogen has many present-day industrial uses that can be found by consulting texts on inorganic chemistry. Biologically speaking, interest in hydrogen centers on its roles in the production of body constituents and the production of energy, as was mentioned previously and will be more fully discussed in the biochemistry section of this text.

OXIDATION-REDUCTION

Oxidation

The term *oxidation* may be defined and illustrated in three ways:

1. Addition of oxygen to a compound or element:

$$4Fe + 3O_2 \rightarrow 2Fe_2O_3$$

Iron   **Ferric oxide (rust)**

2. Removal of electrons from an atom or increase in positive valence:

$$Fe^{+2} - e^{-} \rightarrow Fe^{+3}$$

Ferrous **Ferric**

3. Removal of hydrogen from a compound:

$$2H_2O \rightarrow 2H_2 + O_2$$

Reduction

Chemically, the opposite of oxidation is *reduction,* which can be defined as:

1. Removal of oxygen from a compound:

$$CuO + H_2 \rightarrow H_2O + Cu$$

2. Addition of electrons to an atom or decrease in positive valence:

$$Cu^{+2} + e^{-} \rightarrow Cu^{+}$$

Cupric **Cuprous**

3. Addition of hydrogen to an element or compound:

$$O_2 + 2H_2 \rightarrow 2H_2O$$

Oxidation-reduction reactions

Careful examination of the examples given will show that whenever oxidation occurs in a chemical reaction, reduction must also occur in the same reaction. Furthermore, the oxidizing agent is found to be reduced and the reducing agent is oxidized as in the following example:

CuO	+	H_2	→	H_2O	+	Cu
Cupric oxide						**Copper**
(oxidizing agent)		(reducing agent)		(oxidized reducing agent)		(reduced oxidizing agent)

In CuO the valences can be indicated as $Cu^{+2}O^{-2}$ and in Cu as Cu^0. The copper has decreased in positive valence and has therefore become reduced. The valence of hydrogen in H_2 is 0 and in H_2O is +1 for each hydrogen atom. The hydrogen has increased in positive valence and has therefore been oxidized.

As already mentioned, test-tube oxidation-reduction reactions usually require high temperatures to make them work at a fast rate, but in the body these reactions take place at much lower temperatures with good efficiency because of the presence of enzyme catalysts.

Antiseptic applications. Some oxidizing substances have been found useful as antiseptic agents because of their ability to interfere with the normal metabolism of bacteria, thereby causing their death. Among the most common ones of this type are hydrogen peroxide (H_2O_2), which decomposes readily to give oxygen bubbles; potassium permanganate ($KMnO_4$), which oxidizes organic matter; and iodine (I_2), which, in solution, was the major household antiseptic for many years for all sorts of minor cuts and other external uses. At the present time there are many other effective antiseptic agents that are not oxidizing substances.

Bleaching applications. Many colored substances can be rendered colorless by being oxidized or reduced. It is therefore possible to use both oxidizing and reducing agents as bleaches, depending on the substance being decolorized. One of the most widely used bleaches is Clorox, which contains sodium hypochlorite ($NaClO$), an oxidizing agent that decomposes to give off oxygen. Hydrogen peroxide and potassium permanganate, listed before as antiseptic agents, are also useful as bleaches. Among the reducing type bleaches are oxalic acid [$(COOH)_2$], and sodium thiosulfate ($Na_2S_2O_3$), a very effective agent for removing iodine stains.

SUMMARY

OXYGEN

Occurrence: Discovered in the 1770s, most abundant element, makes up 21% of air where it is found free as oxygen, found combined with many elements in a large variety of compounds, including water, which is eight-ninths oxygen, and makes up about 60% of the weight of the human body.

Preparation: (1) From liquid air, (2) electrolysis of water, (3) heating potassium chlorate ($KClO_3$) in presence of manganese dioxide (MnO_2).

Physical properties: Colorless, odorless, tasteless, slightly heavier than air, only slightly soluble in water.

Chemical properties: Reacts moderately at ordinary temperatures and more readily at higher temperatures with many metallic and non-metallic elements. Forms basic oxides with metals and acidic oxides with nonmetals. Reactions in which oxygen combines with other elements are known as oxidation reactions.

Combustion: Oxygen supports combustion, which is a vigorous and rapid oxidation resulting in the release of heat and light energy. Substances that burn are combustible, flammable, or inflammable. Some substances burn better in pure oxygen than in air. Combustion reactions without oxygen are also possible, e.g., hydrogen burns in chlorine. The body obtains its energy by the oxidation of hydrogen in food chemicals:

$$2H_2 + O_2 \rightarrow 2H_2O + \text{energy}$$

This occurs under mild conditions because of the presence of enzyme catalysts.

Kindling temperature: Minimum temperature that must be reached before combustion of a substance can occur.

Spontaneous combustion: Combustion resulting from accumulation of heat from a slow oxidation.

Extinguishing fires: Fires can be extinguished by deprivation of the oxygen source or lowering the temperature below the kindling point.

Medical uses: Pure oxygen is used for breathing in situations requiring increased concentrations in order to maintain adequate supplies to the tissues. Oxygen is also used during administration of anesthetics.

OZONE

Allotropic modification of oxygen that exists as a gas with the formula O_3.

Preparation: By electric discharge in the presence of oxygen.

Properties: Heavier than oxygen, pale blue gas, has distinctive odor, much more soluble than oxygen in water, much more reactive than oxygen.

HYDROGEN

Occurrence: Discovered in 1776, not found free, but found combined abundantly in water, acids, bases, and many biologically important organic compounds.

Preparation: (1) Electrolysis of water; (2) displacement of hydrogen from acids by metals; (3) many industrial methods.

Properties: Hydrogen is a very light, colorless, odorless, tasteless, combustible gas, quite reactive, combining with many nonmetallic elements. Reactions of hydrogen are susceptible to the influence of platinum catalysts in test tubes and enzyme catalysts in the body. Hydrogen is a reducing agent.

Uses of hydrogen: Has many industrial uses. Biologically useful in the production of energy and body constituents.

OXIDATION-REDUCTION

Oxidation: (1) Addition of oxygen to a compound or element; (2) removal of electrons from an atom or increase in positive valence; (3) removal of hydrogen from a compound.

Reduction: Chemically the opposite of oxidation. (1) Removal of oxygen from a compound; (2) addition of electrons to an atom or decrease in positive valence; (3) addition of hydrogen to an element or compound.

Oxidation-reduction reactions: Whenever oxidation occurs in a chemical reaction, reduction must also occur in the same reaction. The oxidizing agent is reduced and the reducing agent is oxidized. Oxidation-reduction reactions take place in the body with production of energy under mild conditions and with good efficiency because of the presence of enzyme catalysts.

Antiseptic applications: Oxidizing substances often have antiseptic action on the basis of their ability to interfere with the normal metabolism of bacteria. Some commonly used antiseptics are hydrogen peroxide (H_2O_2), potassium permanganate ($KMnO_4$), and iodine (I_2).

Bleaching applications: Many colored substances can be decolorized by being oxidized or reduced. Good oxidizing bleaches are sodium hypochlorite ($NaClO$), potassium permanganate, and hydrogen peroxide. Good reducing bleaches are oxalic acid [$(COOH)_2$] and sodium thiosulfate ($Na_2S_2O_3$), especially useful with iodine stains.

REVIEW QUESTIONS

1. List the more important aspects of the physical and chemical properties of oxygen.
2. What is the relationship of oxygen to combustion? To energy production in the body? How is this related to the medical uses of oxygen?
3. List the more important aspects of the physical and chemical properties of hydrogen. In what body constituents is hydrogen found?
4. Define the terms *oxidation* and *reduction*. For every reaction in which an oxidation takes place, what else must also occur? What else is produced in such reactions in living organisms which is of great importance to the maintenance of life?
5. Define *kindling temperature*. What are the two most important considerations, one physical, one chemical, in putting out a fire?

REFERENCES

Lagowski, J. J.: Modern inorganic chemistry, New York, 1973, Marcel Dekker, Inc.

Masterton, W. L., and Slowinski, E. J.: Chemical principles, ed. 4, Philadelphia, 1977, W. B. Saunders Co.

Pauling, L.: General chemistry, ed. 3, San Francisco, 1970, W. H. Freeman and Co. Publishers.

Petrucci, R. H.: General chemistry: principles and modern applications, ed. 2, New York, 1977, Macmillan Publishing Co., Inc.

Water

7

"Water, water, everywhere . . ." is the beginning of a famous quotation from the poem *The Rime of the Ancient Mariner* by Samuel Taylor Coleridge. This statement is very close to the actual situation because water, the most abundant of all compounds, covers approximately three-fourths of the earth's surface. Because of its chemical stability, however, it was considered an element, not a compound, for many years dating back to the four "elements" of Aristotle (384-322 BC): earth, air, fire, water. Cavendish and others, in the 1770s, found that water was formed when his "inflammable air," recognized as an element and named hydrogen ("water former") by Lavoisier, burned in air. Other experiments by Lavoisier showed that hydrogen was combining with the oxygen of the air as it was burning and formed water, leading to the establishment of water as a compound.

OCCURRENCE

Water is found in, on, or around almost everything on earth. This includes many substances apparently dry until warmed or heated to high temperatures to drive off the water present, resulting in a loss of weight. All plants and animals are largely composed of water, the human body containing about 65% water. All foods usually contain large amounts of water.

PHYSICAL PROPERTIES

Water exists as a solid (ice), liquid (water), and gas (vapor or steam). As a liquid, pure water is odorless, tasteless, and colorless. It has a maximum density (weight per unit volume) at 4° C. Thus, when water freezes at 0° C (32° F), the ice formed is less dense than the surrounding water and consequently it floats on top of the water. This means that large bodies of water will freeze from the top down, and, since ice is a good insulator, it prevents the water under the ice from freezing solid, thereby protecting the plants and animals living below the surface.

At the other end of the scale, water will boil to form a vapor (gas) if heated to a high enough temperature. Liquids reach their boiling point when the vapor pressure (tendency of a liquid to become a gas) is equal to the atmospheric pressure. At sea level, or an atmospheric pressure of 760 mm Hg, water boils at 100° C (212° F). At higher altitudes, such as mountain tops where the atmospheric pressure is lower than at sea level, the boiling point is under 100° C. For instance, at an elevation of 30,000 ft, the boiling point of water is about 70° C. This sometimes leads to difficulties in preparing foods because the rate and efficiency of cooking depend on the temperature and not on the fact that boiling is occurring. In such cases, pressure cookers can be used to raise the boiling point. Pressure cookers are also used in ordinary cooking to raise the boiling point of water above 100° C, thus speeding up the process.

When water boils, each gram of water (1 ml) absorbs 539 calories (cal) of heat energy to provide the molecules with the added kinetic energy required to exist in the gaseous state. This quantity is known as the *heat of vaporization.* At normal body temperature of 37° C (98.6° F), 575 calories of heat are used up in vaporizing 1 g of water. This large quantity of heat required to convert liquid water to gaseous water makes this process very useful in controlling body temperature by way of evaporation of perspiration.

Another very important property of water is its ability to dissolve a large number of substances. This property of water depends to a large extent on its polarity (p. 44). Many of the solid substances that dissolve in water are electrovalent or polar compounds, and when placed in water, the positive ends of these compounds are attracted to the negative ends of the water molecules, and the negative ends are attracted to the positive ends, the net effect being a cancellation of the forces that normally maintain substances in the solid state. The strategic importance of this property of water is emphasized by the fact that most chemical reactions will not take place in the absence of water. The fact that hundreds and thousands of reactions are occurring at any moment in a living organism explains the requirement for the high concentration of water found in all tissues of the body.

Units of measurement based on physical properties of water

Because water is such a common substance, and because it can be highly purified with little difficulty, certain units of measurement are defined in terms of the physical properties of water.

The *Centigrade temperature scale,* devised in 1742 by Celsius, a Swede, has 100 degrees between the freezing point of water, 0° C, and the boiling point of water, 100° C, at standard pressure (760 mm Hg). Thus the centigrade, or Celsius, degree is defined as one one-hundredth of the difference in temperature between the freezing and boiling points of pure water. Using the Celsius scale, the unit of heat, the *calorie,* is defined as the quantity of heat required to raise the temperature of 1 g of water 1 degree from 15° to 16° C.

Another commonly used unit of measurement based on the physical properties of water is known as *specific gravity*. This quantity is determined by comparing the weight of 1 ml of a substance with the weight of 1 ml of water at 4° C, which is 1.000 g.

CHEMICAL PROPERTIES

As already mentioned, water is a relatively stable compound. When heated to 2,000° C, only about 1.5% and at 2,700° C, only 11% is decomposed. This is related to the fact that when one gram molecular weight (1 mole) of water (18 g) is produced by the combination of hydrogen and oxygen, approximately 68,000 cal of heat is liberated. The same large quantity of heat is produced by any reaction that produces water from hydrogen and oxygen, a fact that is the basis for the manner in which the body obtains energy for its requirements. The hydrogen comes from food substances and the oxygen from air breathed in by the body. This will be discussed in more detail in the chapters on metabolism.

The decomposition of water requires the input of at least as much energy as is produced in its synthesis and is most conveniently accomplished by electrolysis (p. 56).

Water reacts with certain metals, or their oxides, producing metallic hydroxides known as *bases:*

$$\underset{\textbf{Sodium}}{2Na} + 2H_2O \rightarrow \underset{\textbf{Sodium hydroxide}}{2NaOH} + H_2\uparrow$$

$$\underset{\textbf{Calcium oxide}}{CaO} + H_2O \rightarrow \underset{\textbf{Calcium hydroxide}}{Ca(OH)_2}$$

Water also reacts with some oxides and nonmetals, producing substances known as *acids:*

$$\underset{\textbf{Sulfur trioxide}}{SO_3} + H_2O \rightarrow \underset{\textbf{Sulfuric acid}}{H_2SO_4}$$

$$\underset{\textbf{Carbon dioxide}}{CO_2} + H_2O \rightarrow \underset{\textbf{Carbonic acid}}{H_2CO_3}$$

Another chemical property of water of great importance to the functioning of living organisms is the process known as *hydrolysis (hydro-* water; *lysis-* breaking apart), breaking apart under the influence of water. A simple example of such a reaction is the hydrolysis of sodium carbonate when dissolved in water:

$$Na_2CO_3 + 2HOH \rightarrow 2NaOH + H_2CO_3$$

Sodium carbonate **Sodium hydroxide** **Carbonic acid**

In the body, the reactions of digestion are mainly hydrolytic, as a result of which large, complex compounds are broken down into smaller components:

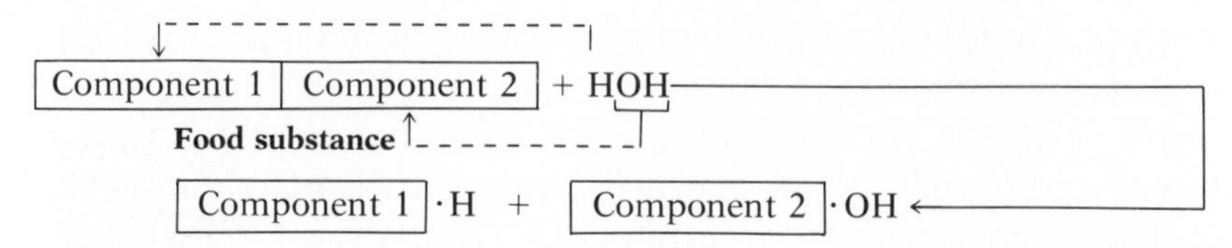

HYDRATES AND THEIR PROPERTIES

When certain substances are allowed to crystallize in the presence of water, combinations known as *hydrates* are formed. In these hydrates, the water molecules are associated in a definite manner with the other substance of the combination but not in true chemical combination. Thus, in a hydrate, both the water molecules and the other substance retain their identities. This relationship is designated by placing a dot after the formula of a substance forming a hydrate, followed by the number of molecules of water associated with the substance. Copper sulfate forms a blue crystalline hydrate in which one molecule of copper sulfate is associated with five molecules of water. Its formula, therefore, is written as $CuSO_4 \cdot 5H_2O$. When hydrates are heated, the water of crystallization is lost, and the substance crumbles to a powder.

Copper sulfate from which the water of crystallization has been removed is called *anhydrous* ("without water") and is written as $CuSO_4$. Some hydrates have a tendency to lose water of crystallization simply on exposure to air (higher vapor pressure than the vapor pressure of air). Such substances are called *efflorescent* substances. Other hydrates have a tendency to absorb water from the air (lower vapor pressure than the vapor pressure of air). These substances are known as *hygroscopic* materials. Hygroscopic substances that tend to absorb water so strongly that a solution is formed are called *deliquescent* substances. A common example of such a substance is calcium chloride, which is frequently used to settle the dust on rural dirt roads or to absorb water from wet basements.

A hydrate of interest in medicine, although it is used mainly in the building industry to make walls and wall boards, is plaster of paris. Plaster of paris is made by heating gypsum, a hydrate of calcium sulfate:

$$2(CaSO_4 \cdot 2H_2O) \rightarrow (CaSO_4)_2 \cdot H_2O + 3H_2O$$

Gypsum **Plaster of paris**

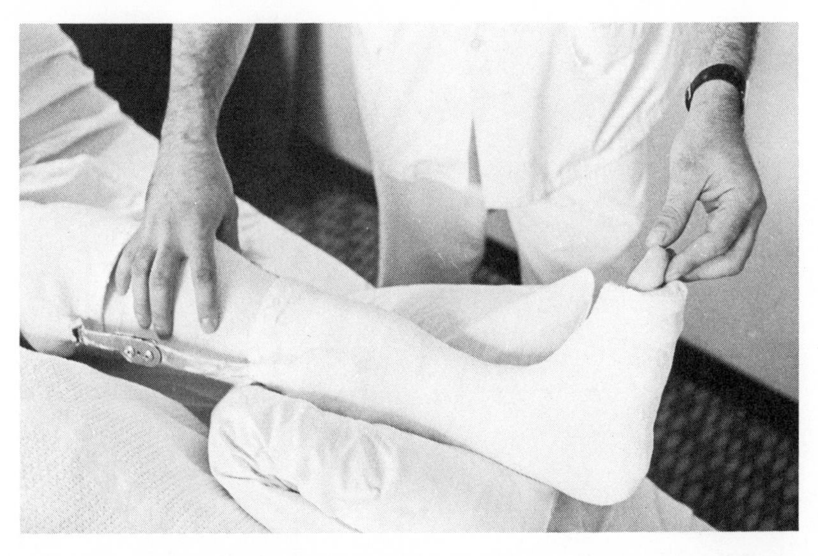

Fig. 7-1. Leg with fractured bone set in cast of plaster of paris, which keeps broken ends rigid and aligned during healing process. (Courtesy Barnes Hospital, St. Louis.)

When plaster of paris is treated with water, it rapidly forms gypsum by a reversal of this reaction and sets in the form of a hard, crystalline hydrate. In setting, or hardening, it expands somewhat, a property that is very useful in making rigid, form-fitting casts around broken bones (Fig. 7-1). Plaster of paris and gypsum represent two different hydrates of the same parent compound. There are other compounds that can form more than one hydrate.

WATER FOR HUMAN USE; PURIFICATION

". . . and not a drop to drink," completes the quotation cited at the beginning of this chapter. This quotation emphasizes the increasing importance of the problems being encountered, in the United States and the rest of the world, in obtaining and purifying sufficient water for human use.

Impurities in water

Natural water supplies may contain any combination or all of the following types of impurities:

1. Suspended substances such as sand, insoluble inorganic or organic matter, bacteria
2. Dissolved inorganic salts, commonly the chlorides, sulfates, and bicarbonates of sodium, potassium, calcium, magnesium, and iron
3. Dissolved organic compounds resulting from decay of once-living matter and industrial wastes

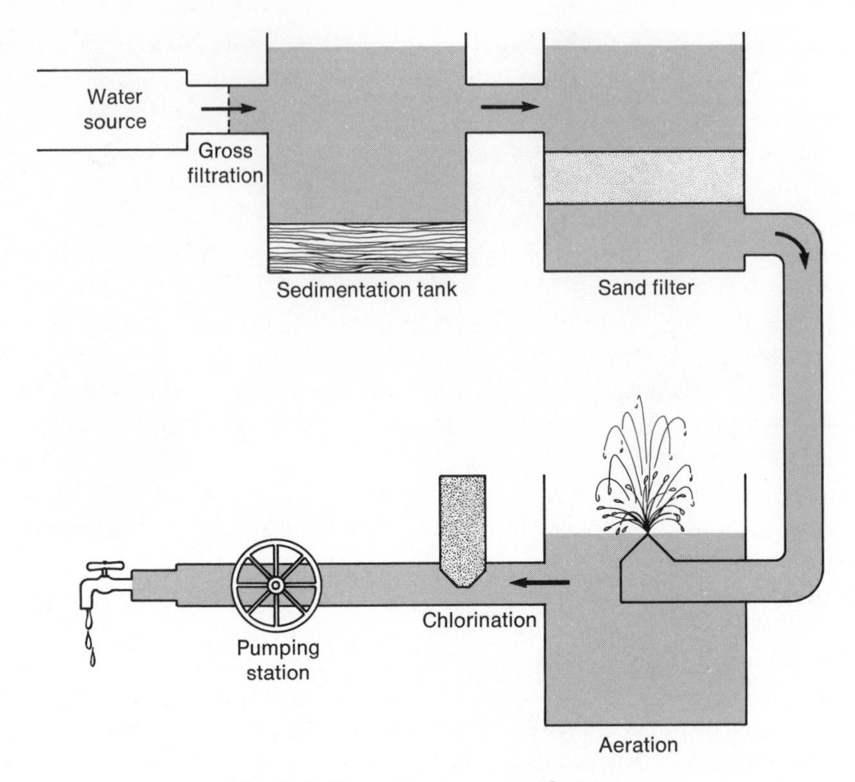

Fig. 7-2. Steps in water purification.

4. Dissolved gases such as carbon dioxide, nitrogen, and ammonia

In the interest of brevity, increasing human contributions to these naturally occurring impurities and the consequences thereof have been omitted here. Regardless of their source, the impurities must still be removed to a sufficient degree to make the water palatable and safe for human consumption (Fig. 7-2).

Methods for removing impurities

Because of the different types of impurities involved, no one method is sufficient to completely purify a water supply. A combination of the methods described here, or possibly a sequence of all methods, is generally required for large water supplies.

Filtration. Most suspended solids can be removed by allowing heavier particles to settle to the bottom in a holding tank and then passing the water through a series of sand beds.

Chlorination. Chlorine added to water is the most commonly used method for killing bacteria in the water because of its effectiveness and economy. A concentration of one part chlorine in one million parts of water is usually sufficient to destroy bacterial contamination in water.

Aeration. Dissolved gases may be removed from water by long exposure to air or by being sprayed into the air. This increased exposure to oxygen also aids in destroying bacteria and oxidizes organic matter.

Ozonation. Although ozone is a more effective germicide than chlorine or oxygen, it is not in general use because of its greater cost. However, for some special applications it may be the method of choice.

Distillation. When water is heated to boiling, steam is formed. The steam can be caused to form liquid water again (distillate) by passing it over a cool surface. Solid materials that cannot vaporize are thus left behind in the boiler, bacteria are killed by the heating, but most gases will still be found in the distillate. As with ozonation, distillation is a costly process and is usually used only for special purposes or when a better method is unavailable.

Removal of salt from sea water. In some areas of the world, the sea would be the major source of water if a suitable method of removing the salt were available. The United States government continues to sponsor research on several different methods of accomplishing this task. The energy requirements, regardless of the method used, are very high, and some of the research is aimed at harnessing atomic or solar energy for this purpose, especially in parts of the world where other sources of energy are lacking.

Deionization. It is sometimes necessary or desirable to purify water by removing all of the components of dissolved salts (ions). This is accomplished by passing the water over columns or beds of appropriate materials (known as ion-exchange resins) that can substitute hydrogen ions for some of the salt ions and hydroxyl ions for others, leaving only water to come off such columns. This water is almost equivalent to distilled water and is sometimes produced in small quantities in the home for use in steam irons.

Hard and soft water

Most purification procedures do not remove dissolved inorganic salts. Certain of these salts form precipitates with soaps, destroying the ability of soaps to produce lather with water. Water containing salts of calcium and magnesium forms such precipitates and is known as *hard water*. Hard water that can be converted to soft water simply by heating is known as *temporary hard water*, as, for example, water containing calcium or magnesium bicarbonates. The bicarbonates are converted to insoluble carbonates by heating, thus removing the calcium or magnesium salt from solution leaving the soap free to lather again:

$$\underset{\text{Calcium bicarbonate}}{Ca(HCO_3)_2} \xrightarrow[\Delta]{} \underset{\text{Calcium carbonate}}{CaCO_3}\downarrow + H_2O + CO_2$$

Water containing the chloride or sulfate salts of calcium or magnesium, which cannot be removed by heating, is known as *permanent hard water.* Such water may be softened by chemical means, such as the addition of sodium carbonate (washing soda). This precipitates the carbonate of calcium or magnesium, again leaving the water free to lather:

$$\underset{\text{Calcium sulfate}}{CaSO_4} + \underset{\text{Sodium carbonate}}{Na_2CO_3} \rightarrow CaCO_3\downarrow + Na_2SO_4$$

Deionization of water, as discussed before, can also soften permanent hard water. For areas with very hard water, tank units are available for home use to soften such water.

Another relatively modern innovation is the use of synthetic detergents for washing purposes. These substances are not inactivated by calcium or magnesium salts in the water and can lather in hard water. However, the use of detergents caused a serious pollution problem because many of these substances were not biodegradable, i.e., they were not destroyed in the natural environment. As a result, many streams ended up as frothy masses of lather. Much progress has been made in synthesizing biodegradable detergents.

SUMMARY

OCCURRENCE Very abundant. All animals and plants are composed largely of water, the human body being about 65% water.

PHYSICAL PROPERTIES Water exists as a solid, liquid, or gas. As a liquid it is odorless, tasteless, colorless, has a maximum density at 4° C, freezes at 0° C (32° F), and boils at 100° C (212° F) at sea level. At higher altitudes (lower atmospheric pressure), the boiling point is lower than 100° C. Heat of vaporization of water is 539 cal/g at 100° C or 575 cal/g at body temperature (37° C, 98.6° F). This provides body with a method for controlling temperature by evaporation of perspiration. Water dissolves a very large number of substances because of its polar nature.

Units of measurement based on physical properties of water:

1. Centigrade temperature scale (Celsius) has 100 degrees between freezing point of water, 0° C, and boiling point of water, 100° C.
2. Calorie—heat required to raise the temperature of 1 g water 1 degree from 15° to 16° C.
3. Specific gravity—weight of 1 ml of a substance compared to the weight of 1 ml of water at 4° C (1.000 g).

CHEMICAL PROPERTIES Water is a stable compound that can be heated to high temperatures with little decomposition. It can be decomposed by electrolysis:

$$2H_2O \xrightarrow[\text{current}]{\text{electric}} 2H_2\uparrow + O_2\uparrow$$

Water reacts with certain metals, or their oxides, to form metallic hydroxides (bases):

$$2Na + 2H_2O \rightarrow 2NaOH + H_2\uparrow$$

Water reacts with some oxides of nonmetals to form acids:

$$SO_3 + H_2O \rightarrow H_2SO_4$$

Water breaks down many substances into smaller components, a process known as hydrolysis:

$$Na_2CO_3 + 2HOH \rightarrow 2NaOH + H_2CO_3$$

The reactions of digestion are mainly hydrolytic, breaking food substances into smaller molecules.

HYDRATES AND THEIR PROPERTIES

Hydrates are formed by the association of certain substances with one or more molecules of water as crystallization occurs. The hydrate of copper sulfate is indicated as $CuSO_4 \cdot 5H_2O$. Heating a hydrate to remove the water produces an anhydrous substance. Hydrates that give up their water on exposure to air are efflorescent substances. Those that absorb water are hygroscopic, and if they absorb enough water to form solutions, are deliquescent substances. The hydrate plaster of paris, when treated with water, is converted to another hydrate, gypsum, which expands as it sets, making it useful for rigid plaster casts around broken bones:

$$(CaSO_4)_2 \cdot H_2O + 3H_2O \rightarrow 2(CaSO_4 \cdot 2H_2O)$$

WATER FOR HUMAN USE; PURIFICATION

Before water is suitable for human use certain impurities must be removed or modified.

Impurities: (1) Suspended solids, sand, bacteria, etc.; (2) dissolved inorganic salts such as chlorides, sulfates, bicarbonates of sodium, potassium, calcium, magnesium, and iron; (3) dissolved organic compounds from decaying matter; and (4) dissolved gases such as carbon dioxide, ammonia.

Methods for removing impurities: (1) Filtration, removes suspended solids, some bacteria; (2) chlorination, kills bacteria; (3) aeration, kills bacteria and removes dissolved gases; (4) ozonation, kills bacteria; (5) distillation, removes dissolved substances; (6) removal of salt from sea water is being investigated; (7) deionization of water is accomplished by passing water through columns of appropriate ion-exchange resins.

Hard and soft water: Hard water is water in which soap cannot form a lather because the soap is precipitated by calcium or magnesium salts. Temporary hard water can be softened by heating:

$$Ca(HCO_3)_2 \xrightarrow[\Delta]{} CaCO_3 \downarrow + H_2O + CO_2$$

Permanent hard water can be softened by chemical treatment (also by deionization):

$$CaSO_4 + Na_2CO_3 \rightarrow CaCO_3 \downarrow + Na_2SO_4$$

Detergents, which are not inactivated by calcium or magnesium salts, can form lather in hard water.

REVIEW QUESTIONS

1. List the more important physical and chemical properties of water. What is the approximate concentration of water in the human body?
2. What is the significance of the high heat of vaporization of water to humans? Is the polar nature of water of significance to humans?
3. Define the process known as hydrolysis. Write an equation or describe the important aspects of the process. What important function in the body occurs mainly as hydrolytic reactions?
4. What types of materials, of possible danger to life, must be removed from water supplies before they are suitable for human use?
5. What is meant by the terms *hard* and *soft water?* What are some of the procedures that may be used to soften hard water?

REFERENCES

Lagowski, J. J.: Modern inorganic chemistry, New York, 1973, Marcel Dekker, Inc.

Masterton, W. L., and Slowinski, E. J.: Chemical principles, ed. 4, Philadelphia, 1977, W. B. Saunders Co.

Pauling, L.: General chemistry, ed. 3, San Francisco, 1970, W. H. Freeman and Co., Publishers.

Petrucci, R. H.: General chemistry: principles and modern applications, ed. 2, New York, 1977, Macmillan Publishing Co., Inc.

Solutions, osmosis, dialysis

8

SOLUTIONS

When granules of sugar are placed in water and stirred, the sugar apparently disappears. The resulting mixture is called a *solution* and is distinguished from another type of mixture, e.g., a mixture of sand and water, in terms of homogeneity (uniformity). In the sugar solution, the molecules of sugar have become so uniformly dispersed among the water molecules that the concentration of sugar in each unit volume of solution is the same as that in any other and remains so. Since molecules cannot be seen by eye, the sugar molecules seem to disappear. In the sand and water mixture, however, the grains of sand are not the same size, and most are heavy enough to settle out almost immediately, so that there is very little uniformity in the various parts of the mixture: the mixture is nonhomogeneous. A solution can be defined, therefore, as a homogeneous mixture of two or more substances at the molecular level.

The components of a solution are known as the *solute* and the *solvent,* the solute being the substance that is dissolved and the solvent being the substance in which the solute is dissolved.

Types of solutions

Because matter can exist in three forms, solid, liquid, and gas, there can be nine different types of solutions. For the study of living organisms, however, the three most important types of solutions are (1) solid in liquid, (2) liquid in liquid, and (3) gas in liquid. Furthermore, the most important liquid solvent is water, and the resultant solutions are known as *aqueous solutions.*

Solutions are significant to life because (1) living tissue is largely composed of water, (2) nutrients are brought to the tissues in the form of solutions and wastes are removed in the form of solutions, and (3) most chemical reactions take place in solution.

Factors influencing solubility

The following considerations apply mainly to the solution of solid solutes in water as the solvent.

Nature of solute and solvent. Water, which is a polar compound, will dissolve most inorganic compounds, which are also polar. Thus follows the generalization that polar solvents tend to dissolve polar solutes. Details of why this is true will be discussed in the chapter on ionization. Organic compounds, which are mostly nonpolar, do not dissolve to any extent in water but do dissolve in organic solvents. Therefore, it is generally found that nonpolar solvents dissolve nonpolar solutes.

Solute surface area. A solute in the form of large crystals or aggregates takes longer to dissolve than does the same substance in a finely powdered form. This is so because in the finely powdered form much more surface of the substance is in contact with the solvent at any one time than is the case with large crystals.

Stirring. A solute added to a solvent and left undisturbed will not dissolve as rapidly as when the mixture is stirred constantly. Such stirring allows fresh solvent molecules to come into more frequent contact with solute surface than if the solute were left undisturbed.

Temperature. In most cases, raising the temperature of the solvent will increase the rate of solution of a solute. There are, however, some instances where the opposite is true.

When gases are dissolved in liquids, the lower the temperature, the greater the solubility. When opened, a cold bottle of soda water does not "fizz" (release carbon dioxide gas) as much as a warm bottle.

Pressure. Changes in pressure have little effect on the solubility of solid solutes. However, in the case of solutions of gases, the higher the pressure, the greater the solubility.

General specification of concentration of solutions

Dilute solutions. When the amount of solute is small compared to the amount of solvent, the solution is known as a *dilute solution.*

Concentrated solutions. When the amount of solute is large compared to the amount of solvent, the solution is a *concentrated* solution.

Saturated solutions. A *saturated* solution is one in which no more solute can be dissolved by the solvent in the presence of some excess solute. Under such conditions with a constant temperature, an equilibrium exists between the dissolved and undissolved solute, i.e., for every portion of the solute that may go into solution, an equal portion of solute will crystallize out of solution.

Supersaturated solutions. Because most solid solutes dissolve to a greater extent at higher temperatures, a saturated solution prepared at a high temperature when allowed to cool slowly and carefully to a lower temperature may, by keeping its excess solute in solution at the lower temperature, form a *supersaturated* solution. If such a solution is jarred sharply or a crystal of solute is added to the solution, the excess solute will generally crystallize out of solution quite suddenly. Honey is a common example of a supersaturated solution that may crystallize slowly on standing. Heating in a hot water bath (honey jar in pot of hot water) will usually redissolve the crystallized sugar.

Exact specification of concentration of solutions

In most chemical and biologic applications, it is usually necessary to know more exactly than is possible with the general designations of dilute or concentrated solutions, how much solute is found in a certain amount of the solution, or its exact concentration. Discussed here are the three specific designations of concentration of solutions used for this purpose. With any of these designations, it becomes a simple matter to calculate exactly how much solute or solution has been or should be used.

Percentage solutions. As indicated by the term, percentage solutions are made up on the basis of 100 parts. To be strictly accurate, the units used should be designated. For example, a 5% solution of sodium chloride (NaCl) in water is usually prepared by dissolving 5 g of sodium chloride in water to make a total volume of 100 ml of solution. This solution, typical of solid solutes in water, should be specified as 5% sodium chloride (w/v), where w indicates weight of sodium chloride and v indicates that the final solution was measured by volume. By strict definition, a percentage solution should be made up on the basis of parts by weight of solute per one hundred parts by weight of final solution, or 5 g sodium chloride in 100 g of final solution in water, designated as 5% sodium chloride (w/w). Percentage solutions of liquids are usually made up on the basis of measured volumes, e.g., a 5% solution of ethyl alcohol (C_2H_5OH) in water is usually prepared by adding 5 ml of alcohol to water to make a final solution of 100 ml, designated as 5% alcohol (v/v). Unless the basis for the percentage solution is indicated as just described, confusion and errors are possible.

In the actual preparation of a w/v percentage solution, the correct amount of solute is weighed and placed in a volumetric flask (a flask with an accurately measured mark to indicate the total volume, e.g., 100 ml) containing less than the full volume of water. The solute is then dissolved and water is added until the measured mark is reached. The solution is then thoroughly mixed and is ready to be used.

Molar solutions. For many purposes, it is desirable to make solutions in such a way that the amount of solute in terms of its molecular weight can easily be calculated from the amount of solution used. Thus a one molar (1M) solution is made by weighing out one gram molecular weight of a solute and dissolving it in water to give a final volume of 1 liter. If 500 ml of such a solution is used, it means that ½ mole (mole = gram molecular weight) of the solute has been used. Some practical examples will illustrate these matters.

OBJECT: To prepare a 1M solution of NaCl.

atomic weight of Na = 23
atomic weight of Cl = 35.5
sum of atomic weights = molecular weight = 58.5

In grams, one gram molecular weight, or 1 mole sodium chloride is 58.5 g. This amount dissolved in water to give a final volume of 1000 ml of solution, makes a 1M sodium chloride solution.

OBJECT: To determine the weight of solute in 250 ml of a 0.5M NaCl solution.

The molecular weight of NaCl, 58.5, is determined as already shown.
1 liter of a 0.5 M NaCl solution contains ½ × 58.5 g NaCl, or 29.3 g NaCl.
250 ml of a 0.5M NaCl solution contains $^{250}/_{1,000}$ × 29.3 g NaCl, or ¼ × 29.3 = 7.3 g NaCl.

Normal solutions. Especially useful for many analytical purposes are solutions prepared in such a way that a definite unit of one solution contains an amount of solute that is chemically equivalent to the amount of any other solute in the same unit of solution. This is achieved by using as the basis for preparing such solutions a quantity known as the *equivalent weight.* An equivalent weight may be defined as that weight of a substance that can liberate or combine with 1 mole of hydrogen (1 g), or 1 mole of hydroxyl ions (17 g). Expressed in grams, this becomes the gram equivalent weight.

The equivalent weight is related to the molecular weight, as the following examples show. One mole of hydrogen chloride (HCl) contains 1 mole of hydrogen, or one equivalent weight. One mole of sulfuric acid (H_2SO_4) contains 2 moles of hydrogen, or two equivalent weights. Thus, the equivalent weight of a substance may also be defined as the molecular weight divided by the appropriate valence:

$$\text{For HCl, the equivalent weight} = \frac{MW}{V} = \frac{36.5}{1} = 36.5$$

$$\text{For } H_2SO_4\text{, the equivalent weight} = \frac{MW}{V} = \frac{98}{2} = 49$$

A one normal (1N) solution of hydrogen chloride would contain 36.5 g hydrogen chloride in water made up to a final volume of 1 liter. A 1N solution of sulfuric acid would contain 49 g of sulfuric acid in 1 liter of solution. To determine the weight of solute in a given amount of a solution of known normality, the same calculations as described under molar solutions are performed using equivalent weights in place of molecular weights.

The particular advantage of preparing solutions on the basis of normality is the ease with which the equivalents, or multiples thereof, used or required, can be calculated. One liter of any 1N solution contains the same number of equivalents, one, as 1 liter of any other 1N solution. The following expressions can be derived from this relationship:

$$\text{Volume in liters} \times \text{Normality} = \text{Number of equivalents}$$

or

$$\text{Volume in ml} \times \text{Normality} = \frac{\text{Number of equivalents}}{1000} =$$

$$\text{milliequivalents (mEq) or simply, ml} \times \text{N} = \text{mEq}$$

To illustrate the application of this equation, suppose that a 10 ml portion of solution A, which is 2N, reacts exactly with, i.e., is chemically equivalent to, 40 ml of solution B of unknown normality. The following relationships hold for this situation:

$$\text{ml solution A} \times \text{N solution A} = \text{mEq A}$$
$$\text{ml solution B} \times \text{N solution B} = \text{mEq B}$$

Since solution A and solution B are equivalent under the conditions listed above:

$$\text{mEq A} = \text{mEq B}$$

$$\therefore \text{ ml solution A} \times \text{N solution A} = \text{ml solution B} \times \text{N solution B}$$

substituting,

$$10 \times 2 = 40 \times \text{N solution B}$$

$$\text{N solution B} = \frac{10 \times 2}{40} = \frac{20}{40} = 0.5$$

The expression:

$$\text{ml}_A \times \text{N}_A = \text{ml}_B \times \text{N}_B$$

applies to any situation where two solutions are found to be chemically equivalent. Therefore, if any three of the factors are known, the fourth can be calculated. Some definite applications will be noted in later discussions.

Physical properties of solutions

In these discussions, attention will be focused on the type of solution of greatest importance to living organisms—solid solutes in water.

Vapor pressure. An important physical property of water is its tendency to evaporate, or escape from the liquid. This escaping tendency is known as the *vapor pressure.* As has already been noted, when the vapor pressure of water becomes equal to the atmospheric pressure (the force tending to prevent the evaporation of water molecules), the water boils. The vapor pressure of water in a solution is less than that of pure water under the same conditions of temperature and pressure. The reason for this may be that in a solution there will be fewer molecules of water at the surface at any time than in pure water and therefore the vapor pressure will be lower (Fig. 8-1).

Evidence that this may be the reason for the decrease in vapor pres-

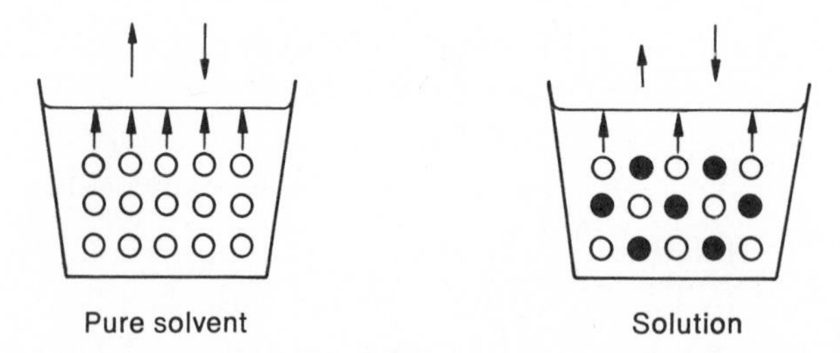

Fig. 8-1. Vapor pressures developed by a solution and the pure solvent. The vapor pressure of the solution is lower than that of the solvent.

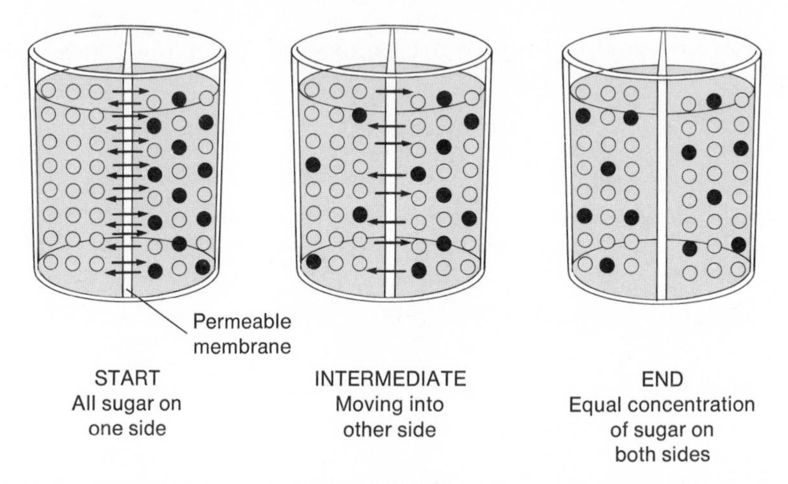

Fig. 8-2. Three stages of the process of diffusion through a *permeable* membrane.

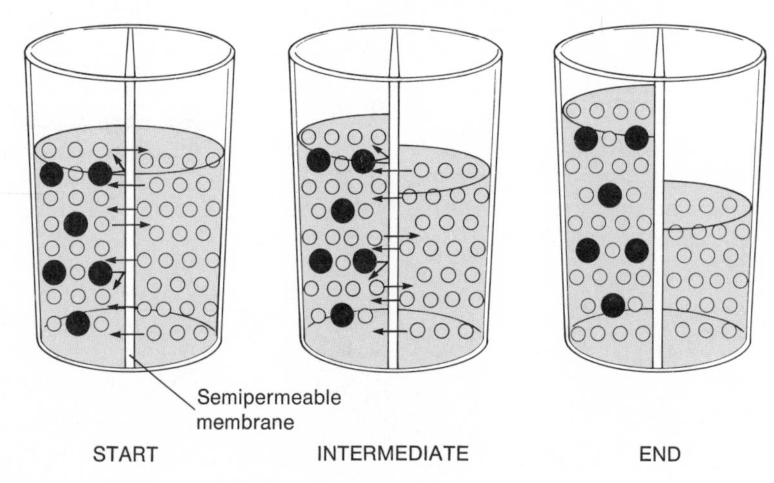

Fig. 8-3. Three stages of the process of osmosis through a *semipermeable* membrane.

sure of solutions compared to that of pure solvent is found in the fact that the type of solid solute used is not important, but the number of particles of solute in a definite amount of solution is important. In other words, if the concentration of one kind of solid solute particles in one solution is the same as the concentration of a different kind of solid solute particles in another solution, the decrease in vapor pressure will be the same for both solutions.

Elevation of the boiling point of the solvent in solution. As already stated, the vapor pressure of the solvent in a solution is lower than that of the pure solvent. This means that the solution must be raised to a higher temperature than the pure solvent before the vapor pressure will be equal to the atmospheric pressure. Therefore the boiling point of water in a solution is higher than the boiling point of pure water.

Lowering of the freezing point of the solvent in solution. Although the explanation is not as simple as in the case of the elevation of the boiling point in solution, it is also found that the freezing point of water in a solution is lower than the freezing point of pure water. Everyone is familiar with the use of antifreezes (these are liquid solutes) to keep the water in car radiators from freezing in the winter. Since these properties are related to changes in the vapor pressure, which are related to the concentration of solute, or ultimately to its molecular weight, the lowering of the freezing point may sometimes be used for the determination of molecular weights.

Diffusion. If a large crystal of blue copper sulfate is placed at the bottom of a long cylindrical container filled with water and left undisturbed for several months, slowly but definitely, a line of blue dissolved copper sulfate will rise in the cylinder. This indicates the process of *diffusion,* or movement of solute in the solvent, resulting from the fact that the molecules of solvent and solute are in constant motion, although the solution appears motionless.

Osmosis. A compartment containing pure water is separated from a compartment containing a solution of sugar in water by a material similar to a common window screen with openings slightly larger than either the water or sugar molecules. Left this way for a time, the sugar molecules will gradually distribute themselves equally in both compartments as a result of diffusion because both water and sugar molecules can pass freely from either compartment to the other (see Fig. 8-2).

If, instead of the sugar solution, a solution containing molecules of a substance too large to pass through the screen is placed in the compartment next to the pure water, something different occurs. Water will now tend to flow from the compartment of pure solvent, or lower concentration of solute, into the compartment of higher concentration of solute, or lower concentration of solvent. This process is known as *osmosis* and is very important in nature because most membranes found in all living cells are similar in action to the imaginary screen used in this discussion. Such membranes are known as *semipermeable membranes,* freely permeable to water and some small dissolved molecules but impermeable to large molecules. As osmosis continues, the volume of solution and the pressure due to the rising column of liquid (hydrostatic pressure) on the side containing the impermeable solute will increase until the tendency of water to pass from one compartment to the other will be equal. At this point, the hydrostatic pressure is equal to the osmotic pressure (see Fig. 8-3).

Although a simple, precise, physical explanation of osmosis is not possible, a practical explanation, related to the principles involved, that will give a working understanding of the process involved is possible and desirable. As stated in the discussion of vapor pressure, the vapor pressure of a solution is lower than that of the pure solvent because fewer molecules of solvent appear at the surface of the solution than in the pure solvent. Similarly, in the case of osmosis, fewer molecules of water will appear at the semipermeable membrane in the compartment containing the impermeable solute than in the compartment containing only water. Water molecules will then have a greater opportunity to go from the pure water compartment into the solution compartment than the reverse.

The simplest way to look at osmosis is as a tendency in nature toward equalization—in this case, the concentration of solute in both compartments. Since the solute molecules are too large to move from one compartment to the other, the burden falls on water, which can try to equalize the concentration of solute by diluting the solute molecules trapped in the one compartment. The water must therefore flow from the compartment with a lower concentration of solute to the compartment with a higher concentration of solute, thereby diluting the solute in the compartment of higher solute concentration and bringing the concentrations in both compartments closer to each other.

In this discussion, the primary concern has been with the impermeable solute concentration. It is also true, when attention is focused on the solvent concentrations, that in osmosis, the flow of water will be from the compartment of higher water (solvent) concentration to the compartment of lower water (solvent) concentration.

Osmosis and the red cell. The wall of the red blood cell acts as a semipermeable membrane. In order for it to maintain its uniformly globular shape, the red cell must be kept in a solution that has the same osmotic pressure as inside the red cell. Under such conditions, there is no net change in volume of the red cell and the solution is said to be an *isotonic* solution. The osmotic pressure of the complex liquid portion of the blood containing water, salts, and other substances can be matched by a 0.9% sodium chloride solution. A red cell placed in such a solution will not be damaged because the 0.9% sodium chloride solution is isotonic.

If, however, red cells are placed in a *hypotonic* solution, a solution containing a lower concentration of solute components than normal blood, osmosis will occur with water moving from the compartment with lower solute concentration, the hypotonic solution, into the compartment with the higher solute concentration, the red cells. Should the process continue sufficiently long, the contents of the red cells increase to the

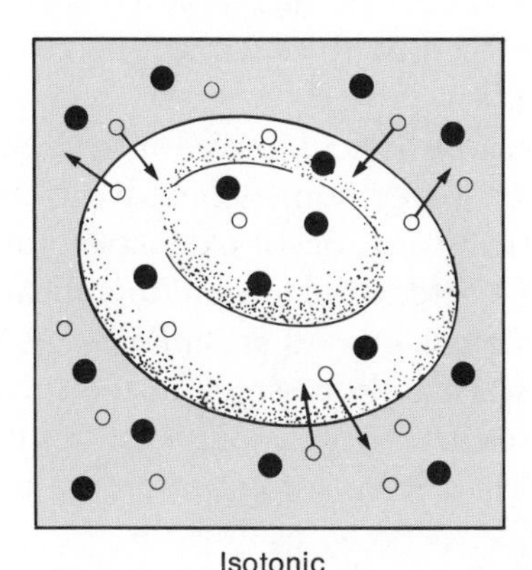

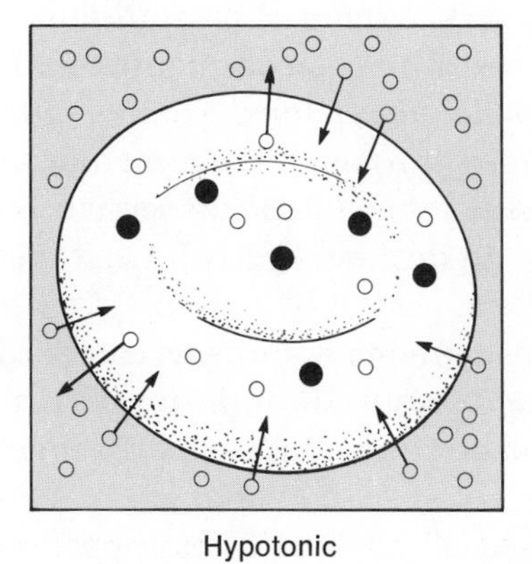

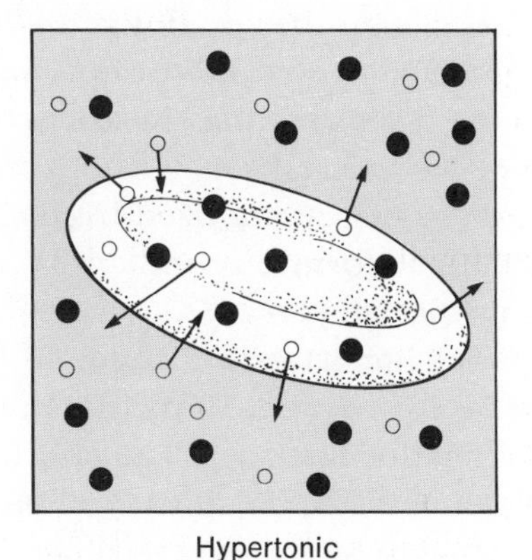

Fig. 8-4. Red cells in isotonic, hypotonic, and hypertonic solutions. No change (net flow of water = 0) in isotonic solution. Lysis occurs in hypotonic solutions (net flow of water is *into* the red cells). Crenation occurs in hypertonic solutions (net flow of water is *out* of red cells).

point at which they burst, or *hemolysis* occurs. Such a situation would occur if red cells were placed in water rather than an isotonic salt solution.

On the other hand, if red cells are placed in a *hypertonic* solution, a solution containing a higher concentration of solute components than normal blood, osmosis will occur in the opposite direction, i.e., water will flow from the red cell, which is now the compartment of lower solute concentration, to the hypertonic solution, which is the compartment of higher solute concentration. This process in which the volume of the red cell decreases and shrinks unevenly is known as *crenation.* These different situations are pictured in Fig. 8-4.

On the basis of these considerations, it is obvious that when injecting substances into the blood great care must be used not to disturb the osmotic equilibrium between the blood and the red blood cells.

COLLOIDS

Thus far the discussions have referred to true solutions, i.e., solutions in which the solute is dissolved homogeneously in the solvent at the molecular level and no settling out occurs. Most of the solutes of this type, which can pass through filters and animal membranes and can be crystallized from solution, are designated *crystalloids.* At the other extreme are substances that do not actually dissolve in the solvent in which they are placed and the particles are heavy enough so that they settle on standing. Such mixtures are called *suspensions,* most paints providing good examples of suspensions.

Intermediate in size between the molecular size of crystalloids and the large size of substances found in suspensions are the *colloids,* substances that, though they do not actually dissolve in the sense of crystalloids, are light enough not to settle out. Such mixtures, although commonly referred to as *colloidal solutions,* are actually *colloidal dispersions.* Solutes of this type may pass through filters but do not pass through animal membranes and are not crystallized from solution. When a colloidal dispersion is in the liquid form, e.g., milk, it is called a *sol.* In the form of a solid, e.g., gelatin, it is called a *gel.*

The importance of colloidal dispersions in living organisms can hardly be overestimated. Many functions in the body result from the properties of colloidal substances. The ability of the blood to regulate the exchange of water between the blood and tissues depends on the colloidal nature of the large plasma protein molecules. When such colloidal particles appear only on one side of a semipermeable membrane, i.e., the blood vessels, particularly the capillaries, an osmotic pressure, known as the *colloid osmotic pressure* of the blood, is produced. The process of blood clotting is based on the formation of a gel. In all cells and tissues are found colloidal substances and many types of semipermeable membranes, the properties of which must be taken into account in order to understand the proper functioning of a living organism.

Properties of colloids

Size. The difference in size between crystalloidal and colloidal substances is most easily demonstrated on the basis of the Tyndall effect. If a beam of light is passed through a true solution, the path of the beam is not visible if viewed at right angles to the direction of the beam because the crystalloidal solute molecules are too small to reflect sufficient light to be seen. However, if the same procedure is carried out with a colloidal solution, the colloidal solute particles are large enough to reflect sufficient light at right angles to the direction of the beam so that the path through the colloidal solution becomes visible (Fig. 8-5).

Osmotic pressure. As has already been stated, the size of colloidal substances makes them unable to pass through semipermeable membranes, and this provides one of the factors necessary for the occurrence of os-

Fig. 8-5. Demonstration of the Tyndall effect. The light source at the left is blocked except for a small aperture that allows a restricted beam of light to pass first through a glass vessel filled with a solution of sodium chloride, a crystalloidal substance, and then through a vessel filled with a colloidal suspension of gelatin. The observer is looking through the solutions from the top of the vessels at right angles to the light beam. No light is seen in the solution to the left, but the beam is obvious in the colloidal suspension to the right and in the glass walls of the vessels, which also contain colloidal molecules large enough to reflect light.

motic pressure. This osmotic pressure may then be used in the control of important functions and processes in the body.

Dialysis. If a mixture of crystalloidal and colloidal solutions is placed in one compartment separated by a semipermeable membrane from another compartment containing only distilled water, it will be found that the small crystalloidal solute molecules will gradually find their way into the distilled water compartment while the colloidal substance remains in its own compartment because it cannot pass through the membrane. Because the crystalloidal solute can also go back to its original compartment, essentially all of it can be coaxed into the distilled water by replacing the distilled water compartment with several changes of fresh distilled water. Such a process is known as *dialysis* (Fig. 8-6).

At the same time, some osmosis, or flow of water into the compartment containing the colloidal substance, will also occur. This is more like the situation as it actually occurs in living tissues such as the intestine during digestion and absorption of food materials, in the bloodstream during distribution of nutrients and collection of waste materials, and in the kidneys as some of the waste materials find their way into the urine. It should always be remembered that the interactions between the various compartments of the body, and between compartments in each cell as well, are mediated by membranes through which both osmosis and dialysis are occurring simultaneously. Some idea of the multiplicity of such membranes may be gained from an examination of Fig. 8-7, showing the membranes to be found in only a small portion of a single cell.

Brownian movement. In contrast to the rapid molecular motion found in true solutions, a much more sluggish movement of colloidal particles was

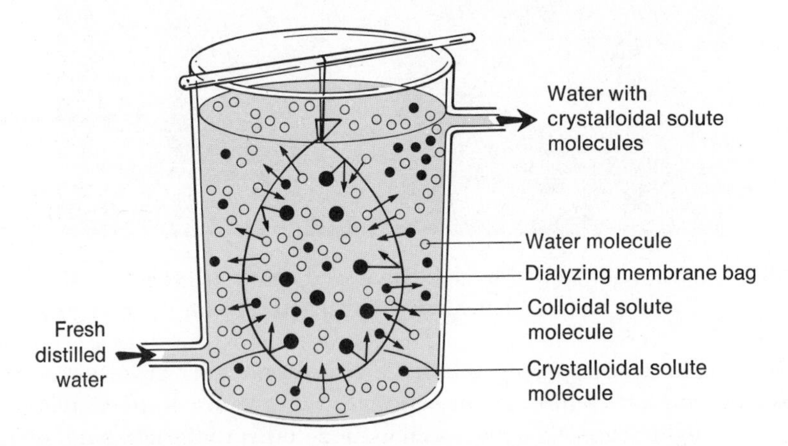

Fig. 8-6. Dialysis. Water and small solute molecules pass through the semipermeable membrane.

first observed under an ultramicroscope by Robert Brown. The motion of the colloidal particles is probably due to bombardment by solvent molecules and the slowness to the relatively larger size of the solute particles.

Electric charge. Colloidal particles in dispersions often carry a characteristic positive or negative charge. As a result, they can be made to move when placed in an electric field, a process known as *electrophoresis*. This process is useful in the separation of colloidal materials for identification and other research purposes.

Adsorption. Because of the large surface area presented by colloids and their ability to take on other substances on such surfaces, colloidal materials are found useful in the separation of other materials in various forms of the process known as *chromatography*. Probably one of the best known uses of colloids for adsorption is the use of charcoal in gas masks for removing poisonous gases from air before it is breathed into the lungs.

Emulsification. If oil and water, which do not dissolve in each other, are shaken vigorously, a cloudy mixture of small droplets of oil mixed in the water, an *emulsion*, is formed but remains as such for only a short time. The droplets of oil then coalesce and a layer of oil forms on top of the wa-

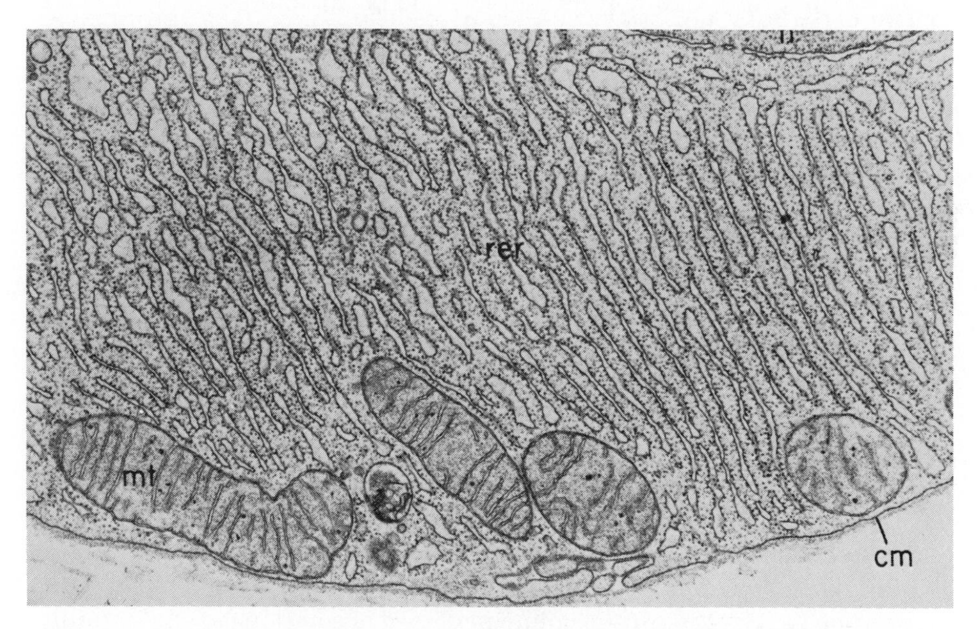

Fig. 8-7. Electron micrograph of basal region in pancreatic exocrine cell (rat) showing the multiplicity of membranes that occur in cells (×20,000). (cm = cell membrane; n = nucleus; mt = mitrochondrion; rer = rough-surfaced endoplasmic reticulum.) (From Palade, G. E.: J.A.M.A. **198**:815, 1966. Copyright 1966, American Medical Association.)

ter, an indication that the emulsion formed was a *temporary emulsion.* Certain colloidal substances, known as *emulsifying agents* or *protective colloids,* added to temporary emulsions can convert them to permanent emulsions, in which there is little or no separation of the oil and water phases. Milk is a good example of a permanent emulsion of butter fat in water with the colloidal protein casein serving as the emulsifying agent.

SUMMARY

SOLUTIONS

Definitions: *True solution* is a homogeneous mixture of two or more substances at the molecular level. *Solute* is the substance dissolved; *solvent* is the substance in which the solute is dissolved.

Types of solutions: Of nine types possible, the three with liquid (water) as the solvent are the most important in living organisms: (1) solid in liquid; (2) liquid in liquid; and (3) gas in liquid. Solutions are important because most chemical reactions take place in solution.

Factors influencing solubility:

Nature of solute and solvent: Polar solvents (e.g., water) dissolve polar solutes; nonpolar solvents dissolve nonpolar solutes.

Solute surface area: Powdered solute dissolves faster than large crystals.

Stirring: The greater the stirring, the faster solution will occur.

Temperature: Most solid solutes dissolve faster at higher temperatures, but there are some exceptions.

Pressure: Important for gases; the higher the pressure, the more gas will dissolve.

General specifications of concentration of solutions:

Dilute solutions: Amount of solute is small compared to solvent.

Concentrated solutions: Amount of solute is large compared to solvent.

Saturated solutions: No more solute can be dissolved by the solvent in the presence of excess solute.

Supersaturated solutions: Excess solute dissolved in solvent at a higher temperature may sometimes remain in the solution when cooled carefully to a lower temperature.

Exact specification of concentration of solutions:

Percentage solutions: Three types: (1) grams of solute per 100 ml of solution (w/v); (2) grams of solute per 100 g of solution (w/w); (3) milliliters of solute per 100 ml of solution (v/v).

Molar solutions: Based on the number of moles (molecular weights) of solute per 1000 ml (1 liter) of solution. A 1M solution contains 1 mole of solute in 1 liter of solution.

Normal solutions: Based on the number of equivalent weights (EW) of solute per liter of solution. EW = molecular weight ÷ valence. A 1N solution contains 1 EW per liter of solution. Using normal solutions A and B:

$$ml_A \times N_A = ml_B \times N_B$$

Physical properties of solutions:

Vapor pressure: Tendency to evaporate or escape. The vapor pressure of aqueous solutions is lower than that of pure water.

Elevation of the boiling point: The boiling point of a solution is higher than for pure water.

Lowering of the freezing point: The freezing point of a solution is lower than for pure water.

Diffusion: Movement of solute molecules in a solution due to constant molecular motion of solute and solvent molecules.

Osmosis: Flow of water, in a two-compartment system separated by a semipermeable membrane, from the compartment of lower impermeable solute concentration into the compartment of higher impermeable solute concentration.

Osmosis and the red cell: The osmotic pressure inside a red cell and in the blood are equal and no osmosis takes place. A solution with the same osmotic pressure as blood, such as 0.9% sodium chloride is an *isotonic solution.* In a *hypotonic solution,* lower solute concentration than blood, water flows into the red cell, which eventually bursts (hemolysis). In a *hypertonic solution,* higher solute concentration than blood, water flows from the red cell, which then shrinks (crenation).

COLLOIDS

Definitions: *Crystalloids* are solid solutes that form true solutions, can pass through filters and animal membranes, can be crystallized from solution. *Suspensions* are solid solutes that do not dissolve and are so heavy that they settle out. *Colloids* are solid solutes, of a size in between crystalloids and suspensions, which do not dissolve as crystalloids but which are light enough so that they do not settle out, may pass through filters but not animal membranes, cannot be crystallized from solution. *Sol* is a liquid colloidal solution; *gel* is a solid colloidal solution. *Colloid osmotic pressure* is the osmotic pressure that occurs as the result of the presence of a colloidal solute (impermeable) on one side of a two-compartment system separated by a semipermeable membrane, such as in the blood capillaries.

Properties of colloids:

Size: The difference in size between crystalloidal and colloidal substances can be demonstrated by way of the Tyndall effect (reflection of light at right angles to the path of light) shown by colloidal solutions only.

Osmosis: Colloids can cause osmotic pressure by supplying the impermeable solute on one side of a two-compartment system separated by a semipermeable membrane.

Dialysis: Removal of crystalloidal solutes from a mixture with colloidal solutes, by allowing the crystalloidal solutes to pass through a semipermeable membrane into distilled water while the colloidal solutes are held back. Actually osmosis and dialysis take place at the same time at many of the membranes found in living tissue.

Brownian movement: Movement of colloidal solute particles resulting from bombardment by the more rapidly moving solvent molecules.

Electric charge: Colloidal particles in dispersion carry electric charges and can be made to move in an electric field.

Adsorption: Colloids have the ability to take on other substances on their large surfaces.

Emulsification: Temporary emulsions, mixtures of liquids that do not mix, may be made permanent by adding protective colloids known as emulsifying agents.

REVIEW QUESTIONS

1. List some important factors influencing the solubility of substances and provide a scientific basis for the effects observed.
2. Define a molar solution. Define a normal solution. What aspect of a solution is specified by its molarity or normality? What is the unique advantage of designating solutions in terms of normality?
3. Define the terms *colloid* and *semipermeable membrane.* In simple terms explain how a difference in concentration of colloidal material on two sides of a semipermeable membrane creates an osmotic pressure. What is the difference between osmosis and dialysis?
4. What happens to a red cell when placed in a hypotonic solution? In a hypertonic solution? Explain on the basis of osmotic pressure.
5. What is the significance of colloids in life processes? Discuss some properties of colloidal materials. What property distinguishes colloids from crystalloids by way of the Tyndall effect?

REFERENCES

Lagowski, J. J.: Modern inorganic chemistry, New York, 1973, Marcel Dekker, Inc.

Masterton, W. L., and Slowinski, E. J.: Chemical principles, ed. 4, Philadelphia, 1977, W. B. Saunders Co.

Pauling, L.: General chemistry, ed. 3, San Francisco, 1970, W. H. Freeman and Co., Publishers.

Petrucci, R. H.: General chemistry: principles and modern applications, ed. 2, New York, 1977, Macmillan Publishing Co., Inc.

Acids, bases, salts

9

In view of the fact that most of the compounds encountered in inorganic chemistry may be classified as acids, bases, or salts, it becomes desirable to learn something of the properties, reactions, uses, and methods of naming these substances.

For defining acids and bases, it is necessary to introduce the concept of *ionization* at this time, although a more detailed discussion is reserved for the next chapter. In simple terms, ionization is the process whereby a compound in solution splits into electrically charged particles that are called *ions*. For example, the gas, hydrogen chloride (HCl), when placed in water, undergoes the following change:

$$HCl \rightarrow H^+ + Cl^-$$

Two ways of defining acids and bases are in general use. Both have their advantages and disadvantages. Rather than stressing one over the other, it is desirable to be aware of both and be able to use either one to its best advantage.

Arrhenius definition of acids and bases

As a consequence of his theory of ionization, Arrhenius defined acids as substances that produce hydrogen ions (H^+) in aqueous solution and bases as substances that produce hydroxyl ions (OH^-) in aqueous solution:

$$\underset{\textbf{Acid}}{HCl} \rightarrow \underset{\textbf{Hydrogen ion}}{H^+} + Cl^-$$

$$\underset{\textbf{Base}}{NaOH} \rightarrow \underset{\textbf{Hydroxyl ion}}{OH^-} + Na^+$$

Brönsted-Lowry definition of acids and bases

This is more general than the Arrhenius definition. According to Brönsted and Lowry, an acid is any substance that can give up or donate a proton, the hydrogen ion, and a base is any substance that can take up or accept a proton:

$$\underset{\textbf{Acid}}{HCl} \rightleftharpoons \underset{\textbf{Proton}}{H^+} + \underset{\textbf{Base}}{Cl^-}$$

Table 9-1. Common acids and bases

Acids		Bases	
Hydrochloric acid	HCl	Sodium hydroxide	$NaOH$
Sulfuric acid	H_2SO_4	Potassium hydroxide	KOH
Carbonic acid	H_2CO_3	Ammonium hydroxide	NH_4OH
Nitric acid	HNO_3	Calcium hydroxide	$Ca(OH)_2$
Boric acid	H_3BO_3	Magnesium hydroxide	$Mg(OH)_2$
Phosphoric acid	H_3PO_4	Aluminum hydroxide	$Al(OH)_3$

A double-headed arrow is used in this case because the reaction can actually go either way. In going to the right, hydrochloric acid is an acid because it gives up a proton and in going to the left, the chloride ion is a base because it accepts a proton to form hydrochloric acid.

Some of the commonly encountered acids and bases are listed in Table 9-1.

ACIDS

Properties of acids

1. Acids are generally described as having a sour taste, e.g., vinegar, which is a dilute solution of acetic acid, and citric acid in fruit juices. (Caution: Taste only those solutions designated for such a purpose.)
2. Acids react with bases (hydroxides) to form water and a salt:

$$HCl + NaOH \rightarrow H_2O + NaCl$$

It should be noted that in this reaction, both the acid and the base are neutralized. Reactions of this type are therefore known as *neutralization reactions.*

Acids also react with metallic oxides to form water and salt in a similar manner:

$$2HCl + \underset{\textbf{Calcium oxide}}{CaO} \rightarrow H_2O + CaCl_2$$

Neutralization reactions find much use in clinical and analytical chemistry in determining the amount of acid or base to be found in various body fluids or other solutions. The procedure for such a determination is known as a *titration* and makes use of the great convenience of normal solutions for simplifying the calculations. A titration is performed as follows (see Fig. 9-1):

STEP 1. *A measured volume, V_1, of a test fluid (containing the acid to be measured) is placed in a beaker.*

STEP 2. *An indicator, a substance that changes color when a certain amount of acid is completely neutralized by a base, is added to the beaker. Litmus, one of the oldest known indicators, is red in acid solution and turns blue in basic (al-*

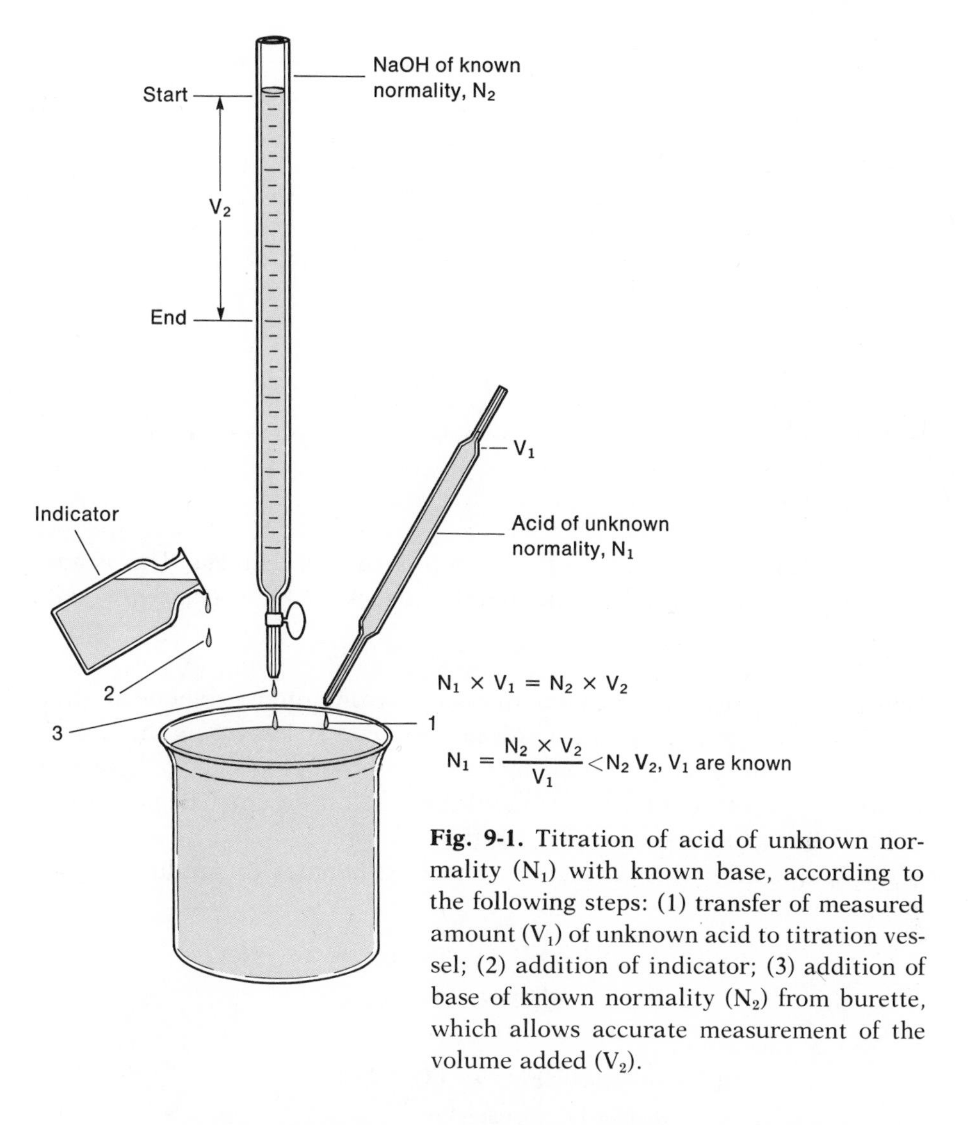

Fig. 9-1. Titration of acid of unknown normality (N_1) with known base, according to the following steps: (1) transfer of measured amount (V_1) of unknown acid to titration vessel; (2) addition of indicator; (3) addition of base of known normality (N_2) from burette, which allows accurate measurement of the volume added (V_2).

kaline) solution. Today, many more sensitive indicators, with a large variety of color changes, are available.

STEP 3. *A base, NaOH, of exact known strength specified in terms of normality, N_2, is added carefully from a burette, a measuring device, to the test fluid until the indicator just changes from the acid color to the alkaline color. The exact volume of NaOH used, V_2, is read from the burette. At this point, three of the four quantities found in the equation under the discussion of normal solutions (Chapter 8) are known, i.e., V_1, N_2, and V_2. The fourth, the normality of the test fluid, can be calculated from the equation:*

$$N_1 \times V_1 = N_2 \times V_2$$

$$N_1 = \frac{N_2 \times V_2}{V_1} \qquad N_2, V_2, V_1 \text{ are known}$$

Table 9-2. Electromotive series

Potassium	K
Sodium	Na
Magnesium	Mg
Zinc	Zn
Iron	Fe
Lead	Pb
Hydrogen	H
Copper	Cu
Silver	Ag
Platinum	Pt
Gold	Au

3. Acids react with some metals to produce hydrogen gas. This is a replacement reaction in which the metal replaces the hydrogen of the acid:

$$2HCl + Zn \rightarrow H_2\uparrow + ZnCl_2$$

The activity of metals in this reaction varies from intense to none at all. If the metals are arranged with hydrogen in what is known as the *electromotive series*, in order of decreasing activity, those metals above hydrogen are found to replace it from acids and those below do not. The partial list in Table 9-2 is arranged in such a series.

4. Acids react with carbonates and bicarbonates to produce carbon dioxide:

$$2HCl + \underset{\substack{\textbf{Sodium carbonate} \\ \text{(Washing soda)}}}{Na_2CO_3} \rightarrow CO_2\uparrow + H_2O + 2NaCl$$

$$HCl + \underset{\substack{\textbf{Sodium bicarbonate} \\ \text{(Baking soda)}}}{NaHCO_3} \rightarrow CO_2\uparrow + H_2O + NaCl$$

Sodium bicarbonate, baking soda, is used in baking powders in conjunction with acidic ingredients that, when wet, produce carbon dioxide gas bubbles in the dough to provide the lightness found in well-baked products. Sodium bicarbonate, as baking soda or in all sorts of patent medicine forms, is also used by many people to neutralize excess acidity in the stomach, a practice that may be dangerous if too much sodium is absorbed from the intestinal tract.

5. Acids react with some salts to produce new acids and new salts:

$$HCl + AgNO_3 \rightarrow HNO_3 + AgCl \downarrow$$

Silver nitrate **Nitric acid**

This is a double displacement or metathetical reaction.

6. Acids damage tissues, fabrics, metals, wood, and other materials. Since a good number of the common acids in concentrated form can be quite injurious to tissues and other substances, it becomes a worthwhile habit to consider all acids at any concentration as potentially dangerous. They should always be handled carefully and with full attention to directions for proper use.

Inevitably, however, some accidents will happen. When an accident does occur, it is important to know the correct procedure to follow as rapidly as possible. All laboratories should be equipped with good water supplies for flushing and with the proper emergency solutions to be used. As soon as possible after spilling acid on the skin or on clothes, the acid should be flushed with water and a solution of a basic substance such as sodium bicarbonate should be applied to help neutralize the acid. In case acid is swallowed, the sodium bicarbonate solution may be taken internally. Solutions of ammonia (NH_4OH) or limewater [$Ca(OH)_2$] may also be used externally. After such steps are completed, the damage can be assessed and medical help sought if necessary. Unfortunately, damage to tissues may be very rapid. Many of the newer fabrics are less susceptible to acid damage.

7. The common laboratory acids are very soluble in water.

Naming acids

The simplest acids are made up of hydrogen and only one other element and are called *binary* acids. These are named by using the prefix *hydro-*, followed by a suitable portion of the nonhydrogen element name, followed by the suffix *-ic:*

HCl	HBr
Hydro-chlor-ic acid	**Hydro-brom-ic acid**

Many of the common acids contain oxygen as a third element and are called *ternary* acids. These acids are named from the element other than hydrogen or oxygen, followed by the suffix *-ic:*

H_2SO_4	HNO_3	H_2CO_3
Sulfur-ic acid	**Nitr-ic acid**	**Carbon-ic acid**

A complication may occur here in that the same three elements may form more than one acid. When this occurs, the one with the higher number of oxygen atoms or oxidation state is given the suffix *-ic* and the one with the lower number of oxygen atoms is given the suffix *-ous:*

HNO_3	HNO_2	H_3PO_4	H_3PO_3
Nitr-ic acid	**Nitr-ous acid**	**Phosphor-ic acid**	**Phosphor-ous acid**

BASES

Properties of bases

The properties listed here are descriptive of bases under the Arrhenius definition, i.e., substances that produce hydroxyl ions in aqueous solution.

1. Bases in solution have a slippery feeling and are described as having a bitter, metallic taste, much as soap does. This is due to the presence of some free base in most soaps. (Caution: Taste only those solutions designated for such a purpose.)

2. Bases react with acids to form water and salt (see Properties of Acids) in a neutralization reaction.

3. Bases react with oxides of nonmetals to form water and a salt:

$$\underset{\text{Calcium hydroxide}}{Ca(OH)_2} + CO_2 \rightarrow H_2O + \underset{\text{Calcium carbonate}}{CaCO_3 \downarrow}$$

This is a common procedure for absorbing the carbon dioxide from a mixture of gases by bubbling the mixture through a basic solution.

4. Bases react with some metals to produce hydrogen and salt in a replacement reaction:

$$6NaOH + \underset{\text{Aluminum}}{2Al} \rightarrow 3H_2 \uparrow + \underset{\text{Sodium aluminate}}{2Na_3AlO_3}$$

5. Bases react with some salts to form new bases and new salts in a double displacement or metathetical reaction:

$$2NaOH + \underset{\text{Barium chloride}}{BaCl_2} \rightarrow \underset{\text{Barium hydroxide}}{Ba(OH)_2 \downarrow} + 2NaCl$$

6. Bases damage tissues, fabrics, and other materials. As in the case of acids, bases should be handled very carefully. If a base is spilled on the skin or clothes, flush the area with water or add a solution of a dilute acid, such as dilute acetic acid (vinegar) or boric acid, to neutralize the base. *Boric acid must be used only externally because it is a poison if taken internally.* Dilute acetic acid may be used to rinse the mouth, if necessary. It is wise to check the presence and location of the emergency solutions in a laboratory before starting experimental work.

7. Sodium, potassium, and ammonium hydroxides are quite soluble in water.

Naming bases

Bases are named as *hydroxides* using the name of the metal found in the compound.

$NaOH$	KOH
Sodium hydroxide	**Potassium hydroxide**

The base ammonium hydroxide, formed when ammonia (NH_3) is added to water, does not contain a metal, but is named the same way:

NH_4OH

Ammonium hydroxide

SALTS

Salts are generally defined on the basis of the reactions in which they are formed, i.e., as the compound other than water that is formed in a neutralization reaction (acid + base) or the compound formed by the replacement of hydrogen of an acid or a base by a metal (see pp. 96, 98).

Properties of salts

1. Tastes of salts or their aqueous solutions cannot be described by any general term. The term *salty* generally refers to the taste of table salt, sodium chloride, or ocean water, which has sodium chloride as a major solute. Other salts have a wide range of tastes. (Caution: Taste only those substances or solutions designated for such a purpose.)

2. Salts can react with other salts to form new salts in double decomposition reactions.

$$Na_2SO_4 + BaCl_2 \rightarrow 2NaCl + BaSO_4 \downarrow$$

3. Salts react with acids to form new salts and acids in double decomposition reactions (see Acids).

If all the hydrogen of an acid is replaced by a metal, the salt formed in such a reaction is called a *normal* salt:

$$2NaOH + H_2SO_4 \rightarrow Na_2SO_4 + 2H_2O$$

Sodium sulfate

(Normal salt)

If at least one of the hydrogen atoms of an acid remains in the salt molecule, it is called an *acid* salt:

$$NaOH + H_2SO_4 \rightarrow NaHSO_4 + H_2O$$

Sodium acid sulfate
or
Sodium bisulfate

(Acid salt)

An acid with three hydrogen atoms, such as phosphoric acid (H_3PO_4), can form two acid salts:

NaH_2PO_4	Na_2HPO_4
Sodium dihydrogen phosphate	**Disodium hydrogen phosphate**

4. Salts react with bases to form new salts and bases in double decomposition reactions (see Bases).

5. Salts may react with metals to form new salts and metals in displacement reactions.

$$CuSO_4 + Fe \rightarrow FeSO_4 + Cu$$

Salt **Metal** **Salt** **Metal**

Reactions of this type occur if the metal, such as iron (Fe), is higher in the electromotive series (see Table 9-2) than the metal, such as copper (Cu), which it replaces from the salt.

6. Salts of sodium, potassium, and ammonium are generally quite soluble in water.

Naming salts

Salts of binary acids are named from the metal, followed by the nonmetal portion to which is added the suffix *-ide.* The ammonium ion is treated as a metal.

$NaCl$	NH_4Cl
Sodium chlor-ide	**Ammonium chlor-ide**

If the metal has more than one valence, the form with the higher positive valence state is given the suffix *-ic,* on the metal, and the form with the lower positive valence state the suffix *-ous:*

$FeCl_3$	$FeCl_2$
Ferr-ic chlor-ide	**Ferr-ous chlor-ide**

In naming salts from ternary acids, if the name of the acid ends in *-ic,* the name of the salt ends in *-ate:*

H_2SO_4	$BaSO_4$
Sulfur-ic acid	**Barium sulf-ate**

When the name of the acid ends in *-ous,* the name of the salt ends in *-ite:*

H_2SO_3	$BaSO_3$
Sulfur-ous acid	**Barium sulf-ite**

If the metal in the salt can have more than one valence, the metal is named in the same way as in the case of the binary acids: the metal is given the suffix *-ic* for the higher positive valence form and *-ous* for the lower positive valence form:

$Hg(NO_3)_2$	$HgNO_3$
Mercur-ic nitr-ate	**Mercur-ous nitr-ate**

Need for salts in the body

All of the body fluids contain some dissolved salts. Some of the better known requirements for salts are the need for iron salts in the production of hemoglobin, required for the transport of oxygen in the blood, and the need for calcium and phosphorus salts for the proper formation of bone and teeth. Those who live in areas with low mineral content in the drinking water or far from sources of salt-water fish may find salts containing iodine added to their table salt because of the need for iodine in the synthesis of thyroid hormone in the body.

Salts are also required for maintenance of the delicate acid-base balance in the body, as will be seen in subsequent chapters. Another delicate balance in the body involves the normal distribution of electrolytes in the various fluids, tissues, and cells. This electrolyte balance involves mainly the sodium, potassium, and magnesium chlorides, bicarbonates, and phosphates. Under various disease conditions these balances may be disturbed and may require the administration of salts as a corrective measure. This may be accomplished orally, intravenously (Fig. 9-2), or possibly by both routes.

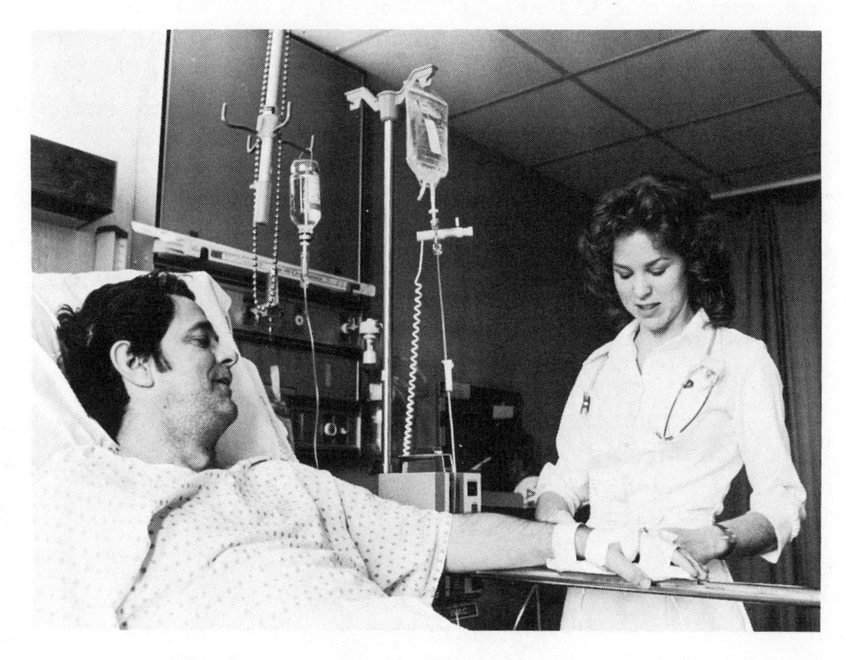

Fig. 9-2. Intravenous administration of salts. (Courtesy Barnes Hospital, St. Louis.)

Some uses of salts in medicine

One of the first encounters with the medical use of salts is at the moment of birth when an antiseptic solution of silver nitrate ($AgNO_3$) is routinely placed in the eyes of the baby to prevent gonorrheal infections, whether an infection is actually suspected or not.

Since sodium bicarbonate may be taken internally, it is often used to neutralize gastric acidity. Other salts useful for this purpose are calcium carbonate, magnesium carbonate, magnesium trisilicate, magnesium hydroxide, and aluminum hydroxide. If it is necessary to generate acidity in the body, ammonium chloride may be administered.

As noted previously (p. 34), suspensions of barium sulfate are used to make x-ray examinations of the intestinal tract much more definitive than is otherwise possible because the barium sulfate is opaque to x-rays, whereas tissues are easily penetrated.

These examples cited include only a small sample of the salts being used in medicine and illustrate only a few of the many purposes for which they are used. Salts have long been of great importance in medicine and continue to be so.

SUMMARY

Definitions:

Arrhenius: An *acid* is a substance that produces hydrogen ions (H^+) in aqueous solutions; a *base* is a substance that produces hydroxyl ions (OH^-) in aqueous solutions.

Brönsted-Lowry: An *acid* is a substance that gives up a proton (H^+ ion); a *base* is a substance that takes up a proton.

ACIDS

Properties of acids:

1. Taste sour.
2. React with bases (hydroxides) to form water and salt, *neutralization reaction:*

$$HCl + NaOH \rightarrow H_2O + NaCl$$

This reaction is the basis for the analytical determination of acids or bases in unknown solutions by a titration procedure.
Also react with metallic oxides to form water and salt:

$$2HCl + CaO \rightarrow H_2O + CaCl_2$$

3. React with some metals to form hydrogen:

$$2HCl + Zn \rightarrow H_2\uparrow + ZnCl_2$$

Metals above hydrogen in the *electromotive series* form hydrogen in this reaction; those below hydrogen do not.

4. React with carbonates and bicarbonates to give carbon dioxide:

$$2HCl + Na_2CO_3 \rightarrow CO_2\uparrow + H_2O + 2NaCl$$

$$HCl + NaHCO_3 \rightarrow CO_2\uparrow + H_2O + NaCl$$

5. React with some salts to produce new acids and new salts:

$$HCl + AgNO_3 \rightarrow HNO_3 + AgCl\downarrow$$

6. Damage tissues, fabrics, metals, wood, and other materials. First aid involves flushing with water and neutralization with weakly basic (alkaline) solutions such as ammonium hydroxide or sodium bicarbonate.
7. Usually very soluble in water.

Naming acids:

Binary: Binary acids (hydrogen plus one other element) are named by using prefix *hydro-* followed by the nonhydrogen element and the suffix *-ic,* e.g., HCl, hydro-chlor-ic acid.

Ternary: Ternary acids (hydrogen, oxygen, and one other element) are named from the element other than hydrogen or oxygen and the suffix *-ic,* e.g., H_2SO_4, sulfur-ic acid. If another acid can be formed from the same elements, the one with less oxygen is given the suffix *-ous,* e.g., H_2SO_3, sulfur-ous acid.

BASES

Properties of bases:

1. Taste metallic, soapy, and feel slippery.
2. React with acids to form water and salt in a *neutralization reaction.*
3. React with oxides of nonmetals to form water and salt:

$$Ca(OH)_2 + CO_2 \rightarrow H_2O + CaCO_3\downarrow$$

4. React with some metals to form hydrogen and salt:

$$6NaOH + 2Al \rightarrow 3H_2\uparrow + 2Na_3AlO_3$$

5. React with some salts to form new bases and new salts:

$$2NaOH + BaCl_2 \rightarrow Ba(OH)_2\downarrow + 2NaCl$$

6. Damage tissues, fabrics, and other materials. First aid involves flushing with water and neutralization with weakly acidic solutions such as acetic or boric acids.
7. Sodium, potassium, and ammonium hydroxides are quite soluble in water.

Naming bases: Bases are named as hydroxides of the metal in the compound, e.g., NaOH, sodium hydroxide.

SALTS

Definition: Salts are compounds other than water, formed in a neutralization reaction or by the replacement of hydrogen of an acid or base by a metal.

Properties of salts:

1. Large range of tastes.
2. React with other salts to form new salts:

$$Na_2SO_4 + BaCl_2 \rightarrow 2NaCl + BaSO_4 \downarrow$$

3. React with acids to form new salts and acids (see under Acids). A *normal salt* is a salt formed when all hydrogen of acid is replaced by metal, e.g., sodium sulfate (Na_2SO_4). An *acid salt* is a salt formed when not all hydrogen is replaced by metal, e.g., sodium acid sulfate ($NaHSO_4$).
4. React with bases to form new salts and new bases (see under Bases).
5. React with some metals to form new salts and new metals:

$$CuSO_4 + Fe \rightarrow FeSO_4 + Cu$$

The free metal iron (Fe) must be higher in the *electromotive series* than the metal in the salt, copper (Cu) in $CuSO_4$, for such a reaction to take place.

6. Sodium, potassium, and ammonium salts are generally quite soluble in water.

Naming salts:

Salts of binary acids: Named from the metal, the nonmetal, and the suffix *-ide;* e.g., NaCl, sodium chlor-ide. If the metal has more than one valence, the form with the higher positive valence has the suffix *-ic* on the metal, the lower has the suffix *-ous;* e.g., $FeCl_3$, ferr-ic chlor-ide; $FeCl_2$, ferr-ous chlor-ide.

Salts of ternary acids: If name of acid ends in *-ic,* name of salt ends in *-ate;* e.g., H_2SO_4, sulfur-ic acid, $BaSO_4$, barium sulf-ate. If name of acid ends in *-ous,* name of salt ends in *-ite;* e.g., H_2SO_3, sulfur-ous acid, $BaSO_3$, barium sulf-ite. Metals with more than one valence are named as in the preceding paragraph, e.g., $Hg(NO_3)_2$, mercur-ic nitr-ate; $HgNO_3$, mercur-ous nitr-ate.

Need for salts in body: All body fluids contain some dissolved salts. Some requirements are iron salts for production of hemoglobin (oxygen transport), calcium and phosphorus salts for bone and teeth, iodine salts for thyroid hormone, and salts for acid-base and electrolyte balances.

Some uses of salts in medicine: Silver nitrate as an antiseptic; sodium bicarbonate, calcium carbonate, magnesium trisilicate, and others as neutralizers of acidity; ammonium chloride to generate acidity; barium sulfate for x-ray purposes.

REVIEW QUESTIONS

1. Is there any significant difference between the Arrhenius and Brönsted-Lowry definitions of an acid? Explain.
2. An acid and what other substance react in in a neutralization reaction? What are the products? What has been neutralized?
3. Name some acids, bases, and salts that have important functions in the body. Where do they function and how do their chemical properties enable them to carry out their functions?
4. In case of an accidental spill of a strong acid (or base) on the skin, should it be neutralized with a strong or weak base (or acid)? Explain why.
5. What is a titration? What is the value of this procedure in a clinical or analytical laboratory?

REFERENCES

Lagowski, J. J.: Modern inorganic chemistry, New York, 1973, Marcel Dekker, Inc.

Masterton, W. L., and Slowinski, E. J.: Chemical principles, ed. 4, Philadelphia, 1977, W. B. Saunders Co.

Pauling, L.: General chemistry, ed. 3, San Francisco, 1970, W. H. Freeman and Co., Publishers.

Petrucci, R. H.: General chemistry: Principles and modern applications, ed. 2, New York, 1977, Macmillan Publishing Co., Inc.

10 Ionization, electrolytes, hydrogen ion concentration (pH)

IONIZATION

In 1833, Michael Faraday, an English scientist, described his investigations of the properties of solutions as conductors, or carriers, of electric current. His studies were a direct outgrowth of the development, in 1800, of a useful electric battery by Alessandro Volta, an Italian physicist, and the electrochemical research of his teacher Sir Humphry Davy, in which Volta's battery was used for the isolation of potassium and sodium in 1807. See Fig. 10-1 for a diagram of a Voltaic cell that can be put together easily in the laboratory.

Faraday explained the conduction of an electric current in certain solutions between two electrodes on the basis of particles, some of which moved toward one electrode while others moved toward the second electrode. He called these particles *ions* (Gr., to go), naming those particles moving toward the positive electrode (anode), *anions,* and those moving toward the negative electrode (cathode), *cations*. He believed that the substance that gave rise to the ions existed in solution as molecules until the potential difference between the two electrodes pulled the molecule apart to form the ions.

Solutions of acids, bases, and salts are generally found to conduct an electric current quite well and as such are called *electrolytes*. Fig. 10-2 shows a simple apparatus that can be used to demonstrate the ability of a solution of an electrolyte to conduct electricity.

Solutions of other substances, such as sugar, that do not conduct an electric current are known as *nonelectrolytes*. Many organic compounds are found in this class of substances.

Arrhenius theory of ionization

A theory of ionization that has stood the test of time with little modification was proposed by Svante Arrhenius in 1887 in his thesis for a doctor's degree:

1. Ions are formed by the dissociation of molecules of acids, bases, and salts in certain solvents, particularly water.
2. Ions carry an electric charge. On dissociation, a molecule forms one or more positive ions and one or more negative ions. The sum of the

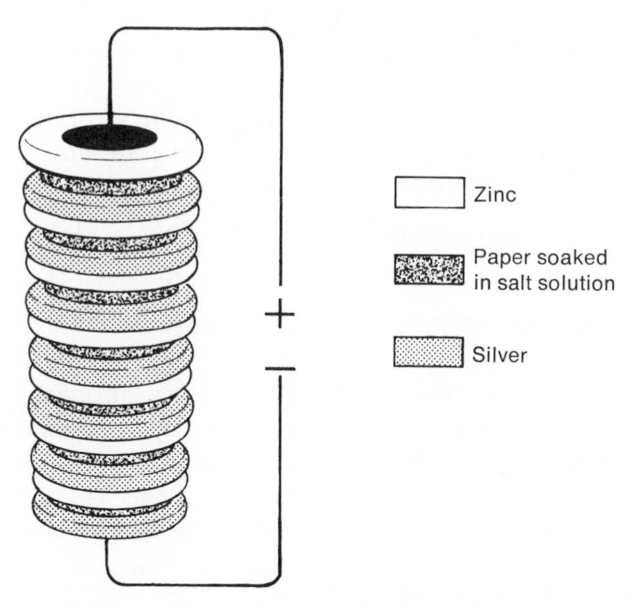

Fig. 10-1. Voltaic pile, the first working battery.

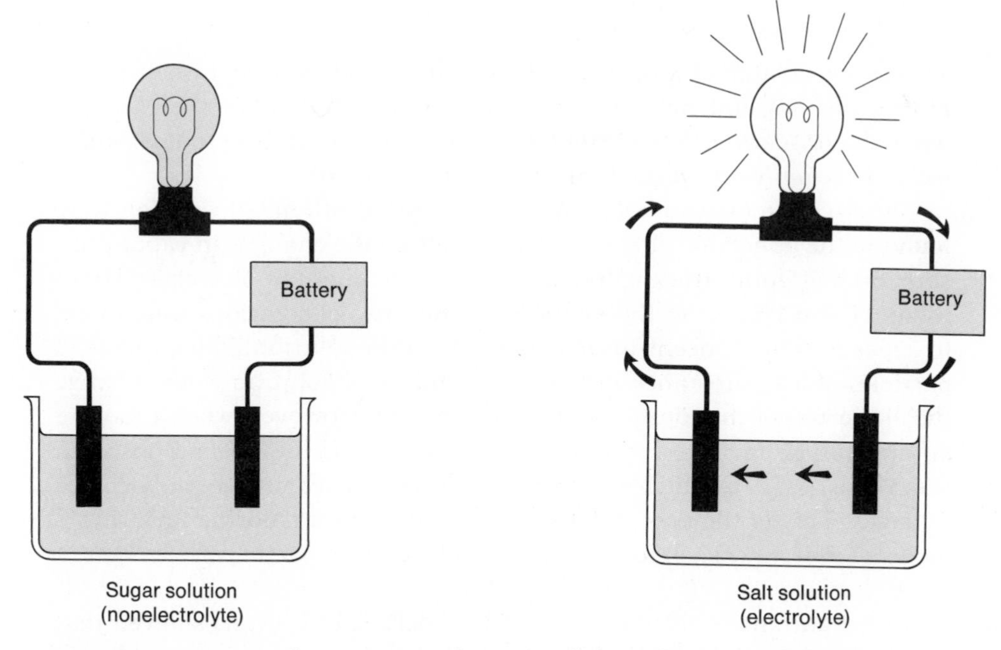

Fig. 10-2. Apparatus for testing conductivity of electrolyte or nonelectrolyte substances.

positive charges must equal the sum of the negative charges, leaving the solution electrically neutral.

3. Nonelectrolytes, aqueous solutions of which do not conduct an electric current, do not produce ions.
4. Complete ionization occurs only in extremely dilute solutions. In ordinary solutions the ions are in equilibrium with undissociated molecules. The situation was pictured as follows:

$$\text{Extremely dilute solution: } NaCl \rightarrow Na^+ + Cl^-$$

$$\text{Ordinary concentrations: } NaCl \rightleftharpoons Na^+ + Cl^-$$

In the more dilute solutions it was considered less likely that recombination of ions to form molecules could occur.

One of the best indications that some substances produce ions in aqueous solution is the observation that although these substances will react readily in solution, no reaction occurs in the dry state. When solutions of sodium sulfate and barium chloride are mixed, there is an instantaneous reaction and a precipitate of barium sulfate is formed. Arrhenius conceived the reaction as occurring in the following manner:

$$Na_2SO_4 \rightarrow 2Na^+ + SO_4^{=}$$

$$BaCl_2 \rightarrow 2Cl^- + Ba^{++}$$

$$\downarrow$$

$$BaSO_4 \downarrow$$

The same substances mixed together as dry powders do not react. It follows from this and many other similar observations that the *chemical properties of solutions are attributable to the ions formed* and that *chemical reactions usually take place between ions, not molecules.*

Other convincing indications of the presence of ions in solution were found in the abnormal effects of electrolytes on the changes in vapor pressure, boiling point, freezing point, and osmotic pressure of such solutions (Chapter 8). Changes in these physical properties of solutions were found to depend on the concentration of *particles* in the solution. Thus, 1 mole of particles of any substance in the same amount of solution would change the properties of the solution to the same extent. However, when 1 mole of an electrolyte such as sodium chloride was used, the change observed was almost twice as much as occurred with 1 mole of a nonelectrolyte such as sugar. This is the case because 1 mole of sodium chloride furnishes 2 moles of ions, and each ion acts as an independent particle.

Modification of Arrhenius theory of ionization: Debye-Hückel theory

Investigations of the structure and properties of crystals showed that some substances are ionized in their crystalline state. Sodium chloride crystals are composed of a regular arrangement of sodium and chloride ions (Fig. 10-3). To take these factors into account, the physical

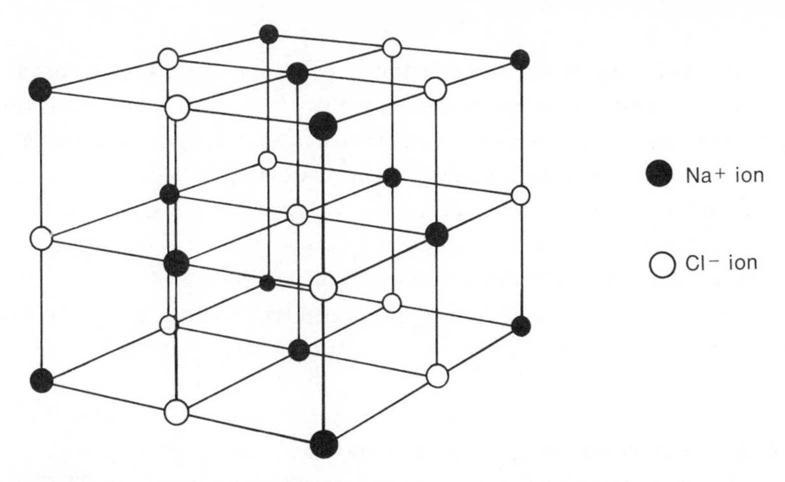

Fig. 10-3. Distribution of sodium (Na^+) and chloride (Cl^-) ions in a crystal of sodium chloride (NaCl).

chemists Debye and Hückel, in 1923, proposed that strong electrolytes, such as sodium chloride, are completely dissociated in water solutions, but that in the more concentrated solutions, the free movement of some of the ions is interfered with by the accumulation of ions of opposite charge (opposite charges attract) around one another. This is in contrast to the Arrhenius theory, which assumed that undissociated molecules were formed under such conditions and an equilibrium was established with the ions in solution. The Arrhenius theory still applies to solutions of weak electrolytes, which ionize to a small extent only, the small number of ions being in equilibrium with a much larger number of undissociated molecules.

In order to distinguish between ionizations of a strong electrolyte and a weak electrolyte, the following notation is often used:

$$\text{Strong electrolyte} \rightarrow \text{positive ions} + \text{negative ions}$$

$$\text{Weak electrolyte} \rightleftharpoons \text{positive ions} + \text{negative ions}$$

The double-headed arrow indicates that an equilibrium exists between undissociated molecules and ions in solution (weak electrolytes), whereas the single-headed arrow indicates relatively complete ionization (strong electrolytes). The terms *strong* and *weak* refer to the extent, or degree, of ionization of acids or electrolytes. In contrast, the designations *concentrated* and *dilute*, as discussed in Chapter 8, refer to the ratio of solute to solvent in solutions. Some of the more common electrolytes are classified as follows:

	Strong	*Weak*
Salts	Almost all	Mercuric salts
Acids	HCl, HNO_3, H_2SO_4	H_2CO_3, H_3BO_3, H_2SO_3
Bases	NaOH, KOH	NH_4OH

At the present time, it is considered that strong electrolytes are largely ionized in aqueous solutions, whereas weak electrolytes remain largely in the form of undissociated molecules. This is a very important factor in the consideration of the properties of solutions of certain salts, as will be discussed later.

ELECTROLYTES

With an understanding of the theory of ionization it is possible to reexamine some aspects of the properties of electrolytes, acids, bases, and salts, for the purpose of providing more accurate explanations of certain phenomena.

Acids

When an acid ionizes in water to produce hydrogen ions these ions attach themselves to water molecules to form *hydronium* ions (H_3O^+).

$$H^+ + H_2O \rightarrow \underset{\textbf{Hydronium ion}}{H_3O^+}$$

However, for convenience in writing equations for the reactions with acids, the water molecule attached to the hydrogen ion is taken for granted and left out of the equation. This directs attention more clearly to what is happening specifically to the hydrogen ion.

Although many electrolytes are electrovalent compounds existing in the form of ions in the crystalline state, other electrolytes are covalent compounds in the dry state. Hydrogen chloride is a covalent, gaseous compound when dry and is a strong electrolyte, ionizing completely when in aqueous solution:

Dry: $H:\ddot{\underset{\cdot\cdot}{Cl}}:$ Hydrogen chloride

In aqueous solution: $H:\ddot{\underset{\cdot\cdot}{Cl}}: \rightarrow H^+ + :\ddot{\underset{\cdot\cdot}{Cl}}:^-$

or,

$$HCl \rightarrow H^+ + Cl^-$$

The ionization of a typical weak acid, carbonic acid, may be written as follows:

$$H_2CO_3 \rightleftharpoons H^+ + HCO_3^-$$

Bases

Bases are compounds that contain both an electrovalent bond and a covalent bond, e.g., the electronic structure of sodium hydroxide may be written as follows:

$$Na^+ :\ddot{\underset{\cdot\cdot}{O}}:H^-$$

Electrovalent bond ↗ ↖ Covalent bond

Since the oxygen atom normally has six electrons in its outer orbit, the hydrogen atom shares its one electron as part of a pair with oxygen to form the covalent bond (oxygen now has seven electrons in its outer orbit), and the sodium atom gives up its one electron to the oxygen atom to form the electrovalent bond (oxygen now has eight electrons in its outer orbit and the hydroxyl radical has a negative charge, the sodium ion a positive charge). In aqueous solution, the electrovalent bond is broken, allowing ions to move about freely but the covalent bond remains intact as part of the hydroxyl ion, or radical. The ionization of sodium hydroxide, a strong base, is written as follows:

$$NaOH \rightarrow Na^+ + OH^-$$

The ionization of a familiar weak base, ammonium hydroxide, is written as follows:

$$NH_4OH \rightleftharpoons NH_4^+ + OH^-$$

In this case, covalent bonds are found in both radicals.

Neutralization reactions. In Chapter 9 the equation for the neutralization reaction between an acid and a base was written as follows:

$$NaOH + HCl \rightarrow H_2O + NaCl$$

In reality, the reaction takes place only between the hydroxyl and hydrogen ions and may be written as follows:

$$NaOH \rightarrow Na^+ + OH^-$$

$$HCl \rightarrow Cl^- + H^+$$

$$\downarrow\uparrow$$

$$H_2O$$

The hydrogen and hydroxyl ions neutralize each other by combining to form undissociated water molecules and the solution is neutral, i.e., it is neither acidic nor basic. The sodium and chloride ions, being constituents of a strong electrolyte, remain in solution as ions.

Salts

Salts do not have any one ion that is common to all members of this class of electrolytes, unlike acids, all of which contain hydrogen ions, and bases, all of which contain hydroxyl ions. Normal salts, which are salts resulting from the neutralization of a strong acid with a strong base, ionize as follows:

$$NaCl \rightarrow Na^+ + Cl^-$$

$$K_2SO_4 \rightarrow 2K^+ + SO_4^=$$

The resulting solution has a neutral reaction because there are no excess hydrogen or hydroxyl ions in the solution.

Salts with acidic reaction. Aqueous solutions of certain salts, other than normal salts, will have an acidic reaction. The two factors responsible for this phenomenon are (1) differences in extent of ionization of the compounds involved, and (2) the fact that although water is not considered an electrolyte, it does ionize to a very small, but measurable, extent:

$$H_2O \rightleftharpoons H^+ + OH^-$$

Thus, when a salt such as ammonium chloride is dissolved in water, the following series of events takes place all at once:

$$NH_4Cl \rightarrow NH_4^+ + Cl^-$$
$$H_2O \rightleftharpoons OH^- + H^+$$
$$\downarrow\uparrow$$
$$NH_4OH$$

Because ammonium hydroxide is a weak electrolyte, ammonium and hydroxyl ions tend to form undissociated molecules when they appear in the same solution. The hydroxyl ions supplied by the water will be removed from the solution while some excess hydrogen ions, also supplied by the water, will remain in solution as ions, because hydrochloric acid is a strong electrolyte (see classification of strong and weak electrolytes in previous list). As a result, the solution has an acidic reaction. In general, *the solution of a salt formed from a strong acid* (the chloride ion of ammonium chloride comes from hydrochloric acid) *and a weak base* (the ammonium ion comes from ammonium hydroxide) *will have an acidic reaction.* This process, in which a small amount of the starting compound, ammonium chloride, may be considered to be decomposed by the action of water, is known as a *hydrolysis* reaction.

Salts with basic reaction. In an analogous manner, it can be shown that *the solution of a salt formed from a strong base and a weak acid will have a basic reaction.* In a solution of sodium bicarbonate, the following takes place:

$$NaHCO_3 \rightarrow Na^+ + HCO_3^-$$
$$H_2O \rightleftharpoons OH^- + H^+$$
$$\downarrow\uparrow$$
$$H_2CO_3$$

When a hydrogen ion meets a bicarbonate ion in solution, the two ions tend to form molecules of carbonic acid, a weak electrolyte. This removes some hydrogen ions, leaving an excess of hydroxyl ions in solution, resulting in a basic reaction.

Importance of ions in life processes

The properties and importance to the living organism of solutions and acids, bases, and salts are discussed in Chapters 8 and 9. These matters should be reviewed at this point in the light of the facts developed in this chapter showing that the properties of solutions, acids, bases, and salts and their reactions are mainly the properties of the constituent ions. Ions, therefore, are intimately involved in all sorts of life processes such as digestion of food, transport of nutrients to tissues and removal of wastes, synthesis (anabolism) and breakdown (catabolism) of body constituents, conduction of nerve impulses, contraction of muscles, and maintenance of electrolyte and acid-base balance in the various compartments of the body. Specific examples will be detailed in the biochemistry chapters.

HYDROGEN ION CONCENTRATION (pH)

Although, as already noted, water ionizes to a very small extent only, this ionization is very important in maintaining the body in a normal functioning state. The hydrogen ion concentration of the different body fluids, such as gastric juice, intestinal juice, and blood, may vary widely as compared to each other, but the range of variation in each specific fluid is usually very small. Much of this requirement for relatively rigid control of variation in hydrogen ion concentration stems from the fact that chemical reactions and the enzymes (protein catalysts) that affect these reactions work best at certain definite hydrogen ion concentrations.

pH system

Pure water ionizes as follows:

$$H_2O \rightleftharpoons H^+ + OH^-$$

One liter of water contains 0.0000001 g of hydrogen ions. This is a rather unwieldy way of expressing such small concentrations. To simplify such numbers, Sørensen devised the pH system. In mathematical shorthand, $0.0000001 = 1 \times 10^{-7}$ or 10^{-7}. By choosing the proper definitions and applying some mathematical manipulations, including the use of logarithms, the simple number 7 indicates the hydrogen ion concentration of a neutral solution:

$$\text{Definition: pH} = \log \frac{1}{\text{H}^+ \text{ concentration}}$$

$$\text{pH of neutral water} = \log \frac{1}{10^{-7}} = \log 10^7$$

Since $\log 10^x = X$
pH of neutral water = 7

For those unfamiliar with the use of logarithms, a more practical explanation may be necessary:

$$0.0000001 = \frac{1}{10{,}000{,}000} = \frac{1}{10 \times 10 \times 10 \times 10 \times 10 \times 10 \times 10} = \frac{1}{10^7}$$

Table 10-1. The Sørensen pH scale: pH of body fluids and solutions

	H^+ concentration (g/liter)		pH	Body fluids	Solutions
Acidic ↑	10^0	1.0	0		
	10^{-1}	0.1	1		0.1N HCl
				Gastric juice	
	10^{-2}	0.01	2		
	10^{-3}	0.001	3		
	10^{-4}	0.0001	4		$0.1N\ H_2CO_3$ / 0.0001N HCl
	10^{-5}	0.00001	5		
↓	10^{-6}	0.000001	6	Urine	
				Saliva	
Neutral	10^{-7}	0.0000001	7		Pure water
				Blood	
Basic ↑	10^{-8}	0.00000001	8	Bile	
				Intestinal juice	
	10^{-9}	0.000000001	9		
	10^{-10}	0.0000000001	10		
	10^{-11}	0.00000000001	11		0.001N NaOH / $0.1N\ NH_4OH$
	10^{-12}	0.000000000001	12		
	10^{-13}	0.0000000000001	13		0.1N NaOH
↓	10^{-14}	0.00000000000001	14		

where 10^7 means 10 multiplied by 10, the answer then multiplied by 10 and repeated until seven 10s have been used and 7 is called the exponent of 10. When a number carrying an exponent is moved from the denominator to the numerator in order to simplify a fraction, the sign of the exponent changes:

$$\frac{1}{10^7} = 10^{-7}$$

Thus, by practical definition, when the hydrogen ion concentration is expressed as 10 with a negative exponent, the pH is equal to the exponent with the negative sign changed to a positive:

$$\text{If } H^+ \text{ concentration} = 10^{-7}$$

$$\text{pH} = +7 = 7 \text{ (pure water)}$$

Acidic solutions will contain higher concentrations of hydrogen ions than that found in pure water, e.g., 0.001 g rather than 0.0000001 g:

$$0.001 = \frac{1}{1000} = \frac{1}{10^3} = 10^{-3}$$

$$\text{pH} = \log\frac{1}{10^{-3}} = \log 10^3 = 3$$

or,

$$\text{If } H^+ \text{ concentration} = 10^{-3}$$

$$pH = +3 = 3$$

Thus, acidic solutions have pH numbers lower than 7 and, in a similar manner, alkaline solutions can be shown to have pH numbers higher than 7. The Sørensen pH scale ranges from 0 to 14. Some of the relationships discussed may become clearer from Table 10-1.

It will be noted from this table that the greater the hydrogen ion concentration (the more acidic the solution), the lower the pH number. This is because the pH is determined by the number of decimal places in the designation of the hydrogen ion concentration and the fact that as the hydrogen ion concentration increases, the number of decimal places decreases, e.g., the pH of a 0.0001N hydrochloric acid solution is 4 whereas the pH of a 0.1N hydrochloric acid solution is 1.

Also from this table, it can be seen that it takes a much higher concentration of a weak acid (one that ionizes to a smaller extent), e.g., carbonic acid, to supply as many hydrogen ions as a lower concentration of a strong acid, e.g., hydrochloric. Note that a 0.1 N carbonic acid solution has the same pH as a 0.0001N hydrochloric acid solution. Because of the logarithmic nature of the pH system, a difference of 1 between any two pH numbers means a difference of ten times in hydrogen ion concentration, e.g., the hydrogen ion concentration corresponding to pH 3 is 0.001 and to pH 4 is 0.0001 g per liter.

Measurement of pH

There are two generally used methods for determination of the pH of solutions or body fluids:

1. Colorimetric: This method is used mainly when an approximate pH of a solution will suffice. pH indicator paper is usually used. The paper, when wet with the test solution, assumes the color characteristic of the pH of the solution. This color is matched against a standard chart of colors listing the corresponding pHs, and the pH of the solution can be estimated. The color of the pH paper results from the response of the indicator dye in the paper to the solution (Fig. 10-4).
2. Electrometric: This method takes advantage of certain electrochemical relationships that make it possible to determine the pH of test solutions by a series of simple manipulations of an electric meter known as a pH meter. Electrodes of the pH meter are placed in the test solution and after the proper adjustments have been made, the pH can be read directly from a dial on the meter (Fig. 10-5). Measurements made with the pH meter are much more accurate than those made with a colorimetric method.

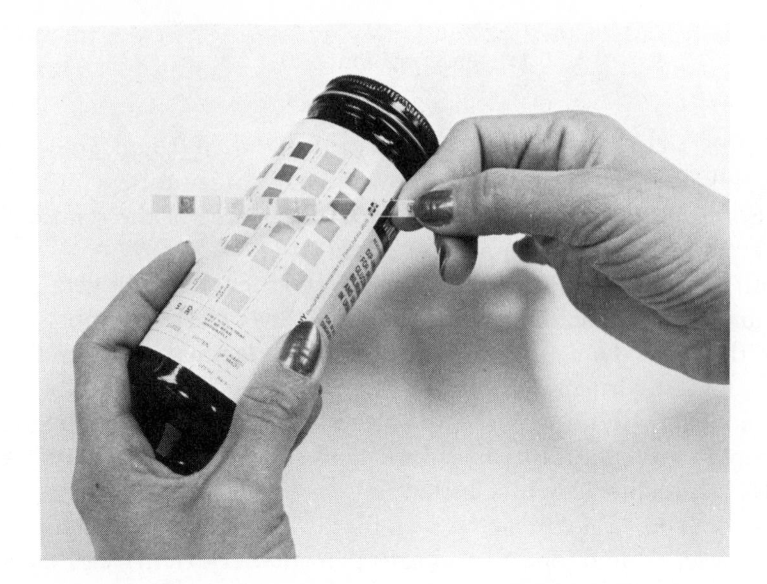

Fig. 10-4. Indicator paper strip being compared with chart of colors for determination of pH of test solution.

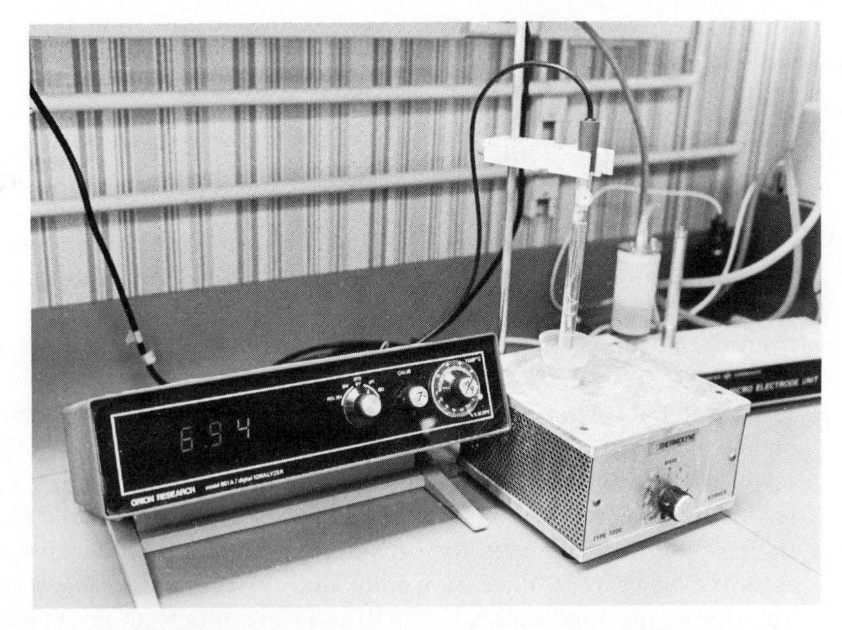

Fig. 10-5. Digital meter for determination of pH of solutions.

Buffer systems

As noted before, in the body and in many laboratory experiments, it is necessary to control the pH within very narrow ranges. One very useful method for accomplishing this is the use of a *buffer system,* which is a system composed of a solution of a weak acid and a salt of the weak acid. Such a combination can withstand changes in pH which otherwise would result from the addition of acids or bases. A buffer system composed of carbonic acid and its salt, sodium bicarbonate, would oppose any changes in pH as follows:

$$NaHCO_3 \rightarrow Na^+ + HCO_3^-$$

$$\text{Added acid} \rightarrow HCl \rightarrow Cl^- + H^+$$

$$\downarrow\uparrow$$

$$H_2CO_3$$

Since bicarbonate and hydrogen ions prefer to form undissociated molecules of carbonic acid, excess hydrogen ions would be removed from the solution, keeping the pH approximately the same as it was before the addition of acid.

$$H_2CO_3 \rightleftharpoons H^+ + HCO_3^-$$

$$\text{Added base} \rightarrow NaOH \rightarrow OH^- + Na^+$$

$$\downarrow\uparrow$$

$$H_2O$$

Since hydrogen and hydroxyl ions prefer to form undissociated molecules of water, excess hydroxyl ions would be removed from the solution, keeping the pH approximately constant.

From these reactions, it can be seen that both a weak acid and its salt must be present in the same solution before buffering can be accomplished —thus a buffering *system.* Confusion is sometimes introduced with the use of such terms as *buffer* or *buffer salt,* implying a requirement for only one substance in order to obtain buffering.

SUMMARY

IONIZATION

Arrhenius theory of ionization:

1. Electrolyte molecules dissociate to form ions.
2. Ions are charged particles. The sum of the positive charges must equal the sum of the negative charges.
3. Nonelectrolytes do not form ions.
4. Complete ionization occurs only in extremely dilute solutions. In more concentrated solutions, the ions are in equilibrium with undissociated molecules.

Solutions of electrolytes react readily, but dry powders do not, indicating that reactions take place between ions, not molecules.

Electrolytes cause abnormal changes in vapor pressure, boiling point, freezing point, and osmotic pressure of solutions because they form more particles in solution when they ionize than do nonelectrolytes.

Modification of Arrhenius theory of ionization: Debye-Hückel theory: Strong electrolytes are ionized in the crystalline state and are completely ionized in solution. Molecules of strong electrolytes do not exist in solution. Instead, the free movement of ions in concentrated solution is hampered by local aggregations of ions of opposite charge around one another. Weak electrolytes do remain largely as undissociated molecules in solution.

ELECTROLYTES

Acids: Hydronium ion is formed when acids ionize in water:

$$H^+ + H_2O \rightarrow \underset{\textbf{Hydronium ion}}{H_3O^+}$$

Strong acids ionize completely:

$$HCl \rightarrow H^+ + Cl^-$$

Weak acids do not ionize completely:

$$H_2CO_3 \rightleftharpoons H^+ + HCO_3^-$$

Bases: Bases contain both electrovalent and covalent bonds:

$$Na^+ \; :\ddot{\underset{\cdot\cdot}{O}}:H^-$$

Electrovalent bond   Covalent bond

When NaOH, a strong base, ionizes, the electrovalent bond is broken, but the covalent bond in the OH^- ion or radical remains intact:

$$NaOH \rightarrow Na^+ + OH^-$$

Ionization of a weak base:

$$NH_4OH \rightleftharpoons NH_4^+ + OH^-$$

Neutralization reactions: The reaction between an acid and a base is actually the reaction between H^+ and OH^- ions:

$$HCl \rightarrow H^+ + Cl^-$$

$$NaOH \rightarrow OH^- + Na^+$$

$$\downarrow\uparrow$$

$$H_2O$$

The hydrogen and hydroxyl ions neutralize each other and form molecules of water. The other ions remain in solution as ions.

Salts: Normal salts, formed from strong acid and strong base, ionize as follows:

$$NaCl \rightarrow Na^+ + Cl^-$$

Salts with acidic reaction: Salts formed from strong acid and weak base give acid reaction:

$$NH_4Cl \rightarrow NH_4^+ + Cl^-$$

$$\begin{array}{ccc} H_2O \rightleftharpoons & OH^- & + \; H^+ \\ & \downarrow\uparrow & \nwarrow \\ & NH_4OH & \text{Causes acidic reaction} \end{array}$$

Salts with basic reaction: Salts formed from strong base and weak acid give basic reaction:

$$NaHCO_3 \rightarrow Na^+ + HCO_3^-$$

$$\begin{array}{ccc} H_2O \rightleftharpoons & OH & + \; H^+ \\ & \nearrow & \downarrow\uparrow \\ \text{Causes basic reaction} & & H_2CO_3 \end{array}$$

Importance of ions in life processes: Ions are involved in digestion of food, transport of nutrients and removal of wastes, anabolism and catabolism of body constituents, conduction of nerve impulses, contraction of muscles, electrolyte and acid-base balance.

HYDROGEN ION CONCENTRATION (pH)

pH system:

$$pH = \log \frac{1}{H^+ \text{ concentration}}$$

If the hydrogen ion concentration of water (neutral) is 10^{-7},

$$pH = +7 = 7$$

The Sørensen pH scale ranges from 0 to 14: 0 to 6 = acidic solutions, 7 = neutrality, 8 to 14 = basic solutions. The lower the pH number, the higher the H^+ ion concentration. A weak acid (less ionization) requires a higher concentration to give the same pH as a more dilute solution of a strong acid (more ionization). A difference of one pH number, e.g., from pH 3 to 4, means a difference of ten times in the hydrogen ion concentration.

Measurement of pH:

1. Colorimetric. Using, for example, pH indicator paper, an approximate pH value is determined according to the change in color of the paper.
2. Electrometric. Using an electric pH meter, an accurate pH value is obtained.

Buffer systems: Composed of a solution of weak acid, e.g., carbonic acid (H_2CO_3), and a salt of the weak acid, e.g., sodium bicarbonate ($NaHCO_3$), to withstand changes in pH due to added acid or base:

$$NaHCO_3 \rightarrow Na^+ + HCO_3^-$$

$$\begin{array}{c} \text{Added acid} \rightarrow HCl \rightarrow Cl^- + H^+ \\ \downarrow\uparrow \\ H_2CO_3 \end{array}$$

Hydrogen ion removed as H_2CO_3, therefore no change in pH.

$$H_2CO_3 \rightleftharpoons H^+ + HCO_3^-$$

$$\text{Added base} \rightarrow NaOH \rightarrow \underset{\underset{H_2O}{\downarrow\uparrow}}{OH^-} + Na^+$$

OH^- removed as H_2O, therefore no change in pH.

REVIEW QUESTIONS

1. What are the major differences between the Arrhenius and Debye-Hückel theories of ionization? How are the ionizations of strong and weak electrolytes indicated in equations in this text?
2. Using the concepts of ionization, indicate what actually takes place in a neutralization reaction.
3. List the important aspects of the pH system. Why is a lower pH number indicative of a higher hydrogen ion concentration? What is the actual difference in concentration of hydrogen ion if the pH numbers differ by 1?
4. What is a buffer system? How does it operate? What is the significance of buffers in the body?
5. Show with equations how the slight ionization of H_2O becomes an important factor in the chemical properties of electrolytes.

REFERENCES

Lagowski, J. J.: Modern inorganic chemistry, New York, 1973, Marcel Dekker, Inc.

Masterton, W. L., and Slowinski, E. J.: Chemical principles, ed. 4, Philadelphia, 1977, W. B. Saunders Co.

Pauling, L.: General chemistry, ed. 3, San Francisco, 1970, W. H. Freeman and Co., Publishers.

Petrucci, R. H.: General chemistry: principles and modern applications, ed. 2, New York, 1977, Macmillan Publishing Co., Inc.

UNIT THREE

ORGANIC CHEMISTRY

General principles of organic chemistry

11

Interest in the large variety of common and exotic compounds of carbon that could be obtained from plant and animal sources dates back thousands of years to the very beginnings of chemistry. By the second half of the seventeenth century, attempts were being made to separate organic from inorganic chemistry by classifying substances according to source as mineral, animal, or vegetable. Because of the difficulties involved in working with organic compounds and the inability to synthesize such compounds from materials not obtained from living organisms, the belief grew that organic compounds could be formed only in living organisms under the influence of a "vital force."

The vital force theory did not come under serious challenge until 1828 when Friedrich Wöhler, a German scientist, accidentally prepared urea, a common product of animal life, by heating the inorganic compound, ammonium cyanate:

$$\underset{\textbf{Ammonium cyanate}}{NH_4OCN} \underset{\Delta}{\rightarrow} \underset{\textbf{Urea}}{CO(NH_2)_2}$$

As other successful syntheses of organic compounds followed, it became clear that a "vital force" or a living organism was not an absolute requirement for the synthesis of such compounds. Further study of these compounds and their reactions showed that the same chemical forces and laws applied to organic compounds as well as to inorganic compounds, and that organic compounds always contained the element carbon in their structures. On this basis, it has been determined that the differences in properties and reactions between organic and inorganic compounds can be ascribed mainly to the properties of carbon. Thus, *organic chemistry* is generally defined, at the present time, simply as *the study of the compounds of carbon.* A few substances such as carbon itself, carbon monoxide and dioxide, and carbonic acid and its salts are not included in this definition and are usually treated as inorganic compounds.

Although the vital force theory no longer exists, the importance of organic compounds to life has not diminished since these compounds play key roles in all types of plant and animal life. In order to attempt to understand life processes, normal and abnormal, it is essential to acquire an understanding of the properties and reactions of organic compounds.

In modern times, the ingenuity of human beings in the laboratory has sometimes matched that of nature in the living world. The number of organic compounds that have been reported to date, mostly synthetic, is in the millions. The achievements of the organic chemist are well known in the preparation of fibers such as nylon, all sorts of new plastic materials, synthetic natural products such as vitamins and antibiotics, many new drugs, and hosts of other products that make life more comfortable. In spite of such an impressive list of accomplishments, nature still has a few challenges remaining in the almost incredibly complicated structures of some naturally occurring compounds and in their interrelationships—secrets that are yielding only very grudgingly, if at all, to massive research efforts in many laboratories.

COMPARISON OF ORGANIC AND INORGANIC COMPOUNDS

As stated before, the same basic principles of chemistry that apply to inorganic compounds and their reactions also apply to organic compounds. Why then is a division usually made between the study of inorganic and organic chemistry? It is done mainly on the basis of convenience. Inorganic chemistry concerns itself with the study of a relatively small number of compounds composed of various combinations of one or more of 104 out of 105 elements (e.g., in July, 1978, less than 4% of 4.3 million chemical substances recorded in the Chemical Abstracts Service Chemical Registry System were inorganic compounds). Organic chemistry concerns itself with the study of compounds of the remaining element, carbon, generally in combination with hydrogen and oxygen, and occasionally with nitrogen, phosphorus, and sulfur. The large number of organic compounds (in the millions range) becomes manageable on the basis that carbon forms many large series of related compounds with similar properties. Thus the major differences between inorganic and organic compounds are attributable to the differences in properties of carbon compounds in comparison to the properties of inorganic compounds.

One of the major differences is the type of bond formed by these two classes of compounds. As stated in previous chapters, most inorganic compounds are electrovalent, whereas organic compounds are usually covalent. This leads to several important consequences. Inorganic compounds are generally quite soluble in water, a polar solvent that tends to pull the ions apart in solution. Organic compounds, which usually do not form any ions, are relatively insoluble in water but soluble in nonpolar organic solvents. Inorganic compounds are not soluble in organic solvents. Thus, polar solvents tend to dissolve polar or ionic compounds, and nonpolar solvents tend to dissolve nonpolar compounds.

Another consequence of the difference in bonds found in organic and inorganic compounds follows from the fact that reactions usually occur between ions and not molecules. This means that inorganic reactions generally take place much more rapidly and under much less drastic conditions than organic reactions. Fortunately, most of the burden thus imposed on organic reactions is met by nature in the form of enzymes, protein catalysts that facilitate almost all of the reactions involving organic compounds in all living organisms.

One of the most interesting properties of carbon compounds is the ability of carbon atoms to combine with each other in long chains, rings, and almost infinite variations. This leads not only to endless opportunities to form large numbers of compounds but also molecules with very high molecular weights. In contrast, although inorganic compounds can be made from 104 elements, the possible combinations are quite limited and the molecular weights are generally much lower than those found in organic compounds.

IMPORTANT PROPERTIES OF CARBON

Tetracovalency

Carbon, with an atomic number of 6, has two electrons in its first orbit and four in its second or outer orbit. Thus, to achieve the stable configuration of the nearest noble gas, carbon might have the choice of losing four electrons (with a resulting configuration like helium) or gaining four electrons (with a resulting configuration like neon). Instead of giving up electrons, as metallic elements do, or gaining electrons, as nonmetallic elements do, carbon shares its electrons, in the form of pairs, with four other electrons contributed by other carbon atoms or the other atoms, such as hydrogen and oxygen, usually found in organic compounds. Therefore, *carbon invariably has a valence of four in organic compounds on the basis of four covalent bonds.* For example, the simplest organic compound of carbon can be formed from one carbon atom and four hydrogen atoms. This compound is methane (CH_4), and the four covalent bonds may be shown as follows:

```
  H                 H
  ••                |
H:C:H     or    H — C — H
  ••                |
  H                 H
```

For greater convenience in writing structural organic formulas, the pair of electrons in the covalent bond is usually represented by a short, straight line connecting the two atoms contributing one electron each to the covalent bond. The complete electronic configuration for methane is shown in Fig. 11-1. To know that carbon always has a valence of four is of great help in the study of organic chemistry.

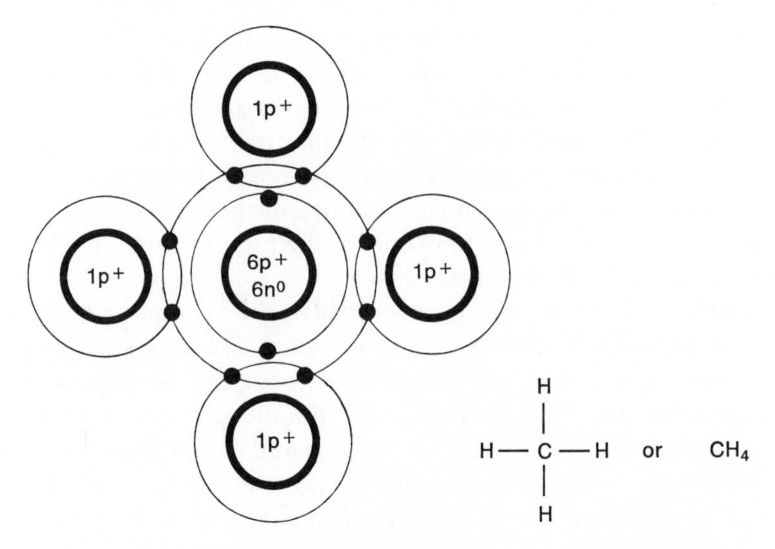

Fig. 11-1. Electron configuration of methane (CH_4). Note that carbon forms four covalent bonds, as it does almost invariably in all organic compounds.

Equivalence of carbon valences

All four valence bonds of the carbon atom are equivalent because the replacement of one of the four hydrogen atoms in methane always produces only one form of the new compound. If the valence bonds of the carbon atom were different, it would be expected that more than one new compound could be formed in single replacement reactions. Although attempts were made, this has never been accomplished. In other words, if a bromine atom replaces one hydrogen atom in methane, only one methyl bromide is formed and it makes no difference which hydrogen atom is replaced. In representing the structure of methyl bromide, each one of the following formulas is equivalent:

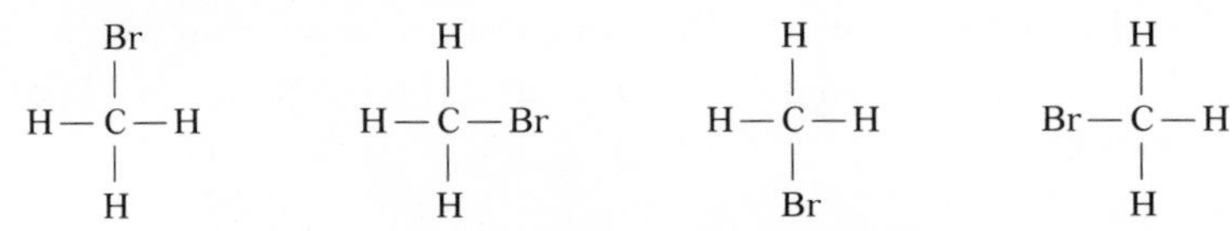

Equivalent forms of methyl bromide

Regular tetrahedron

On the basis of the equivalence of the valence bonds of the carbon atom, it would be expected that these bonds would be distributed around the carbon atom in a regular manner. X-ray and other studies show that in three-dimensional form, the carbon atom occupies the center of a regular tetrahedron with the four valence bonds directed toward the four points (Fig. 11-2). The hydrogen atoms of methane are considered to be attached at the points of the tetrahedron. There are several models available for classroom use that make these relationships more easily understandable (Fig. 11-3).

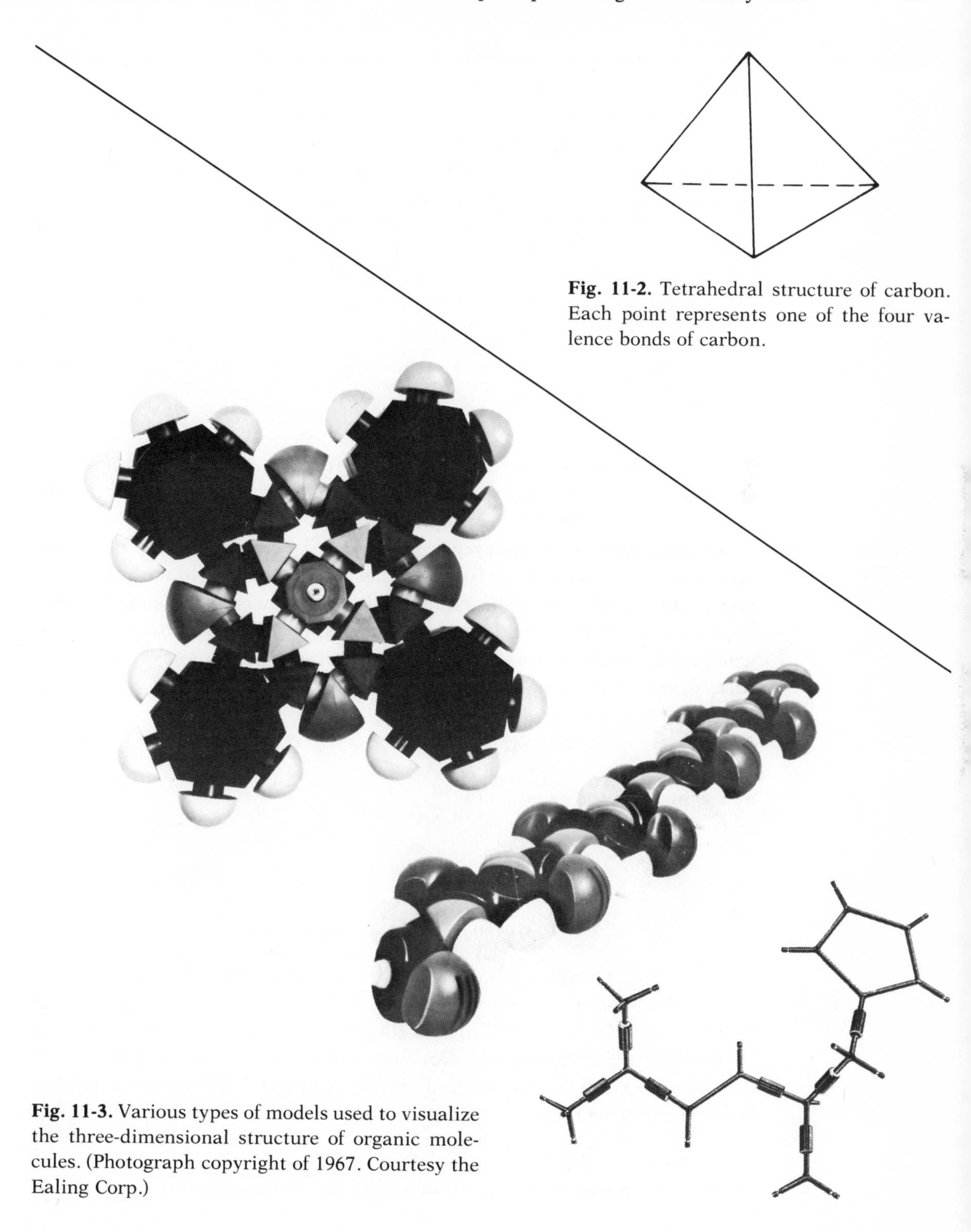

Fig. 11-2. Tetrahedral structure of carbon. Each point represents one of the four valence bonds of carbon.

Fig. 11-3. Various types of models used to visualize the three-dimensional structure of organic molecules. (Photograph copyright of 1967. Courtesy the Ealing Corp.)

Self-combining capacity

A very important reason for the large number of organic compounds is the ability of the carbon atom to combine with itself.

Open-chain compounds (aliphatic or fatty compounds). Various sizes of open chains of carbon atoms attached to each other are found in nature and are easily synthesized in the laboratory. The electron structures of a simple open-chained and a branched-chain compound are:

```
 ·· ·· ··          ·· ·· ··
:C:C:C:           :C:C:C:
 ·· ·· ··          ·· ·· ··
                     :C:
                      ··
```

The corresponding structural formulas are:

```
  |   |   |           |   |   |
— C — C — C —       — C — C — C —
  |   |   |           |   |   |
                        — C —
                          |
```

The branch of organic chemistry dealing with such open-chain compounds is known as *aliphatic organic chemistry* because of the fatty nature of the members of this class that were first studied.

Closed-chain compounds (cyclic compounds). Carbon atoms may also combine with each other in the form of rings. When they do so with only carbon atoms in the ring, the compounds are known as *carbocyclic compounds.* If an atom other than carbon also appears in the ring, e.g., nitrogen, these compounds are known as *heterocyclic compounds.* The structural formulas of these types are as follows:

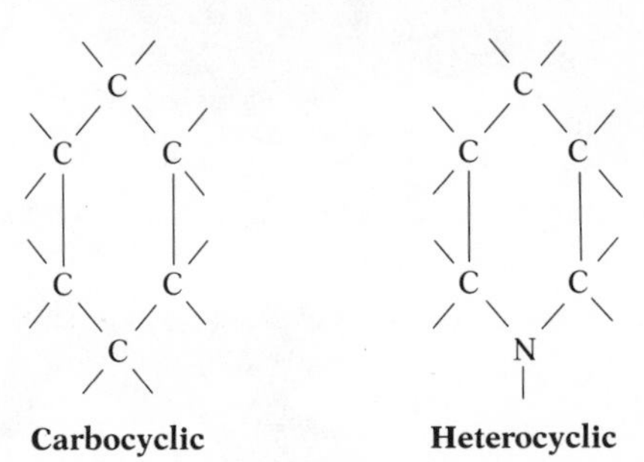

Carbocyclic Heterocyclic

The branch of organic chemistry concerned with carbocyclic compounds is sometimes known as *aromatic organic chemistry* because many of the first compounds discovered in this class have pleasant aromas.

Single, double, triple bonds. Not only can carbon atoms combine with each other readily in the form of chains or rings, but they can combine with one, two, or three bonds between any two carbon atoms, as follows:

```
 ·· ··        ·  ·
:C:C:       :C::C:       :C:::C:
 ·· ··
```

Single bond **Double bond** **Triple bond**

or,

$$\begin{array}{c}|\;\;\;|\\ -\mathrm{C}-\mathrm{C}-\\ |\;\;\;|\end{array}\qquad\qquad \begin{array}{c}\diagdown\;\;\;\;\diagup\\ \mathrm{C}=\mathrm{C}\\ \diagup\;\;\;\;\diagdown\end{array}\qquad\qquad -\mathrm{C}\equiv\mathrm{C}-$$

Those compounds in which all the carbon atoms are connected with each other by single bonds only are called *saturated compounds,* whereas compounds containing double or triple bonds between carbon atoms are called *unsaturated compounds.* Unsaturated compounds are usually much more reactive than saturated compounds.

Single-bonded carbon atoms are considered to be joined by single points of the two tetrahedra representing the carbon atoms. Double-bonded carbon atoms are joined by two points along one edge of the tetrahedra and triple-bonded carbon atoms are joined by three points along one face of the tetrahedra. These relationships are shown in Fig. 11-4.

Homologous series. If, for example, the series of compounds formed from carbon and hydrogen containing one, two, three, and four carbon atoms is examined, it is found that each two consecutive members of the series differ by a factor of CH_2:

$$\begin{array}{c}\mathrm{H}\\ |\\ \mathrm{H}-\mathrm{C}-\mathrm{H}\\ |\\ \mathrm{H}\end{array}\qquad \begin{array}{c}\mathrm{H}\;\;\;\mathrm{H}\\ |\;\;\;\;|\\ \mathrm{H}-\mathrm{C}-\mathrm{C}-\mathrm{H}\\ |\;\;\;\;|\\ \mathrm{H}\;\;\;\mathrm{H}\end{array}\qquad \begin{array}{c}\mathrm{H}\;\;\;\mathrm{H}\;\;\;\mathrm{H}\\ |\;\;\;\;|\;\;\;\;|\\ \mathrm{H}-\mathrm{C}-\mathrm{C}-\mathrm{C}-\mathrm{H}\\ |\;\;\;\;|\;\;\;\;|\\ \mathrm{H}\;\;\;\mathrm{H}\;\;\;\mathrm{H}\end{array}\qquad \begin{array}{c}\mathrm{H}\;\;\;\mathrm{H}\;\;\;\mathrm{H}\;\;\;\mathrm{H}\\ |\;\;\;\;|\;\;\;\;|\;\;\;\;|\\ \mathrm{H}-\mathrm{C}-\mathrm{C}-\mathrm{C}-\mathrm{C}-\mathrm{H}\\ |\;\;\;\;|\;\;\;\;|\;\;\;\;|\\ \mathrm{H}\;\;\;\mathrm{H}\;\;\;\mathrm{H}\;\;\;\mathrm{H}\end{array}$$

or,

$$\begin{array}{ccccccc} CH_4 & & C_2H_6 & & C_3H_8 & & C_4H_{10}\\ & + \;\; /\!\!/ & & + \;\; /\!\!/ & & + \;\; /\!\!/ & \\ & CH_2 & & CH_2 & & CH_2 & \end{array}$$

Such a series is known as a *homologous series,* and the members are called *homologs* of each other. Other types of homologous series also differ by the constant factor of CH_2. Members of homologous series usually differ from

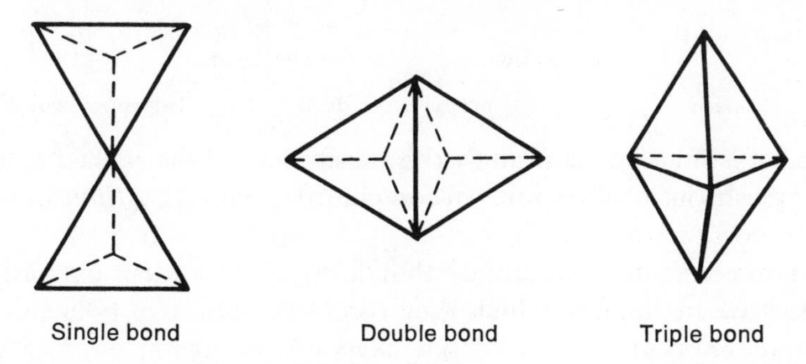

Fig. 11-4. Representation of single (two points), double (one edge), and triple (one face) bonds between two carbon atoms.

each other in various properties in a regular manner, making it easier to study large groups of compounds together rather than one at a time.

ISOMERS

Since carbon atoms can combine in so many different ways, there are many examples of two or more compounds that have the same molecular (empirical) formula but differ in structure or arrangement of the atoms or radicals. Such compounds are called *isomers.*

Chain or nuclear isomerism

Turning again to the simple compounds of carbon and hydrogen, it is found that when a compound containing four carbon atoms in an open chain is considered, there are two ways in which the four carbon atoms can be arranged:

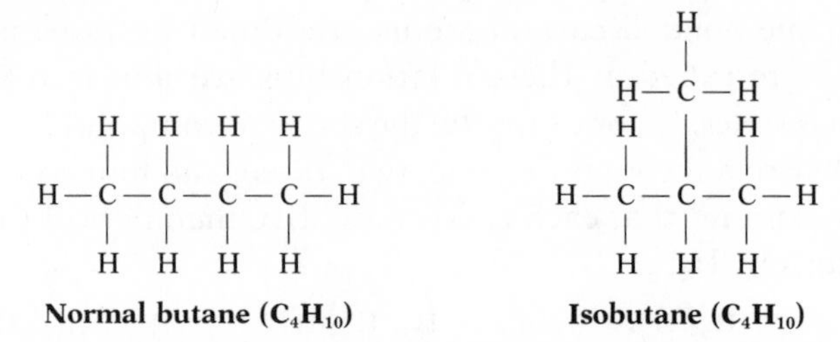

Normal butane (C_4H_{10}) **Isobutane (C_4H_{10})**

These two compounds are isomers, and this type of isomerism involving a different structure due to the manner of linking the carbon chain is termed *chain* or *nuclear isomerism.* Note that the molecular formulas give no indication of the difference in structure of the isomers.

Position isomerism

If in a compound such as propane (C_3H_8) one of the hydrogen atoms is replaced by another atom, e.g., bromine, two different products are possible:

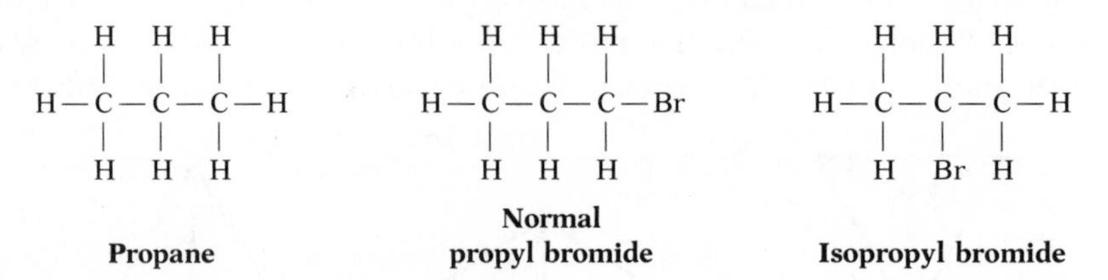

Propane **Normal propyl bromide** **Isopropyl bromide**

This type of isomerism, caused by the attachment of the replacing atom in different positions of the same carbon chain, is called *position isomerism.*

Optical isomerism

There are certain compounds that have almost identical properties except for the manner in which they rotate the plane of polarized light. Compounds containing at least one *asymmetric* carbon atom, a carbon atom to which are attached four different other atoms, will exhibit this type of isomerism known as *optical isomerism.* Examination of Fig. 11-5 will show that the two carbon atoms, with atoms a, b, c, d attached, can-

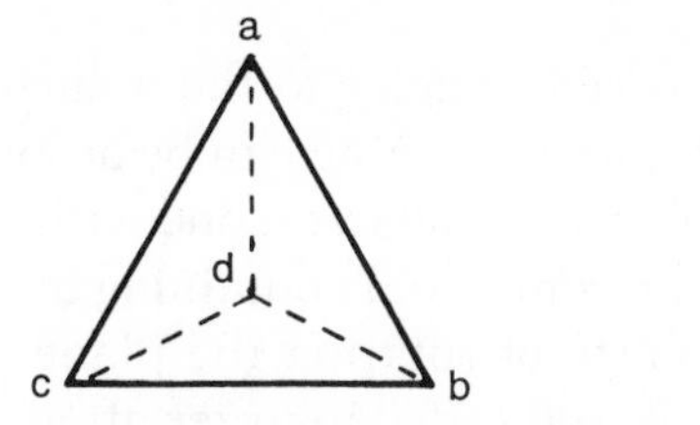

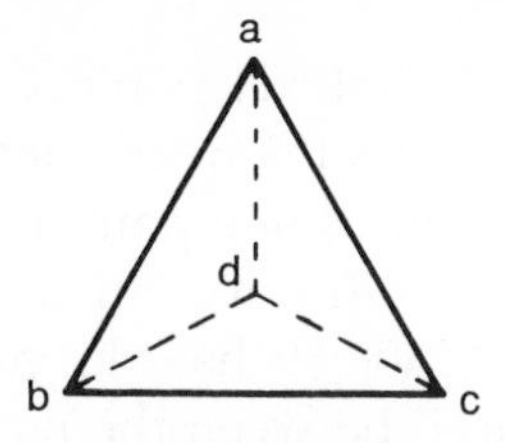

Fig. 11-5. Mirror image configurations around an asymmetric carbon atom.

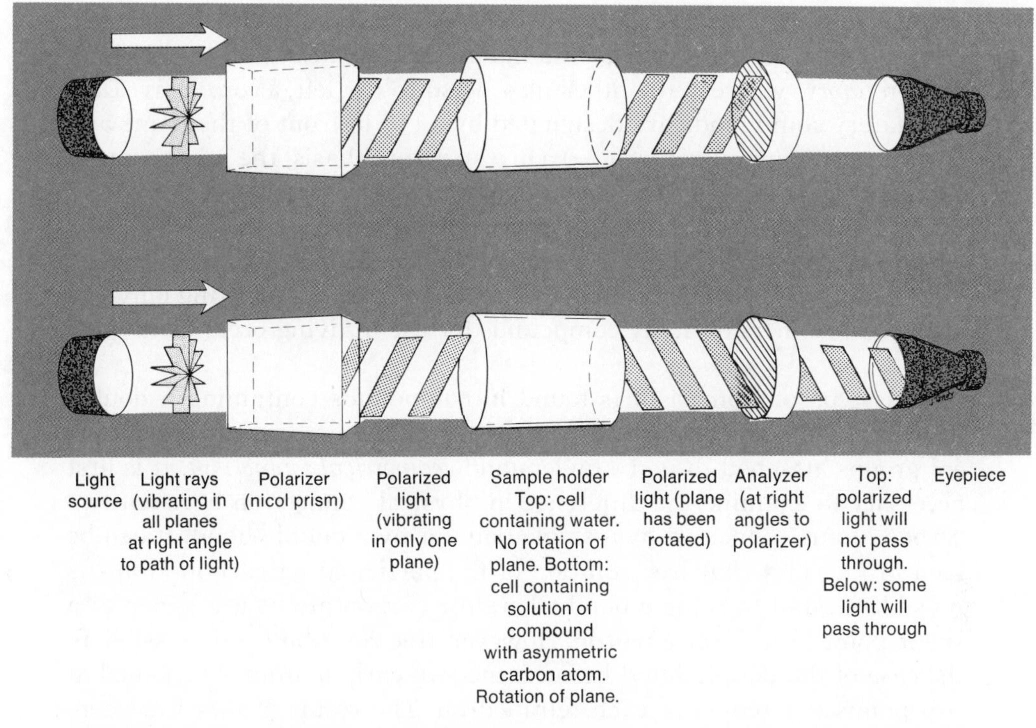

Fig. 11-6. Representation of a polarimeter, an instrument used to determine the optical rotation produced by substances containing asymmetric carbon atoms. When turned to right angles from each other, no light will pass through the combination of polarizer and analyzer. A solution containing asymmetric carbon atoms has been placed in the lower sample holder, with the results as diagrammed.

not be superimposed on each other as long as they are kept in the plane of the paper. That these two structures cannot be superimposed on each other may be more readily seen when shown in three-dimensional form. Such optical isomers are called *enantiomorphs* and are mirror images of each other. Fig. 11-6 explains, in a simple manner, the phenomenon of the polarization of light. An instrument embodying these principles is the *polarimeter,* which contains two polarizing units known as the polarizer and the analyzer, respectively.

When the analyzer is placed at right angles to the polarizer, no light will pass through the instrument (the same thing can be accomplished by rotating the lenses of a pair of polaroid sunglasses, one within the other). If, however, a solution of an organic compound containing an asymmetric carbon atom, which has the property of rotating the plane of polarized light, is placed between the polarizer and the analyzer after the analyzer has been positioned to allow no light to pass through, some light does come through. The polarimeter allows the exact determination of the degree and direction of rotation produced by the test substance. One of a pair of optical isomers may rotate the plane of polarization to the right, *dextrorotatory,* whereas the other may do so to the left, *levorotatory.* Dextrorotatory compounds are designated by a (+) in front of the name and levorotatory compounds by a (−). On a structural basis, the mirror-image compounds are designated by D- and L- prefixes, respectively. The D- and L- designations are not related to the (+) and (−) notations (p. 184).

This type of isomerism is very important because, as will be discussed later, nature chose to be very specific in many cases by selecting only one of the two isomers of many compounds for use in living organisms.

Geometrical isomerism

This type of isomerism is found in compounds containing a double bond between two carbon atoms, each of which has the same two different groups attached to it. To understand *geometrical isomerism,* it is first necessary to examine the difference in physical connection between two carbon atoms connected by a single bond or by a double bond. It can be seen in Fig. 11-4 that free rotation of the participating carbon atoms is possible around the single bond where the carbon atoms are joined at a single point of each tetrahedron. However, free rotation is not possible in the case of the double bond because the two carbon atoms are joined at two points (one edge) of each tetrahedron. The result of this fixed configuration of the carbon atoms is that when each carbon atom has the same two different groups attached to it, two different arrangements are possible (Fig. 11-7). The *cis*-form has similar groups on each side of the double bond whereas the *trans*-form has the similar groups on opposite sides of the double bond.

Fig. 11-7. Geometric isomers, with *cis*-form showing similar groups on the same side of the double bond, and *trans*-form showing similar groups on opposite sides of the double bond.

NAMING ORGANIC COMPOUNDS

Because of the many simple and complicated compounds possible in organic chemistry, the problem of naming these compounds became very serious. The first of a series of international meetings on the subject was held in Geneva in 1892. Certain rules and numbering systems were agreed on for simplification. However, the names of the simpler compounds, which had not been named in a systematic manner by the original discoverers, were usually retained. Also, it became more obvious as time went on that no single system could cover all possibilities. Therefore, after many international meetings, the names generally in use today are made up of trivial names for the simpler compounds, names under the Geneva rules for the larger and more complicated compounds, and other types of names as the need develops for them. The different types of names will be discussed with the specific classes of compounds.

SUMMARY

According to the "vital force" theory the synthesis of organic compounds was possible only by living organisms. Since the formation of the organic compound, urea, from the inorganic compound, ammonium cyanate, by Wöhler in 1828, the vital force theory has been invalidated. *Organic chemistry* is now defined simply as the *chemistry of the compounds of carbon.*

COMPARISON OF ORGANIC AND INORGANIC COMPOUNDS

	Organic compounds	Inorganic compounds
Number of compounds	Very large	Small
Constituent elements	C, H, O, N, S, P, plus a few others	Any of 104 elements
Bonds	Covalent	Electrovalent
Solubility	Insoluble in water (polar solvents), soluble in organic solvents (nonpolar)	Soluble in water, insoluble in organic solvents
Rates of reaction	Slow	Rapid (ionic)
Molecular weights	High	Low

IMPORTANT PROPERTIES OF CARBON

Tetracovalency: Carbon always has a valence of four on the basis of four covalent bonds.

Valences are equivalent: Each of the four valences of carbon are equivalent. Therefore, if one hydrogen atom of methane (CH_4), is replaced by another atom, e.g., Br, only one form of the product, CH_3Br, is possible.

Regular tetrahedron: The carbon atom can be considered as occupying the center of a regular tetrahedron with the four valences directed toward the four points.

Self-combining capacity: Almost unlimited, forming:

Open-chain compounds (aliphatic or fatty compounds):

Aliphatic organic chemistry: Chemistry of open-chain compounds.

Closed-chain compounds (cyclic compounds):

Carbocyclic compounds: Ring compounds containing only carbon atoms in ring.

Aromatic organic chemistry: Chemistry of carbocyclic compounds.

Heterocyclic compounds: Ring compounds containing, in the ring, other atoms in addition to carbon.

Single, double, triple bonds: Any two adjacent carbon atoms may be joined together with a single (one shared pair of electrons), double (two shared pairs of electrons), or triple bond (three shared pairs of electrons) between them.

Saturated compounds: Compounds with only single bonds between carbon atoms.

Unsaturated compounds: Compounds with double or triple bonds between carbon atoms.

Homologous series: Series of related compounds in which each succeeding member differs from the others by the constant factor of CH_2.

ISOMERS Compounds that have the same molecular (empirical) formula but differ in structure or arrangement of the atoms or radicals.

Chain or nuclear isomerism: Involves a different structure, in two or more compounds, because of the manner of linking of the carbon chain.

Position isomerism: Involves a different structure, in two or more compounds, because of the attachment of a replacing atom, e.g., Br, to a different carbon atom in the same chain.

Optical isomerism: Isomers that are identical in properties except for their effect on the plane of polarization of light. Optical isomers must contain at least one asymmetric carbon atom.

Asymmetric carbon atom: Carbon atom to which four different atoms or groups are attached.

Enantiomorphs: Optical isomers that are mirror images of each other, and identified by D- and L- prefixes.

Dextrorotatory: Molecules that rotate plane of polarization of light to the right, designated by the prefix (+).

Levorotatory: Molecules that rotate plane of polarization of light to the left, designated by the prefix (−).

Geometrical isomerism: Found in compounds containing a double bond between two carbon atoms, each of which has the same two different groups attached to it.

Cis-Form: Similar atoms on same side of double bond.
Trans-Form: Similar atoms on different sides of double bond.

NAMING ORGANIC COMPOUNDS

The names generally in use today are made up of trivial names for the simpler compounds, names under the Geneva rules for the larger and more complicated compounds, and other types of names as the need develops for them. No single system covers all the possibilities.

REVIEW QUESTIONS

1. What are the important chemical properties of carbon?
2. What are some reasons for separating the study of carbon compounds from other compounds? What is the "vital force" theory and what is its current status?
3. What are isomers? What are some forms of isomerism? What is the significance of isomers to organic chemistry?
4. Review the specific differences between chain isomerism and position isomerism.
5. What is the difference between D- and L-isomers? What is the significance with respect to the physical properties of compounds designated with the prefixes (+) or (−)?

REFERENCES

Allinger, N. L., and others: Organic chemistry, ed. 2, New York, 1976, Worth Publishers, Inc.
Geissman, T. A.: Principles of organic chemistry, ed. 4, San Francisco, 1977, W. H. Freeman and Co., Publishers.
Holum, J. R.: Organic chemistry: a brief course, New York, 1975, John Wiley & Sons, Inc.
Morrison, R. T., and Boyd, R. N.: Organic chemistry, ed. 3, Boston, 1973, Allyn & Bacon, Inc.
Weininger, S. J.: Contemporary organic chemistry, New York, 1972, Holt, Rinehart and Winston, Inc.

12 Open-chain (aliphatic) organic compounds

In the next few chapters the structures and properties of the organic compounds that are of direct and indirect interest to students of the life sciences are discussed. Emphasis is placed, as much as possible, on the relationships to the living organism rather than on industrial applications. Further information on industrial applications, undoubtedly of great importance to the welfare of modern living, is available in many good texts and reference books.

FUNCTIONAL GROUPS

Fortunately for all students of organic chemistry, the existence of homologous series of compounds makes it possible to study the general properties of a large number of compounds as a group rather than the specific properties of each compound one by one. Another simplifying factor is the observation that certain atoms, radicals, or groups, referred to as *functional groups,* are largely responsible for the properties of the compounds in which they appear. Thus, it is desirable to study the general properties of the functional groups rather than the individual compounds that may contain such groups.

HYDROCARBONS

The hydrocarbons are the simplest organic compounds from the point of view of atomic constituents because these compounds, as the name implies, consist of only hydrogen and carbon.

Saturated hydrocarbons (alkanes)

The homologous series, starting with methane (CH_4), is known as the saturated hydrocarbon series because, except for the valences required in connecting the chain of carbon atoms, all other carbon valences are filled, or *saturated,* with hydrogen atoms. This series is also designated *alkane* in the modern nomenclature system, the suffix *-ane* indicating complete saturation. On the basis of the fact that saturated hydrocarbons are characteristically inert, showing very reluctant tendencies to combine with any reagents, they are frequently referred to as *paraffin* hydrocarbons (*L., parvum affinis,* little affinity). It is also common practice to name a homologous series after the first member. Therefore, the saturated hydrocarbons are sometimes referred to as the *methane* series.

The first four members of this series, methane (CH_4), ethane (CH_3CH_3), propane ($CH_3CH_2CH_3$), and butane ($CH_3CH_2CH_2CH_3$), retain their original, nonsystematic names. Starting with the five-carbon member, pentane, the names of the saturated hydrocarbons are derived mainly from the Greek numerals, with some Latin prefixes also being used.

As noted in Chapter 11, butane is the first member of the methane series to show chain isomerism. The two possible forms for a four-carbon saturated hydrocarbon are written as follows:

$$\begin{array}{ccccccccc} & & H & & H & & H & & H & & \\ & & | & & | & & | & & | & & \\ H & - & C & - & C & - & C & - & C & - & H \\ & & | & & | & & | & & | & & \\ & & H & & H & & H & & H & & \end{array} \qquad \begin{array}{ccccccc} & & H & & H & & H & & \\ & & | & & | & & | & & \\ H & - & C & - & C & - & C & - & H \\ & & | & & | & & | & & \\ & & H & & | & & H & & \\ & & H & - & C & - & H & & \\ & & & & | & & & & \\ & & & & H & & & & \end{array}$$

n-Butane **Isobutane**

or,

$$CH_3(CH_2)_2CH_3 \qquad (CH_3)_3CH$$

Possibilities for isomerism rise rapidly with an increasing number of carbon atoms, as seen from the list below:

No. of carbon atoms	*Possible isomers*
5	3
6	5
7	9
8	18
9	35
10	75
15	4,347
20	366,319
30	4,111,846,763

Although most of the vast number of possible isomers have not been isolated or synthesized, this list emphasizes the tremendous potentialities that exist for the production of new organic compounds.

Occurrence. Saturated hydrocarbons are found in large abundance as the major constituents of natural gas and petroleum. Natural gas is made up largely of methane with much smaller amounts of the next few members of the series, all gases. The hydrocarbons in petroleum range in carbon content from one to forty, most existing as liquids and solids in the pure state.

Chemical properties. As noted before, hydrocarbons of the methane series are generally unreactive substances. Those reactions that do take place usually depend on special structures, high temperatures, or powerful catalysts.

Chlorination, bromination (halogenation). An example of the unusual conditions required for reaction is the chlorination of methane. Mixtures of chlorine and methane do not react in the dark, even after long periods of storage. On exposure to sunlight, however, a rapid reaction occurs that may become explosive:

$$\begin{array}{c} H \\ | \\ H-C-H \\ | \\ H \end{array} + Cl_2 \xrightarrow[\text{light}]{\text{ultraviolet}} HCl + \begin{array}{c} H \\ | \\ H-C-Cl \\ | \\ H \end{array}$$

Methane **Methyl chloride**

or,

$$CH_4 + Cl_2 \xrightarrow[\text{light}]{\text{ultraviolet}} HCl + CH_3Cl$$

Once formed, the methyl chloride reacts more easily with more chlorine to give methylene chloride (CH_2Cl_2), which in turn reacts further until chloroform ($CHCl_3$) and carbon tetrachloride (CCl_4) are also produced. In spite of this series of reactions, much of the original methane does not react. Bromine reacts with saturated hydrocarbons the same way as chlorine but less vigorously.

Oxidation (burning). The output of energy is very high when the saturated hydrocarbons burn, or combine with oxygen. Advantage has been taken of this property in the use of gasoline (a mixture of hydrocarbons) as fuel in automobile engines and in industrial applications of other petroleum products. Normal octane, $CH_3(CH_2)_6CH_3$, produces 11.43 Cal (11,430 cal) per g when completely oxidized. The other saturated hydrocarbons release a similar amount of energy when oxidized. As will be seen later, it is oxidation processes of this type, under much less drastic conditions (under the influence of enzymes), that produce the energy required by living organisms.

Unsaturated hydrocarbons: alkenes (ethylenic hydrocarbons)

Hydrocarbons of this series are characterized by a double bond between two carbon atoms. Since only one bond is ordinarily required to link two carbon atoms, this situation provides a pair of electrons (one bond) for potential combination with other atoms. Thus, unsaturated hydrocarbons containing a double bond are much more reactive than the corresponding saturated compounds, i.e., the alkanes or paraffin hydrocarbons, even though the physical properties of the two series are very similar.

The first few members of the *ethylenic* hydrocarbon or *alkene* series are known by their common names: ethylene, propylene, butylene, amylene.

By the Geneva system, the *-ane* ending for the methane series is changed to *-ene* for the ethylene series.

The double bond may be considered as a functional group because it conveys specific chemical properties to the compounds containing it. As a result, starting with the butenes, there may be position isomers, as follows:

$$\begin{array}{ccccccccc} & & H & & H & & H & & H \\ & & | & & | & & | & & | \\ H & - & C & - & C & - & C & = & C \\ & & | & & | & & & & | \\ & & H & & H & & & & H \end{array} \qquad \begin{array}{ccccccccccc} & & H & & H & & H & & H & & \\ & & | & & | & & | & & | & & \\ H & - & C & - & C & = & C & - & C & - & H \\ & & | & & & & & & | & & \\ & & H & & & & & & H & & \end{array}$$

Butene-1 **Butene-2**

or,

$$CH_3CH_2CH{=}CH_2 \qquad CH_3CH{=}CHCH_3$$

Under the Geneva system, the position of the double bond is designated by a number, the number 1 indicating that the double bond is between carbon atoms 1 and 2, the number 2 indicating that the double bond is between carbon atoms 2 and 3, and so on.

Chemical properties. The more important reactions of the alkenes can be considered as *addition reactions,* i.e., one of the two bonds between the unsaturated carbon atoms opens and a new atom is added to each of these carbon atoms.

Halogenation. The alkenes react so readily with chlorine and bromine (halogens) that the decolorization of a red bromine solution (in carbon tetrachloride) is used as a test for the presence of double bonds in compounds:

$$\begin{array}{ccccccc} & & H & & H & & \\ & & | & & | & & \\ H & - & C & = & C & - & H \end{array} + Br_2 \rightarrow \begin{array}{ccccccc} & & H & & H & & \\ & & | & & | & & \\ H & - & C & - & C & - & H \\ & & | & & | & & \\ & & Br & & Br & & \end{array}$$

Ethylene **Ethylene dibromide**

or,

$$CH_2{=}CH_2 + Br_2 \rightarrow CH_2BrCH_2Br$$

Addition of hydrogen halides. Hydrogen halides can add to the double bonds in the same way as the halogens themselves but the order of reactivity is reversed: HI > HBr > HCl.

$$\begin{array}{ccccccc} & & H & & H & & \\ & & | & & | & & \\ H & - & C & = & C & - & H \end{array} + HBr \rightarrow \begin{array}{ccccccc} & & H & & H & & \\ & & | & & | & & \\ H & - & C & - & C & - & H \\ & & | & & | & & \\ & & H & & Br & & \end{array}$$

Ethylene **Ethyl bromide**

or,

$$CH_2{=}CH_2 + HBr \rightarrow CH_3CH_2Br$$

Hydrogenation (reduction). Under the influence of suitable catalysts, hydrogen can be made to react with alkenes to form the corresponding alkanes:

$$\begin{array}{c}\text{H}\quad\text{H}\\ |\quad\;|\\ \text{H}-\text{C}{=}\text{C}-\text{H}\end{array} + H_2 \xrightarrow{\text{catalyst}} \begin{array}{c}\text{H}\quad\text{H}\\ |\quad\;|\\ \text{H}-\text{C}-\text{C}-\text{H}\\ |\quad\;|\\ \text{H}\quad\text{H}\end{array}$$

Ethylene **Ethane**

or,

$$CH_2{=}CH_2 + H_2 \xrightarrow{\text{catalyst}} CH_3CH_3$$

Unsaturated hydrocarbons: alkynes (acetylenic hydrocarbons)

The unsaturated hydrocarbons containing a triple bond between two carbon atoms are known as acetylenic hydrocarbons, from acetylene, the first member of the series. Under the Geneva system, this series is called the *alkynes.* Because of the greater degree of unsaturation, the alkynes are somewhat more reactive than the alkenes, but their reactions are quite similar, being mainly addition reactions (see reactions of alkenes). The alkynes are of no importance in the biochemistry of living organisms.

Alkyl radicals

Because several series of compounds with different functional groups can be regarded as derivatives of the methane series by replacement of one hydrogen atom, these new compounds are generally named after the parent compound to show this relationship. Thus, the original alkane hydrocarbons minus one hydrogen atom are known as *alkyl radicals,* which are named according to the following examples:

Alkane hydrocarbons	*Alkyl radicals*
Methane, CH_4	Methyl, CH_3-
Ethane, CH_3CH_3	Ethyl, CH_3CH_2-
Propane, $CH_3CH_2CH_3$	Propyl, $CH_3CH_2CH_2-$

A useful generalization, especially when writing long formulas or reactions, is to substitute the letter R for the entire alkyl group when attention is being focused on what is happening to the functional group of a compound or a series. In this designation, RH is the parent alkane (C_nH_{2n+2}) and R— is the alkyl radical (C_nH_{2n+1}).

ALCOHOLS

Compounds of the general formula ROH may be considered either as hydroxyl derivatives of the alkanes or as alkyl derivatives of water and are known as *alcohols:*

$$R{\cdot}H \xrightarrow[\text{with OH}]{\text{replace H}} R{\cdot}OH$$

Alkane

$$H{\cdot}OH \xrightarrow[\text{with R}]{\text{replace H}} R{\cdot}OH$$

Water

Because of their close relationship to water, the lower members of the alcohols are soluble in water. However, as the molecules become larger, and the portion represented by the hydroxyl group becomes proportionally smaller, the solubility in water decreases rapidly.

As usual, the lower members are often referred to by their common names, e.g., methyl, ethyl, and propyl alcohols. Under the Geneva system the *-ane* ending of the alkanes is changed to *-anol,* the corresponding names being methanol, ethanol, and propanol. Position isomerism in the alcohols starts with the propyl alcohols, which have two possible forms:

```
    H   H   H                  H   H   H
    |   |   |                  |   |   |
H — C — C — C — OH         H — C — C — C — H
    |   |   |                  |   |   |
    H   H   H                  H   O   H
                                   H
```

Normal propyl alcohol
Propanol-1

Isopropyl alcohol
Propanol-2

or,

$CH_3CH_2CH_2OH$ $\qquad$ $CH_3CHOHCH_3$

The number of position isomers increases rapidly with the number of carbon atoms in the chain.

There are three different types of alcohol functional groups determined by the differences in structure of the carbon atoms to which the hydroxyl group is attached. In n-propyl alcohol (see above), the hydroxyl group is attached to a straight chain at the end carbon atom that in turn is attached to only one other carbon atom. This is termed a primary alcohol. The hydroxyl group of isopropyl alcohol (see above) is attached to a carbon atom that in turn is attached to two other carbon atoms. This is a secondary alcohol. Of the four possible butyl alcohols, one has the hydroxyl group attached to a carbon that in turn is attached to three other carbon atoms, making this compound a tertiary alcohol:

```
       CH3
       |
CH3 —  C — OH
       |
       CH3
```

tert-Butyl alcohol

These designations are important because the differences in properties of primary, secondary, and tertiary alcohols are related to the differences in structure of these compounds.

Chemical properties

Although the alcoholic hydroxyl group and the hydroxyl group encountered in inorganic bases appear to be similar, the alcoholic hydroxyl group does not ionize to give hydroxyl ions. Alcohols, therefore, do not have the alkaline properties associated with inorganic bases that do produce hydroxyl ions in aqueous solution.

From the biochemical standpoint, the most important chemical property of the alcohols is their ability to undergo oxidation reactions. When subjected to mild oxidation, primary alcohols yield aldehydes:

$$\begin{array}{ccccc} & H & & H & \\ & | & & | & \\ H- & C & - & C & -OH \\ & | & & | & \\ & H & & H & \end{array} \xrightarrow{\text{oxidation}} \begin{array}{ccccc} & H & & H & \\ & | & & | & \\ H- & C & - & C & =O \\ & | & & & \\ & H & & & \end{array}$$

Ethyl alcohol (Primary alcohol) **Acetaldehyde**

or,

$$CH_3CH_2OH \xrightarrow{\text{oxidation}} CH_3CHO$$

Under similar conditions, secondary alcohols form ketones:

$$\begin{array}{ccccccc} & H & & H & & H & \\ & | & & | & & | & \\ H- & C & - & C & - & C & -H \\ & | & & | & & | & \\ & H & & O & & H & \\ & & & H & & & \end{array} \xrightarrow{\text{oxidation}} \begin{array}{ccccccc} & H & & & & H & \\ & | & & & & | & \\ H- & C & - & C & - & C & -H \\ & | & & \| & & | & \\ & H & & O & & H & \end{array}$$

Isopropyl alcohol (Secondary alcohol) **Acetone** (Ketone)

or,

$$CH_3CHOHCH_3 \xrightarrow{\text{oxidation}} CH_3COCH_3$$

Again, oxidations of this type, under controlled enzymatic conditions, produce the energy required by living organisms.

Tertiary alcohols are broken down to smaller compounds when subjected to oxidation. Aldehydes and ketones can undergo further oxidation. Details of these reactions will be discussed under the respective functional groups.

Alcohols also react with organic acids to form esters. This reaction will be discussed under organic acids.

Methyl alcohol. When wood is heated to high temperatures without burning (destructive distillation) one of the products given off is methyl alcohol. Originally this was the best method for the preparation of methyl alcohol, which became known as wood alcohol. At present, methyl alcohol is usually produced by synthetic processes. It is a colorless liquid with a characteristic odor and relatively unpleasant taste.

Because methyl alcohol is an important industrial material for the production of many other substances and finds its way into the home by way of antifreeze products for the automobile radiator and solvents for paints and varnishes, it should be noted that it is poisonous when taken internally, even in relatively small amounts, and can cause blindness or death. Advantage is taken of its unpleasant taste and poisonous nature to make industrial ethyl alcohol unfit to drink by adding some methyl alcohol to it to produce what is known as *denatured alcohol.*

Ethyl alcohol. The history of ethyl alcohol dates back to prehistoric times when wines were produced by fermentation of grapes just as today. The distillation of fermented products to produce alcohol of higher concentrations was discovered about 900 AD. Fermentation of grains, grapes, and other sugar-containing fruits is still used to produce much of the industrial alcohol supplies.

Ethyl alcohol is a colorless liquid with a relatively pleasant odor. Industrially it is used for many manufacturing purposes ranging from the production of many chemicals and drugs to its use, because of its high volatility, in perfumes. It is widely used in hospitals as an antiseptic substance, as a skin-bathing liquid for lowering body temperatures rapidly, and for many other medicinal purposes. In fact, it is such a commonly used substance that the term alcohol, by itself, is usually understood to mean ethyl alcohol.

Another important use of ethyl alcohol is as an ingredient of beverages of many types. In contrast to methyl alcohol, ethyl alcohol may be taken internally, generally without ill effects unless relatively large amounts are involved. The percentage of alcohol in beverages is often designated by the term *proof,* which is twice the percentage by volume. Thus, a whiskey containing 50% alcohol is said to be 100 proof.

Polyhydroxylic alcohols. Thus far only alcohols containing one hydroxyl alcohol group have been discussed. Such compounds are called *monohydroxylic* alcohols. There are many natural and synthetic substances known which contain more than one alcohol group per molecule. These compounds are known as *polyhydroxylic* alcohols or by a name using a prefix to indicate the specific number of alcohol groups in the molecule.

A trihydroxylic alcohol, glycerol (glycerine), is an important constituent of fats (p. 210). It is also used in various medications, cosmetics, and lotions.

$$\begin{array}{c} \mathrm{H} \\ | \\ \mathrm{H-C-OH} \\ | \\ \mathrm{H-C-OH} \\ | \\ \mathrm{H-C-OH} \\ | \\ \mathrm{H} \end{array}$$

Glycerol

ALDEHYDES

Compounds of the general formula RCHO are known as aldehydes. As noted above in the discussion of alcohols, aldehydes are readily produced by the mild oxidation of primary alcohols:

$$\begin{array}{ccccc} & \mathrm{H} & & \mathrm{H} & \\ & | & & | & \\ \mathrm{H-} & \mathrm{C} & \mathrm{-} & \mathrm{C} & \mathrm{-OH} \\ & | & & | & \\ & \mathrm{H} & & \mathrm{H} & \end{array} \xrightarrow{\text{oxidation}} \begin{array}{ccccc} & \mathrm{H} & & \mathrm{H} & \\ & | & & | & \\ \mathrm{H-} & \mathrm{C} & \mathrm{-} & \mathrm{C} & \mathrm{=O} \\ & | & & & \\ & \mathrm{H} & & & \end{array}$$

Ethyl alcohol **Acetaldehyde**

or,

$$CH_3CH_2OH \xrightarrow{\text{oxidation}} CH_3CHO$$

Thus the functional group of aldehydes may be written as:

$$\begin{array}{cc} & \mathrm{H} \\ & | \\ \mathrm{-} & \mathrm{C=O} \end{array} \quad \text{or} \quad \mathrm{-CHO}$$

Aldehyde group

Chemical properties

Aldehydes, formed by the mild oxidation of primary alcohols, may themselves be oxidized further to produce organic acids:

$$\begin{array}{ccccc} & \mathrm{H} & & \mathrm{H} & \\ & | & & | & \\ \mathrm{H-} & \mathrm{C} & \mathrm{-} & \mathrm{C} & \mathrm{=O} \\ & | & & & \\ & \mathrm{H} & & & \end{array} \xrightarrow{\text{oxidation}} \begin{array}{ccccc} & \mathrm{H} & & \mathrm{O} & \\ & | & & /\!/ & \\ \mathrm{H-} & \mathrm{C} & \mathrm{-} & \mathrm{C} & \mathrm{-OH} \\ & | & & & \\ & \mathrm{H} & & & \end{array}$$

Acetaldehyde **Acetic acid**

or,

$$CH_3CHO \xrightarrow{\text{oxidation}} CH_3COOH$$

Under the proper conditions, aldehydes may also be reduced to form primary alcohols:

$$\begin{array}{ccccc} & \mathrm{H} & & \mathrm{H} & \\ & | & & | & \\ \mathrm{H-} & \mathrm{C} & \mathrm{-} & \mathrm{C} & \mathrm{=O} \\ & | & & & \\ & \mathrm{H} & & & \end{array} \xrightarrow{\text{reduction}} \begin{array}{ccccc} & \mathrm{H} & & \mathrm{H} & \\ & | & & | & \\ \mathrm{H-} & \mathrm{C} & \mathrm{-} & \mathrm{C} & \mathrm{-OH} \\ & | & & | & \\ & \mathrm{H} & & \mathrm{H} & \end{array}$$

Acetaldehyde **Ethyl alcohol**

or,

$$CH_3CHO \xrightarrow{\text{reduction}} CH_3CH_2OH$$

Aldehydes are very reactive compounds that may undergo polymerization, a type of reaction in which single unit molecules condense to form larger units known as polymers:

$$3CH_3CHO \rightarrow (CH_3CHO)_3$$

Acetaldehyde **Paraldehyde**
(Polymer of acetaldehyde)

Formaldehyde. Formaldehyde (HCHO) is a colorless gas with a distinctive, relatively unpleasant odor that is never forgotten by any biology student after dissection of a preserved specimen. A 40% solution of formaldehyde, known as formalin, is widely used for the preservation of all sorts of biologic materials. It also finds much use as an antiseptic and as a starting material for many industrial processes.

Acetaldehyde. In addition to its many industrial uses, acetaldehyde forms the polymer paraldehyde that is sometimes used to induce sleep in hospital patients.

KETONES

Compounds of the general formula $R_2C{=}O$ are known as ketones. Ketones are produced by the oxidation of secondary alcohols:

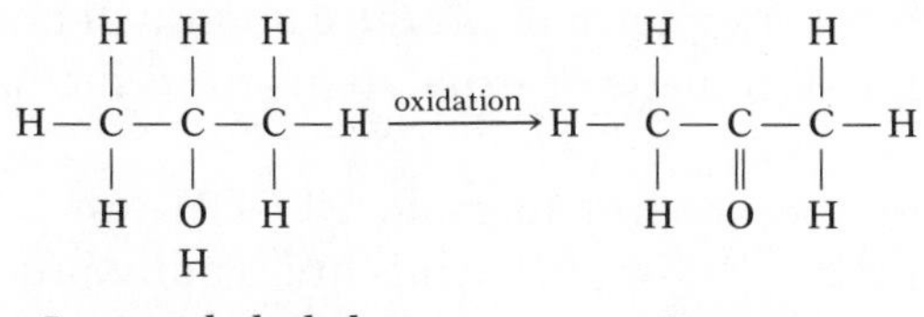

Isopropyl alcohol (Secondary alcohol) **Acetone** (Ketone)

or,

$$CH_3CHOHCH_3 \xrightarrow{\text{oxidation}} CH_3COCH_3$$

The functional group of ketones, known as the carbonyl group, may be written as:

$$\rangle C{=}O$$

Ketone, or carbonyl, group

The similarity and differences between aldehyde and ketone groups should be noted. The ketone group consists of a carbonyl group that is connected to two alkyl groups and has no hydrogen on the carbon atom. The aldehyde group also has a carbonyl group but is attached to only one alkyl group, leaving one hydrogen on the carbon atom.

It should also be noted that with two R groups connected to the carbonyl group, the possibility exists for two similar R groups, as in symmetrical ketones, or two different R groups, as in mixed ketones.

Chemical properties

Under the proper conditions, ketones may be reduced to form secondary alcohols:

$$\begin{array}{ccccc} & H & & H & \\ & | & & | & \\ H- & C & - C - & C & -H \\ & | & \| & | & \\ & H & O & H & \end{array} \xrightarrow{\text{reduction}} \begin{array}{ccccccc} & H & & H & & H & \\ & | & & | & & | & \\ H- & C & - & C & - & C & -H \\ & | & & | & & | & \\ & H & & O & & H & \\ & & & H & & & \end{array}$$

Acetone **Isopropyl alcohol**
(Secondary alcohol)

or,

$$CH_3COCH_3 \xrightarrow{\text{reduction}} CH_3CHOHCH_3$$

Oxidation of ketones produces mixtures of products with carbon chains of smaller size than the original ketone.

Acetone. Acetone is a colorless liquid with a pleasant, fruity odor in low concentration. It is a very good solvent for many organic compounds and is a key chemical for the production of many other substances. Patients suffering from diabetes sometimes produce abnormal amounts of acetone that may be excreted in urine or exhalations from the lungs (p. 325).

ORGANIC ACIDS

Compounds of the general formula RCOOH are known as organic acids. As explained before, organic acids are readily produced by the oxidation of aldehydes.

$$\begin{array}{ccccc} & H & & H & \\ & | & & | & \\ H- & C & - & C & =O \\ & | & & & \\ & H & & & \end{array} \xrightarrow{\text{oxidation}} \begin{array}{ccccc} & H & & O & \\ & | & & \# & \\ H- & C & - & C & -OH \\ & | & & & \\ & H & & & \end{array}$$

Acetaldehyde **Acetic acid**

or,

$$CH_3CHO \xrightarrow{\text{oxidation}} CH_3COOH$$

The functional group of organic acids, known as the carboxyl group, may be written as:

$$\begin{array}{c} O \\ \# \\ -C-OH \end{array} \quad \text{or} \quad -COOH$$

Organic acid, or carboxyl, group

Compounds containing the carboxyl group are known as acids because the hydrogen atom in the carboxyl group can ionize to give hydrogen ions in aqueous solution:

$$\begin{array}{c} \text{H} \quad \text{O} \\ | \quad // \\ \text{H—C—C—OH} \\ | \\ \text{H} \end{array} \underset{\text{solution}}{\overset{\text{aqueous}}{\rightleftharpoons}} H^+ + \left[\begin{array}{c} \text{H} \quad \text{O} \\ | \quad // \\ \text{H—C—C—O} \\ | \\ \text{H} \end{array}\right]^-$$

Acetic acid **Acetate ion**

or,

$$CH_3COOH \underset{\text{solution}}{\overset{\text{aqueous}}{\rightleftharpoons}} H^+ + CH_3COO^-$$

Since the extent of ionization is usually small, most organic acids are weak acids, although some strong acids are also encountered.

Organic acids may have more than one carboxyl group per molecule, in which case the acids may be referred to as *monobasic* (one carboxyl per molecule), *dibasic* (two carboxyls), or *tribasic* (three carboxyls) *acids.* Other important natural organic acids frequently have functional groups in addition to the carboxyl group. Among these are *hydroxy acids,* containing one or more hydroxyl groups, *keto acids,* containing a ketone group, and *amino acids,* containing an amine group (to be discussed below) in addition to the carboxyl group.

Chemical properties

Because of their acidic properties, organic acids can react with inorganic bases, in neutralization reactions, to form water and organic salts:

$$\begin{array}{c} \text{H} \quad \text{O} \\ | \quad // \\ \text{H—C—C—O}\boxed{\text{H} + \text{Na}}\boxed{\text{OH}} \\ | \\ \text{H} \end{array} \rightarrow H_2O + \begin{array}{c} \text{H} \quad \text{O} \\ | \quad // \\ \text{H—C—C—ONa} \\ | \\ \text{H} \end{array}$$

Acetic acid **Sodium acetate** (Organic salt)

or,

$$CH_3COOH + NaOH \rightarrow H_2O + CH_3COONa$$

In a reaction that is superficially similar to the one with sodium hydroxide, organic acids react with alcohols to form esters:

$$\begin{array}{c} \text{H} \quad \text{O} \\ | \quad // \\ \text{H—C—C—} \\ | \\ \text{H} \end{array}\boxed{\text{OH} + \text{H}}\begin{array}{c} \text{H} \quad \text{H} \\ | \quad | \\ \text{O—C—C—H} \\ | \quad | \\ \text{H} \quad \text{H} \end{array} \rightarrow \begin{array}{c} \text{H} \quad \text{O} \quad \text{H} \quad \text{H} \\ | \quad // \quad | \quad | \\ \text{H—C—C—O—C—C—H} \\ | \quad\quad\quad | \quad | \\ \text{H} \quad\quad\quad \text{H} \quad \text{H} \end{array} + H_2O$$

Acetic acid **Ethyl alcohol** **Ethyl acetate** (Ester)

or,

$$CH_3COOH + CH_3CH_2OH \rightarrow CH_3COOCH_2CH_3 + H_2O$$

In the process, a molecule of water is split out.

Formic acid. The ingredient in bee and other insect bites that causes irritation and discomfort is formic acid, HCOOH. Some people develop such great sensitivity to this substance that insect bites may on rare occasions prove fatal.

Acetic acid. Acetic acid, as ethyl alcohol, is a natural product that was recognized and used by humans from the earliest times. It is the active ingredient of vinegar, which contains approximately 4% acetic acid.

Pyruvic acid; lactic acid. Pyruvic acid is a keto acid formed in the body as a result of the oxidation of glucose, a sugar. Pyruvic acid and lactic acid, a hydroxy acid, are related as oxidation-reduction products of each other under varying conditions during the oxidation of glucose (p. 295):

```
    H   H       O               H       O
    |   |      //               |      //
H — C — C — C — OH  →  H — C — C — C — OH + [2H]
    |   |                       |   ‖
    H   O                       H   O
        H
```

Lactic acid (Reduced form) **Pyruvic acid** (Oxidized form)

or,

$$CH_3CHOHCOOH \rightarrow CH_3COCOOH + [2H]$$

ESTERS Compounds of the general formula RCOOR are known as *esters* and are generally formed by the reaction of an organic acid and alcohol, with the splitting out of water:

```
    H     O            H   H              H     O       H   H
    |    //            |   |              |    //       |   |
H — C — C — ┆OH + H┆O — C — C — H  →  H — C — C — O — C — C — H + H₂O
    |                  |   |              |             |   |
    H                  H   H              H             H   H
```

Acetic acid **Ethyl alcohol** **Ethyl acetate**

or,

$$CH_3COOH + CH_3CH_2OH \rightarrow CH_3COOCH_2CH_3 + H_2O$$

The functional group of esters may be written as:

```
     O       |                    /
    //       |                   
— C — O — C —        or     — COOC —
             |                   
             |                    \
```

Ester group

Esters characteristically have pleasant, fruity odors and tastes. In fact, naturally occurring esters are the active ingredients supplying the taste and odor of many fruits. Because of this, esters are used to a large extent in the flavoring and perfume industries as well as for the production of some medicinals.

Chemical properties

The ease of formation of esters from organic acids and alcohols is matched by the ease with which esters can be decomposed to their constituents by hydrolysis, indicating that there is little difference in energy levels of the reactants and products:

$$
\begin{array}{c}
\text{H}-\overset{\displaystyle \text{H}}{\underset{\displaystyle \text{H}}{\text{C}}}-\overset{\displaystyle \text{O}\,/\!\!/}{\text{C}}-\text{O}-\overset{\displaystyle \text{H}}{\underset{\displaystyle \text{H}}{\text{C}}}-\overset{\displaystyle \text{H}}{\underset{\displaystyle \text{H}}{\text{C}}}-\text{H} + \text{H}_2\text{O} \rightarrow \text{H}-\overset{\displaystyle \text{H}}{\underset{\displaystyle \text{H}}{\text{C}}}-\overset{\displaystyle \text{O}\,/\!\!/}{\text{C}}-\text{OH} + \text{H}-\overset{\displaystyle \text{H}}{\underset{\displaystyle \text{H}}{\text{C}}}-\overset{\displaystyle \text{H}}{\underset{\displaystyle \text{H}}{\text{C}}}-\text{OH}
\end{array}
$$

Ethyl acetate **Acetic acid** **Ethyl alcohol**

or,

$$CH_3COOCH_2CH_3 + H_2O \rightarrow CH_3COOH + CH_3CH_2OH$$

ETHERS

Compounds of the general formula ROR are known as *ethers* and may be considered to be derivatives of the alcohols in which the hydrogen of the alcoholic hydroxyl group is replaced with an R group. As with the ketones, the two R groups of an ether may be the same, in symmetrical ethers, or they may be different, as in mixed ethers. Ethers may be prepared from alcohols by a reaction with concentrated sulfuric acid in which the net result is the splitting out of water from two molecules of alcohol:

$$
\text{H}-\overset{\displaystyle \text{H}}{\underset{\displaystyle \text{H}}{\text{C}}}-\overset{\displaystyle \text{H}}{\underset{\displaystyle \text{H}}{\text{C}}}-\fbox{OH + H}\,\text{O}-\overset{\displaystyle \text{H}}{\underset{\displaystyle \text{H}}{\text{C}}}-\overset{\displaystyle \text{H}}{\underset{\displaystyle \text{H}}{\text{C}}}-\text{H} \xrightarrow{H_2SO_4} \text{H}-\overset{\displaystyle \text{H}}{\underset{\displaystyle \text{H}}{\text{C}}}-\overset{\displaystyle \text{H}}{\underset{\displaystyle \text{H}}{\text{C}}}-\text{O}-\overset{\displaystyle \text{H}}{\underset{\displaystyle \text{H}}{\text{C}}}-\overset{\displaystyle \text{H}}{\underset{\displaystyle \text{H}}{\text{C}}}-\text{H} + \text{H}_2\text{O}
$$

2 Ethyl alcohol **Diethyl ether** (Ether)

or,

$$2CH_3CH_2OH \xrightarrow{H_2SO_4} CH_3CH_2OCH_2CH_3 + H_2O$$

The functional group of ethers may be written as:

$$
-\overset{|}{\underset{|}{\text{C}}}-\text{O}-\overset{|}{\underset{|}{\text{C}}}- \quad \text{or} \quad {}^{\diagdown}_{\diagup}\!\!-\text{COC}-\!\!{}^{\diagup}_{\diagdown}
$$

Ether group

The ethers are generally found to be insoluble in water but are very useful as solvents for organic compounds of most types.

Chemical properties

Although ethers may be induced to react under special circumstances, they are considered to be inert substances in the same sense as the paraffin hydrocarbons.

Diethyl ether (ether). The use of diethyl ether, more commonly known simply as ether, for anesthesia marked an important medical milestone. It was first used in a surgical procedure by Dr. Crawford Long in Georgia in 1842 but he did not publish his results. The use of ether as an anesthetic was rediscovered by a Boston dentist, William Morton, who demonstrated its use at Massachusetts General Hospital in 1846. This demonstration led to its more general use and to interest in developing other anesthetic agents because of the unpleasant side effects that sometimes accompany the use of ether.

AMINE DERIVATIVES

Amines

Amines may conveniently be considered as derivatives of ammonia (NH_3) in which one or more hydrogen atoms are replaced by R groups. The general formulas for amines may then be written as RNH_2 for primary amines, R_1R_2NH for secondary amines, and $R_1R_2R_3N$ for tertiary amines. Examples of these types are written as:

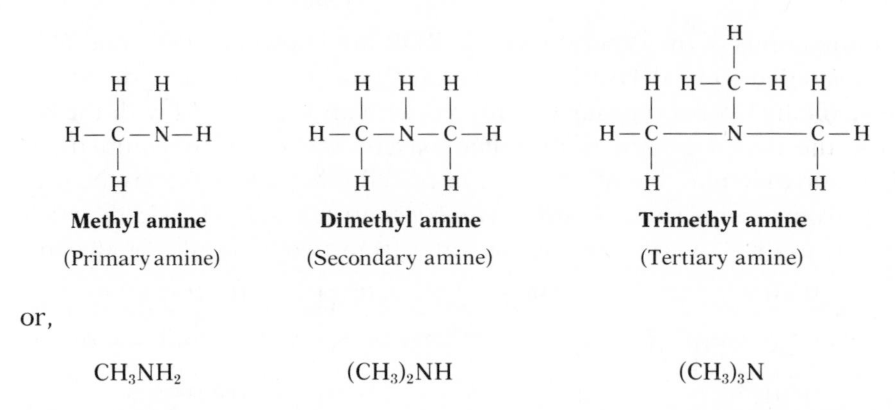

or,

CH_3NH_2 $(CH_3)_2NH$ $(CH_3)_3N$

The R groups may be the same or different. The characteristic groupings for amine groups may then be written as:

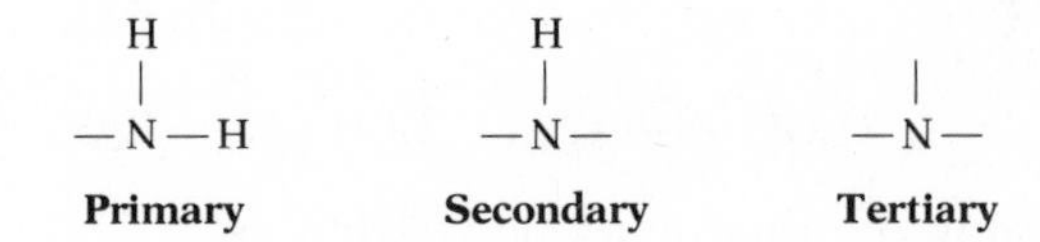

Amines are generally recognized by their unpleasant, fishy odors. The characteristic odor of decaying fish is due to the amines that are produced in the process of decay. Chemically, the amines are best characterized by their strong basic properties, which enable them to react with inorganic acids:

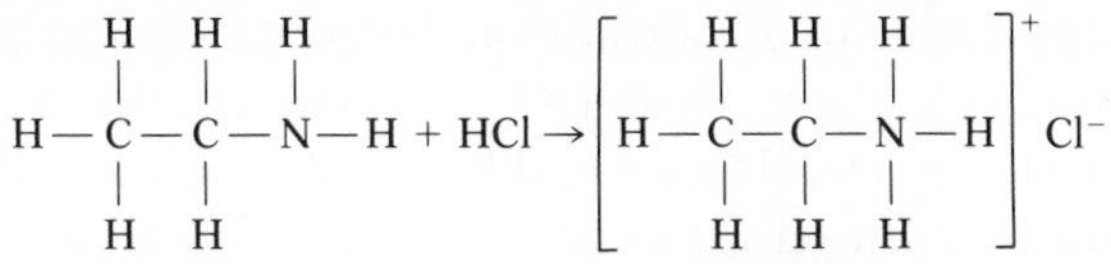

$$
\begin{array}{ccccccc}
 & & H & & H & & H \\
 & & | & & | & & | \\
H & - & C & - & C & - & N-H \\
 & & | & & | & & \\
 & & H & & H & &
\end{array}
+ HCl \rightarrow
\left[
\begin{array}{ccccccc}
 & & H & & H & & H \\
 & & | & & | & & | \\
H & - & C & - & C & - & N-H \\
 & & | & & | & & | \\
 & & H & & H & & H
\end{array}
\right]^{+} Cl^{-}
$$

Ethyl amine **Ethyl amine hydrochloride**

or,

$$CH_3CH_2NH_2 + HCl \rightarrow [CH_3CH_2NH_3]^{+}Cl^{-}$$

or,

$$CH_3CH_2NH_2 \cdot HCl$$

Amines have long been of strategic industrial importance for the manufacture of dyes, explosives, propellants, and other substances.

Amino acids

Amino acids are organic compounds containing two functional groups, an amine group and a carboxyl group. The simplest amino acid, glycine, may be written as:

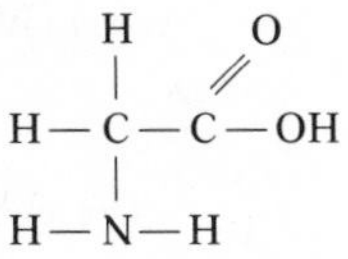

$$
\begin{array}{ccccc}
 & & H & & O \\
 & & | & & /\!\!/ \\
H & - & C & - & C-OH \\
 & & | & & \\
 & H- & N & -H &
\end{array}
\qquad \text{or} \qquad CH_2NH_2COOH
$$

Glycine

The most important amino acids, the variety largely found in nature, are the α-amino acids. The designation α indicates that the amine group is on the carbon atom that is connected to the carboxyl group. The α-amino acids are the basic units from which protein molecules are built, the proteins being essential components of living organisms (p. 216). A discussion of the properties of amino acids is deferred until the chemistry of proteins is taken up.

Amides

Compounds of the general formula $RCONH_2$ are known as *amides* and may be considered as being derived from an organic acid by replacement of the hydroxyl group of the carboxyl group by an amine group. The structure of a typical amide may be written as:

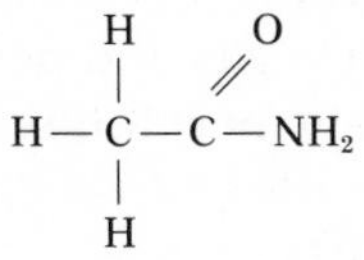

$$
\begin{array}{ccccc}
 & & H & & O \\
 & & | & & /\!\!/ \\
H & - & C & - & C-NH_2 \\
 & & | & & \\
 & & H & &
\end{array}
\qquad \text{or} \qquad CH_3CONH_2
$$

Acetamide

(Primary amide)

Much of the importance of amides lies in the relationship of the structure of secondary amides to the structure formed when amino acids combine in peptide bonds to form proteins, as will be discussed later. The structure of a secondary amide is illustrated here:

$$H-\overset{\displaystyle H}{\underset{\displaystyle H}{C}}-\overset{\displaystyle O}{\overset{\|}{C}}-\underset{\displaystyle H}{N}-\underset{\displaystyle H}{\overset{\displaystyle H}{C}}-\underset{\displaystyle H}{\overset{\displaystyle H}{C}}-H \qquad \text{or} \qquad CH_3CONHCH_2CH_3$$

N-Ethyl acetamide

ALKYL HALIDES The replacement of one or more hydrogen atoms from hydrocarbons by halogen atoms (fluorine, chlorine, bromine, iodine) produces compounds known as *alkyl halides*. The structures of some alkyl halides that have been found useful for medical purposes are listed here:

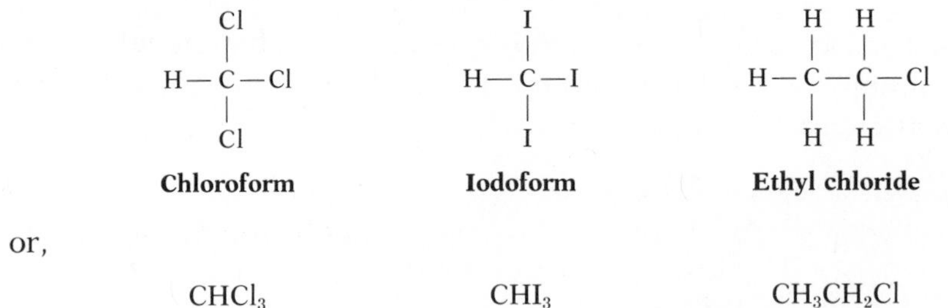

Chloroform **Iodoform** **Ethyl chloride**

or,

$CHCl_3$ CHI_3 CH_3CH_2Cl

Chloroform has been used as an anesthetic, iodoform as an antiseptic, the odor of which used to be recognized as the "hospital smell," and ethyl chloride as a local anesthetic that performs its function by freezing nerve endings in a local area because of its rapid evaporation.

SUMMARY

FUNCTIONAL GROUPS Atoms, radicals, or groups of atoms that are largely responsible for the properties of the compounds in which they appear.

HYDROCARBONS Compounds of hydrogen and carbon only.

Saturated hydrocarbons: Compounds in which all carbon-to-carbon links are single bonds. Also known as alkanes, paraffin, or methane hydrocarbons. Chain isomerism starts with butanes.

Occurrence: Major constituents of natural gas and petroleum.

Chemical properties: Alkanes are relatively inert but react under special conditions.

Chlorination, bromination (halogenation):

$$CH_4 + Cl_2 \xrightarrow[\text{light}]{\text{ultraviolet}} HCl + CH_3Cl$$

Oxidation (burning): Alkanes give off large amounts of energy when burned. Normal octane produces 11.43 Cal (11,430 cal) per g when completely oxidized or burned.

Unsaturated hydrocarbons: alkenes (ethylenic hydrocarbons): Characterized by the presence of a double bond between two carbon atoms, therefore much more reactive than the corresponding saturated compounds. Position of the double bond in alkenes is designated by the number of the carbon atom where the double bond starts. Position isomerism starts with butenes.

Chemical properties: Mainly addition reactions.

Halogenation (Cl_2, Br_2):

$$CH_2{=}CH_2 + Br_2 \rightarrow CH_2BrCH_2Br$$

Addition of hydrogen halides (HI, HBr, HCl):

$$CH_2{=}CH_2 + HBr \rightarrow CH_3CH_2Br$$

Hydrogenation (reduction):

$$CH_2{=}CH_2 + H_2 \rightarrow CH_3CH_3$$

Unsaturated hydrocarbons: alkynes (acetylenic hydrocarbons): Characterized by the presence of a triple bond between two carbon atoms. Alkynes are more reactive than alkenes in similar reactions but are not important in biochemistry.

Alkyl radicals: Alkane minus one hydrogen atom, e.g., methane, CH_4; methyl, CH_3—. The letter R is sometimes used to indicate alkyl radicals.

ALCOHOLS

Compounds of the general formula ROH. Lower members are soluble in water. Position isomerism starts with propanols. *Primary alcohol*—alcohol group attached to carbon that is attached to one other carbon atom. *Secondary alcohol*—alcohol group attached to carbon that is attached to two other carbon atoms. *Tertiary alcohol*—alcohol group attached to carbon that is attached to three other carbon atoms.

Chemical properties: Alcoholic hydroxyl group does not ionize. Biochemically, the most important reactions of alcohols are the oxidation reactions:

$$\text{Primary alcohol: } CH_3CH_2OH \xrightarrow{\text{oxidation}} CH_3CHO$$

$$\text{Secondary alcohol: } (CH_3)_2CHOH \xrightarrow{\text{oxidation}} CH_3COCH_3$$

Oxidations of this type produce the energy required by living organisms.

$$\text{Tertiary alcohol: } (CH_3)_3COH \xrightarrow{\text{oxidation}} \text{breakdown products}$$

Alcohols also react with organic acids to form esters (see Organic Acids).

Methyl alcohol: Prepared from wood, therefore known as wood alcohol. Poisonous when taken internally, therefore used for denaturing ethyl alcohol.

Ethyl alcohol: Known as a fermentation product for many centuries. Safe when taken internally unless in large amounts.

Polyhydroxylic alcohols: Compounds containing more than one alcoholic hydroxyl group. Glycerol, a trihydroxylic alcohol, is an important constituent of fat.

ALDEHYDES Compounds of the general formula RCHO. Produced by oxidation of primary alcohols.

Chemical properties: May be oxidized to organic acids:

$$CH_3CHO \xrightarrow{\text{oxidation}} CH_3COOH$$

May be reduced to primary alcohols:

$$CH_3CHO \xrightarrow{\text{reduction}} CH_3CH_2OH$$

May polymerize:

$$3CH_3CHO \rightarrow (CH_3CHO)_3$$

Formaldehyde: Gas, with distinctive unpleasant odor. In 40% solution as formalin, used as tissue preservative and antiseptic.

Acetaldehyde: Widely used industrially and, as the polymer paraldehyde, is sometimes used as a sedative.

KETONES Compounds of the general formula $R_2C{=}O$. Produced by oxidation of secondary alcohols.

Chemical properties: May be reduced to form secondary alcohols:

$$CH_3COCH_3 \xrightarrow{\text{reduction}} CH_3CHOHCH_3$$

Oxidation of ketones produces breakdown products.

Acetone: Colorless liquid with pleasant fruity odor, good solvent, key industrial chemical. Produced in abnormal amounts by diabetic patients.

ORGANIC ACIDS Compounds of the general formula RCOOH. Produced by the oxidation of aldehydes. Have acidic properties because the hydrogen atom in the carboxyl group ionizes in aqueous solution, generally to a small extent only; therefore weak acids. *Monobasic acid:* one carboxyl group per molecule; *dibasic acid:* two carboxyl groups per molecule; *tribasic acid:* three carboxyl groups per molecule. *Hydroxy acids:* one or more hydroxyl groups plus carboxyl group; *keto acids:* keto group plus carboxyl group; *amino acids:* amine group plus carboxyl group.

Chemical properties: React with inorganic bases in a neutralization reaction to give water and an organic salt:

$$CH_3COOH + NaOH \rightarrow H_2O + CH_3COONa$$

React with alcohols to form esters:

$$CH_3COOH + CH_3CH_2OH \rightarrow CH_3COOCH_2CH_3 + H_2O$$

Formic acid: Active ingredient of insect stings.

Acetic acid: Active ingredient of vinegar.

Pyruvic: lactic acid: Pyruvic acid is a keto acid formed in the body by oxidation of glucose (a sugar) and is related to lactic acid, a hydroxy acid:

$$CH_3CHOHCOOH \rightarrow CH_3COCOOH + [2H]$$

ESTERS

Compounds of the general formula RCOOR. Formed by the reaction of alcohols with organic acids. Have pleasant, fruity odors and tastes, therefore used in flavorings and perfumes.

Chemical properties: Hydrolyzed by water to form constituent alcohol and organic acid:

$$CH_3COOCH_2CH_3 + H_2O \rightarrow CH_3COOH + CH_3CH_2OH$$

ETHERS

Compounds of the general formula ROR. Prepared from alcohols:

$$2CH_3CH_2OH \xrightarrow{H_2SO_4} CH_3CH_2OCH_2CH_3 + H_2O$$

Ethers are insoluble in water but excellent solvents for organic compounds.

Chemical properties: Rather inert substances.

Diethyl ether (ether): First substance used for anesthesia in 1842 by surgeon Crawford Long, in Georgia. Demonstrated in hospital in 1846 by dentist William Morton.

AMINES AND DERIVATIVES

Amines:

Primary amines: RNH_2

Secondary amines: R_2NH

Tertiary amines: R_3N

The R groups may be the same or different. Have unpleasant, fishy odors and have strong basic properties; therefore react with inorganic acids:

$$CH_3CH_2NH_2 + HCl \rightarrow CH_3CH_2NH_2 \cdot HCl$$

Amino acids: Contain an amine group and a carboxyl group on same molecule. Those with amine group on carbon connected to carboxyl carbon, α-amino acids, are the basic units of protein molecules.

Amides: Compounds of the general formula $RCONH_2$ are primary amides. Structure of secondary amides RCONHR is related to the structure of the peptide bond in proteins.

ALKYL HALIDES

Compounds in which one or more hydrogen atoms of hydrocarbons are replaced by halogen atoms (fluorine, chlorine, bromine, iodine). Many alkyl halides have been used for medical purposes.

REVIEW QUESTIONS

1. What are functional groups? What is the significance of functional groups? What is a radical?
2. What is the structure of a primary open-chain alcohol? How does it differ from an inorganic base, e.g., sodium hydroxide? How does it differ from a secondary alcohol in structure and chemical properties?
3. In the reaction between an alcohol and an organic acid, what is the major product formed? What else is produced and where does it come from? Is the reaction between the alcohol and acid reversible?
4. What is an amine group? If an amine group is found on the carbon atom next to the carboxyl group of an organic acid, what is this compound called? Of what important class of naturally occurring substances is this compound a fundamental unit?
5. What is an unsaturated compound? What are the different degrees of unsaturation and what are the relationships to reactivity? Which form(s) appear in biologic substances?

REFERENCES

Allinger, N. L., and others: Organic chemistry, ed. 2, New York, 1976, Worth Publishers, Inc.

Geissman, T. A.: Principles of organic chemistry, ed. 4, San Francisco, 1977, W. H. Freeman and Co., Publishers.

Holum, J. R.: Organic chemistry: a brief course, New York, 1975, John Wiley & Sons, Inc.

Morrison, R. T., and Boyd, R. N.: Organic chemistry, ed. 3, Boston, 1973, Allyn & Bacon, Inc.

Weininger, S. J.: Contemporary organic chemistry, New York, 1972, Holt, Rinehart and Winston, Inc.

Cyclic (aromatic) organic compounds

13

Cyclic organic compounds form the same functional groups as open-chain organic compounds. However, the properties of the two large classes of organic compounds are sufficiently different to make it convenient to study them separately. The differences in properties between these groups of compounds must be caused by the differences in properties of open chains of carbon atoms as compared to cyclic arrangements of carbon atoms.

Many of the cyclic compounds first isolated in the 1800s were found to contain a six-carbon cyclic unit in their structures and possessed aromatic odors. As a result, the chemistry of the cyclic compounds became known as *aromatic organic chemistry*. It was subsequently determined that the six-carbon parent compound was benzene, first isolated by Michael Faraday in 1825. In 1834, Eilhardt Mitscherlich prepared benzene from benzoic acid and determined its structure to be C_6H_6, a structure that, by analogy, would indicate a degree of unsaturation comparable to that of acetylene (an alkyne), C_2H_2. These cyclic compounds, however, were much less reactive than the alkynes and not much more reactive than the alkanes. Thus the properties of the *aromatic compounds*, a designation now reserved for benzene and related compounds, must be accounted for on the basis of the structure of benzene, as will be discussed later.

OCCURRENCE

Coal tar, obtained from bituminous coal by heating it at high temperatures in the absence of air, is the major source of many cyclic compounds or the starting material from which these compounds are synthesized. Many of the tremendous technologic advances that have been made in the last 50 years are connected with products from the coal tar industry. More details about these accomplishments may be found in general organic chemistry textbooks or books written specifically about the coal tar industry.

CARBOCYCLIC COMPOUNDS

Carbocyclic compounds may be defined as compounds containing a ring structure in which all of the atoms in the ring are carbon atoms.

Carbocyclic aromatic hydrocarbons

Carbocyclic compounds containing only carbon and hydrogen atoms in their structures are known as *carbocyclic aromatic hydrocarbons.*

Benzene. Benzene (C_6H_6) is the simplest member of the carbocyclic hydrocarbon series. As noted before, the discrepancies in the observed properties of benzene and the properties usually exhibited by compounds of the general formula C_nH_n posed difficult problems in attempting to determine the structure of benzene. In 1865, 40 years after the first isolation of benzene, Friedrich August Kekulé proposed a structure that accounted for the observed properties of benzene. In this structure, there are six carbon atoms joined together in the form of a regular hexagon, with alternating single and double bonds, each carbon atom having one hydrogen atom attached to it. This structure may be written as follows:

```
        H                              H
        ::                             C
        C                            //  \
     .·   ·.                        HC     CH
  H:C       C:H                     |      ||
    ::      ::          or          HC     CH
  H:C       C:H                      \\   /
     ·.   .·                           C
        C                              H
        ::
        H
```

Benzene (C_6H_6)

A further simplification for purposes of convenience takes the following form:

In this form, it is understood that at each point there is a carbon atom to which is attached a hydrogen atom.

An examination of this proposed structure would indicate that two different forms should be possible:

In addition, each of these forms should have properties similar to those exhibited by the alkenes, but this is not the case. To explain this, Kekulé further proposed that the two forms shown are in a state of dynamic oscillation between the two forms, a phenomenon now referred to as *resonance:*

resonance

As a result of this resonance the distance between any adjacent carbon atoms in the ring, as determined by x-ray studies, is intermediate between the distances normally found for the C—C and C=C bonds. In simple terms this means that the six bonds between the six carbon atoms are equivalent in properties, which on a chemical basis places their reactivity somewhere between that of an open-chain single or double bond. Although these bonds are never actually single or double, in written formulas either one of the equivalent static forms shown was used with the understanding that it represented one of the extreme forms of the resonating structures. In recent years, the equivalent bonds in benzene, intermediate between single and double bonds, have been represented as a circle within a hexagon:

Whenever a resonating structure is possible, the equal distribution (on a time basis) of the involved electrons is indicated by a circle inside the appropriate figure. The most important thing to remember is that all of these forms in print are static representations of a dynamic structure.

Benzene is a colorless liquid with an aromatic odor. It is insoluble in water but soluble in organic solvents. It is a good organic solvent and is sometimes used as a fuel and as the starting material in the synthesis of many important substances.

Chemical properties. Benzene can be nitrated by heating with nitric acid in the presence of sulfuric acid, with the formation of water:

$$C_6H_6 \text{ (Benzene)} + HNO_3 \xrightarrow{H_2SO_4} C_6H_5NO_2 \text{ (Nitrobenzene)} + H_2O$$

Benzene **Nitrobenzene**

or,

$$C_6H_6 + HNO_3 \xrightarrow{H_2SO_4} C_6H_5NO_2 + H_2O$$

Benzene may be sulfonated by heating with concentrated sulfuric acid:

$$C_6H_6 + H_2SO_4 \rightarrow C_6H_5SO_3H + H_2O$$

Benzene **Benzene sulfonic acid**

or,

$$C_6H_6 + H_2SO_4 \rightarrow C_6H_5SO_3H + H_2O$$

Treatment of benzene with bromine in the presence of a catalyst produces bromobenzene:

$$C_6H_6 + Br_2 \xrightarrow{\text{catalyst}} C_6H_5Br + HBr$$

Benzene **Bromobenzene**

This is a substitution reaction in which only one bromine atom is attached to the ring. If this were an addition reaction, characteristic of carbon double bonds, two bromine atoms would be on the ring and no HBr would be formed. This is taken as evidence that chemically the bonds between carbon atoms of the benzene ring are not as reactive as the double bonds found in alkenes, e.g., ethylene.

Naming benzene derivatives. In the reactions used as examples, only one substituent is shown attaching itself to the benzene ring. It is a relatively simple process to name such compounds. However, as soon as two substituents must be accounted for, complications arise. The locations of two substituents on the benzene ring are usually designated by the following scheme:

1. If the two substituents are located on *adjacent* carbon atoms, they are said to be *ortho* to each other.
2. If the two substituents are located on carbon atoms *separated by one other carbon atom,* they are said to be *meta* to each other.
3. If the two substituents are located on carbon atoms *opposite* each other in the ring (separated by two other carbon atoms), they are said to be *para* to each other.

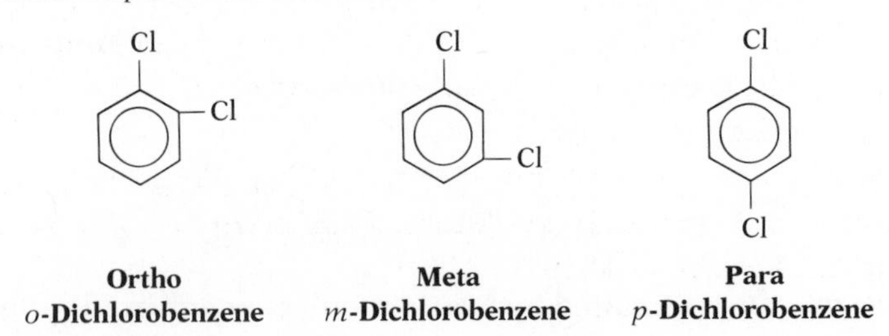

Ortho *o*-**Dichlorobenzene** **Meta** *m*-**Dichlorobenzene** **Para** *p*-**Dichlorobenzene**

Because of the equivalence of the bonds in the benzene ring, the particular locations of the substituents on the printed page is of little importance except in relation to each other. As an example, all of the structures listed here for *o*-dichlorobenzene are equivalent:

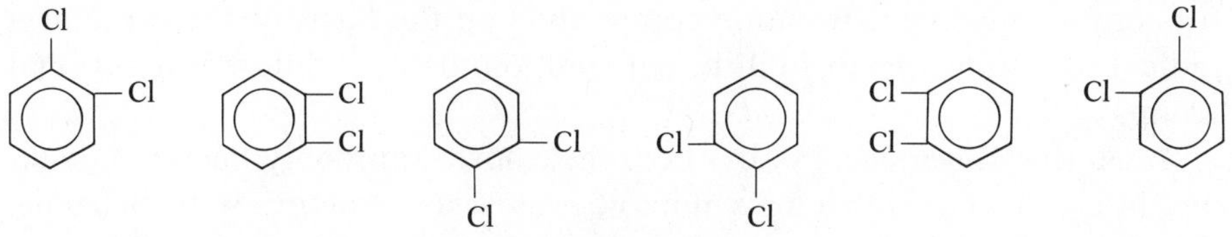

o-**Dichlorobenzene**

The significant factor is that in each case the substituents are on adjacent carbon atoms. The same reasoning holds true for the meta and para compounds and other naming systems.

Introduction of a third substituent, whether all the same or all different, brings up more complications in naming, which are generally solved by using a numbering system in which the carbon atoms of the benzene ring are numbered from one to six:

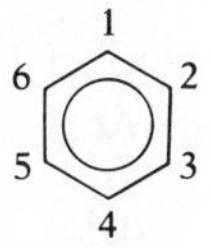

Benzene

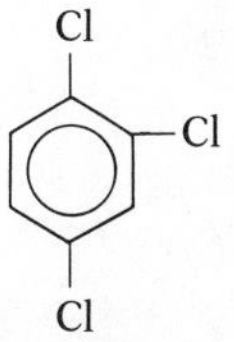

1,2,4-Trichlorobenzene

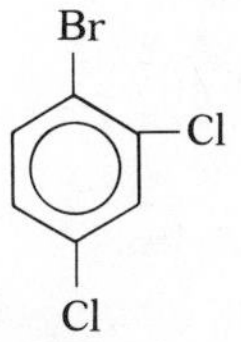

1-Bromo-2,4-dichlorobenzene

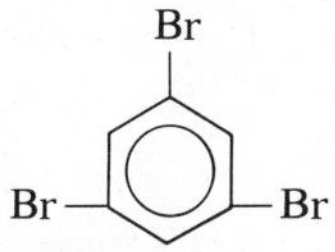

1,3,5-Tribromobenzene
(symmetrical tribromobenzene)

Any atoms, groups, or radicals attached to the carbon atoms in the benzene ring are known as *side chains*. When naming compounds under the systems discussed, it becomes a simple matter to indicate the exact location and nature of each component.

Aryl radicals. Aryl radicals may be defined as the parent carbocyclic hydrocarbon minus one hydrogen atom. Thus benzene C_6H_6, would form the aryl radical known as phenyl, C_6H_5-:

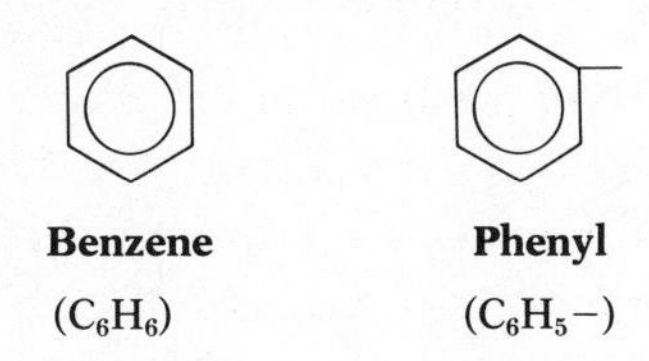

Benzene (C_6H_6)

Phenyl (C_6H_5-)

It is often convenient to name compounds on the basis of the same aryl radical that may appear in different compounds with different functional groups.

Homologs of benzene. Methyl benzene, more commonly known as *toluene,* is the next member of a homologous series starting with benzene. Toluene is probably most noted for its role as starting material for the production of the explosive TNT (trinitrotoluene). It is a colorless liquid used in the manufacture of many other substances, as a constituent of fuels, and is a good organic solvent.

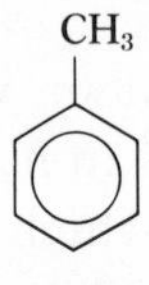

CH_3

Toluene

Addition of a second methyl group to the benzene ring leads to the three *xylenes,* which are identified by the ortho, meta, para system:

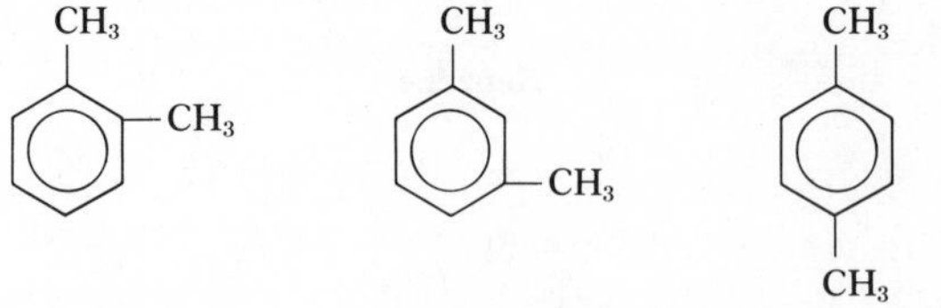

CH_3 CH_3 CH_3 CH_3 CH_3 CH_3

***o*-Xylene** ***m*-Xylene** ***p*-Xylene**

The xylenes are useful in many of the same ways as toluene.

Condensed ring hydrocarbons. The simplest condensed ring system is found in *naphthalene,* the structure of which contains two benzene rings fused so as to share two common carbon atoms:

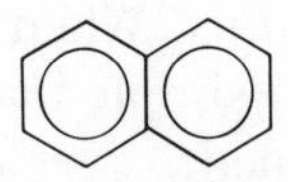

Naphthalene

Three benzene rings condense in two different ways, forming anthracene and phenanthrene:

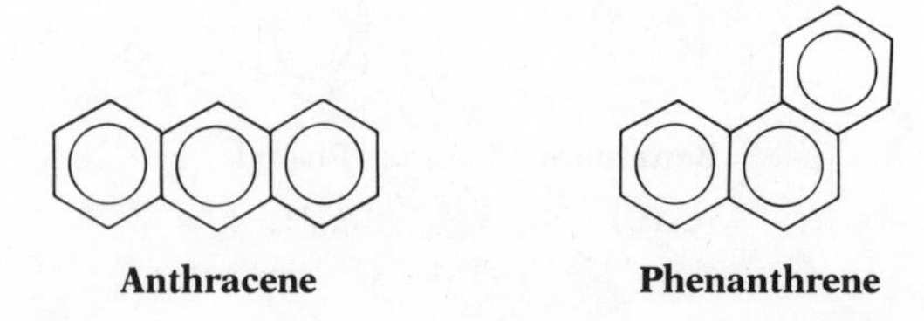

Anthracene **Phenanthrene**

These compounds are important in the dye and drug industries. The phenanthrene structure, as will be shown later, forms the major part of the structure of the steroid hormones, cholesterol, and vitamin D.

Aromatic alcohols

The aromatic alcohols may be considered to be composed of a benzene ring with an aliphatic alcohol as a side chain. *Benzyl alcohol,* the simplest member of this series, appears structurally as a phenyl derivative of methyl alcohol (CH_3OH).

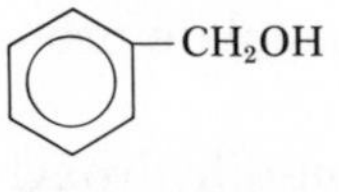

$C_6H_5-CH_2OH$ $C_6H_5-CH_2CH_2OH$

Benzyl alcohol 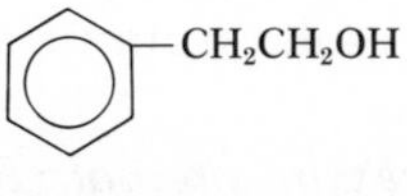 **Phenyl ethyl alcohol**

The next member of this homologous series is *phenyl ethyl alcohol.* Note that in these compounds the alcoholic hydroxyl groups are on an aliphatic side chain, in contrast to the structure of phenols, discussed next.

Because the alcohol functional group is on the aliphatic side chain, the chemical properties of the aromatic alcohols are very similar to those of the aliphatic alcohols, i.e., they are oxidized first to aldehydes, then to acids, react with organic acids to form esters, and may be prepared by reduction of an aldehyde (see chemical properties of aliphatic alcohols and aldehydes). Aromatic alcohols are reactive compounds with pleasant odors and are very useful in the drug and perfume industries.

Phenols

Compounds in which the functional hydroxyl group is connected directly to a benzene ring carbon atom are known as *phenols.* The simplest member of this series is known as *phenol* or *carbolic acid:*

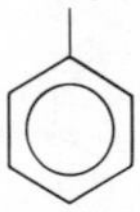

C_6H_5-OH

Phenol

The most distinguishing characteristic of the phenols is their weak acidity caused by the slight ionization of the hydrogen of the hydroxyl group. Thus, phenols react with sodium to form salts known as *phenolates:*

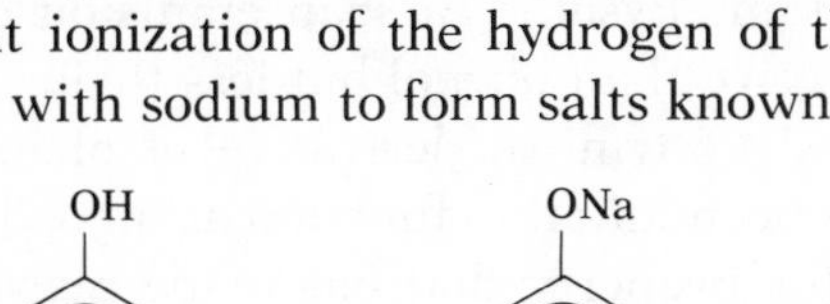

$$C_6H_5OH + NaOH \rightarrow C_6H_5ONa + H_2O$$

Phenol **Sodium phenolate**

Phenols are very reactive compounds and serve as the starting material for many types of synthetic processes. In reactions that show their similarities to alcohols, phenols can form esters and ethers.

Important phenols

Phenol. A colorless crystalline solid with a strong odor, phenol is the oldest known disinfectant. It was first used as a bactericidal agent by Joseph Lister in 1867 for surgical applications. Although phenol is a strong antiseptic, its uses are limited because it is poisonous when taken internally and produces severe burns on contact with skin. Its bactericidal activity probably results from its ability to coagulate proteins, in this case bacterial proteins.

Resorcinol; hexylresorcinol. Resorcinol, *m*-dihydroxybenzene, is less toxic than phenol and also less effective as an antiseptic agent. Introduction of alkyl side chains in the four-position *(para)* increases the effectiveness of such compounds. One of these, *hexylresorcinol,* was used in 1924 as a urinary antiseptic but has been largely supplanted by the sulfa drugs for that purpose. Hexylresorcinol is still used in many preparations as a general disinfectant.

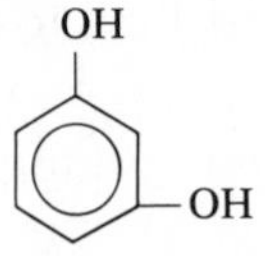

OH OH

OH OH

$CH_2(CH_2)_4CH_3$

Resorcinol 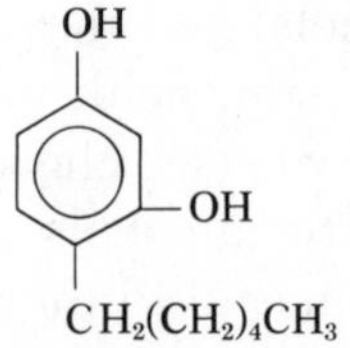 **Hexylresorcinol**

Cresols. Among other phenol derivatives that have been investigated for antiseptic properties are the *cresols,* or hydroxytoluenes:

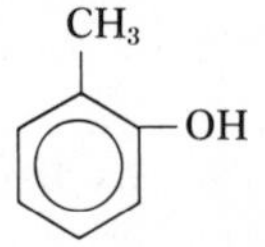

CH_3 CH_3 CH_3

OH OH OH

***o*-Cresol** 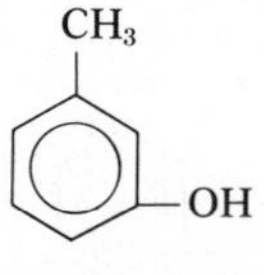 ***m*-Cresol**  ***p*-Cresol**

A common disinfectant, Lysol, is a soap emulsion of the three cresols, which are more effective than phenol but less toxic.

Picric acid. The 2,4,6-trinitro derivative of phenol, *picric acid,* is a strong acid that has been found to function as a good precipitating agent for proteins. It is this property that has made picric acid useful in the treatment of burns. When applied to an area where the skin has been destroyed, picric acid precipitates a film of protein that acts to prevent infection or loss of tissue fluids.

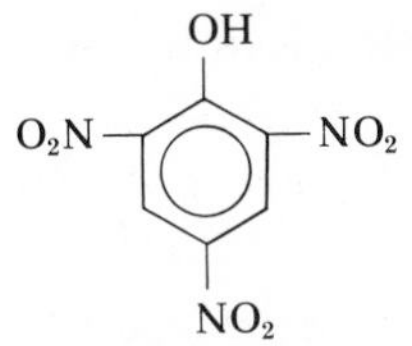

Picric acid
(2,4,6-Trinitrophenol)

Aldehydes

The simplest aromatic aldehyde, *benzaldehyde*, has the aldehyde group (—CHO) attached to a ring carbon. It is found in bitter almonds and may be prepared by oxidation of toluene under proper conditions:

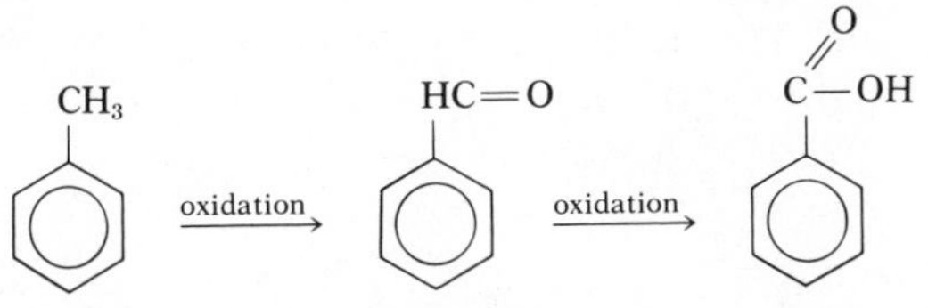

$$\text{Toluene} \xrightarrow{\text{oxidation}} \text{Benzaldehyde} \xrightarrow{\text{oxidation}} \text{Benzoic acid}$$

Toluene **Benzaldehyde** **Benzoic acid**

The aldehyde group may also be located on an aliphatic side chain. In any case, aromatic aldehydes have properties quite similar to those of the aliphatic aldehydes. Thus, benzaldehyde may be further oxidized to benzoic acid, as just shown. The aromatic aldehydes are very reactive compounds that are often used as intermediates in the production of drugs, dyes, and other substances. Because of their pleasant odor and taste, they are also widely used in perfumes and flavorings.

Ketones

Since the functional group of a ketone, the carbonyl group ($>C{=}O$), must have two radicals attached to it, an aromatic ketone may be fully or partially aromatic. The simplest fully aromatic ketone is *benzophenone* and the simplest mixed aromatic-aliphatic ketone is *acetophenone:*

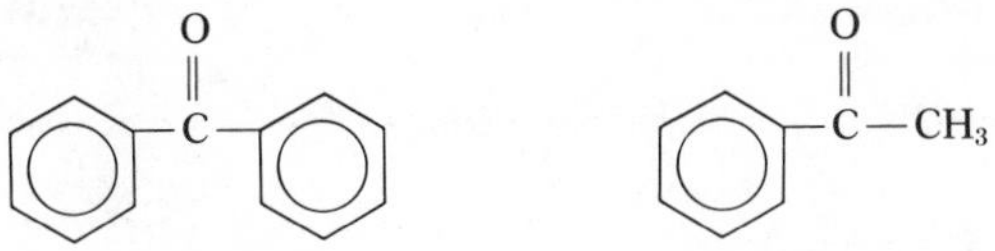

Benzophenone **Acetophenone**

The aromatic ketones are reactive compounds that are used in a variety of synthetic processes.

Acids

Aromatic acids may have the functional carboxyl group (—COOH) on a ring carbon or on a side chain. The simplest aromatic acid is *benzoic acid,* which, as described before, may be prepared by the oxidation of toluene:

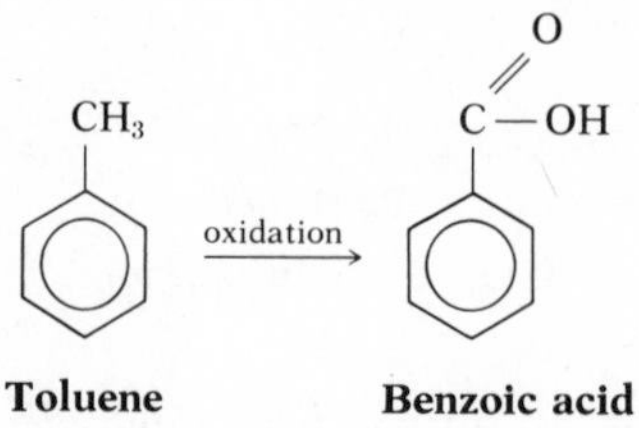

$$C_6H_5CH_3 \xrightarrow{\text{oxidation}} C_6H_5COOH$$

Toluene **Benzoic acid**

As an acid, benzoic acid may enter into neutralization reactions with inorganic bases. With sodium hydroxide, the salt *sodium benzoate* is formed:

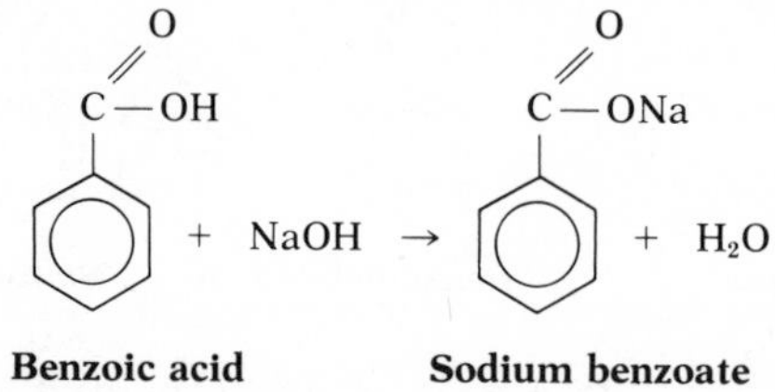

$$C_6H_5COOH + NaOH \rightarrow C_6H_5COONa + H_2O$$

Benzoic acid **Sodium benzoate**

Sodium benzoate is very frequently used in low concentrations as a food preservative.

Important aromatic acids

Salicylic acid and derivatives. The *ortho*-hydroxy derivative of benzoic acid is salicylic acid, which itself gives rise to a number of important derivatives. When salicylic acid is treated with sodium hydroxide, *sodium salicylate,* a salt, is formed (reaction with carboxyl group), and when treated with acetic acid, the ester *acetylsalicylic acid* is formed (reaction with phenolic hydroxyl group):

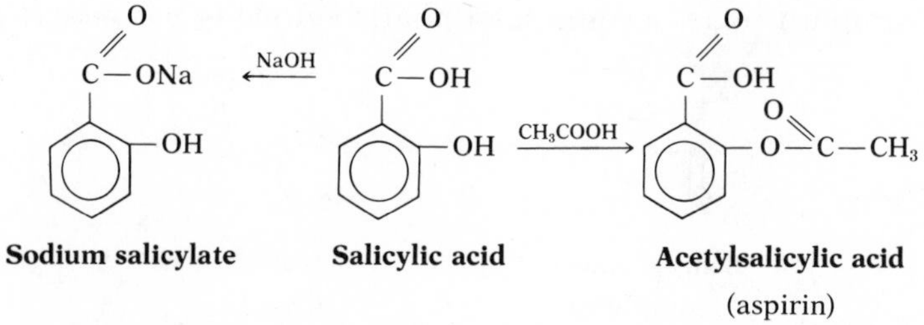

$$C_6H_4(OH)COONa \xleftarrow{NaOH} C_6H_4(OH)COOH \xrightarrow{CH_3COOH} C_6H_4(OCOCH_3)COOH$$

Sodium salicylate **Salicylic acid** **Acetylsalicylic acid** (aspirin)

Aspirin, acetylsalicylic acid, is the very widely used drug for the relief of fever (antipyretic) and pain (analgesic). Sodium salicylate has similar properties. Recent research indicates that aspirin plays an important role in controlling the blood-clotting process. Reaction of salicylic acid with methyl alcohol produces the ester *methyl salicylate,* commonly known as oil of wintergreen:

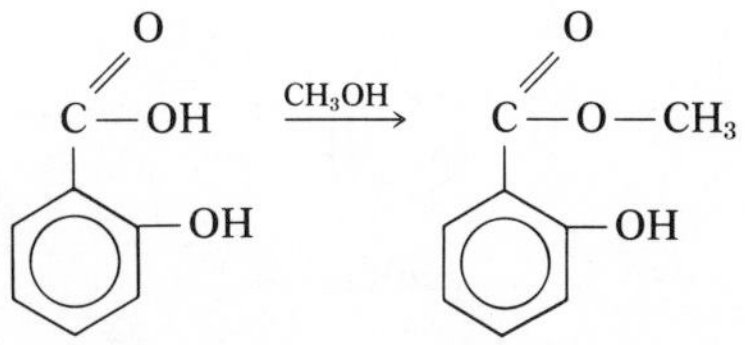

$$\text{C}_6\text{H}_4(\text{OH})\text{COOH} \xrightarrow{CH_3OH} \text{C}_6\text{H}_4(\text{OH})\text{C(=O)—O—CH}_3$$

Salicylic acid **Methyl salicylate**

Methyl salicylate is often found in liniments, perfumes, and flavorings.

Esters

The production and uses of two representative aromatic esters, acetylsalicylic acid (aspirin) and methyl salicylate, are discussed under Acids. The importance of aspirin is attested to by the astronomic figure for the annual consumption of this drug in the United States, now estimated to be in the billions of pounds.

Amines

The functional amine group, $—NH_2$, may be located on an aliphatic side chain or directly on the benzene ring. Two of the lower members of the aliphatic side chain type are *benzylamine* and *phenylethylamine:*

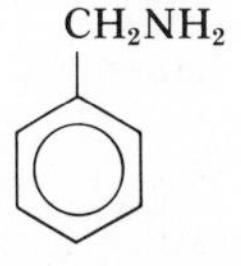

CH_2NH_2 $CH_2CH_2NH_2$

Benzylamine 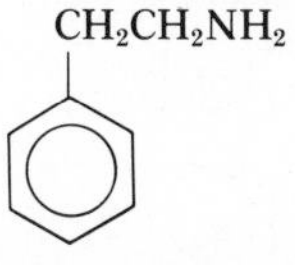 **Phenylethylamine**

Phenylethylamine forms an important part of the structure of the hormone *epinephrine,* or adrenaline, and some drugs such as *benzedrine:*

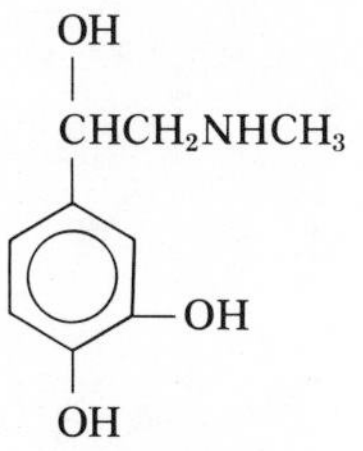

OH | $CHCH_2NHCH_3$ — OH — OH

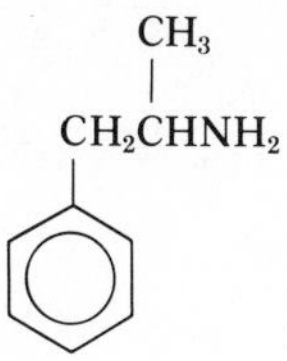

CH_3 | CH_2CHNH_2

Epinephrine (Adrenaline) **Benzedrine**

Epinephrine is a hormone produced in the adrenal gland (adrenal medulla) and has an important function in the metabolism of carbohydrates. Benzedrine acts by stimulating the central nervous system.

The simplest amine with the functional group directly on the ring is *aniline:*

NH_2

Aniline

Aniline has long been one of the most strategic commercial organic chemicals because of the many synthetic processes in which it is used. Of greater importance to the health sciences are some derivatives of aniline.

Important aniline derivatives

Sulfonamides. A great milestone in medical history was passed in the middle 1930s when it was discovered that *sulfanilamide* was effective in the treatment of many dreaded bacterial infections such as pneumonia and streptococcal meningitis, where no effective treatments existed before. The dramatic decreases in mortality rates, from 95% to 5% in some cases, earned the sulfa drugs the designation of "wonder drugs." As is often the case with such discoveries, many derivatives of sulfanilamide have subsequently been investigated. Among these are *sulfadiazine, sulfathiazole,* and *Sulfasuxidine,* all of which have slightly different properties, effectiveness, and toxicities.

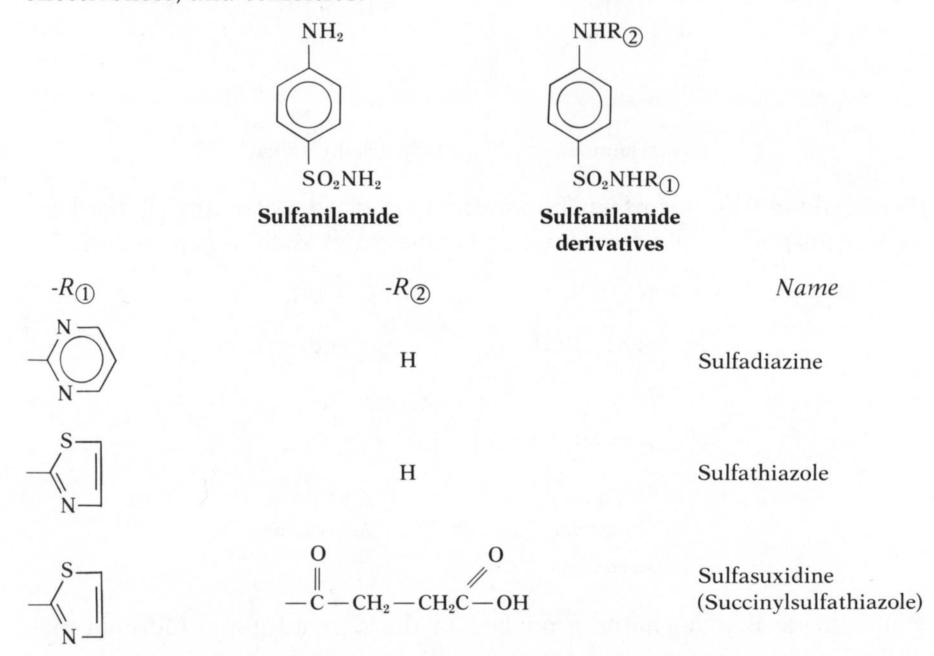

Acetanilid; phenacetin. The N-acetyl derivative of aniline, *acetanilid,*

and one of its derivatives, *phenacetin,* have been used as antipyretics and analgesics:

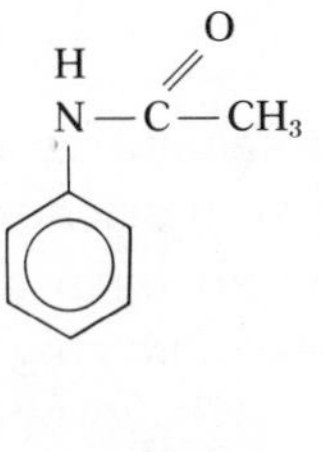

Acetanilid

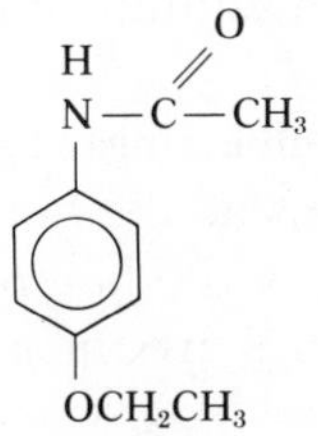

Phenacetin

HETEROCYCLIC COMPOUNDS

The heterocyclic compounds contain ring structures in which at least one atom is of some element other than carbon. The atom or atoms that make the ring structures heterocyclic will be referred to as the *hetero* atoms in the following discussions. Oxygen, nitrogen, and sulfur, alone or in combinations, are usually the hetero atoms in naturally occurring heterocyclic compounds of biochemical interest. Many of the vitamins contain heterocyclic rings in their structures. The heterocyclic rings are usually five- or six-membered rings, which sometimes combine with each other or in various combinations to form condensed ring structures.

Location of substituents on heterocyclic rings is usually accomplished by one of two methods. In the numbering system, the hetero atom becomes atom 1 and the other atoms are numbered in order. Sometimes a lettering system is used in which the atoms adjacent to the hetero atom are designated as the α atoms, and the next atoms as the β atoms. These numbering systems will be illustrated when appropriate.

Oxygen heterocyclics

The unit structures containing oxygen as the hetero atom are *furan,* a five-membered ring, and *pyran,* a six-membered ring:

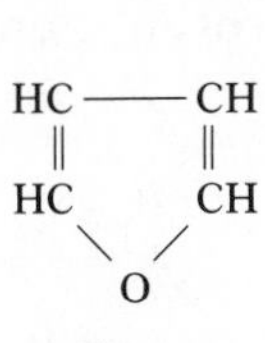

Furan

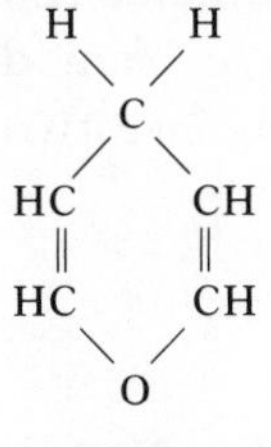

Pyran

or,

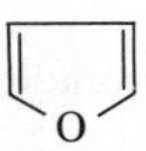

Furan

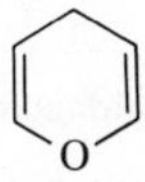

Pyran

Both structures are found in the carbohydrates. A furanose ring is also found in vitamin C and a condensed ring system containing a ring similar to pyran is found in vitamin E (p. 428).

Nitrogen heterocyclics

Pyrrole and pyridine rings. Heterocyclic structures containing one or more nitrogen atoms as the hetero atoms are more numerous than those containing oxygen as the hetero atom. The nitrogen counterparts of the furan and pyran rings are the *pyrrole* and *pyridine* rings:

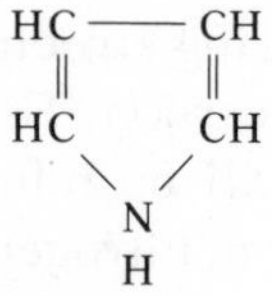

```
                                H
                                C
                              //  \
   HC ---- CH              HC      CH
   ||      ||              |       ||
   HC      CH              HC      CH
     \    /                  \\   /
       N                       N
       H
```

Pyrrole 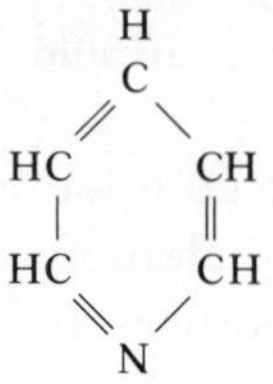 **Pyridine**

or,

```
   [pyrrole ring]        [pyridine ring]
       N                        N
       H
```

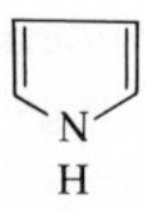

The pyrrole ring is a basic unit in the biosynthesis of the porphyrins, which are found in animal hemoglobin (p. 377) and vitamin B_{12} (p. 436); cytochromes; plant chlorophyll; and also the condensed ring compound, indole. The pyridine ring is found in nicotinic acid and pyridoxine, both members of the B complex of vitamins, and also in isonicotinic acid hydrazide (isoniazid), a drug that is used in the treatment of tuberculosis.

Imidazole and pyrimidine rings. Thus far, ring systems containing only one hetero atom have been discussed. However, in the *imidazole* and *pyrimidine* rings there are two hetero nitrogen atoms separated by one carbon atom:

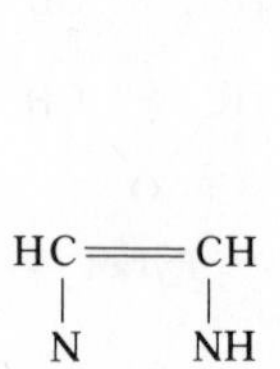

```
                                H
                                C
                              / 6 \\
   HC ==== CH              N1      5 CH
   |       |               ||        |
   N       NH              HC2     4 CH
    \\    /                  \  3  //
      C                         N
      H
```

Imidazole 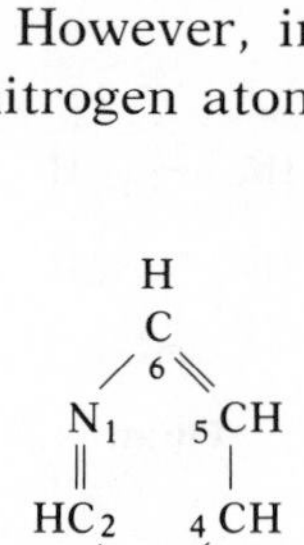 **Pyrimidine**

or,

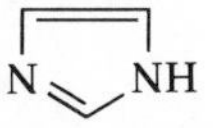

```
 ┌────┐          N⁄‾‾\
 N╲  ╱NH         │ ( ) │
   ‾‾             ╲__╱
                   N
```

The imidazole ring is found in the amino acid histidine (p. 218) and the pyrimidine ring is found in several important compounds that contribute to the structure of the nucleic acids (p. 238), ribonucleic acids (RNA) and deoxyribonucleic acids (DNA).

The parent compound of the barbiturates, *barbituric acid,* may be considered to be a triketo derivative of pyrimidine:

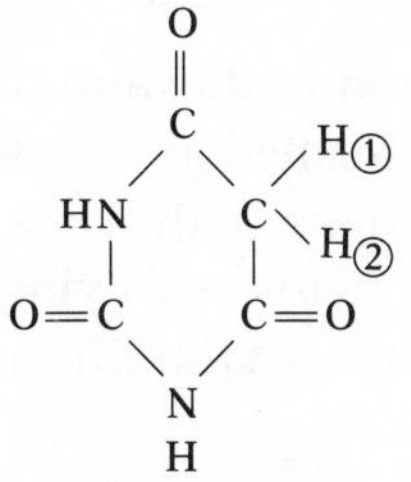

```
          O
          ‖
          C
        /   \   / H①
     HN       C
      |       | \ H②
    O=C       C=O
        \   /
          N
          H
```

Barbituric acid

If the two hydrogen atoms on the ring carbon are replaced as indicated in the chart, a variety of barbiturates may be obtained, all of which have been used for sedative and hypnotic purposes.

Replacement of		*Resulting barbiturate*
H①	*H②*	
$CH_3CH_2—$	$CH_3CH_2—$	Barbital
$CH_3CH_2—$	$C_6H_5—$	Phenobarbital
$CH_2{=}CHCH_2—$	$CH_3CH_2CH_2CH—$ (with CH_3 on the CH)	Seconal

Condensed ring systems

Purines. Condensation of pyrimidine and imidazole rings forms the ring system of the *purines:*

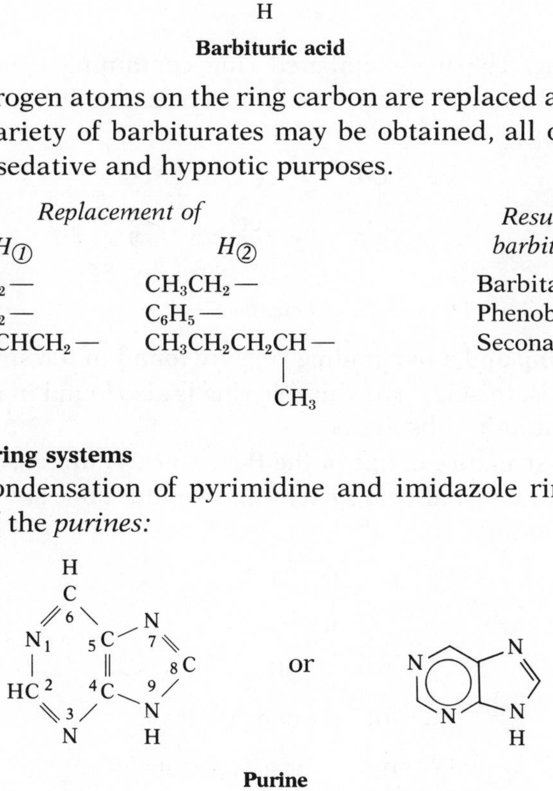

```
          H
          C
       // 6 \      N
    N1      5C⁄  7 \\
    |        ‖       8C        or      N⁄‾‾\___N
   HC2      4C\  9 /                   │ ( ) │   \\
      \\ 3 /       N                    \__/ ‾‾‾N /
          N        H                     N      N
                                                H
```

Purine

The purines as well as the pyrimidines are important constituents of the nucleic acids (p. 238), RNA and DNA.

Indole. The *indole* ring system is formed by the condensation of a pyrrole ring with a benzene ring:

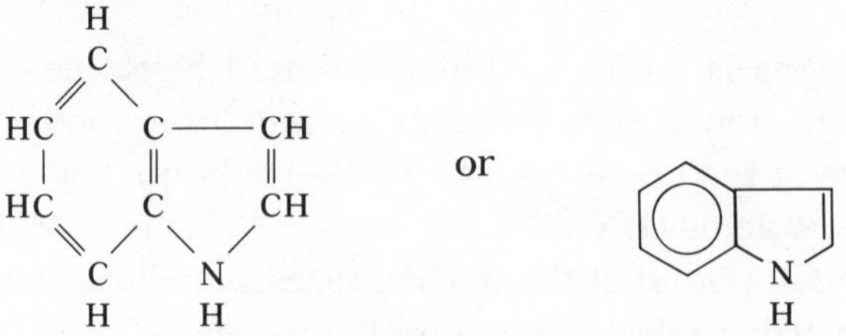

```
         H
         C
       //  \
    HC      C——CH
    |       ‖   ‖        or
    HC      C   CH
      \\  /  \  /
        C     N
        H     H
```

N
H

Indole

The indole ring system is found in the amino acid, tryptophan. Indole and skatole (3-methylindole) are formed as a result of the putrefaction of proteins in the large intestine and are mainly responsible for the odor of feces.

Other condensed ring systems involving hetero nitrogen atoms are found in several of the B-complex vitamins, riboflavin (B_2), folic acid, and cyanocobalamin (B_{12}).

Sulfur and nitrogen heterocyclics

Thiazole ring. The five-membered ring containing a hetero nitrogen atom separated by one carbon atom from a hetero sulfur atom is known as the *thiazole* ring:

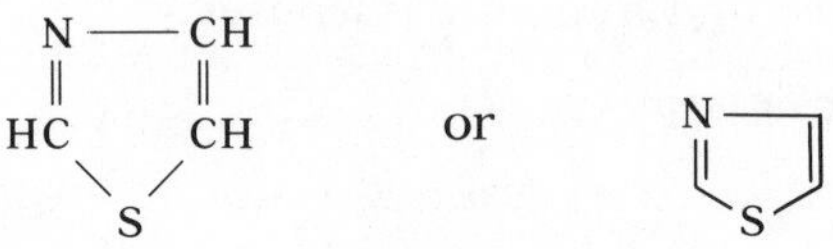

```
    N——CH
    ‖   ‖
   HC   CH        or
     \ /
      S
```

N
S

Thiazole

The thiazole ring and a pyrimidine ring are found in the structure of thiamine, vitamin B_1 (p. 432). The thiazole ring is also found in the penicillins, important antibiotic substances.

Biotin. The structure of one of the B-complex vitamins, *biotin,* is made up largely of an unusual ring system that includes one sulfur and two nitrogen hetero atoms:

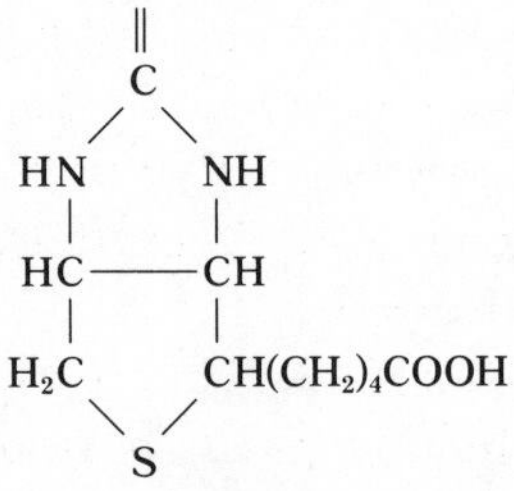

```
          O
          ‖
          C
        /   \
      HN     NH
      |       |
      HC—————CH
      |       |
     H2C     CH(CH2)4COOH
        \   /
          S
```

Biotin

Alkaloids

As their name implies, *alkaloids* are basic substances containing nitrogen as the hetero atom in ring systems, which are generally very complex. These substances are isolated from plants and have long been used as drugs because of their useful but strong physiologic effects. Many are toxic or habit-forming, requiring caution in their prescription and use as drugs. For purposes of illustration, several alkaloids and their uses are listed here.

Atropine. *Atropine* is isolated from the root of the plant, *Atropa belladonna,* commonly known as the deadly nightshade, or belladonna. It acts on the central nervous system and inhibits smooth muscle contraction. It has been used to dilate the pupil of the eye to make examination easier and to diminish spasm (antispasmodic) and excess motor activity in gastrointestinal disorders. *Cocaine,* which is quite similar to atropine in structure, is found in leaves of the coca plant *(Erythroxylon coca)* and is used as a local anesthetic, particularly of the eye, nose, throat, and urethra.

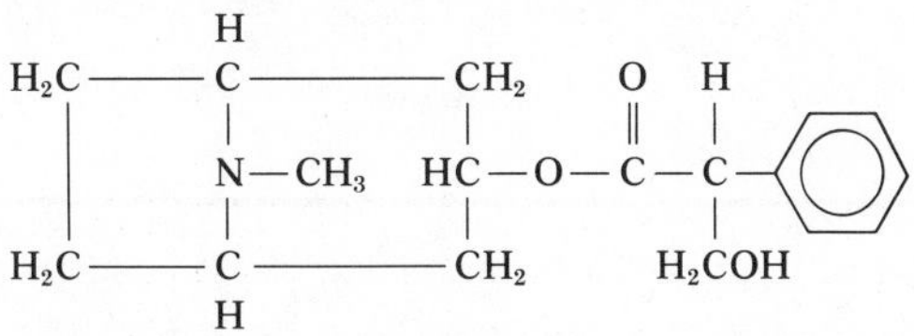

```
            H
H2C ——————— C ——————————— CH2     O   H
 |          |              |      ||   |
 |          N—CH3         HC—O—C—C—⟨benzene⟩
 |          |              |           |
H2C ——————— C ——————————— CH2        H2COH
            H
```

Atropine

Morphine. The most important alkaloid obtained from the unripe seed capsules of the Oriental poppy, *Papaver somniferum,* is *morphine.* The phenanthrene condensed ring system is found in the structure of morphine. The outstanding therapeutic property of morphine is its ability to

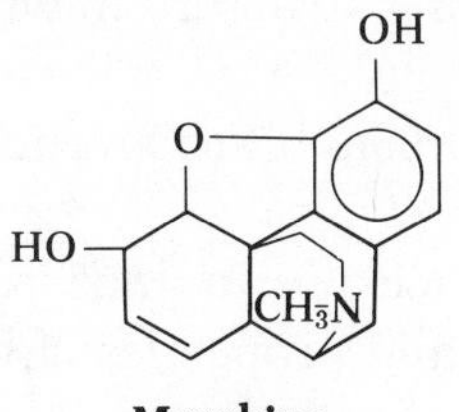

OH

O

HO

CH_3N

Morphine

relieve pain. Its major drawback is that it is a habit-forming drug. *Codeine,* the methyl ester (on the benzene ring) of morphine, is used mainly for the relief of coughing. Many unsuccessful attempts have been made to synthesize compounds with the analgesic effects of the opium alkaloids but without their habit-forming properties.

Quinine. Until World War II, the only available antimalarial drug was *quinine,* an alkaloid obtained from the bark of the cinchona tree.

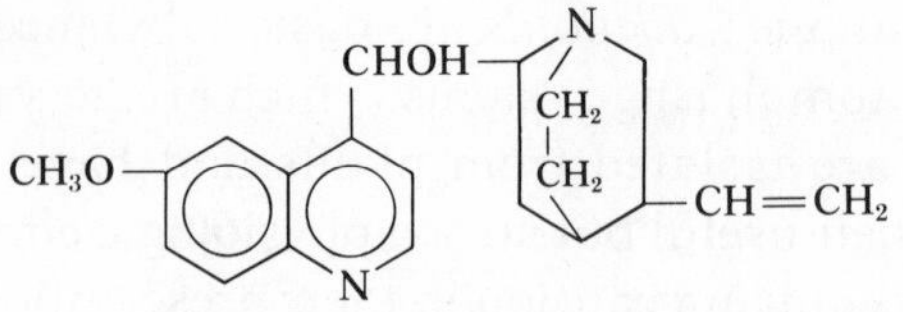

CH_3O — CHOH — N — CH_2 — CH_2 — $CH{=}CH_2$ — N

Quinine

As a result of intensive research, a number of possible substitutes for quinine were synthesized, of which *atabrine* became the best known because of its tendency to color the skin yellow. Troops returning from malaria infested areas could be recognized by the color of their skin.

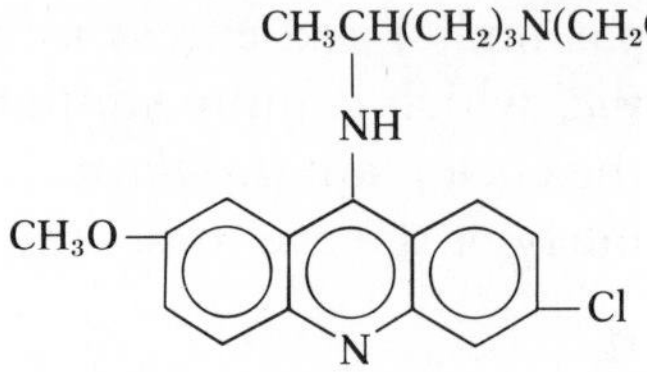

$CH_3CH(CH_2)_3N(CH_2CH_3)_2$ — NH — CH_3O — N — Cl

Atabrine

SUMMARY

Cyclic organic compounds have the same functional groups as open-chain compounds. Differences in properties are mainly because of differences in properties of cyclic benzene structure versus open-chain structure.

OCCURRENCE Coal tar is the major source of many cyclic compounds.

CARBOCYCLIC COMPOUNDS Compounds containing a ring structure in which all atoms in the ring are carbon atoms.

Carbocyclic aromatic hydrocarbons: Carbocyclic aromatic compounds of carbon and hydrogen only.

Benzene: Proposals by Kekulé in 1865 resolved the major questions about the structure and properties of benzene. Proposed structure:

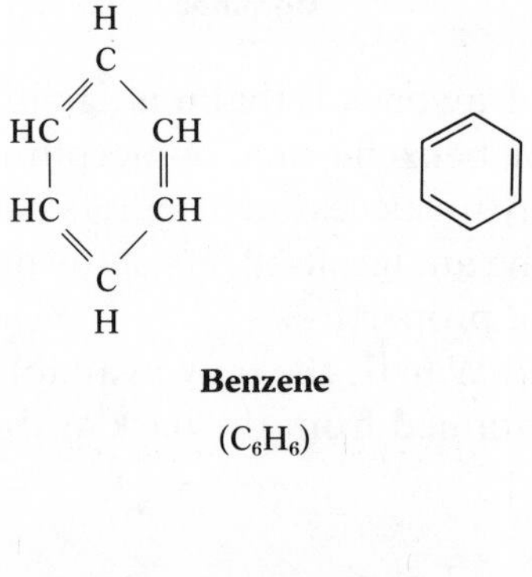

H C HC CH HC CH C H

Benzene

(C_6H_6)

Resonance between the two possible forms accounts for the properties and equivalence of all the bonds in the ring:

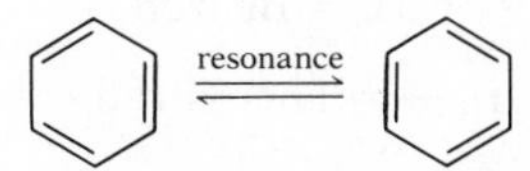

Another way of representing the equivalent bonds in benzene as something intermediate between single and double bonds is as follows:

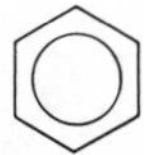

Chemical properties: Benzene can be nitrated:

$$C_6H_6 + HNO_3 \xrightarrow{H_2SO_4} C_6H_5NO_2 + H_2O$$

Benzene can be sulfonated:

$$C_6H_6 + H_2SO_4 \rightarrow C_6H_5SO_3H + H_2O$$

Benzene can react with bromine in substitution reactions:

$$C_6H_6 + Br_2 \rightarrow C_6H_5Br + HBr$$

Naming benzene derivatives: Two substituents on adjacent carbons are *ortho* to each other, on carbons separated by one carbon atom are *meta,* on carbons opposite each other (separated by two carbons) are *para:*

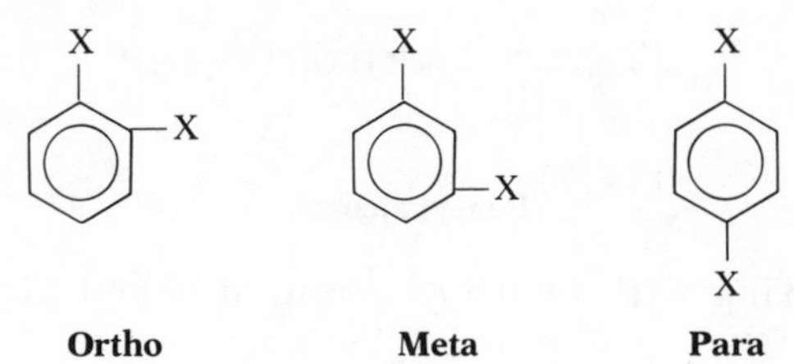

Ortho **Meta** **Para**

With three or more substituents, a numbering system is usually used:

Br
Cl
1 2 3 4 5 6
Cl

1-Bromo-2,4-dichlorobenzene

Aryl radicals: Parent carbocyclic hydrocarbon minus one hydrogen atom:

$$C_6H_6 = \text{Benzene}$$

$$C_6H_5— = \text{Phenyl radical}$$

Homologs of benzene: Methyl benzene, $C_6H_5CH_3$, or toluene, is the next higher homolog of benzene. The dimethyl benzenes, or xylenes, are found in the ortho, meta, and para forms.

Condensed ring hydrocarbons:

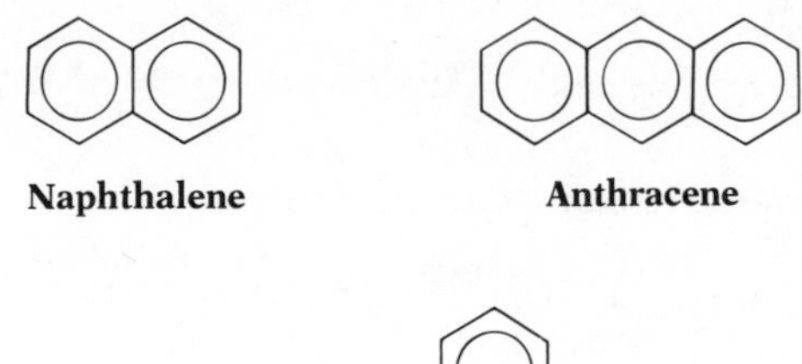

Naphthalene **Anthracene**

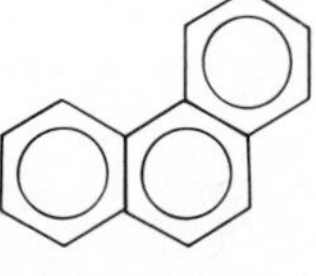

Phenanthrene

Phenanthrene structure is found in steroid hormones, cholesterol, vitamin D, and other important compounds.

Aromatic alcohols: Benzene ring with aliphatic alcohol side chain; therefore, have properties similar to aliphatic alcohols, i.e., can be oxidized to aldehyde, then acid, can react with organic acids to form esters:

CH_2OH

Benzyl alcohol

Phenols: Benzene ring with hydroxyl group attached directly to ring:

OH

Phenol

Phenols are weakly acidic because of the slight ionization of the hydrogen of the hydroxyl group, therefore can form salt with sodium hydroxide. Phenols also react to form esters and ethers.

Important phenols:

Phenol: Oldest known disinfectant, first used by Joseph Lister in

1867 in surgery. Coagulates proteins thereby destroying bacteria. Produces severe burns on contact with skin.

Resorcinol; hexylresorcinol: Resorcinol, *m*-dihydroxybenzene, less toxic than phenol but less effective. Hexylresorcinol has been used as a urinary disinfectant.

Cresols: Lysol is a soap emulsion of a mixture of the *o*-, *m*-, and *p*-cresols (methyl phenols).

Picric acid (2,4,6-Trinitrophenol): Strong acid and good precipitating agent for proteins. Used in treatment of burns to form film of precipitated protein that protects against infection and loss of tissue fluid.

Aldehydes: Functional group —CHO may be directly on benzene ring or on aliphatic side chain, and have properties similar to aliphatic aldehydes, i.e., may be oxidized to acids.

Ketones: Functional group $>C=O$ may connect two aromatic radicals or one aromatic and one aliphatic.

Acids: Functional group —COOH may be directly on benzene ring or on side chain. React with inorganic bases in neutralization reaction to form salts and water.

Important aromatic acids:

Salicylic acid and derivatives: Salicylic acid, *o*-hydroxybenzoic acid, forms important derivatives. Acetylsalicylic acid is the very widely used antipyretic known as *aspirin*. Methyl salicylate is used in liniments, perfumes, flavorings.

Esters: Compounds formed by reaction of alcohol and acid. See esters, acetylsalicylic acid, and methyl salicylate in the discussion of acids.

Amines: Functional group—NH_2 may be located directly on benzene ring or on aliphatic side chain. Phenylethylamine ($C_6H_5CH_2CH_2NH_2$) is in the structure of the hormone epinephrine, or adrenaline. Aniline ($C_6H_5NH_2$) has some important derivatives:

Important aniline derivatives:

Sulfonamides: Sulfanilamide and its derivatives were found to be effective in the treatment of some bacterial diseases where no

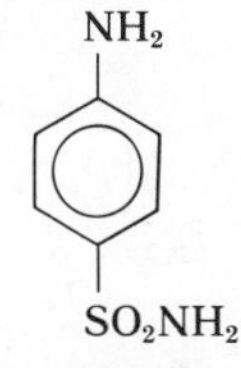

Sulfanilamide

treatment had been available (1935). Some other sulfonamides are sulfadiazine, sulfathiazole, Sulfasuxidine.

Acetanilid; phenacetin: Used as antipyretics and analgesics.

HETEROCYCLIC COMPOUNDS Compounds containing a ring structure in which at least one atom is of some element other than carbon. The different atom is called the hetero atom and is usually oxygen, nitrogen, or sulfur.

Oxygen heterocyclics: Furan or pyran rings are found in carbohydrates, vitamins C and E.

Furan **Pyran**

Nitrogen heterocyclics:

Pyrrole and pyridine rings:

Pyrrole **Pyridine**

Pyrrole ring is found in porphyrins, vitamin B_{12}, indole. Pyridine ring is found in nicotinic acid and pyridoxine (B-complex vitamins) and isonicotinic acid hydrazide used in treatment of tuberculosis.

Imidazole and pyrimidine rings: Two hetero nitrogen atoms per ring:

Imidazole **Pyrimidine**

Imidazole ring is found in amino acid histidine. Pyrimidine ring is found in nucleic acids, RNA and DNA, and in barbiturates, used as sedatives.

Condensed ring systems:

Purines: Condensation of pyrimidine and imidazole rings:

Purine

Purines, as well as pyrimidines, found in nucleic acids RNA and DNA.

Indole: Condensation of pyrrole ring with benzene ring:

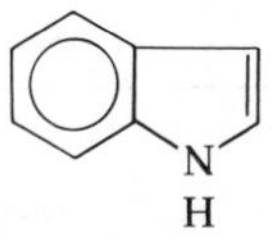

N
H

Indole

Indole ring is found in amino acid tryptophan.

Sulfur and nitrogen heterocyclics:

Thiazole ring:

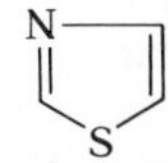

Thiazole

The thiazole ring and pyrimidine ring are found in thiamine, vitamin B_1. Thiazole ring also found in antibiotic penicillins:

Biotin:

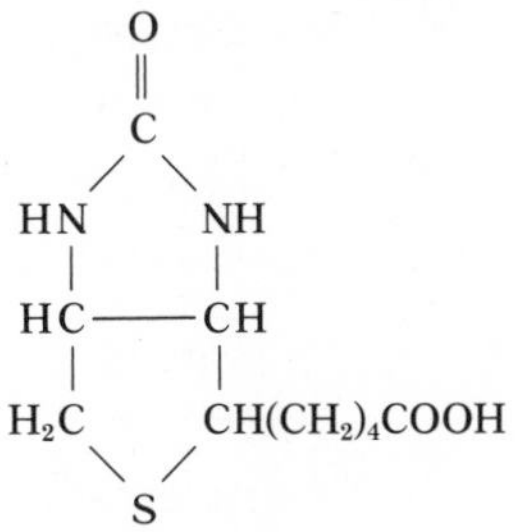

```
        O
        ‖
        C
       / \
    HN     NH
    |      |
    HC ——— CH
    |      |
   H2C     CH(CH2)4COOH
      \   /
        S
```

Biotin

Biotin is one of the B-complex vitamins.

Alkaloids: Basic substances with nitrogen as the hetero atom in complicated ring structures. Alkaloids are isolated from plants and are used as drugs because of their useful physiologic effects. Many are toxic or habit-forming.

Atropine: Obtained from root of belladonna plant, it is used to dilate the pupil of the eye or as an antispasmodic in gastrointestinal disorders. Cocaine, which is similar to atropine, is used as a local anesthetic for the eye, nose, throat, and urethra.

Morphine: Most important of the opium alkaloids containing a phenanthrene ring system. Morphine has outstanding ability to relieve pain but is habit-forming. Codeine, the methyl ester of morphine, is used to control coughing.

Quinine: Obtained from bark of cinchona tree and used in treatment of malaria. Other drugs, such as atabrine, have been developed for the treatment of malaria.

REVIEW QUESTIONS

1. What are the major differences in physical and chemical properties between open-chain and carbocyclic compounds? Do they have similar functional groups?
2. What is the important difference in chemical properties between aromatic alcohols and phenols? For what purposes are some important phenol derivatives used?
3. What is the difference between heterocyclic and carbocyclic compounds? What are the most common hetero atoms in biologic compounds?
4. What are condensed ring systems? Which one is found in important steroid hormones?
5. What are the structural differences between the pyrimidine and purine ring systems? In what important biologic substances are they found?

REFERENCES

Allinger, N. L., and others: Organic chemistry, ed. 2, New York, 1976, Worth Publishers, Inc.

Geissman, T. A.: Principles of organic chemistry, ed. 4, San Francisco, 1977, W. H. Freeman and Co., Publishers.

Holum, J. R.: Organic chemistry: a brief course, New York, 1975, John Wiley & Sons, Inc.

Morrison, R. T., and Boyd, R. N.: Organic chemistry, ed. 3, Boston, 1973, Allyn & Bacon, Inc.

Weininger, S. J.: Contemporary organic chemistry, New York, 1972, Holt, Rinehart and Winston, Inc.

Chemistry of carbohydrates

14

Carbohydrates—sugars, starches, and cellulose—may be defined as polyhydroxylic (polyhydric) aldehydes and ketones and their derivatives. These substances were named carbohydrates as a result of the observation that the empirical formula for many members of this class of substances was $C_nH_{2n}O_n$, which can be rearranged to the form $C_n(H_2O)_n$. Thus, most carbohydrates are composed of only three elements, carbon, hydrogen, and oxygen, in which the ratio of hydrogen to oxygen is the same as that found in water (2:1), but in which the hydrogen and oxygen atoms do not appear in the form of water molecules.

PHOTOSYNTHESIS

The central role of carbohydrates as the major source of energy for the support of life in the animal organism has already been noted (p. 9). As will be shown subsequently in discussing the metabolism of carbohydrates, the light energy trapped by photosynthesis in green plants in the form of chemical energy in carbohydrates can be released in the body, as required, by a series of enzymatic oxidation reactions. The overall reactions are summarized as follows:

$$6CO_2 + 6H_2O + \text{light energy} \xrightarrow[\text{photosynthesis}]{\text{chlorophyll}} \underset{\textbf{Sugar}}{C_6H_{12}O_6} + 6O_2$$

$$C_6H_{12}O_6 + 6O_2 \xrightarrow[\text{by body}]{\text{oxidation}} 6CO_2 + 6H_2O + \text{energy}$$

The dependence of animals on plants for photosynthesis, a process that animals cannot carry out, and the energy cycle are illustrated in Chapter 2, Fig. 2-1. In addition to serving as a source of energy, carbohydrates are also used in the biosynthesis of some of the fatty acids (p. 292) and amino acids (p. 363).

Although the end result of photosynthesis, the capture of light energy and its conversion into the chemical energy of carbohydrates, has been known for many years, the exact manner in which this is accomplished is still under intensive investigation. Much of the research into the mechanism of photosynthesis has been possible only because of the availability of radioactive *tracers* (p. 32), which are suitable radioactive chemical compounds that can be traced from one point to another on synthetic pathways. Use of these tracers has enabled investigators to show that plants exposed to radioactive carbon 14–labeled CO_2 ($^{14}CO_2$) incorporate the carbon 14 into carbohydrates in a very short time and that within a few minutes carbon 14 can also be found in amino acids and lipids. Photosynthesis is divided into two major phases, the first being the absorption of energy, and the second, the use of this energy to synthesize the carbohydrates and then the other metabolites. The first phase requires light and is sometimes referred to as the *light reaction;* the second phase does not require further light and may be termed the *dark reaction.*

CLASSIFICATION OF CARBOHYDRATES

Carbohydrates are generally classified on the basis of three structural considerations: (1) the number of simple molecular units to be found in each carbohydrate on hydrolysis, (2) the number of carbon atoms in the molecular unit, and (3) the identifying functional group in the molecular unit.

Simple carbohydrates consisting of a single molecule of sugar are known as *monosaccharides.* The names of the monosaccharides, which always end in *-ose,* generally indicate their origin, as in *fructose,* for fruit sugar. Monosaccharide units are readily combined by splitting out a molecule of water between each two units. If two monosaccharide units are combined in this way, the resulting compound is known as a *disaccharide.* Three units form a *trisaccharide,* and a large number of units form a *polysaccharide.* Di- and trisaccharide names also end in *-ose* but polysaccharides are named systematically according to the constituent monosaccharide with the ending *-an* or some trivial name, e.g., *starch* (trivial name), a polysaccharide of *glucose* units, would be known as a *glucosan* (systematic name). On complete hydrolysis, all of the saccharides, from di- to polysaccharides, split into simple monosaccharides.

The monosaccharides are subdivided on the basis of the number of carbon atoms in the unit molecules. Thus, a three-carbon carbohydrate is called a *triose,* a five-carbon carbohydrate is a *pentose,* and a six-carbon carbohydrate is a *hexose.*

The carbohydrates from trioses up can be further subdivided on the basis of the presence of an aldehyde or ketone group in the molecule. Trioses with aldehyde groups are known as *aldoses,* or *aldotrioses;* those with ketone groups, as *ketoses,* or *ketotrioses.* The same nomenclature applies to the larger monosaccharides.

This classification system is in the summary at the end of this chapter, including the physiologically important compounds of each class.

STRUCTURAL CONSIDERATIONS

In order to facilitate the following discussions, the simplified structural formulas for the important monosaccharides are presented. Also included here is a shorthand version of these structural formulas that is very convenient for writing purposes.

H
C=O
H—C—OH
H—C—OH
H

○ *

D-Glyceraldehyde

H
H—C—OH
C=O
H—C—OH
H

Dihydroxyacetone

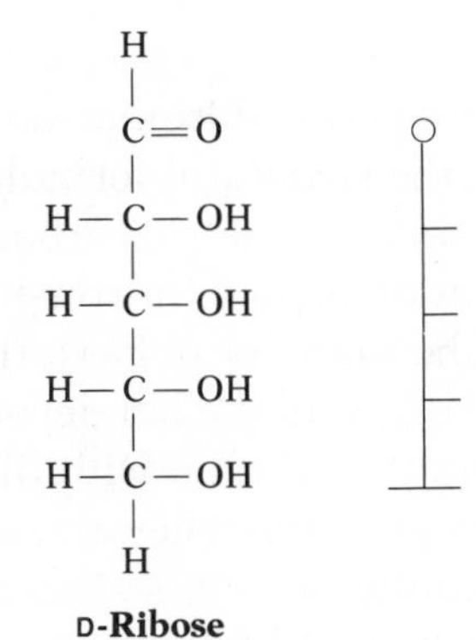

D-Ribose

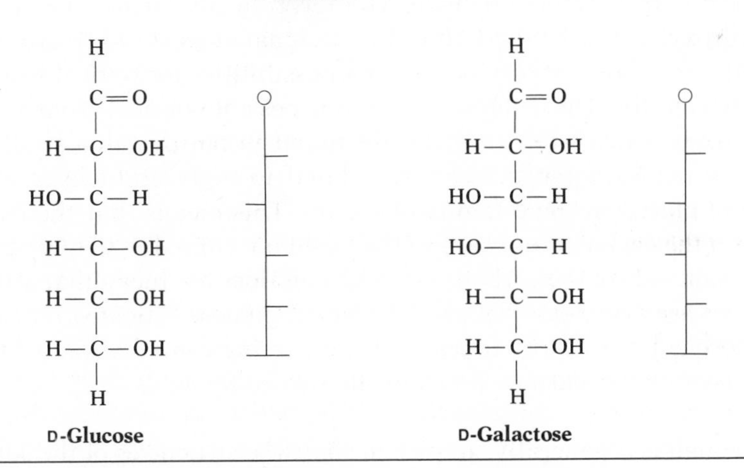

D-Glucose **D-Galactose**

---

*○ = terminal —C(H)=O or interior C=O

— = OH group (H understood to be on opposite side)

T or ⊥ = terminal —CH_2OH group.

```
          H
          |
     H—C—OH
          |
          C=O
          |
   HO—C—H
          |
     H—C—OH
          |
     H—C—OH
          |
     H—C—OH
          |
          H
```

D-Fructose

Optical isomerism from asymmetric carbon atoms

At this point, the section on optical isomerism in Chapter 11 should be reviewed. It can be seen in the structural formula of glyceraldehyde that the middle carbon atom is an asymmetric carbon atom. This means that there are two possible enantiomorphic (mirror-image) forms of glyceraldehyde—one that rotates the plane of polarization of light to the right, dextrorotatory (+), and the other, to the left, levorotatory (−). By custom, the form that is dextrorotatory is written with the hydroxyl group on the asymmetric carbon atom to the right when the aldehyde group is on top and is said to be in the D-family of sugars. The complete designation for this form is D(+)-glyceraldehyde. The levorotatory form is written with the hydroxyl group to the left and is designated as L(−)-glyceraldehyde.

With the more complex sugars, the possibilities for optical isomerism increase rapidly. Thus, there are sixteen possible optical isomers of the aldohexoses. Only a few of these are found in nature; almost all of the naturally occurring sugars can be related to D-glyceraldehyde and are therefore placed in the D-family of sugars. This means that the hydroxyl group on the carbon atom next to the bottom terminal —CH_2OH group is on the right side of the carbon skeleton. This does not mean that all D-family sugars are also dextrorotatory. In fact, D-glucose is dextrorotatory and D-fructose is levorotatory because of the varying contributions of the different asymmetric carbon atoms in the two compounds.

Cyclic structure

Although it is generally simpler to show the structures of the sugars as open-chain compounds, the naturally occurring sugars exist mainly in a ring form as the result of the reaction of an alcoholic hydroxyl group with the aldehyde or ketone group, for example, in the case of glucose, the following occurs:

```
 H     O←-┐                         H    OH
  \   //  ¦                          \  /
   C←--┐  ¦                           C ─────────┐
   |   ¦  ¦                           |          |
 HCOH  ¦  ¦                         HCOH         |
   |   ¦  ¦                           |          |
HOCH   ¦  ¦          ⇌             HOCH          O
   |   ¦  ¦                           |          |
 HCOH  ¦  ¦                         HCOH         |
   | ┌-┘  ¦                           |          |
 HCOH --┘                            HC ─────────┘
   |                                  |
 HCOH                               HCOH
  H                                   H
```

Open-chain form **D-Glucose** **Pyranose ring form**

The ring structure formed in glucose is derived from the heterocyclic pyran ring and the glucose in this form is referred to as D-glucopyranose. If a five-membered heterocyclic ring is formed, as may occur with fructose, the ring is called a furanose ring because of its relation to the heterocyclic furan ring:

```
   H                    H                    H
 HCOH                 HCOH                 HCOH
   |                    |                    |
   C ──────┐           C=O                   C ──────┐
   |       |            |                    |       |
 HOCH      |          HOCH                 HOCH      |
   |       |     ⇌      |          ⇌         |       O
 HCOH      O          HCOH                 HCOH      |
   |       |            |                    |       |
 HCOH      |          HCOH                  HC ──────┘
   |       |            |                    |
  HC ──────┘          HCOH                 HCOH
   H                    H                    H
```

D-Fructopyranose **D-Fructose** **D-Fructofuranose**

An important consideration here is the fact that although in solution the ring structures are the more favored forms, a small amount of the open-chain forms do exist in equilibrium with the ring forms so that the aldehyde and ketone properties of the sugars can be expressed in their chemical reactions.

A further refinement in representing the actual structure of the sugars in printed formulas was proposed by Haworth. In the Haworth formulas, the heterocyclic ring is represented in perspective to be at right angles to the plane of the paper with the groups below the ring and those above the ring corresponding to the groups on the right and the left, respectively, in the open-chain formula. The Haworth structure for glucose is written as shown on p. 186.

D-Glucose

The various representations of the structures are used interchangeably depending on the particular purposes for which they are being used.

MONOSACCHARIDES

Trioses

D-Glyceraldehyde and dihydroxyacetone. D-Glyceraldehyde and dihydroxyacetone occur as intermediates along the pathway of breakdown of carbohydrates (p. 295), the process by which energy is made available to the living organism.

Pentoses

D-Ribose and D-2-deoxyribose. Ribose and deoxyribose are important constituents of ribonucleic acids (RNA, p. 240) and deoxyribonucleic acids (DNA, p. 239), substances that are essential in genetic processes and protein synthesis. The structures of ribose and deoxyribose are the same except that the 2-carbon atom of deoxyribose does not have a hydroxyl group attached to it.

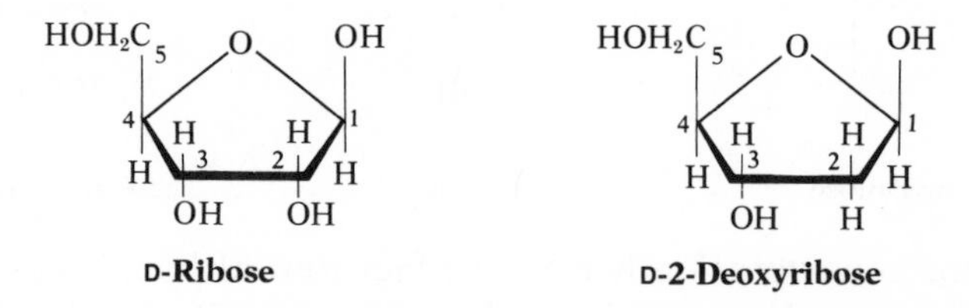

D-Ribose **D-2-Deoxyribose**

Hexoses

Glucose, fructose, galactose. Glucose and fructose are abundant in nature, occurring free in some foodstuffs and combined with each other, as the disaccharide sucrose, in others. Glucose also furnishes the basic unit from which the polysaccharides—starch, glycogen, and cellulose—are formed. Fructose also forms a polysaccharide, inulin, from which it may be prepared commercially by acid hydrolysis. Glucose usually has only the pyranose ring structure, but fructose generally has the pyranose ring when free and the furanose structure when combined in the higher saccharides.

Glucose is the sugar normally found in the blood. Medically, it is therefore important as a source of carbohydrates that can be administered intravenously for patients unable to take nourishment by mouth (Fig. 14-1).

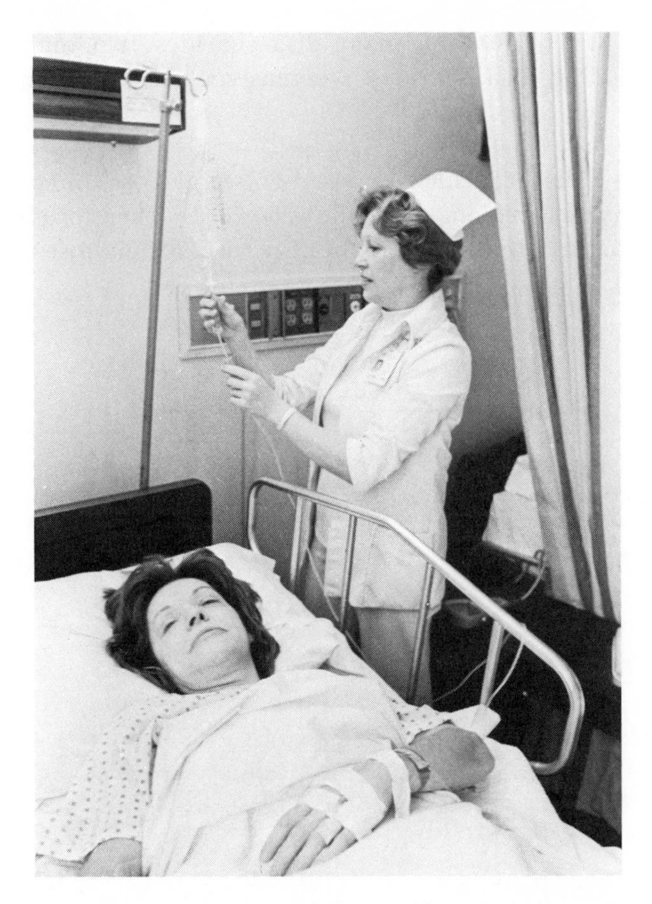

Fig. 14-1. Intravenous administration of glucose (dextrose). (Courtesy Barnes Hospital, St. Louis.)

Some of the older names by which glucose is known are *dextrose* and *grape sugar*. Fructose is sometimes called *levulose* or *fruit sugar*.

Galactose does not occur free, but does occur combined with glucose in the disaccharide lactose (milk sugar). Galactose is also in nerve tissues, combined with lipids in compounds known as cerebrosides.

DISACCHARIDES

As noted before, disaccharides are formed by the combination of two monosaccharides with the splitting out of a molecule of water. This results from the reaction of the aldehyde group of one monosaccharide with an alcoholic hydroxyl or ketone group of the second monosaccharide. The exact type of reaction is important because the *reducing power* of the resulting disaccharide depends on the availability of a free aldehyde or ketone group. Thus, of the three major disaccharides, two, maltose and lactose, are reducing disaccharides, whereas sucrose is a nonreducing disaccharide.

Reducing disaccharides: maltose and lactose

Maltose is a disaccharide composed of two glucose units combined in such a way as to leave one aldehyde group free. Therefore, maltose is a reducing sugar. It will be noted in the structures for maltose that the alde-

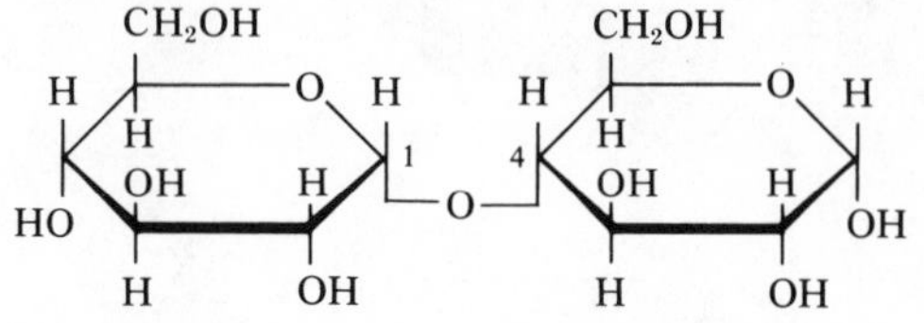

Maltose

(Glucose-oxygen-glucose)

or,

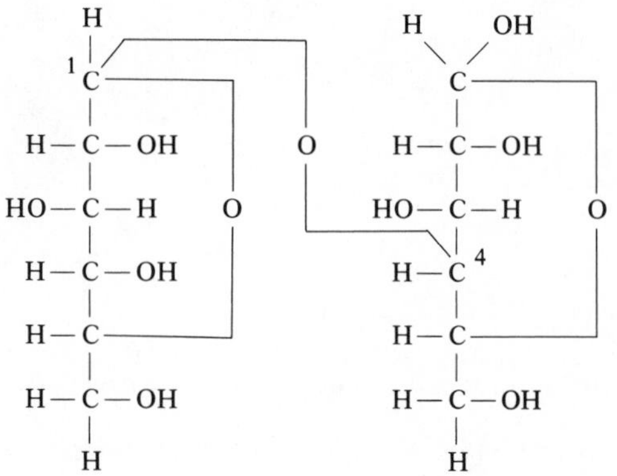

hyde group on the glucose unit to the right is not involved in the combining link and is able to give the disaccharide its reducing properties. This may be a little more obvious from the linear structure. The linkage between the two units is from the 1-carbon of one unit to the 4-carbon of the second unit and is referred to as a 1,4-linkage.

Maltose received its name from the observation that the enzymes found in malt hydrolyze starch to the disaccharide stage. The same is true of the amylase enzymes in the human intestinal tract. On further hydrolysis, two glucose units are obtained.

Lactose, the disaccharide found in milk, is composed of one glucose and one galactose unit. Examination of the structural formula shows that the 1,4-linkage is from the 1-carbon of galactose to the 4-carbon of glucose, leaving the aldehyde group on the 1-carbon of glucose free. As a result, lactose is a reducing disaccharide.

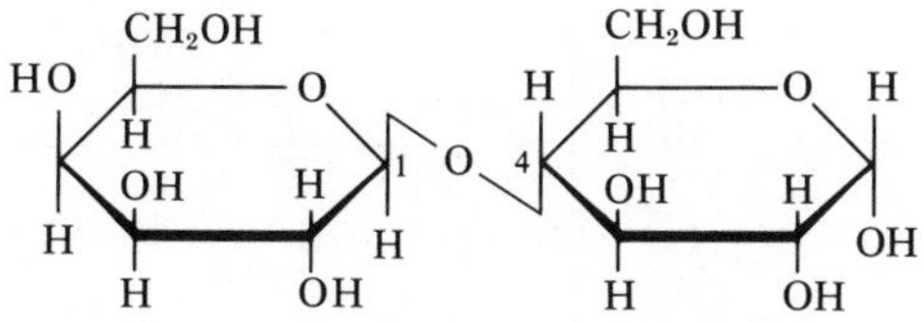

Lactose

(Galactose-oxygen-glucose)

or,

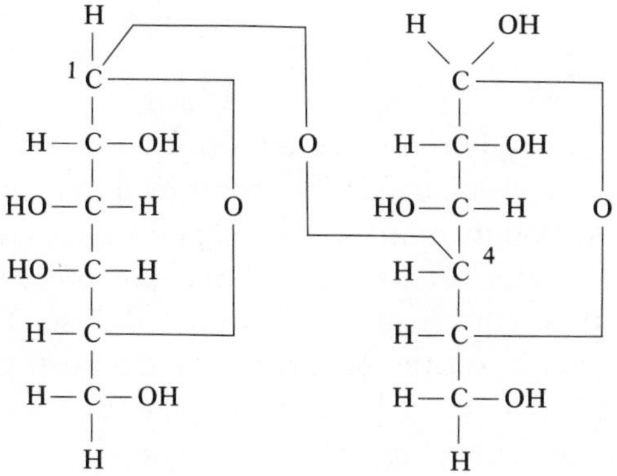

Nonreducing disaccharide: sucrose

The disaccharide sucrose, obtained mainly from sugar cane and sugar beets, is an important ingredient of the human diet, by itself or in combination in many different food preparations. Sucrose is composed of one glucose and one fructose unit, the 1,2-linkage involving the aldehyde group of the glucose unit and the ketone group of the fructose unit, leaving sucrose with no free reducing group. On hydrolysis, the furanose ring (five-membered) of fructose as it occurs in sucrose becomes the pyranose ring (six-membered) in the free fructose.

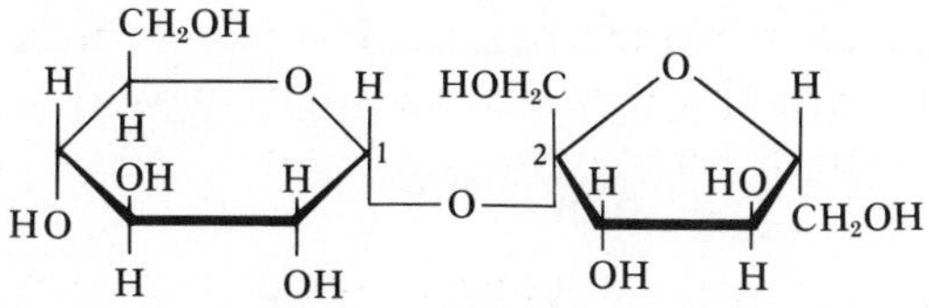

Sucrose

(Glucose-oxygen-fructose)

or,

POLYSACCHARIDES Polysaccharides are high–molecular weight compounds composed of large numbers of monosaccharide units combined to form one large molecule or polymer. Although many types of polysaccharides are possible, attention will be directed to the three important polysaccharides composed of glucose units: starch, glycogen, and cellulose. These compounds may be referred to as hexosans or glucosans. Because of their high molecular weights and manner of linkage, polysaccharides are insoluble in water and tasteless and have no reducing power.

Starch Starch represents the storage form of carbohydrates in plants and, as such, is an important source of carbohydrates for the diet of animals. There are two types of starch molecules: amyloses, composed of straight chains of 1,4-linked glucose units, and amylopectins, a branched type molecule of straight 1,4-linked units, as in amylose, that form branches by way of 1,6-linked units:

Amylose

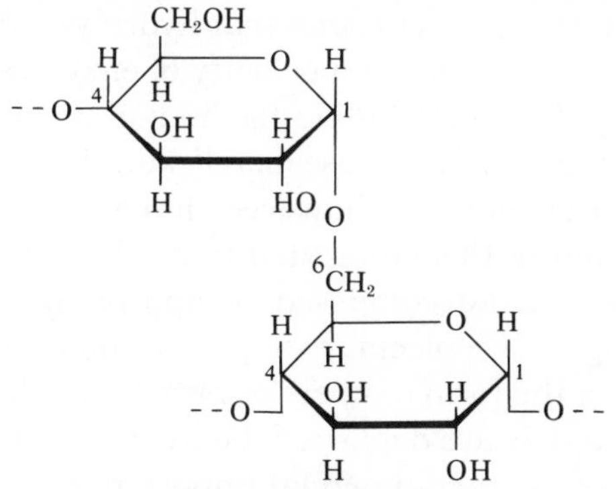

1,6 Branching linkage in amylopectin and glycogen

Amyloses have been reported with molecular weights ranging from 69,000 to 1,000,000. The amylopectins, with a branch for every 24 to 30 glucose units, are larger molecules with molecular weights ranging from 200,000 to the millions.

The enzymatic hydrolysis of starch by amylases in saliva and pancreatic juice produces smaller polysaccharides known as dextrins and ends with the formation of the disaccharide, maltose, as noted before.

Glycogen

Glycogen serves as the storage form of carbohydrates in animals, being found in liver and muscle tissue where enzyme systems are available to break down the glycogen to glucose units as required for the production of energy or for synthetic purposes. Glycogen is a branched polysaccharide resembling amylopectin but with a greater degree of branching (1 branch per 12 glucose units) and larger molecules. Molecular weights for glycogen have been reported in the range of 1 to 200 million.

Cellulose

Cellulose is an important constituent of the supporting structure of plants. It is similar in structure to amylose, being composed of glucose units combined in the form of straight chains. However, the 1,4-linkage in cellulose is arranged differently from that in amylose:

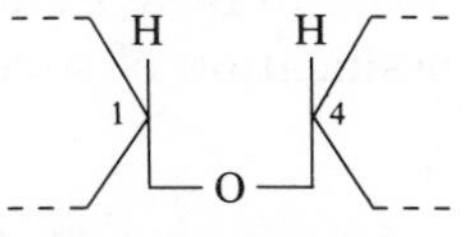

Amylose 1,4-linkage
(α-linkage)

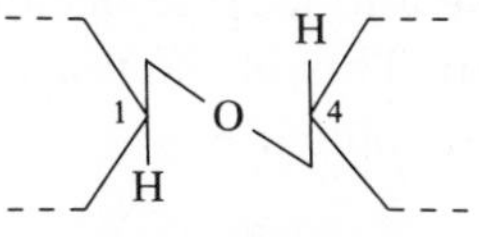

Cellulose 1,4-linkage
(β-linkage)

Because of this difference, the enzymes that hydrolyze amylose cannot act on cellulose. This results from the specificity of enzymes, a property to be discussed later. Therefore, cellulose in the human diet cannot supply carbohydrates in a usable form. Cellulose merely adds bulk or roughage to the intestinal contents. In ruminants, however, the cellulose is digested by the microorganisms found in their intestinal tracts.

Cellulose appears to be composed of approximately 3,000 glucose units, corresponding to a molecular weight of about 500,000. The molecules are arranged in the form of fibers or bundles of chains, giving structural strength where they are deposited. Cellulose and the many products formed from it are of great commercial importance.

CHEMICAL PROPERTIES OF CARBOHYDRATES

Reducing sugars

The term *reducing sugars* is applied to sugars that are capable of reducing certain reagents, particularly the cupric ion in alkaline solutions containing a complexing agent. In order to exhibit this reducing power, a sugar must have a free or potential aldehyde or ketol (ketone with adjacent alcoholic hydroxyl group, as in fructose) group in its structure. Heating such a sugar with an appropriate alkaline copper solution such as Benedict's reagent results in the reduction of the cupric ion (Cu^{++}) to the cuprous state (Cu^{+}) in cuprous oxide (Cu_2O), which settles out as a red precipitate. At the same time the sugar itself is oxidized to an acid:

$$\underset{\text{Reducing sugar}}{\mathrm{H{-}C{=}O}} + \underset{\substack{\text{in Cu}^{++}\text{ complex} \\ \text{(Blue solution)} \\ \text{Cupric ion}}}{Cu^{++}} \xrightarrow{\Delta} \underset{\substack{\text{in Cu}_2\text{O} \\ \text{(Red precipitate)} \\ \text{Cuprous ion}}}{Cu^{+}} + \underset{\text{Sugar acid}}{\mathrm{C({=}O){-}OH}}$$

This reaction provides a very useful qualitative test for the determination of the presence or absence of reducing sugars and by careful standardization of the procedure has also been used in a quantitative manner to determine the amount of sugar present. According to the requirements for a positive test, all of the free monosaccharides discussed here should give this test, but in the case of the disaccharides, sucrose can be distinguished from maltose and lactose because sucrose does not have a free aldehyde or ketone group. This is the reason for the classification of sucrose as a nonreducing disaccharide.

Oxidation

Under the proper conditions the 1-carbon atom, the 6-carbon atom, or both the 1- and the 6-carbon atoms of glucose may be oxidized to give the products listed here:

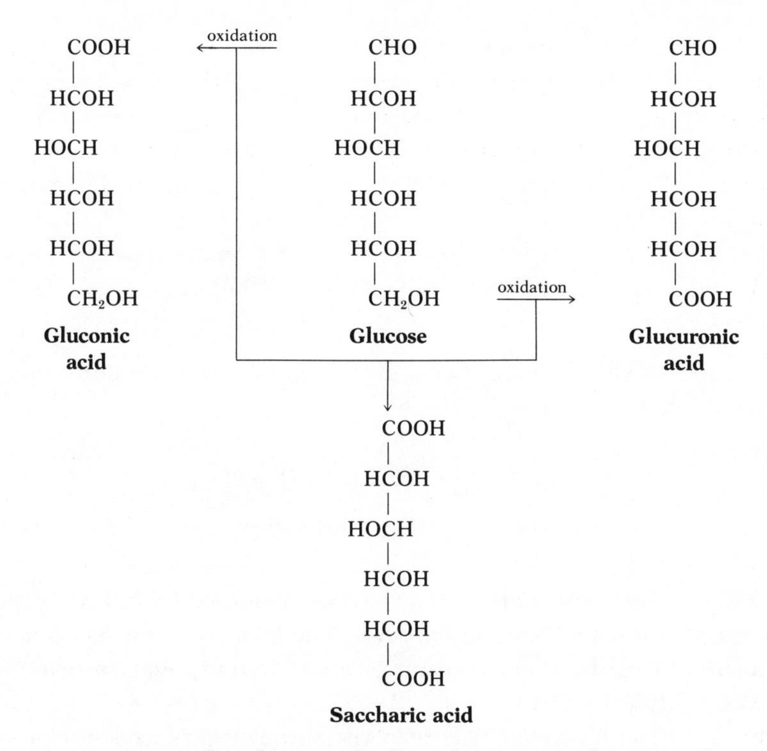

The same reactions take place with the other sugars forming the general classes of aldonic, uronic, and saccharic acids, respectively.

As noted before, the Benedict's test for reducing sugars is not selective. If it is desired to analyze a biologic sample specifically for glucose, advantage is taken of the availability of an enzyme, glucose oxidase, found in some molds. This enzyme acts only on glucose, oxidizing it to gluconic acid and hydrogen peroxide (H_2O_2). The hydrogen peroxide is then made to react with appropriate reagents to produce a color, the intensity of which is directly related to the amount of glucose in the sample.

Reduction

Sugars may also be reduced to form polyhydroxylic alcohols, e.g., glucose would be converted to sorbitol, and ribose to ribitol (found as a constituent of important coenzymes):

$$\begin{array}{ccc} \text{CHO} & & \text{CH}_2\text{OH} \\ | & & | \\ \text{HCOH} & & \text{HCOH} \\ | & & | \\ \text{HOCH} & & \text{HOCH} \\ | & \xrightarrow{\text{reduction}} & | \\ \text{HCOH} & & \text{HCOH} \\ | & & | \\ \text{HCOH} & & \text{HCOH} \\ | & & | \\ \text{CH}_2\text{OH} & & \text{CH}_2\text{OH} \\ \textbf{Glucose} & & \textbf{Sorbitol} \end{array}$$

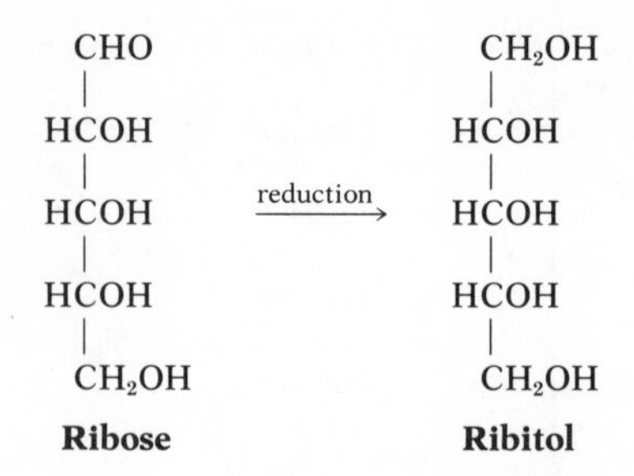

Fermentation

In a process known as fermentation, bakers' yeast can convert glucose to ethyl alcohol:

$$\underset{\textbf{Glucose}}{C_6H_{12}O_6} \xrightarrow[\text{yeast}]{\text{bakers'}} \underset{\textbf{Ethyl alcohol}}{2CH_3CH_2OH} + 2CO_2$$

Fructose, maltose, and sucrose can also be fermented by bakers' yeast, but galactose, lactose, and pentoses are not. The fermentation reaction can be used as an aid to the identification of a particular sugar in urine and in other body fluids.

Why a particular yeast will ferment some sugars and not others involves the presence or absence of appropriate enzymes and their specificity. Disaccharides must be broken down to their monosaccharide components before the alcoholic fermentation can take place. This is noted with bakers' yeast, which can ferment glucose but cannot ferment lactose because it lacks an enzyme to split lactose into glucose and galactose.

Iodine color reaction

Characteristic colors are obtained when some polysaccharides are treated with an iodine solution. Amylose gives a very deep blue color, amylopectin a reddish-purple color, and glycogen a reddish-brown color. Although starch is a mixture of amylose and amylopectin, the deep blue of the amylose prevails when starch is treated with iodine. As starch is broken down by hydrolysis to the lower–molecular weight dextrins, the color reaction with iodine changes from blue to violet to red-brown (erythrodextrins) to colorless (achroodextrins). This provides a simple way to follow the course of digestion of starch in the laboratory.

Furfural color tests

Furfural (from pentoses) and hydroxymethyl furfural (from hexoses) are easily formed by dehydrating carbohydrates by heating with nonoxidizing acids. The furfurals couple with aromatic amines and phenols to give colored compounds that can be used for qualitative or quantitative tests for carbohydrates.

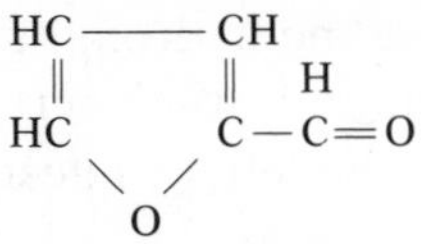

```
HC ——— CH                     HC ——— CH
||      || H                  ||      || H
HC      C—C=O      HOH2C—C      C—C=O
  \   /                          \   /
    O                              O
```

Furfural 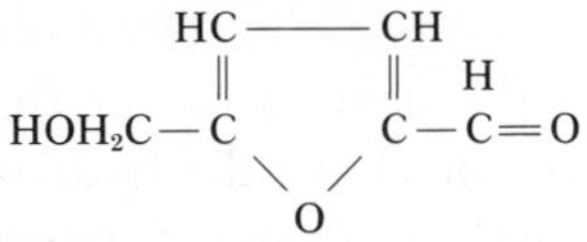 **Hydroxymethyl furfural**

Among the tests that operate in this way are the Molisch reaction for all carbohydrates, the Selivanoff test for ketoses, and the Bial, Tollen, and Tauber tests for pentoses.

SUMMARY

DEFINITION

Carbohydrates—sugars, starches, and cellulose—are polyhydroxylic aldehydes or ketones.

PHOTOSYNTHESIS

Process by which light energy is trapped and converted to chemical energy in the form of carbohydrates:

$$6CO_2 + 6H_2O + \text{light energy} \xrightarrow{\text{chlorophyll}} \underset{\textbf{Sugar}}{C_6H_{12}O_6} + 6O_2$$

CLASSIFICATION

Monosaccharides:

Unit molecules:

Trioses: three carbon atoms per monosaccharide.

Aldotriose: triose with aldehyde group, e.g., D-glyceraldehyde.

Ketotriose: triose with ketone group, e.g., dihydroxyacetone.

Pentoses: five carbon atoms per monosaccharide.

Aldopentose: D-ribose.

Deoxypentose: D-2-deoxyribose.

Hexoses: six carbon atoms per monosaccharide.

Aldohexoses: D-glucose, D-galactose.

Ketohexose: D-fructose.

Disaccharides:

Two monosaccharides chemically combined:

Reducing: maltose (two glucose units), lactose (glucose, galactose).

Nonreducing: sucrose (glucose, fructose).

Polysaccharides:

Many monosaccharide units chemically combined:

Hexosans (glucosans): starch, glycogen, cellulose.

STRUCTURAL CONSIDERATIONS

Optical isomerism from asymmetric carbon atoms: Optical isomerism should be reviewed in Chapter 11. D-Glyceraldehyde, with one asymmetric carbon atom, serves as the model compound to which all other carbohydrates are related. D-Carbohydrates have the hydroxyl group to the right on the next to the last carbon atom, as in D-glyceraldehyde; L-carbohydrates have the hydroxyl group to the left. (+) designates dextrorotatory; (−) designates levorotatory.

Cyclic structure: Carbohydrates exist mainly in the form containing ring structures. The six-membered heterocyclic (hetero oxygen) ring is the pyranose form. The five-membered ring is the furanose form. Haworth structures show carbohydrates in their true ring form.

MONOSACCHARIDES

Trioses:

D-*Glyceraldehyde and Dihydroxyacetone.* Occur as intermediates in breakdown of carbohydrates.

Pentoses:

D-*Ribose and* D-*2-Deoxyribose:* Important constituents of RNA and DNA, respectively, which are essential in genetic processes and protein synthesis.

Hexoses:

Glucose, Fructose, Galactose: Glucose and fructose occur free in foods and also combined as the disaccharide sucrose. Glucose is the basic unit of the polysaccharides—starch, glycogen, and cellulose. Glucose is the sugar normally found in the blood and may be given intravenously when food cannot be taken by mouth. Glucose has also been called dextrose or grape sugar. Fructose has also been called levulose or fruit sugar. Galactose occurs combined with glucose in the disaccharide lactose and in cerebrosides (nerve tissue).

DISACCHARIDES

Reducing disaccharides:

Maltose: Composed of two glucose units in 1,4-linkage with one aldehyde group free; therefore, has reducing power. Starch is hydrolyzed to the maltose stage by amylase enzymes.

Lactose: Composed of glucose and galactose in 1,4-linkage with one aldehyde group free, therefore has reducing power. Lactose occurs in milk.

Nonreducing disaccharides:

Sucrose: Composed of glucose and fructose in 1,2-linkage with no free aldehyde or ketol group; therefore has no reducing power. Sucrose occurs in cane sugar and is important in the human diet.

POLYSACCHARIDES

High–molecular weight compounds made up of many monosaccharide units; are insoluble in water, tasteless; and have no reducing power.

Starch: Storage form of carbohydrates in plants. There are two types of starch molecules: (1) Amylose = straight chains of glucose units in 1,4-

linkage; (2) Amylopectin = branched type with chains of 1,4-units joined at branch points in 1,6-linkage.

Glycogen: Storage form of carbohydrates in animals, in liver and muscle tissue. Resembles amylopectin.

Cellulose: Forms supporting structure of plants. Cannot be digested by humans because the 1,4-linkage is different than the ones found in starch and glycogen.

CHEMICAL PROPERTIES OF CARBOHYDRATES

Reducing sugars: Refers to ability of sugars with free or potential aldehyde or ketol groups to reduce alkaline copper solutions, e.g., Benedict's reagent:

$$\underset{\text{Sugar}}{\text{H}-\text{C}{=}\text{O}} + \underset{\text{in Cu}^{++}\text{ complex}}{Cu^{++}} \rightarrow \underset{\text{in Cu}_2\text{O (Red precipitate)}}{Cu^{+}} + \underset{\text{Sugar acid}}{\text{O}{=}\text{C}-\text{OH}}$$

Used as qualitative and quantitative test.

Oxidation: Oxidation of aldehyde group produces aldonic acid:

$$\text{CHO} \cdots \text{CH}_2\text{OH} \xrightarrow{\text{oxidation}} \text{COOH} \cdots \text{CH}_2\text{OH} \quad \text{aldonic acid}$$

Oxidation of $-CH_2OH$ group produces uronic acid:

$$\text{CHO} \cdots \text{CH}_2\text{OH} \xrightarrow{\text{oxidation}} \text{CHO} \cdots \text{COOH} \quad \text{uronic acid}$$

Oxidation of both groups produces saccharic acid:

$$\text{CHO} \cdots \text{CH}_2\text{OH} \xrightarrow{\text{oxidation}} \text{COOH} \cdots \text{COOH} \quad \text{saccharic acid}$$

A specific test for glucose is the use of the enzyme, glucose oxidase, which oxidizes glucose to gluconic acid and hydrogen peroxide (H_2O_2). The hydrogen peroxide then reacts with appropriate reagents to produce a color, from the intensity of which the quantity of glucose present can be determined.

Reduction: Sugars may be reduced to form polyhydroxylic alcohols, e.g.,

$$\text{glucose} \xrightarrow{\text{reduction}} \text{sorbitol}$$

$$\text{ribose} \xrightarrow{\text{reduction}} \text{ribitol}$$

Fermentation: Bakers' yeast ferments glucose to ethyl alcohol:

$$C_6H_{12}O_6 \xrightarrow[\text{yeast}]{\text{bakers'}} 2CH_3CH_2OH + 2CO_2$$

Fructose, maltose, sucrose also fermented but not galactose, lactose, pentoses. Lack of fermentation means lack of appropriate enzyme. Disaccharides must be broken down to monosaccharides before fermentation can take place.

Iodine color reactions: Iodine added to starch causes deep blue color; with amylopectin, reddish-purple; with glycogen, reddish-brown. As starch breaks down, dextrins give red-brown color (erythrodextrins), then no color (achroodextrins). Starch digestion can be followed with these color reactions.

Furfural color tests: Furfurals, formed from carbohydrates, couple with aromatic amines and phenols to give colored compounds used as qualitative and quantitative tests for carbohydrates: Molisch test for all carbohydrates; Selivanoff test for ketoses; Bial, Tollen, Tauber tests for pentoses.

REVIEW QUESTIONS

1. Of what significance to human life is photosynthesis? What are the key factors involved in the process? Describe the food chain.
2. What are the characteristic functional groups of carbohydrates? What is the importance of these groups in the formation of disaccharides? Polysaccharides?
3. Discuss the chemistry involved in determining the reducing power of a carbohydrate. Do all monosaccharides exhibit reducing power? Disaccharides? Polysaccharides? Explain.
4. What is the animal storage form of carbohydrates? The plant storage form? Are they similar in structure? Explain.
5. Which monosaccharide is normally found in the blood? Does it exist in open chain or ring form in solution? Does it exhibit optical isomerism? Explain.
6. Describe some chemical tests for the detection of carbohydrates.

REFERENCES

Advances in carbohydrate chemistry (series), New York, Academic Press, Inc.

Annual review of biochemistry, Palo Alto, Calif., Annual Reviews.

Barry, J. M., and Barry, E. M.: An introduction to the structure of biological molecules, Englewood Cliffs, N.J., 1969, Prentice-Hall, Inc.

Florkin, M., and Stotz, E. H. editors: Comprehensive biochemistry, Vol. 5, Amsterdam, 1963, Elsevier.

Guthrie, R. D.: Introduction to carbohydrate chemistry, ed. 4, New York, 1974, Oxford University Press, Inc.

Pigman, W. W., and Horton, D.: The carbohydrates, 4 vols., ed. 2, New York, 1970-1978, Academic Press, Inc.

Shreeve, W. W.: Physiological chemistry of carbohydrates in mammals, Philadelphia, 1974, W. B. Saunders Co.

Stoddart, J. F.: Stereochemistry of carbohydrates, New York, 1971, John Wiley & Sons, Inc.

Whistler, R. L., and Wolfram, M. L., editors: Methods in carbohydrate chemistry, 7 vols., New York, 1962-1971, Academic Press, Inc.

Chemistry of lipids

15

Lipids cannot be defined in as simple a manner as were the carbohydrates. Probably the most general statements that may be made about lipids are that fatty acids (open-chain carboxylic acids) usually constitute an important part of the lipid molecules and that lipids are generally insoluble in water but soluble in organic solvents. Lipids are composed mostly of carbon, hydrogen, and oxygen, with nitrogen and phosphorus appearing in some types of lipids. Lipids have long been considered quantitatively important because of their characteristics as fuel for energy purposes. Not only can lipids be stored in the body to an unlimited extent (a sorry fact for many dieters), but on complete oxidation they yield twice as much energy per gram as do carbohydrates or protein (p. 316). Certain *essential fatty acids* must be supplied by the diet because they cannot be synthesized in the body. Among their functions, lipids are known to be good insulators against excessive heat exchanges and to form good padding to protect internal organs from severe shock. Although certain lipids are present in high concentration in nerve tissue, the functions of these lipids are largely unknown. One of the most interesting developments in recent research is the investigation of the function of lipoproteins (compounds of lipids with proteins) as constituents of cell membranes. Since most body functions depend on reactions that take place in, on, or around the many membranes that are found in tissue cells, the potential importance of such functions for lipids can hardly be exaggerated. Shown in Fig. 15-1 is a model representation of a membrane composed of a phospholipid matrix in which protein molecules are embedded, some extending through the membrane on both sides. This model has been used to explain some important properties and functions of membranes.

CLASSIFICATION OF LIPIDS

The lipids are discussed according to the classification as summarized at the end of this chapter.

The simple lipids consist of fatty acids esterified with an alcohol compound, which is glycerol in the fats and a substance with a higher molecular weight, such as cholesterol, in the waxes. Compound lipids con-

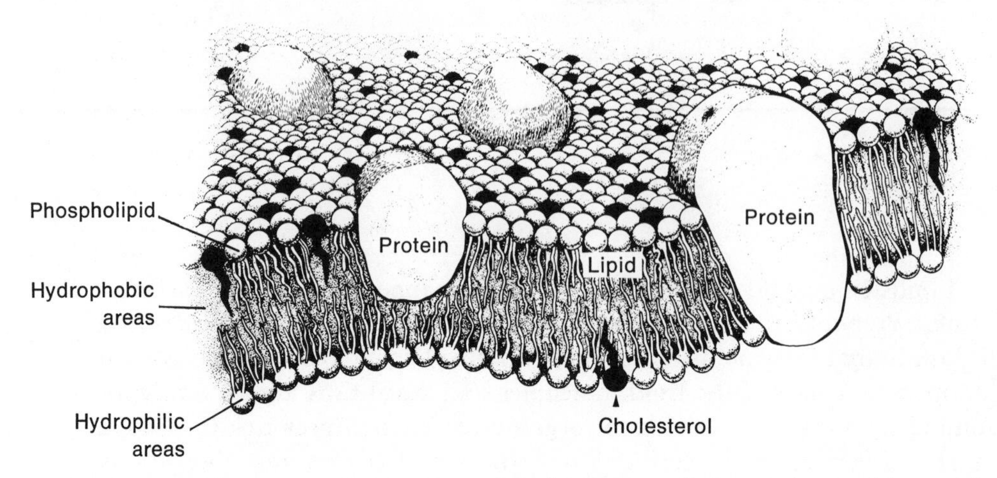

Fig. 15-1. Plasma membrane. Note globular (potato-like) proteins in this model, embedded in bimolecular layer of lipids (including cholesterol). (From Smith, A. L.: Microbiology and pathology, ed. 12, The C. V. Mosby Co., 1980.)

tain, in addition, some other groups such as phosphoric acid and basic nitrogen components in phospholipids, carbohydrate in glycolipids, and protein in lipoproteins. The steroids, though not always included with the lipids, are substances that are usually associated with lipids in nature because of the similarity of their solubility properties.

SIMPLE LIPIDS

Fats

A molecule of fat consists of glycerol, a trihydroxylic alcohol, esterified with three open-chain fatty acids:

$$\begin{array}{l} \quad\ \ \mathrm{H} \\ \quad\ \ | \qquad \mathrm{O} \\ \mathrm{H-C-O-\overset{\|}{C}-R_1} \\ \quad\ \ | \qquad \mathrm{O} \\ \mathrm{H-C-O-\overset{\|}{C}-R_2} \\ \quad\ \ | \qquad \mathrm{O} \\ \mathrm{H-C-O-\overset{\|}{C}-R_3} \\ \quad\ \ | \\ \quad\ \ \mathrm{H} \end{array}$$

Fat or glyceride

The naturally occurring fats are generally of the *mixed glyceride* type, meaning that the fatty acids represented by R_1, R_2, and R_3 are different. If the three fatty acids in a molecule of fat were oleic, palmitic, and stearic acids, the name of the fat would be *oleopalmitostearin.* There may, of course, be molecules with two or three acids the same and these would be named appropriately, e.g., *oleodistearin* and *tripalmitin.* Tripalmitin, with all the acids the same, is termed a *simple glyceride.*

The fats are insoluble in water but soluble in organic solvents such as carbon tetrachloride, benzene, and ether. An interesting property of the fats is the fact that their melting points vary according to the type of fatty acids in the molecule: the saturated acids, with relatively longer chains, impart a higher melting point (more solid) to the fats, and the unsaturated acids, with relatively shorter chains, impart a lower melting point (more liquid). Those fats that remain liquid at room temperature are referred to as oils and are generally found in vegetables. Animal fats, which contain a greater proportion of saturated fatty acids, are usually solids and represent the storage lipid in the body as well as a major part of the lipid intake of the human diet.

Glycerol. Glycerol, in contrast to fats, of which it is an important component, is soluble in water and insoluble in organic solvents that ordinarily dissolve lipids. However, when two of the fatty acids are hydrolyzed off leaving a monoglyceride, the monoglyceride acts as a good detergent and emulsifying agent because it now has a water soluble end—the singly esterified glycerol portion—and a lipid soluble end—the fatty acid portion. This property is important in the process of digestion of fats. It also led to the development of synthetic detergents, which do not form insoluble substances with the calcium or magnesium ions of hard water.

Fatty acids. Except for the lower fatty acids, which are soluble in water, naturally occurring fats, with longer chains, are insoluble in water but soluble in organic solvents. The sodium and potassium salts of the fatty acids, which are known as soaps, are soluble in water and insoluble in organic solvents. The shorter chain unsaturated fatty acids melt at a lower temperature than the longer, saturated acids.

Of the fatty acids found in nature, almost all are of the straight-chain type and have an even number of carbon atoms. In the box on p. 202 are included the more important fatty acids found in animals—palmitic, stearic, and oleic acids being the most commonly occurring ones. For example, human depot fat (storage fat) contains approximately 25% palmitic acid, 6% stearic acid, and 50% oleic acid.

In names of the unsaturated acids, the positions of the double bonds are indicated by a number, the numbering starting from the carboxyl carbon as number 1. Thus, oleic acid, in which the double bond is between the number 9 and 10 carbon atoms, has the systematic name of *9-octadecenoic acid* (9 for the position of the double bond, *octa*-[8], *-dec*-[10] to indi-

FATTY ACIDS

Saturated

Butyric (n-butanoic)	$CH_3(CH_2)_2COOH$
Caproic (hexanoic)	$CH_3(CH_2)_4COOH$
Caprylic (octanoic)	$CH_3(CH_2)_6COOH$
Capric (decanoic)	$CH_3(CH_2)_8COOH$
Lauric (dodecanoic)	$CH_3(CH_2)_{10}COOH$
Myristic (tetradecanoic)	$CH_3(CH_2)_{12}COOH$
Palmitic (hexadecanoic)	$CH_3(CH_2)_{14}COOH$
Stearic (octadecanoic)	$CH_3(CH_2)_{16}COOH$
Lignoceric (tetracosanoic)	$CH_3(CH_2)_{22}COOH$

Unsaturated

Palmitoleic (9-hexadecenoic)	$CH_3(CH_2)_5CH{=}CH(CH_2)_7COOH$
Oleic (9-octadecenoic)	$CH_3(CH_2)_7CH{=}CH(CH_2)_7COOH$
Nervonic (15-tetracosenoic)	$CH_3(CH_2)_7CH{=}CH(CH_2)_{13}COOH$
Linoleic (9,12-octadecadienoic)	$CH_3(CH_2)_4CH{=}CHCH_2CH{=}CH(CH_2)_7COOH$
Linolenic (9,12,15-octadecatrienoic)	$CH_3CH_2CH{=}(CHCH_2CH{=})_2CH(CH_2)_7COOH$
Arachidonic (5,8,11,14-eicosatetraenoic)	$CH_3(CH_2)_4CH{=}(CHCH_2CH{=})_3CH(CH_2)_3COOH$

Cyclic

Chaulmoogric (13-[2-cyclopentenyl] tridecanoic)

```
HC═══CH
|      |
H2C    CH(CH2)12COOH
  \   /
    C
    H2
```

cate a total of 18 carbon atoms, *-en-* to indicate a double bond, *-oic* as the systematic ending for a carboxylic acid). The essential fatty acids, those that must be supplied by the diet because they cannot be synthesized by the body, are the unsaturated acids—linoleic, linolenic, and arachidonic.

Chaulmoogric acid is a cyclic fatty acid of plant origin that has been used in the treatment of leprosy.

Prostaglandins, thromboxanes. Until recently, the nature of the essentiality of some fatty acids was entirely unknown. However, it has now been established that essential fatty acids are precursors in the biosynthesis of a class of naturally occurring substances known as *prostaglandins* (p. 322), the general structure of which is illustrated by the compound designated as PGE_2:

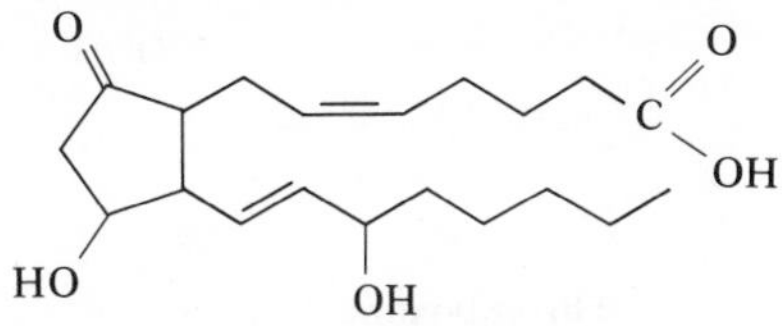

PGE_2

The nomenclature of the prostaglandins derives from differences in the ring structure, substituents, unsaturation in the ring, long side chains, and the spatial arrangements of the atoms and substituents. Four basic ring structures in the prostaglandins are indicated here:

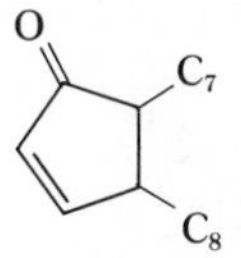

A 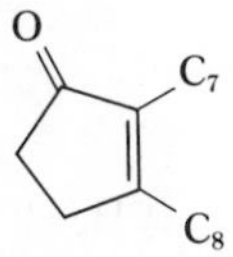 **B** 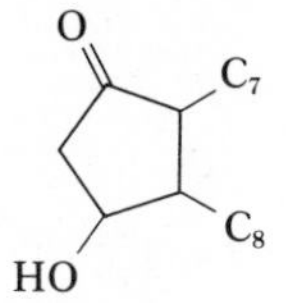 **E** 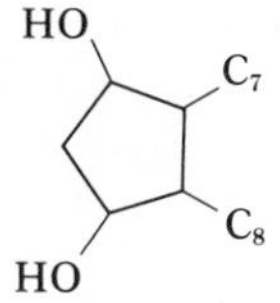 **F**

It is obvious that there are many opportunities for isomerism and the number of compounds described, natural and synthetic, is continually increasing.

Although first named "prostaglandins" by Ulf Svante von Euler (a Swedish Nobel prize–winning scientist) in 1935, these important substances have been the subject of intensive research mainly in recent years. The prostaglandins are made by many cell types and the list of physiologic effects attributed to them is still growing (p. 322).

One function in particular has been receiving increasing attention and that is the effect of the prostaglandins on blood clotting (p. 372). As a result of such studies, other derivatives of the prostaglandins, *prostaglandin endoperoxides, prostacyclin,* and *thromboxanes,* have been discovered. The thromboxanes are so named because they were first isolated from blood platelets, or thrombocytes, and contain an oxane ring. Examples of structures of these compounds are shown here:

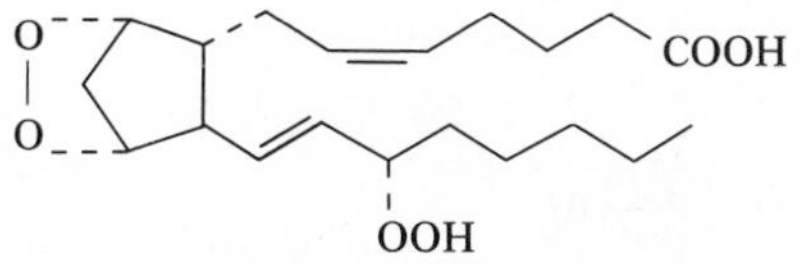

Prostaglandin endoperoxide
(PGG_2)

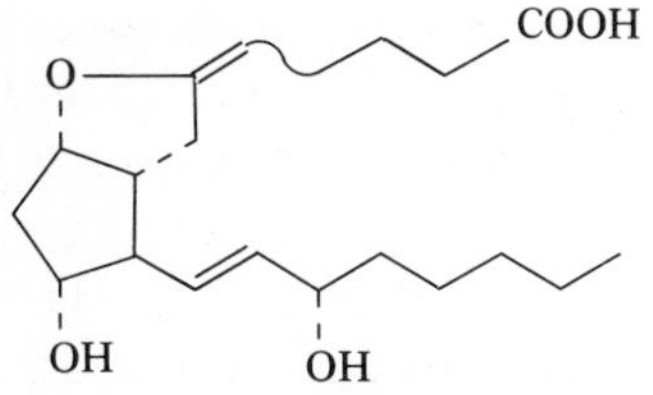

Prostacyclin
(PGI_2)

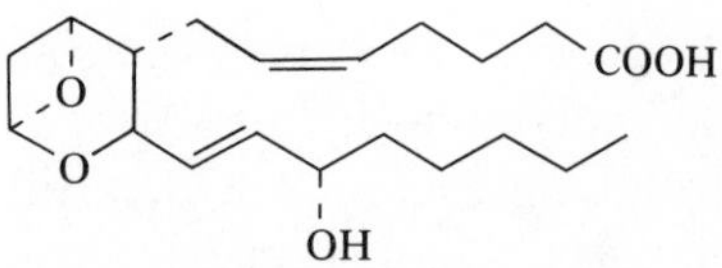

Thromboxane A_2
(TXA_2)

Waxes

The waxes are esters of fatty acids and alcohols other than glycerol and have higher molecular weights. A wax found in blood plasma, *cholesteryl palmitate,* contains the alcohol cholesterol, a member of the steroid series to be discussed later:

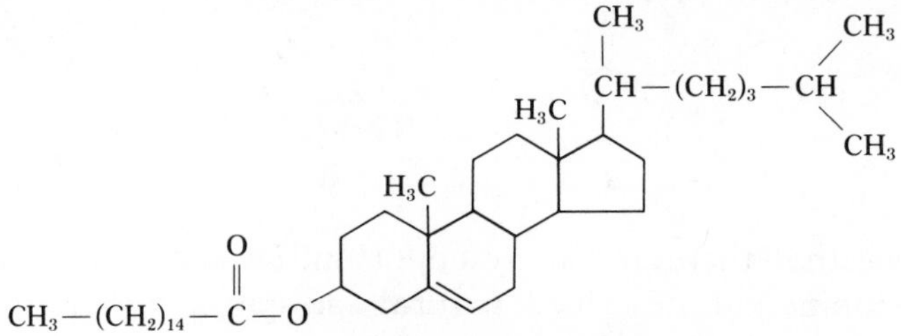

Cholesteryl palmitate

Some of the waxes found in skin are esters of hydroxylated fatty acids and open-chain alcohols.

COMPOUND LIPIDS
Phospholipids (phosphatides)

The phospholipids are composed of glycerol, fatty acids, phosphoric acid, and a basic nitrogen compound. These substances are also known as phosphatides and are sometimes named as derivatives of the parent compound, a *phosphatidic acid,* which is missing only the nitrogen portion of the phospholipids:

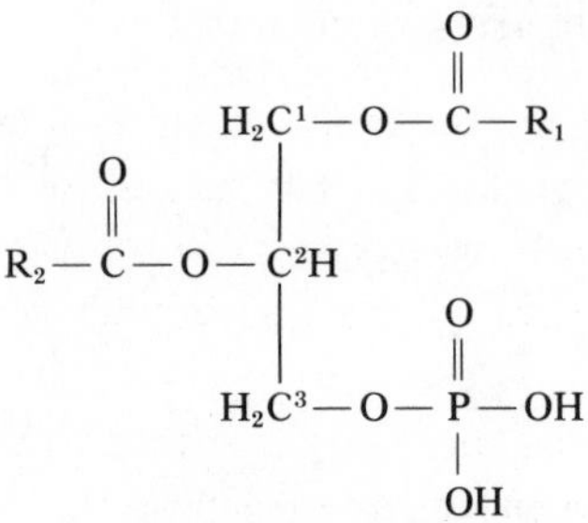

Phosphatidic acid

Examination of this general formula shows that the middle carbon atom (C-2) of the glycerol portion is asymmetric. The phosphatides are related to L-glycerolphosphate, the form in which the hydroxyl group on the 2-carbon is shown to the left (p. 326). Therefore, the group on the 2-carbon atom of the phosphatides should be written to the left when the phosphate group is at the bottom.

Lecithins. Lecithins are phosphatides of choline, which, under physiologic conditions, have the following general structure:

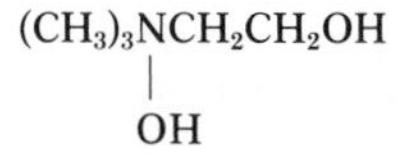

$$\begin{array}{l} (CH_3)_3NCH_2CH_2OH \\ \quad\quad\;\; | \\ \quad\quad\; OH \end{array}$$

Choline

$$\begin{array}{rcl} & & \quad\quad\quad\quad\; O \\ & & \quad\quad\quad\quad\; \| \\ & & H_2C-O-C-R_1 \\ O & & \;\;\, | \\ \| & & \;\;\, | \\ R_2-C-O- & & CH \\ & & \;\;\, | \quad\quad\quad\; O \\ & & \;\;\, | \quad\quad\quad\; \| \\ & & H_2C-O-P-OCH_2CH_2\overset{+}{N}(CH_3)_3 \\ & & \quad\quad\quad\quad\; | \\ & & \quad\quad\quad\quad\; O^- \end{array}$$

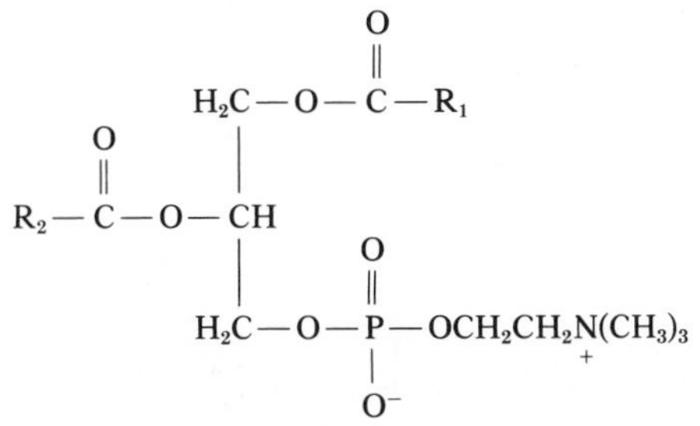

A lecithin

(Phosphatidyl choline)

Only a few of the fatty acids, such as palmitic, stearic, oleic, linoleic, linolenic, and arachidonic, are usually found in the lecithins. R_1 (in the general structure) is usually a saturated fatty acid, and R_2 is usually an unsaturated fatty acid.

Lecithins probably occur in all cells, and are found in especially high concentration in egg yolk, from which its name is derived (Gr., *lekithos,* yolk). The specific functions of the lecithins are under investigation with much interest centered on activities occurring in membranes, such as secretion of products by cells and transport of other substances into and out of cells. The lecithins are used commercially in chocolate candies and other products as emulsifying agents.

Certain enzymes in snake venom can cause the hydrolysis of the unsaturated fatty acids on the 2-carbon atom of phospholipids, resulting in the production of compounds known as *lysolecithins* (and *lysocephalins,* see later), substances that have strong hemolytic (red cell–destroying) action. Death may result if the hemolysis in the victim of a snake bite is extensive enough.

Cephalins. The cephalins are very similar to the lecithins, being phosphatides containing ethanolamine or serine (an α-amino acid) in place of choline.

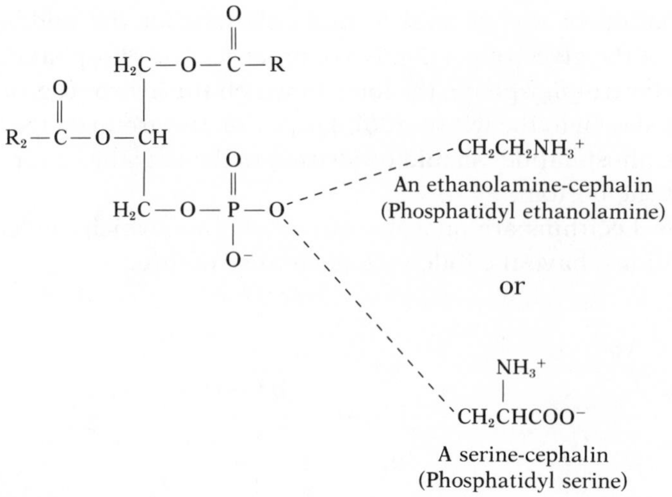

$$\begin{array}{l} \qquad\qquad\qquad\qquad O \\ \qquad\qquad\qquad\qquad \| \\ \qquad\qquad H_2C-O-C-R \\ \quad O \qquad\quad | \\ \quad \| \qquad\quad | \\ R_2-C-O-CH \qquad\qquad\qquad\qquad CH_2CH_2NH_3^+ \\ \qquad\qquad\quad | \qquad O \\ \qquad\qquad\quad | \qquad \| \\ \qquad\qquad H_2C-O-P-O- \\ \qquad\qquad\qquad\qquad | \\ \qquad\qquad\qquad\qquad O^- \end{array}$$

An ethanolamine-cephalin
(Phosphatidyl ethanolamine)

or

$$\begin{array}{c} NH_3^+ \\ | \\ CH_2CHCOO^- \end{array}$$

A serine-cephalin
(Phosphatidyl serine)

The cephalins were named because of their prevalence in brain tissue (Gr., *kephale,* head). Stearic, oleic, linoleic, and arachidonic acids have been found as fatty acid constituents of the cephalins. Enzymes in snake venom can cause the formation of lysocephalins (see lysolecithins, before). The cephalins have been implicated in the process of blood coagulation (clotting, p. 372).

Sphingomyelins. The sphingomyelins are composed of a complex basic amino alcohol, *sphingosine,* with a fatty acid in amide linkage on the amino group and the phosphorylcholine group attached by way of the terminal alcohol group:

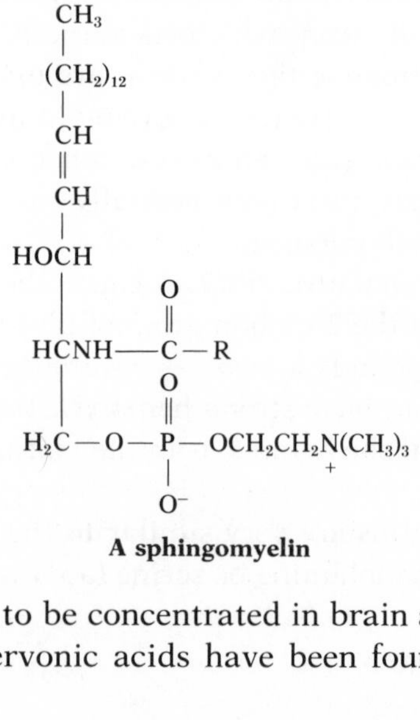

$$\begin{array}{l} CH_3 \\ | \\ (CH_2)_{12} \\ | \\ CH \\ \| \\ CH \\ | \\ HOCH \qquad O \\ | \qquad\qquad \| \\ HCNH-C-R \\ | \qquad\qquad O \\ | \qquad\qquad \| \\ H_2C-O-P-OCH_2CH_2\overset{+}{N}(CH_3)_3 \\ \qquad\qquad | \\ \qquad\qquad O^- \end{array}$$

A sphingomyelin

Sphingomyelins seem to be concentrated in brain and nerve tissue. Stearic, lignoceric, and nervonic acids have been found in sphingomyelins.

Glycolipids

The glycolipids are so named because a portion of the molecule is formed by a carbohydrate.

Cerebrosides. Cerebrosides are sphingoglycolipids that contain only simple sugars, usually galactose, in place of the phosphorylcholine of sphingomyelin. The portion of the molecule leaving out the sugar is known as *ceramide:*

```
                                       O           Fatty  ⎫
                                   H   ‖          / acid  ⎪
                             OH    N — C — R ────/        ⎬ Ceramide
                             |     |                      ⎪
CH3(CH2)12CH = CH — CH — CH — CH2O ───────────────────────⎭
                                                 |
                                                HC ──────────┐
                                                 |           |
                                                HCOH         |
                                                 |           |
                                               HOCH          O
                                                 |           |
                                               HOCH          |
                                                 |           |
                                                HC ──────────┘
                                                 |
                                               H2COH
```

A cerebroside (sphingoglycolipid)

(Ceramide-galactose)

The cerebrosides are found in relatively large concentration in white matter of brain and in the myelin sheaths of nerves. The base in most cerebrosides is sphingosine, but dihydrosphingosine (saturated sphingosine) is sometimes found. The fatty acid components usually found are lignoceric acid in *kerasin(e) (cerasine)*, α-hydroxylignoceric acid in *phrenosin(e) (cerebron)*, nervonic acid in *nervon*, and α-hydroxynervonic acid in *hydroxynervon (oxynervon)*. Palmitic and stearic acids have also been found. The carbohydrate portion may consist of a single carbohydrate molecule of galactose or glucose or a chain of these sugars. The more complicated types of glycolipids have various derivatives of the sugars attached to the ceramide portion in place of the simple sugars.

Lipoproteins

There are certain substances in which it appears that lipids are combined with proteins. However, the nature of the linkage between the two portions is still under active investigation. Studies with these substances have been greatly hampered by the difficulties encountered in attempts to obtain pure samples or to handle these compounds in the natural state.

The various types of lipids found in the bloodstream are believed to be carried as lipoprotein complexes. The plasma lipoproteins have been separated into two subgroups, the low-density lipoproteins, with densities below 1.063, and the high-density lipoproteins, with densities between 1.063 and 1.20. Of interest is the observation that the concentration of some of these lipoproteins changes depending on the clinical state of the individual. Whether or not these changes are directly related to specific disease conditions is not presently known.

Also under continuing investigation is the relationship between the lipoproteins and the development of atherosclerosis or the susceptibility to heart attacks.

Much of the current interest in lipoproteins is related to the possible function of these substances in cell membranes and other intracellular components such as the nucleus and the mitochondria.

STEROIDS

Compounds containing in their structure the cyclopentanoperhydrophenanthrene condensed ring system are known as *steroids:*

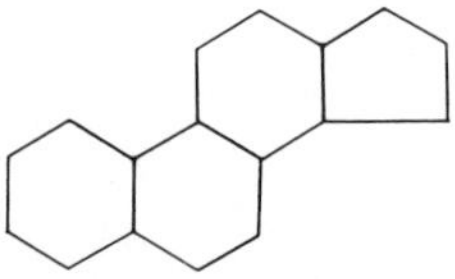

Steroid nucleus

(Cyclopentanoperhydrophenanthrene)

Among the different types of substances containing this structure are vitamins, hormones, drugs and poisons, bile acids, and the sterols. Although the steroid nucleus is common to the groups named, the physiologic properties differ widely. The sterols, to be discussed next, are generally associated with lipids in nature, possibly because of their solubility characteristics.

Sterols

Probably the most talked about representative of the sterols, high molecular weight cyclic alcohols, is cholesterol, already discussed as a component of waxes. Its recent notoriety is derived from its relationship to atherosclerosis (hardening of the arteries), a disease in which it is found deposited in the walls of the arteries, destroying the elasticity normally present. Cholesterol is also well known as a constituent of gallstones.

Cholesterol, which is synthesized by the animal body from small units such as acetic acid, is itself used for the synthesis of steroid hormones (p. 451) by certain endocrine glands. It can also be converted to 7-dehydrocholesterol in the animal body, and when it reaches the skin it is changed to vitamin D_3, cholecalciferol, by ultraviolet radiation from sunlight. Ergosterol, a plant sterol similar to cholesterol, is converted to calciferol, vitamin D_2, by irradiation.

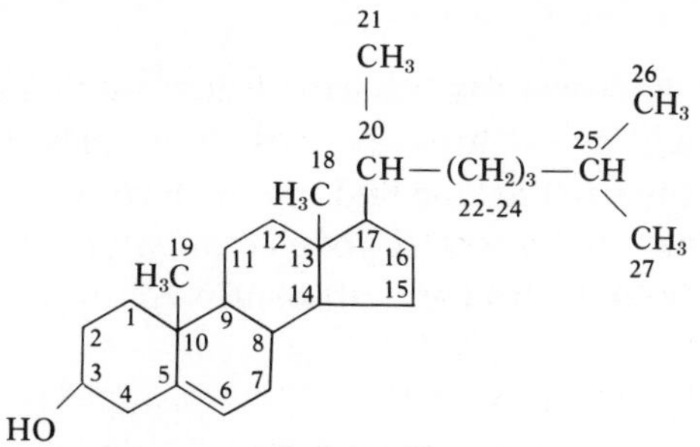

Cholesterol

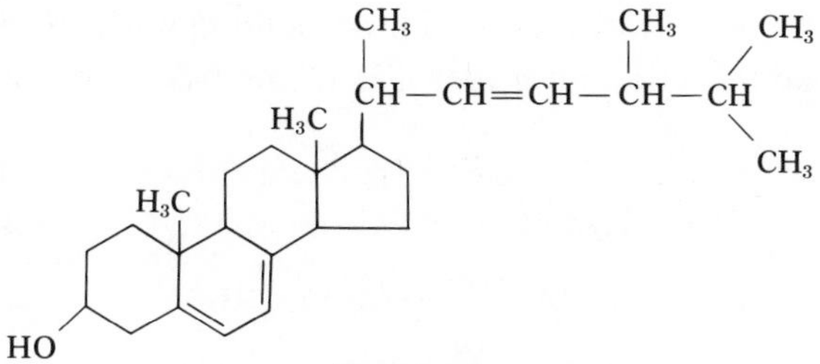

Ergosterol

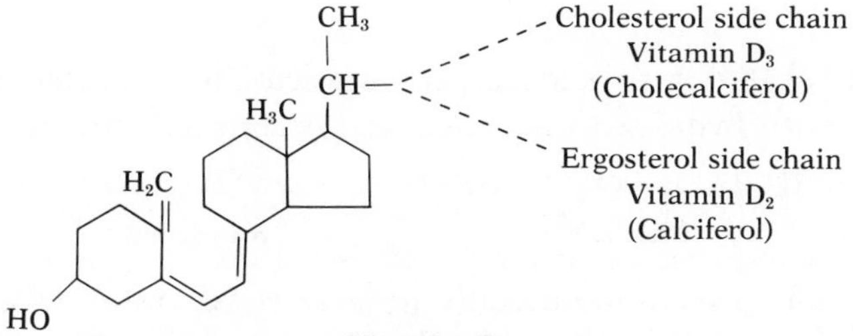

Cholesterol side chain
Vitamin D_3
(Cholecalciferol)

Ergosterol side chain
Vitamin D_2
(Calciferol)

Vitamins D

Bile acids

The bile acids have a carboxyl group on the side chain and additional hydroxyl groups:

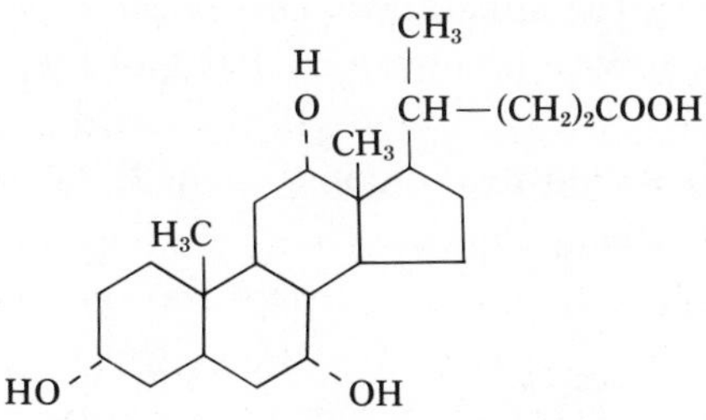

Cholic acid

These acids are found in bile combined (conjugated) by way of their carboxyl groups with the amino acid glycine (glycocholic acids) or a derivative of the amino acid cysteine, taurine (taurocholic acids). At the pH of the bile, the bile acid conjugates are present as salts. Thus, with an ionic group (water-soluble salt) at one end and a sterol nucleus (soluble in lipids) at the other end, the bile salts act as emulsifying agents and, as such, are important in the digestion and absorption of lipids and their hydrolysis products.

CHEMICAL PROPERTIES OF LIPIDS

Acrolein formation

When glycerol is heated in the presence of a dehydrating agent such as sodium bisulfate, an unsaturated aldehyde, *acrolein*, is produced. Acrolein has the characteristic unpleasant odor associated with the burning of fats. The reaction occurs whether glycerol is in the free or combined state.

$$\begin{matrix} H_2COH \\ | \\ HCOH \\ | \\ H_2COH \end{matrix} \xrightarrow[\Delta]{NaHSO_4} \begin{matrix} H \\ C{=}O \\ | \\ CH \\ \| \\ CH_2 \end{matrix} + 2H_2O$$

Glycerol **Acrolein**

Hydrogenation

The unsaturated fatty acids can be converted to saturated compounds by reaction with hydrogen, e.g., oleic acid is converted to stearic acid:

$$\underset{\textbf{Oleic acid}}{CH_3(CH_2)_7CH{=}CH(CH_2)_7COOH} + H_2 \rightarrow \underset{\textbf{Stearic acid}}{CH_3(CH_2)_{16}COOH}$$

This reaction is used commercially to convert vegetable oils (high concentration of unsaturated fatty acids, therefore low melting points) to solid forms (shortenings).

Iodine number

It should be recalled that halogens also can add to double bonds. Therefore the amount of iodine that reacts with a lipid can be a measure of the unsaturation of the substance. The *iodine number* is defined as the number of grams of iodine taken up by 100 g of fat.

Hydrolysis

Fats can be hydrolyzed to the constituent fatty acids and glycerol by acids, bases, or enzymes:

$$\begin{matrix} H_2C{-}O{-}\overset{O}{\overset{\|}{C}}{-}R_1 \\ | \\ HC{-}O{-}\overset{O}{\overset{\|}{C}}{-}R_2 \\ | \\ H_2C{-}O{-}\overset{O}{\overset{\|}{C}}{-}R_3 \end{matrix} + 3H_2O \rightarrow \begin{matrix} H_2COH \\ | \\ HCOH \\ | \\ H_2COH \end{matrix} + \begin{matrix} R_1COOH \\ \\ R_2COOH \\ \\ R_3COOH \end{matrix}$$

Fat **Glycerol** **Fatty acids**

Saponification. The alkaline hydrolysis of a fat is known as *saponification* because among the products formed are the alkali metal salts of the fatty acids, known as soaps. The reaction may be considered as occurring in two stages, first the hydrolysis to give the free fatty acids, as shown, then the neutralization reaction between the fatty acids and the alkali to form the soap:

$$\mathrm{RC(=O)-OH} + \mathrm{NaOH} \rightarrow \underset{\textbf{Soap}}{\mathrm{RC(=O)-ONa}} + H_2O$$

In practice, the two stages cannot be separated. The sodium salts produce hard soaps, and the potassium salts produce soft soaps. Soaps owe their cleansing action to their emulsifying properties, which are the result of the water-soluble nature of the salt end of the molecule and the lipid-soluble nature of the hydrocarbon end of the molecule. The insoluble soaps of calcium and magnesium are formed in hard water, thereby destroying the cleansing action of soap.

The *saponification number* for a fat is defined as the number of milligrams of potassium hydroxide required to completely hydrolyze 1 g of fat. On this basis, the higher the saponification number, the smaller will be the fatty acid chains found in the molecule.

Rancidity. Fats may become *rancid* on standing, i.e., they develop an unpleasant odor or taste. This may result from either or both of two degradative processes, *hydrolytic* or *oxidative.* In the hydrolytic process, enzymes from microorganisms in the fat hydrolyze the fat, forming free fatty acids and glycerol. If there is a sufficiently high concentration of the lower fatty acids, such as butyric acid, in rancid butter, a rather unpleasant odor develops. In the oxidative process, exposure to the oxygen of air results in the formation of short-chain aldehydes or ketones with unpleasant odors and tastes. Antioxidants, i.e., substances that will undergo oxidation more easily than the components of fat, protect the fats from this type of rancidity. Commercially, the tocopherols, E vitamins, are used for such a purpose in many modern food preparations.

Miscellaneous tests

There are specialized tests available for the detection and quantitative measurement of choline, cholesterol, and the bile acids that may be found in appropriate reference books.

SUMMARY

Lipids are composed of carbon, hydrogen, and oxygen with nitrogen and phosphorus occurring in some types. Fatty acids constitute an important

the structure of most lipids. Lipids are generally insoluble in water
luble in organic solvents.

Simple lipids
A. Fats (glycerides)
B. Waxes
I. Compound lipids
A. Phospholipids (phosphatides)
1. Lecithins
2. Cephalins
3. Sphingomyelins
B. Glycolipids
1. Cerebrosides
C. Lipoproteins
III. Steroids
A. Sterols
B. Bile acids

SIMPLE LIPIDS

Fats (glycerides): Fats consist of glycerol esterified with three open-chain fatty acids. *Mixed glycerides* have different acids; *simple glycerides* have all three acids the same. Saturated fats have higher melting points than unsaturated fats.

Glycerol: Soluble in water, insoluble in lipid solvents. Monoglycerides (glycerol esterified with one fatty acid) make good emulsifying agents.

Fatty acids: Most insoluble in water, soluble in lipid solvents. Almost all natural fatty acids are open-chain compounds and have an even number of carbon atoms. Palmitic, stearic, and oleic acids are the most common ones in animals. The essential fatty acids, linoleic, linolenic, arachidonic, all unsaturated, must be obtained in the diet.

Prostaglandins, thromboxanes: Prostaglandins are synthesized by most cells from essential fatty acids. Structure includes five-membered ring and two long side chains. Many physiologic functions are being attributed to prostaglandins. Thromboxanes and prostacyclin are involved in the blood clotting process.

Waxes: Esters of fatty acids and alcohols other than glycerol and have higher molecular weight, e.g., cholesterol.

COMPOUND LIPIDS

Phospholipids (phosphatides): Composed of glycerol, fatty acids, phosphoric acid, basic nitrogen compound. Parent compound is phosphatidic acid, esterified with fatty acids on the 1- and 2-carbons of glycerol and with phosphoric acid on the remaining carbon. Phosphatides are related to L-glycerolphosphate.

Lecithins: Phosphatides of choline. Present in all cells and may be important in membrane functions. Enzymes in snake venom can hydrolyze fatty acid on the middle carbon (C-2) producing a *lysolecithin,* a substance that causes hemolysis (destroys red cells).

Cephalins: Phosphatides of ethanolamine or serine. Found in brain tissue and function in blood clotting. *Lysocephalins* can also be formed by snake venom.

Sphingomyelins: Composed of amino alcohol, sphingosine, fatty acid in amide linkage on amino group, and phosphorylcholine on the terminal alcohol group. Concentrated in brain and nervous tissue.

Glycolipids: Lipids combined with carbohydrate portion.

Cerebrosides: Sphingoglycolipids that contain simple sugars (glucose or galactose) in place of phosphorylcholine of sphingomyelin. Cerebrosides are found in white matter of brain and myelin sheaths of nerves. More complicated types have derivatives of sugars and chains of sugars in place of the simple sugars.

Lipoproteins: Lipids combined with protein portion. Believed to be involved in transport of lipids and in membrane functions. Blood concentration of lipoproteins changes in disease conditions.

STEROIDS

Compounds containing the cyclopentanoperhydrophenanthrene condensed ring system:

Steroid nucleus

Steroid structure is found in vitamins, hormones, drugs, bile acids, and sterols.

Sterols: Cyclic alcohols with high molecular weight, e.g., cholesterol. Cholesterol is found deposited in arterial walls in atherosclerosis and forms gallstones in bile. Cholesterol is synthesized by animals from small units such as acetic acid and is used to synthesize steroid hormones and vitamin D_3, cholecalciferol, by the body. Vitamin D_2, calciferol, is made from the plant sterol, ergosterol.

Bile acids: Have more hydroxyl groups than cholesterol and a carboxyl group on side chain. In bile, are conjugated with glycine or taurine. At pH of bile, bile acids are in the form of salts and are good emulsifying agents, aiding in the digestion and absorption of lipids.

CHEMICAL PROPERTIES OF LIPIDS

Acrolein formation: Acrolein, $CH_2{=}CHCHO$, an unsaturated aldehyde with unpleasant odor, is formed when free or combined glycerol is heated with dehydrating agent ($NaHSO_4$).

Hydrogenation: Unsaturated fats will add hydrogen across double bonds and will become more solid (higher melting point).

Iodine number: Number of grams of iodine taken up by 100 g of fat. This is a measure of the unsaturation of the fat.

Hydrolysis: Fats are hydrolyzed by acids, bases, and enzymes.

Saponification: Alkaline hydrolysis produces soaps, i.e., sodium and potassium salts of the fatty acids. Soaps have cleansing action because of their emulsifying properties. Saponification number = mg KOH required to saponify 1 g of fat. The higher the saponification number, the smaller the fatty acids in the fat.

Rancidity: Two major types: (1) hydrolytic—bacterial enzyme action liberates short chain fatty acids with unpleasant odors and tastes; and (2) oxidative—exposure to oxygen of air results in breakdown to short chain aldehydes or ketones with unpleasant odors and tastes. Antioxidants such as vitamin E are used to prevent oxidative rancidity.

Miscellaneous tests: Tests are also available for the detection and quantitative measurement of choline, cholesterol, and bile acids.

REVIEW QUESTIONS

1. Define a fatty acid in chemical terms. Why are some fatty acids termed essential fatty acids? What are prostaglandins?
2. What are the similarities in structure of the lipids and phospholipids? Differences? Can they be distinguished by the acrolein test?
3. What type of structure characterizes the steroids? How are the locations of substituents indicated? Name some classes of biologically important compounds that contain the steroid nucleus.
4. Where are lipoproteins found in the body? What might be their possible significance in the various locations?
5. What type of substance is responsible for the unpleasant odor of rancid fats? How do these substances arise from fats? Can this process be prevented? Explain.

REFERENCES

Advances in lipid research (series), New York, Academic Press, Inc.

Annual review of biochemistry, Palo Alto, Calif., Annual Reviews.

Barry, J. M., and Barry, E. M.: An introduction to the structure of biological molecules, Englewood Cliffs, N.J., 1969, Prentice-Hall, Inc.

Florkin, M., and Stotz, E. H., editors: Comprehensive biochemistry, Vols. 6, 10, 18, Amsterdam, 1963-1971, Elsevier.

Holmes, W. L., and Bortz, W. M., editors: Biochemistry and pharmacology of free fatty acids, Basel, Switzerland, 1971, S. Karger.

Marinetti, G. V., editor: Lipid chromatographic analysis, New York, 1976, Marcel Dekker, Inc.

Ramwell, P. W., editor: The prostaglandins, New York, 1973-1977, Plenum Publishing Corp.

Chemistry of proteins

16

The proteins are complex compounds with high molecular weights. They are made up of approximately twenty naturally occurring amino acids acting as the basic building units. These amino acids are usually composed of carbon, hydrogen, oxygen, and nitrogen, with sulfur and iodine appearing in a few. Every protein has a very specific order in which the unit amino acids must be linked together and may consist of one or more chains per functioning molecule. Drawing from a pool of some twenty amino acids, which theoretically may be put together in an almost limitless number of ways, it should be obvious that the possible number of compounds and the complexity of structures are much greater for the proteins than they are for the carbohydrates or lipids. Many of these possibilities have been achieved by nature in the form of molecules having molecular weights that vary from the tens of thousands to many millions and vary in shape from spheres to long, thin strands.

On a functional basis, the proteins have long been recognized to be of prime importance, as the derivation of the name implies (Gr., *proteios*, primary). Proteins have important roles in cell walls and other membranes, in the liquid portion of the cells as well as the various particles and other structures in the cell, in blood, in connective tissue, in muscles, as enzyme catalysts, and as hormones that regulate various processes occurring in the body.

Because of their large size, proteins are colloidal substances that do not pass through semipermeable membranes (p. 87). Most proteins are insoluble in organic solvents, while a few are soluble in water and others are soluble in the presence of some salts, acids, or bases. Other important properties will be discussed later.

NITROGEN CYCLE

As one of the constituent elements of amino acids, nitrogen takes on strategic importance because of the relative scarcity of sources available to plants, which produce some of the essential amino acids that cannot be synthesized in the animal body. Dietary requirements for these important compounds can be met by eating plants (or animals that have eaten plants) that do produce these essential amino acids. The recycling of nitro-

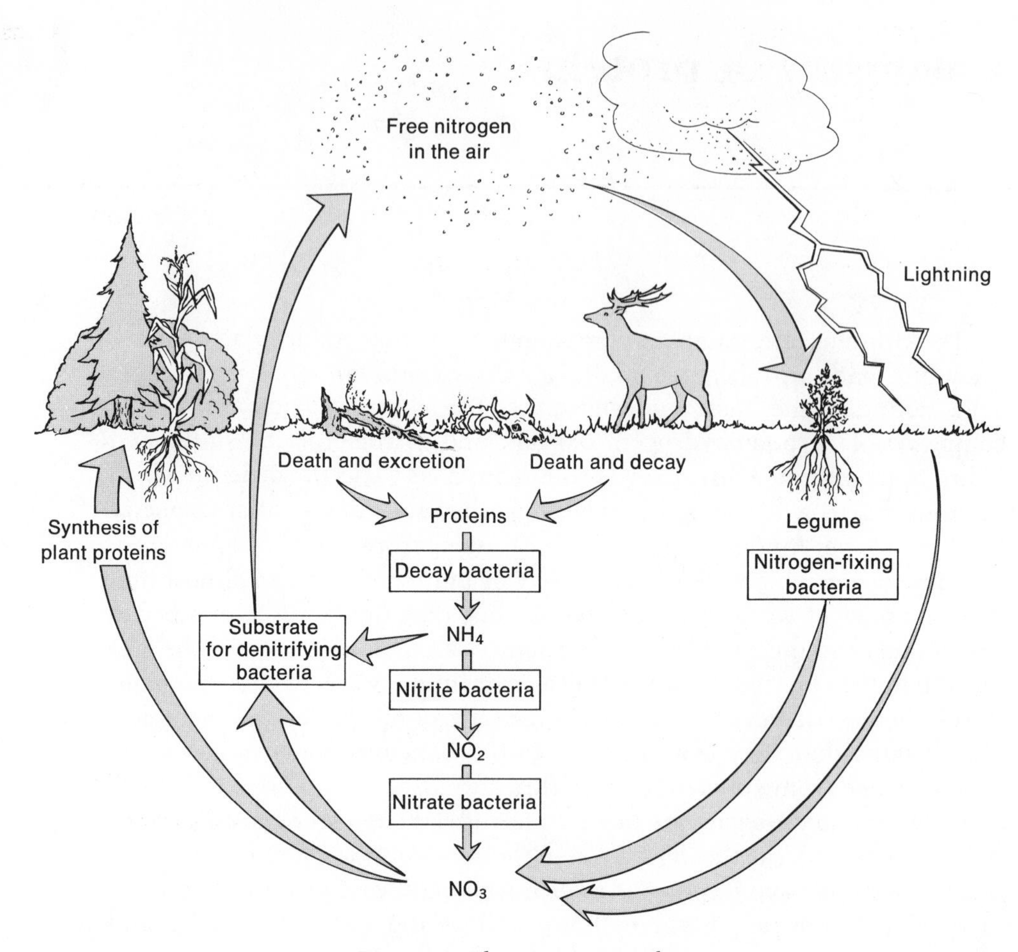

Fig. 16-1. The nitrogen cycle.

gen in nature is illustrated in Fig. 16-1. The important role of certain types of bacteria in the nitrogen cycle should be noted.

AMINO ACIDS

As already noted, proteins are made up of many amino acid units combined to form large molecules. The constituent amino acids may be obtained by hydrolysis of a protein. The specific amino acids usually found in proteins are classified here. Examination of the structural formulas will show that all of these are *α-amino acids,* i.e., the amino group is on the α-carbon atom, the carbon next to the carboxyl group. Only rarely is a β-amino acid encountered in nature.

Classification

I. Aliphatic amino acids
 A. Neutral (one carboxyl, one amino group)
  1. Simple amino acids

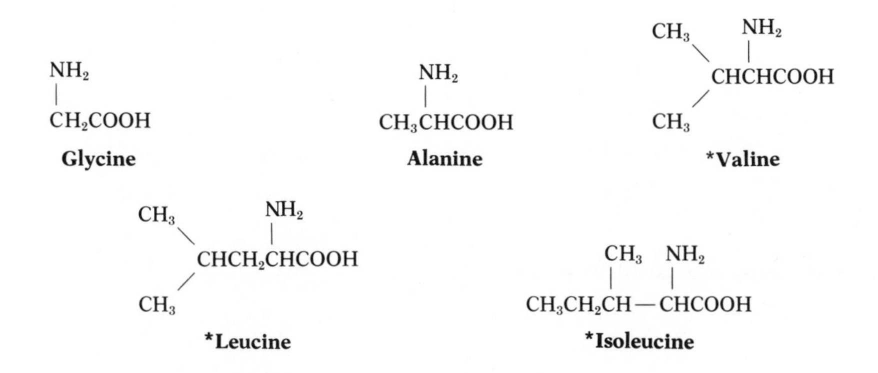

  2. Sulfur-containing amino acids

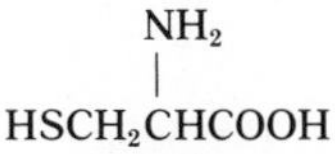

NH_2
|
$HSCH_2CHCOOH$

Cysteine

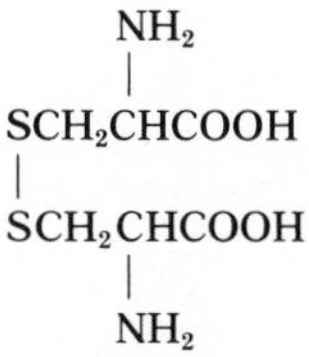

NH_2
|
$SCH_2CHCOOH$
|
$SCH_2CHCOOH$
|
NH_2

Cystine

NH_2
|
$CH_3SCH_2CH_2CHCOOH$

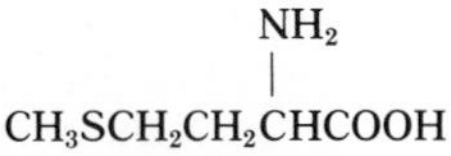

***Methionine**

  3. Hydroxyl-containing amino acids

 B. Acidic amino acids (two carboxyls, one amino group)

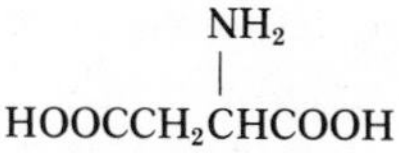

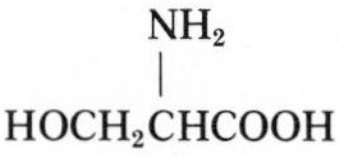

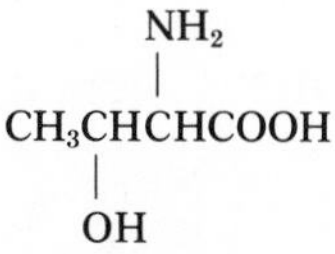

NH_2
|
$HOOCCH_2CHCOOH$

Aspartic acid

NH_2
|
$HOOCCH_2CH_2CHCOOH$

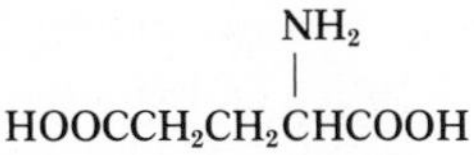

Glutamic acid

---

*Essential amino acids that must be obtained in the diet.

C. Basic amino acids (one carboxyl, two amino groups)

$$H_2N-\overset{\overset{\displaystyle NH}{\|}}{C}-NH-CH_2CH_2CH_2\overset{\overset{\displaystyle NH_2}{|}}{C}HCOOH$$

***Arginine**

$$H_2NCH_2CH_2CH_2CH_2\overset{\overset{\displaystyle NH_2}{|}}{C}HCOOH$$

***Lysine**

$$H_2NCH_2\overset{\overset{\displaystyle OH}{|}}{C}HCH_2CH_2\overset{\overset{\displaystyle NH_2}{|}}{C}HCOOH$$

Hydroxylysine

II. Aromatic amino acids

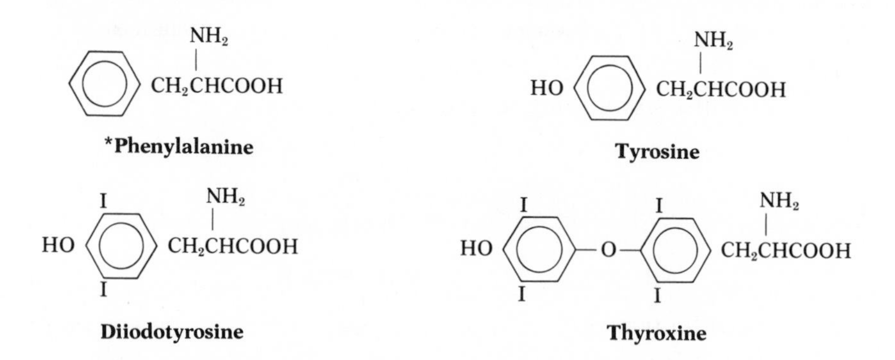

III. Heterocyclic amino acids

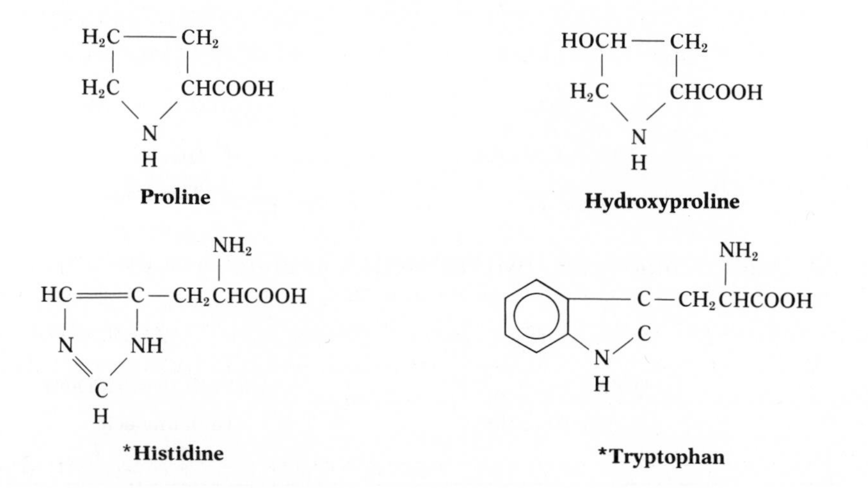

*Essential amino acids must be obtained in the diet.

Amphoterism of amino acids—buffer action

Although a simple amino acid such as glycine, in a solution of approximately neutral pH is electrically neutral, the molecule actually has two regions of charge rather than no charge. In this form, the carboxyl group is negatively charged and the amino group is positively charged, giving the molecule a net charge of zero:

$$\begin{array}{l} CH_2COO^- \\ | \\ NH_3^+ \end{array}$$

Dipolar ions of this type are known as *zwitterions*. In this form, the amino acid will not migrate in an electrical field, and the pH at which this situation occurs is known as the *isoelectric point* (pI).

Because of their zwitterion structure, amino acids can react both as weak acids and as weak bases. Substances that can react this way are said to have *amphoteric* properties. The ionic changes that take place in such reactions may be written as follows, using glycine as an example:

$$\begin{array}{lclcll} CH_2COOH & & CH_2COO^- & & CH_2COO^- & \\ | & \xleftarrow{H^+} & | & \xrightarrow{OH^-} & | & + H_2O \\ NH_3^+ & & NH_3^+ & & NH_2 & \\ \textbf{Glycine (+)} & & \textbf{Glycine (±)} & & \textbf{Glycine (−)} & \\ \text{(acidic to pI)} & & \text{(at pI)} & & \text{(basic to pI)} & \end{array}$$

Thus, at a pH acidic to the isoelectric point, the glycine becomes positively charged because the ionization of the carboxyl group is suppressed, and at a pH basic to the isoelectric point the glycine becomes negatively charged as the titratable hydrogen ion is removed from the ammonium group of the zwitterion to form water. In an electrical field, the positively charged form will move toward the cathode, and the negatively charged form will move toward the anode. These reactions can be reversed as follows:

$$\begin{array}{lclcl} CH_2COOH & \overset{H_2O}{\uparrow} & CH_2COO^- & & CH_2COO^- \\ | & \xrightarrow{OH^-} & | & \xleftarrow{H^+} & | \\ NH_3^+ & & NH_3^+ & & NH_2 \\ \textbf{Glycine (+)} & & \textbf{Glycine (±)} & & \textbf{Glycine (−)} \end{array}$$

As a result of these reactions, amino acids act as buffers (p. 119), the more positive form representing the weak acid and the more negative form representing the salt of the weak acid. The buffering properties of proteins derive from such properties of the individual amino acids and are of strategic importance to the well-being of the animal.

Asymmetry of amino acids

Examination of the structural formulas of the amino acids will show that in all cases, except for glycine, the α-carbon atom has four different atoms or groups attached to it, i.e., the α-carbon is asymmetric (p. 130).

L-Serine, which serves as the reference compound for the amino acids, has the same configuration as L-glyceraldehyde. This means that when the carboxyl group is placed on top, the amino group of the L-amino acids is written to the left and for the D-amino acids, the amino group goes to the right:

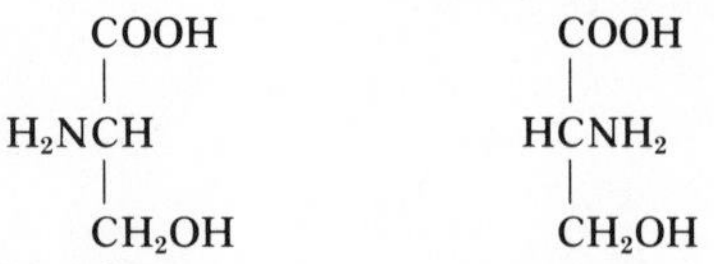

$$\begin{array}{c} COOH \\ | \\ H_2NCH \\ | \\ CH_2OH \end{array} \qquad \begin{array}{c} COOH \\ | \\ HCNH_2 \\ | \\ CH_2OH \end{array}$$

L-Serine **D-Serine**

As with the carbohydrates, the direction of optical rotation is indicated by (+) or (−) and is not directly related to the L- or D-configuration. Almost all of the amino acids found in nature are of the L-series. Some D-amino acids are found in products synthesized by bacteria and used as antibiotics.

PROTEINS

Peptide bond

The type of bond that unites the constituent amino acids of large protein molecules is known as the *peptide bond.* It is formed between the carboxyl group of one amino acid and the amino group of another amino acid, with the splitting out of water. As can be seen from the following equations, the peptide bond resembles an amide bond formed by the reaction of ammonia (or substituted ammonia) with a carboxylic acid:

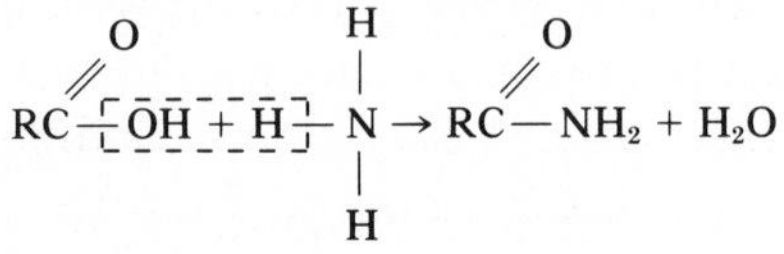

$$\underset{\text{Carboxylic acid}}{RC(=O)-OH} + \underset{\text{Ammonia}}{H-N(H)-H} \rightarrow \underset{\text{Amide}}{RC(=O)-NH_2} + H_2O$$

$$\underset{\text{Amino acid 1}}{R_1-C(NH_2)(H)-C(=O)-OH} + \underset{\text{Amino acid 2}}{H-N(H)-C(H)(R_2)-C(=O)-OH} \rightarrow R_1-C(NH_2)(H)-\underbrace{C(=O)-N(H)}_{\text{Peptide bond}}-C(H)(R_2)-C(=O)-OH + H_2O$$

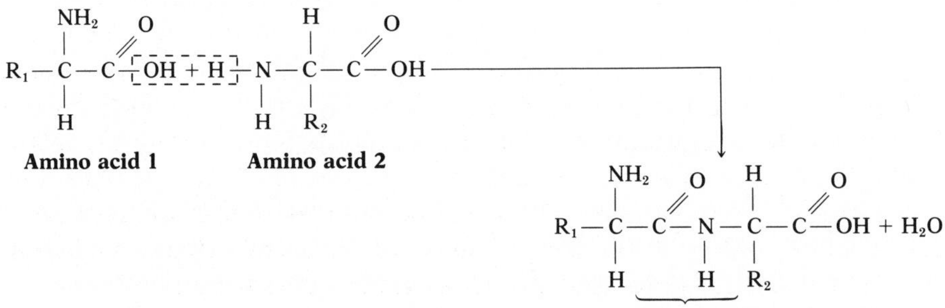

(Dipeptide)

By convention, peptides are written with the free amino end to the left. In naming peptides, the name of the amino acid residue (the amino acid as it exists in the peptide) with the free amino group comes first, with its ending changed to *-yl,* followed in order by any other amino acids named the same way, until the end amino acid with the free carboxyl group is reached, this amino acid getting its full name as the ending to the peptide name. The naming of the two peptides of glycine and alanine will make this clear:

```
   NH2  O   H    O                  NH2  O   H    O
   |   //   |   //                  |   //   |   //
H—C—C—N—C—C—OH              CH3C—C—N—C—C—OH
   |       |   |                    |       |   |
   H       H  CH3                   H       H   H
```

Glycylalanine **Alanylglycine**

Peptides formed from two amino acids are known as *dipeptides.* A confusing factor here is that a dipeptide has only one peptide bond. It must be remembered that the designations di-, tri-, etc., peptide, refer to the number of amino acid residues in the molecule and not the number of peptide bonds. The larger molecules made up of great numbers of amino acids are called *polypeptides.*

Classification

Proteins are such complex polypeptides that their classification has been and still is a difficult task. One system in general use is based on composition and solubility and will be used here. In this system the proteins are divided into three main groups: (1) simple proteins, (2) conjugated proteins, and (3) derived proteins. Other systems based on molecular size, shape, or biologic function are also possible.

Simple proteins. On hydrolysis, the simple proteins are broken down to amino acids only. Further subdivision of the simple proteins is on the basis of solubility characteristics.

Albumins and globulins. Some of the best known proteins are the albumins and globulins, e.g., plasma albumin and globulins, egg albumin (egg white), and lactalbumin (in milk). The albumins are more soluble in pure water than the globulins, which may require the presence of some salt to keep them in solution. Albumins and globulins are sometimes defined on the basis of the concentration of salt required to precipitate them from solution. Globulins are precipitated from solution by lower concentrations of salts such as ammonium sulfate or sodium sulfate than are required to precipitate albumins. The separation is not completely effective and some overlap of the albumins and globulins usually does occur.

Glutelins. Glutelins are insoluble in pure water and dilute salt solutions but are soluble in dilute acids or bases. An example of this group is glutenin from wheat.

Prolamines. Prolamines are insoluble in water and absolute alcohol but are soluble in 70% to 80% aqueous alcohol. Zein from corn and gliadin from wheat are examples of this group.

Scleroproteins (albuminoids). Scleroproteins are generally insoluble in the usual solvents and when solution does take place, it probably results from the breakdown of the native structure of the scleroproteins by the solvent. Examples of this group are keratin from nails, hair, horn, and feathers; collagen from tendons, skin, and bone; and elastin from ligaments.

Conjugated proteins. Conjugated proteins are composed of simple proteins combined with a nonprotein portion. This class is subdivided on the basis of the nature of the nonprotein portion, usually known as the *prosthetic group.*

Nucleoproteins. The prosthetic groups of the nucleoproteins are the nucleic acids, which will be discussed in detail in the next chapter. Among the proteins to which the nucleic acids are attached are the basic proteins, the protamines and the histones, which are discussed in the section on derived proteins. The nucleoproteins are in all cells—in the cytoplasm as well as the nuclei.

Phosphoproteins. The phosphoric acid found in phosphoproteins is linked to the protein as an ester by way of the hydroxyl group of serine. Milk casein is an example of this group of proteins.

Porphyrinoproteins (chromoproteins). The prosthetic group of the porphyrinoproteins is the complex heterocyclic (hetero nitrogen) porphyrin group, which is usually associated with a metallic atom. In hemoglobin (the oxygen-transporting compound of the blood), the porphyrin contains iron (p. 377), whereas in chlorophyll (the green photosynthesizing pigment of plants), the porphyrin contains magnesium in its structure.

Glycoproteins. Glycoproteins are proteins combined with polysaccharides (heterosaccharidic, amino-sugar type) and are often referred to as mucopolysaccharides. The mucins in saliva and associated with mucous membranes are examples of this group.

Lipoproteins. Lipoproteins are proteins combined with lipids and are discussed in Chapter 15. Good examples of this group are the lipoproteins of the blood plasma.

Derived proteins. Derived proteins result from the splitting of conjugated proteins, hydrolysis of simple proteins, and the alteration of the structure of native proteins—a process known as denaturation. It should be clear, then, that all derived proteins are obtained by some type of manipulation of naturally occurring proteins.

Protamines and histones. The protamines and histones result from the splitting of the nucleoproteins into their major components. The protamines are basic, water-soluble, relatively low–molecular weight proteins. The histones differ from the protamines on the basis of higher molecular weights.

Proteoses, peptones, peptides. Proteoses and peptones are mixtures of substances intermediate in weight between the proteins and smaller peptides and result from progressive enzymatic or chemical hydrolysis of proteins. The dividing line between proteoses (molecular weight, approximately 5000) and peptones (molecular weight, approximately 2000) is sometimes set on the basis of the precipitability of proteoses with ammonium sulfate. Some of the smaller peptides can be identified as individual chemical compounds.

Denatured proteins. Because of the relatively delicate nature of proteins and their complicated structures, it is a simple matter, in most instances, to change the structure of, or denature, naturally occurring proteins. In fact, in working with proteins, many precautions must be taken to ensure against denaturation of proteins when it is important to study the properties of a protein in its native state. Denatured proteins are usually significant only in a negative sense, i.e., they are unwanted.

STRUCTURAL CONSIDERATIONS

Levels of organization

The study of proteins and their functions in the animal body has developed, with convincing evidence, the hypothesis that the composition, organization, and shape of a protein molecule are directly related to its function. Research on this hypothesis has progressed relatively rapidly in recent years and some of the mysteries hidden in the complicated structures of proteins are beginning to be unraveled. Because of the complicated structures involved, a new terminology referring to levels of organization of protein molecules has been developed. These levels of organization are designated as primary, secondary, tertiary, and quaternary structures of proteins.

Primary structure. The primary structure of a protein involves the amino acid composition in a quantitative sense, i.e., how much of each different amino acid is present, and the specific sequence in which they appear in the molecule, linked in peptide bonds. With appropriate combinations of methods (some of which will be discussed later), such as chemical and enzymatic hydrolysis, chromatography, electrophoresis, and Sanger reaction for determining the end amino acid with the free amino group, the difficult task of determining the primary structure of a protein has been accomplished rather completely for insulin, hemoglobin, and a few other proteins. To gain an insight into the magnitude of the problems to be encountered in research of this type, it may be noted that hemoglobin (molecular weight, 68,000) is made up of four chains, two α chains and two β chains. The α chains each contain 141 amino acid residues and the β chains each contain 146 amino acid residues.

Secondary structure. Studies of structure usually involve determination of some aspects of the manner in which the amino acids are arranged in the protein chains. Some parts of the chain may exist in extended form

A

B

Fig. 16-2. A, Three-dimensional models (tertiary structure) of the α chain (white) and β chain (black) of hemoglobin. **B,** Three-dimensional model (quaternary structure) of hemoglobin, with the heme groups of the front white and black peptide (globin) chains indicated by gray discs. (**A,** from Perutz, M.: Science **140:**863, 1963. Also in *Nobel Lectures, Chemistry; 1942-1962.* Courtesy of the Nobel Foundation, Stockholm, Sweden, published by Elsevier, Amsterdam, 1964. **B,** from Perutz, M., and others: Nature **185:**416, 1960.)

while others exist in the form of coiled helices (as a spiral staircase). Study of this aspect of the structure of proteins involves intricate physical-chemical measurements with rather sophisticated and expensive equipment.

Tertiary structure. The three-dimensional picture, or the actual shape or conformation of the protein chain, and the atomic forces that keep the chain in its specific shape are the main concern of studies of tertiary structure. Secondary and tertiary structural considerations often overlap and may not always be differentiated one from the other. The tertiary structure of a hemoglobin chain is illustrated in Fig. 16-2. Much of the study of tertiary structure involves x-ray studies that enable the investigators to determine the exact position of each atom as a result of the deflection of the x-rays by the constituent atoms. This work involves hundreds of thousands of measurements and calculations and awaited the development of the high-speed computers before it could progress at a reasonable rate.

Quaternary structure. In a molecule such as hemoglobin, the three-dimensional picture of the total molecule involves not only the conformation or tertiary structure of each individual chain, but also the placement of each chain with respect to the other chains and the forces involved in keeping the chains in position, which is the concern of studies of quaternary structure. Much of this work also involves x-ray studies. The quaternary structures of the hemoglobin molecule are also shown in Fig. 16-2.

COLLOIDAL PROPERTIES OF PROTEINS

Because of their large size, proteins are colloidal particles. Thus, they have certain properties common to all colloidal substances as discussed in Chapter 8. The ability of colloids to exert an osmotic pressure through semipermeable membranes, the inability to pass through such membranes, the ability to adsorb smaller molecules, and the property of having electric charges are very important in enabling proteins to carry out their biologic functions. These properties should be reviewed at this point in preparation for the discussions in the enzyme and metabolism chapters that will point out specific functions for which these properties have made proteins so well suited.

SEPARATION OF AMINO ACIDS AND PROTEINS

Before any one of the approximately twenty amino acids or the thousands of specific proteins can be studied in the laboratory, it is almost inevitable that some types of separation procedures must be carried out. Some of the more widely used methods are described here briefly.

Chromatography

Chromatography, originally used to separate colored products, has been developed to a very precise state for the separation of amino acids and proteins under the proper conditions. Chromatography is generally carried out in columns or tubes or on flat supporting materials such as filter paper. For column chromatography, finely divided substances, such as powdered silica or aluminum oxide, which have large surface areas for adsorption of solutes passing by, are used. The solution containing the solutes to be separated is placed on top of the column and allowed to percolate slowly through the length of the column. Since the adsorption affinity of the column material is likely to be different for most solutes, the solutes will spread out in bands down the column, each band, in an ideal separation, being washed off the column by the solvent into a different container placed at the bottom of the column (Fig. 16-3). With the appropriate choice of conditions, adsorbent, solvent, percolation rate, and column size, many separations can be accomplished that might not be possible otherwise and can generally occur at a much faster rate than by strictly chemical methods.

Another type of column chromatography involves ion-exchange resins that exchange ions with the solutes passing by. Amino acids carry electrical charges depending on the pH of their solution and can be picked up by the ion-exchange resins under conditions which are different for the different amino acids. An amino acid that stays on the column under certain specific conditions while the others pass through can then be washed off the column by an appropriate change in the pH of the solvent.

In gel filtration chromatography, the gel particles can be considered as hollow particles that allow solute molecules of certain sizes to enter the hollow spaces while larger molecules are excluded and are washed off the column.

Another technique, gas chromatography, involves the volatilization of heat-stable substances that can be differentially absorbed along a packed tube in much the same manner as in the liquid method of column chromatography. All of these column chromatographic methods are now completely automated.

A more recent chromatographic technique, known as affinity chromatography, has rapidly become a very effective means of concentrating and purifying proteins and other substances. This procedure takes advantage of the high degree of specific affinity that some pairs of substances have for one another (e.g., enzymes-inhibitors, p. 254, or antigens-antibodies, p. 352). Typically, one of the substances is coupled to or immobilized on the column material and solutions of the other substance are passed through the column. As a molecule of the dissolved member of the pair meets a molecule of the immobilized member, they combine and remain on the column. The column can then be washed free of contaminating substances, after which the conditions can be changed so that the desired material can be dissociated from the immobilized partner and washed off

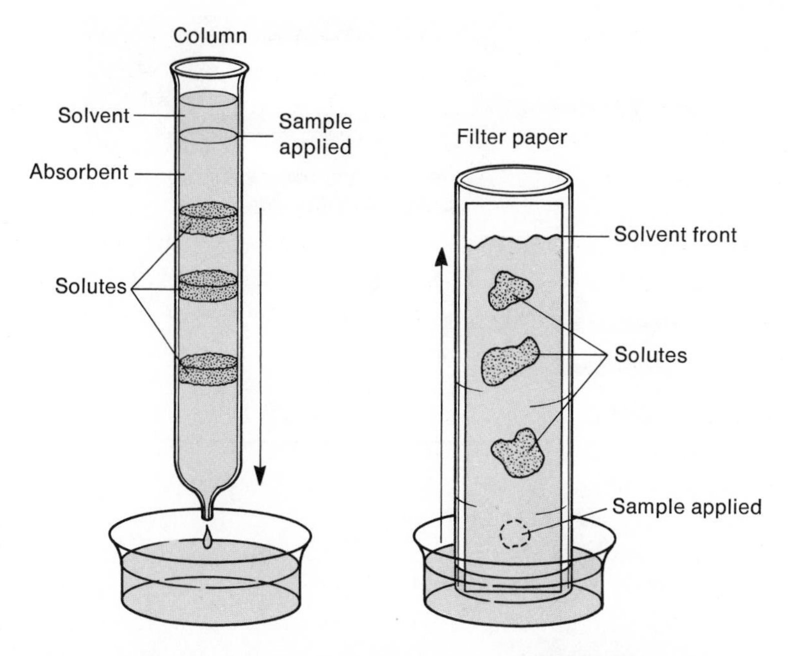

Fig. 16-3. Column and paper chromatography.

the column in much higher concentration and purity than in the original solution.

In paper chromatography, a mixture of amino acids or small peptides in solution may be spotted on a piece of filter paper that is placed in a solvent (Fig. 16-3). As the solvent front moves past the solute spot, the different solutes will be carried to different places on the paper depending mainly on their solubilities. Their positions on the paper can be determined by using an appropriate color-forming reagent, which for amino acids would be triketohydrindene hydrate (Ninhydrin). For any specific solvent, the relative positions of the amino acids or small peptides remain constant and may be identified on the basis of their R_f numbers, the R_f number being the ratio of the distance traveled by a particular solute to the distance traveled by the solvent front. Sometimes it is useful to chromatograph on a large square of filter paper in one direction using one solvent and then at right angles using a second solvent. This is known as two-dimensional chromatography and produces what may be termed an amino acid map. In thin-layer chromatography, which is similar to paper chromatography, some of the adsorbents used in column chromatography are spread out in thin layers on glass plates.

Electrophoresis

Because of their amphoteric properties, amino acids and proteins in solution bear a net electric charge except at their isoelectric points. There-

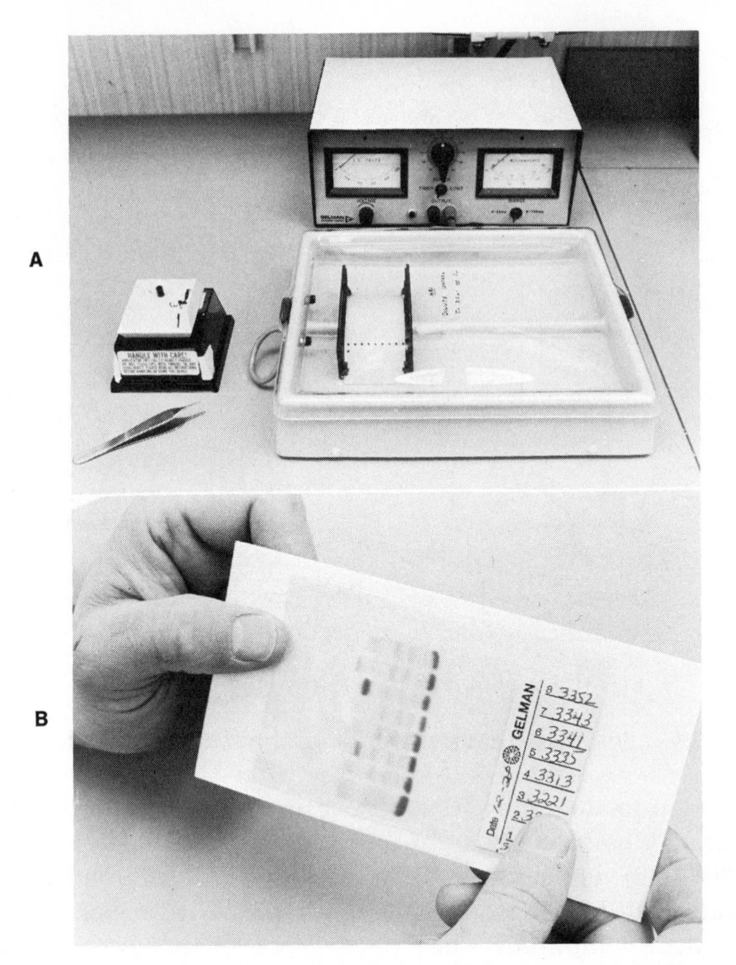

Fig. 16-4. A, Apparatus used for paper electrophoresis. **B,** Results from tests.

fore, under the proper conditions they can be made to migrate in an electrical field, a process known as *electrophoresis.* Since the isoelectric points of the various amino acids and proteins are different, the net charges under specific conditions will differ and the rates of migration will also differ. This makes it possible to effect a separation of components of a mixture.

Electrophoresis has been used with great success to separate the various components of plasma or serum proteins for both clinical and research purposes. By adjusting the pH of a particular sample until it is alkaline to the isoelectric points of all the protein fractions, each component will carry a net negative charge and will migrate toward the anode. The simplest form of electrophoresis uses a strip of filter paper or cellulose acetate as the supporting medium. The paper strip is wet with the appropriate buffer, the test solution is applied toward one end, the ends of the strip are placed in electrode vessels filled with the same buffer, and the electric field is applied. After an appropriate time, the paper strip is removed and stained to show the location of the components. An example of a separation of normal human serum components and the apparatus used in the process are shown in Fig. 16-4. Starch cakes, starch gel, and other gels are sometimes used instead of paper as supporting media. The procedure may also be carried out as *free* or *moving boundary* electrophoresis in a buffer solution, but this involves much more intricate and expensive equipment.

Ultracentrifugation

The high molecular weights of the proteins make it possible, by applying strong gravitational forces in the ultracentrifuge, to cause them to sediment. Because the rate at which sedimentation occurs is related to the molecular weight of a particular molecule, ultracentrifugation can be used not only to separate proteins but also to determine molecular weights.

Precipitation methods

Advantage has been taken of the differences in solubilities of various proteins when treated with suitable reagents to devise useful methods for separating such proteins. As already mentioned in connection with the criteria for classification of proteins, certain salts, such as sodium sulfate and ammonium sulfate, can be used to precipitate certain proteins while others remain in solution. Similar procedures using alcohol at low temperatures and heavy metals such as zinc have also been used successfully.

Dialysis

The discussion of the process of dialysis, used for the removal of small, crystalloidal contaminants from colloidal substances such as proteins, should be reviewed in Chapter 8.

DETECTION AND QUANTITATION OF AMINO ACIDS AND PROTEINS

Ninhydrin color reaction

The Ninhydrin color reaction is the test most generally used for the detection of amino acids. The color produced by amino acids varies from blue to violet. Although the Ninhydrin color reaction test is positive for all free amino groups in amino acids, peptides, or proteins, the test is much weaker for peptides or proteins because not as many free amino groups are available as in amino acids. Automated quantitative methods have been developed for amino acid analysis based on the Ninhydrin color reaction test:

$$\text{Ninhydrin } \left(C_6H_4(C{=}O)_2C(OH)_2\right) + R\overset{H}{\underset{NH_2}{C}}COOH \rightarrow C_6H_4(C{=}O)_2C{=}N{-}C(C{=}O)(C{-}O^-)C_6H_4 \text{ (Ninhydrin-color compound)}$$

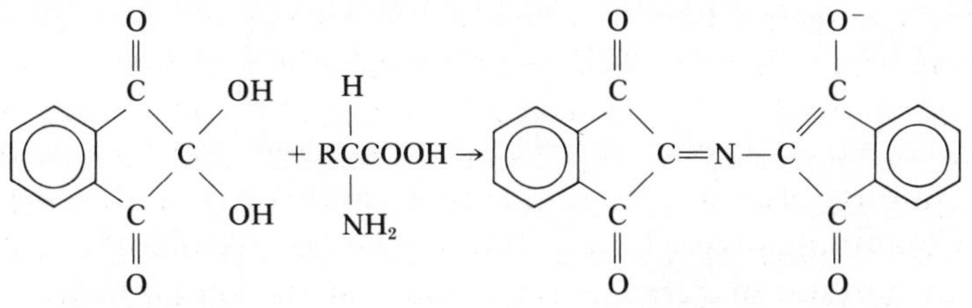

Ninhydrin **Ninhydrin-color compound**

Heat coagulation

Because of the ease with which most proteins can be coagulated, e.g., the white of a boiled egg, the formation of a coagulum on simple heating is often used as a presumptive test for the presence of a protein. By proper drying and weighing of the protein precipitate, the amount of protein may be determined.

Biuret reaction

With substances containing two or more peptide linkages, the biuret reagent (dilute copper solution in strong alkali) gives a blue-violet color. This reaction as well as various modifications have been used for quantitative determinations of proteins.

Total nitrogen—Kjeldahl method

The Kjeldahl method determines the nitrogen content of amino acids or proteins by digesting or converting the amino acid or protein in the presence of concentrated sulfuric acid, forming ammonium sulfate. The ammonia is forced out of solution by the addition of sodium hydroxide, and the amount of ammonia produced may then be measured accurately by titration with hydrochloric acid or by an appropriate color reaction:

$$\text{Protein} \xrightarrow{H_2SO_4} (NH_4)_2SO_4 + NaOH \downarrow NH_3 + HCl \rightarrow NH_4Cl$$

Because the average protein contains approximately 16% nitrogen, the protein content of a test sample can be determined by multiplying the percent nitrogen by the factor 6.25 (i.e., $100/16 = 6.25$).

Xanthoproteic reaction

When protein is treated with nitric acid and then made alkaline, the initial yellow color, due to the presence of tyrosine or tryptophan (cyclic or heterocyclic rings), is converted to an orange color. The same colors result when nitric acid comes in contact with the skin and soap is used in washing it off.

Millon test

A red color is produced when substances containing phenolic groups, as in tyrosine, are heated with Millon reagent, which is a mixture of mercuric and mercurous nitrates and nitrites. This test is used to verify the presence of free tyrosine and tyrosine-containing proteins.

Hopkins-Cole ring test

In a test that is specific for tryptophan (indole ring system) or tryptophan-containing proteins, sulfuric acid is carefully layered under the test solution containing glyoxylic acid (CHO—COOH). A violet ring appears at the interface in the presence of tryptophan or its compounds.

Heller ring test

A precipitate is formed at the interface when concentrated nitric acid is layered over a protein-containing solution. Concentrated hydrochloric and sulfuric acids may also precipitate proteins.

Sulfur test

When cysteine or cystine is heated with alkali in the presence of lead acetate, a black precipitate of lead sulfide forms. This test is negative for methionine, another sulfur-containing amino acid.

Anionic precipitants (alkaloidal reagents)

Proteins in the cationic form (+ charge) are precipitated by alkaloidal and other reagents in the anionic form (− charge). Substances such as picric, tannic, and tungstic acids and metaphosphate and trichloroacetate anions all form precipitates with protein. The formation of such a precipitate, which acts as a protective film, is the basis for the use of picric and tannic acids in the treatment of skin burns.

Cationic precipitants (heavy metals)

Proteins in the anionic form may be precipitated by heavy metal ions, such as silver and mercury. Preparations containing heavy metals are sometimes used as topical antiseptics that precipitate bacterial proteins.

OTHER IMPORTANT REACTIONS OF AMINO ACIDS AND PROTEINS

Reaction with fluorodinitrobenzene (Sanger reagent)

Fluorodinitrobenzene was used by Sanger in determining the identity of the terminal amino acid residue of a protein that had the free amino group:

$$R-\underset{COOH}{\overset{H}{C}}-NH_2 + F-C_6H_3(NO_2)-NO_2 \rightarrow R-\underset{COOH}{\overset{H}{C}}-NH-C_6H_3(NO_2)-NO_2 + HF$$

This was an important reaction in the determination of the total structure of the peptide hormone, insulin. Because the reaction takes place only with free amino groups, it is very useful in determining the number of chains present in a protein molecule.

Carbamino reaction

The free amino groups of proteins can condense with carbon dioxide to form carbamino compounds. This reaction is important in the transport of carbon dioxide in blood:

$$R-NH_2 + CO_2 \rightarrow R-\overset{H}{N}-\overset{O}{\overset{\|}{C}}-O^- + H^+$$

Carbamino compound

SUMMARY

Proteins are complex, high–molecular weight compounds made up of about twenty amino acids composed of carbon, hydrogen, oxygen, nitrogen, and in a few cases, sulfur and iodine. They are very important in many biologic processes.

NITROGEN CYCLE Important because of scarcity of sources of nitrogen.

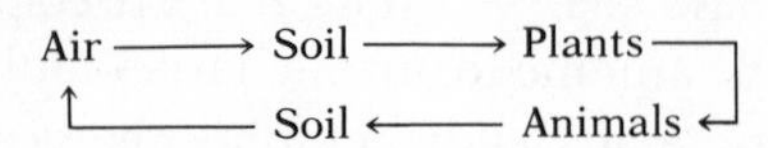

AMINO ACIDS Almost all naturally occurring amino acids are α-amino acids.

Classification: For formulas, see text.

I. Aliphatic amino acids
  A. Neutral (one carboxyl, one amino group)
    1. Simple amino acids
       Glycine
       Alanine
       *Valine
       *Leucine
       *Isoleucine
    2. Sulfur-containing amino acids
       Cysteine
       Cystine
       *Methionine
    3. Hydroxyl-containing amino acids
       Serine
       *Threonine
  B. Acidic amino acids (two carboxyls, one amino group)
     Aspartic acid
     Glutamic acid
  C. Basic amino acids (one carboxyl, two amino groups)
     *Arginine
     *Lysine
     Hydroxylysine
II. Aromatic amino acids
     *Phenylalanine
     Tyrosine
     Diiodotyrosine
     Thyroxine

*Essential amino acids, which must be obtained in the diet.

III. Heterocyclic amino acids
- Proline
- Hydroxyproline
- *Histidine
- *Tryptophan

Amphoterism of amino acids—buffer action: Amino acids can react both as weak acids, i.e., give off hydrogen ions, and as weak bases, i.e., take up hydrogen ions. At their isoelectric point, pI (net charge of zero), amino acids exist as dipolar zwitterions:

$$\begin{array}{l} CH_2COO^- \\ | \\ NH_3^+ \end{array}$$

Because of the ability of amino acids to give up hydrogen ions to neutralize hydroxyl ions in solution or take up excess hydrogen ions, the pH of the solution remains relatively constant. Therefore, amino acids and proteins are good buffers.

Asymmetry of amino acids: All amino acids, except glycine, have an asymmetric α-carbon atom and are optically active. L-Serine, the reference amino acid, has the same configuration as L-glyceraldehyde. Almost all natural amino acids are L-amino acids.

PROTEINS

Peptide bond: Resembles an amide bond, being formed between the carboxyl group of one amino acid and the amino group of another amino acid:

$$\begin{array}{ccccccccccc} & & NH_2 & & O & & H & & O & & \\ & & | & & /\!/ & & | & & /\!/ & & \\ R_1 & - & C & - & C & - & N & - & C & - & C-OH \\ & & | & & & & | & & | & & \\ & & H & & & & H & & R_2 & & \end{array}$$

Peptide bond

A dipeptide, as shown, has one peptide bond and two amino acid residues. Larger molecules are called polypeptides.

Classification: Often based on composition and solubilities.

Simple proteins: Give only amino acids on hydrolysis.

Albumins and globulins: Albumins are more soluble in pure water than globulins and require more salt to precipitate them from solution than do the globulins. Examples: plasma albumin and globulins.

Glutelins: Insoluble in water and dilute salt solutions but soluble in dilute acids or bases. Glutenin in wheat.

---

*Essential amino acids, which must be obtained in the diet.

Prolamines: Insoluble in water and absolute alcohol, but soluble in aqueous alcohol. Zein from corn.

Scleroproteins (albuminoids): Insoluble in all usual solvents. Keratin from nails, collagen from tendon.

Conjugated proteins: Composed of simple proteins combined with a nonprotein portion known as a prosthetic group.

Nucleoproteins: Prosthetic groups are the nucleic acids; protein portion includes protamines and histones (see below). Nucleoproteins found in all cells.

Phosphoproteins: Simple protein plus phosphorus as in milk casein.

Porphyrinoproteins (chromoproteins): Colored proteins containing heterocyclic (hetero nitrogen) porphyrin group as in hemoglobin.

Glycoproteins: Proteins combined with polysaccharides as in mucins of saliva.

Lipoproteins: Proteins combined with lipids as in the lipoproteins of blood plasma.

Derived proteins: Result from splitting conjugated proteins, hydrolysis of simple proteins, and denaturation of proteins.

Protamines and histones: Basic, water-soluble proteins found in nucleoproteins.

Proteoses, peptones, peptides: Mixtures of peptides resulting from progressive hydrolysis of proteins.

Denatured proteins: Result from changes in the natural structure of proteins, e.g., the effect of heat on proteins.

STRUCTURAL CONSIDERATIONS

Levels of organization:

Primary structure: Amino acid composition in a quantitative sense and the specific sequence in which the amino acids are linked in the protein molecule.

Secondary structure: The arrangement of the chain in the form of coiled helices or extended chains.

Tertiary structure: The complete three-dimensional picture of the protein chain.

Quaternary structure: The three-dimensional arrangement of multiple chains in a functioning molecular unit; e.g., hemoglobin consists of four chains arranged together as a unit.

COLLOIDAL PROPERTIES OF PROTEINS

Proteins are colloidal and therefore exhibit the properties of colloids—produce osmotic pressure, adsorb, and carry electric charges. (See Chapter 8 for a review of these properties.)

SEPARATION OF AMINO ACIDS AND PROTEINS

Chromatography: Column and paper chromatographic methods separate amino acids on the basis of differences in adsorption, ionic charges, size, affinity, and solubility of molecules.

Electrophoresis: Effects separations in an electrical field on the basis of differences in charges carried by amino acids and proteins under specific conditions.

Ultracentrifugation: Effects separation on the basis of differences in molecular weights when large gravitational forces are applied in the ultracentrifuge.

Precipitation methods: Salts, such as sodium sulfate and ammonium sulfate, at specific concentrations, precipitate some proteins while others remain in solution. Alcohol at low temperatures and zinc have been used in a similar way.

Dialysis: For the removal of small, crystalloidal molecules from protein solutions (Chapter 8).

DETECTION AND QUANTITATION OF AMINO ACIDS AND PROTEINS

Ninhydrin reaction: Produces blue to violet color with free amino acids. Color may be measured quantitatively.

Heat coagulation: Proteins coagulate easily on heating, e.g., egg white. Precipitate may be weighed after drying.

Biuret reaction: Relatively specific for compounds containing two or more peptide links. Dilute copper solution in strong alkali gives blue-violet color with peptides. May be used quantitatively.

Total nitrogen—Kjeldahl method: Converts amino nitrogen to ammonia, which can be measured accurately. Protein contains 16% nitrogen; therefore $\%N \times 6.25$ = protein content of sample.

Xanthoproteic reaction: Protein and nitric acid yield yellow color; subsequent addition of alkali results in orange color.

Millon test: Red color in presence of protein. Detects phenolic groups as in tyrosine.

Hopkins-Cole ring test: Violet ring forms at interface. Detects indole ring as in tryptophan.

Heller ring test: Forms precipitate at interface.

Sulfur test: Black precipitate of lead sulfide forms when cysteine or cystine is heated with alkaline lead acetate. Methionine does not react.

Anionic precipitants (alkaloidal reagents): Picric, tannic, and tungstic acids in the form of negative ions (anions) precipitate proteins in the form of positive ions (cations).

Cationic precipitants (heavy metals): Heavy metal ions, such as silver and mercury, in positive form (cations) precipitate proteins in the form of negative ions (anions).

OTHER IMPORTANT REACTIONS OF AMINO ACIDS AND PROTEINS

Reaction with fluorodinitrobenzene (Sanger reagent): Free amino groups react with fluorodinitrobenzene to form dinitrophenyl derivatives. Used in analysis of peptide structure.

Carbamino reaction: Free amino groups condense with carbon dioxide forming carbamino compounds, important in transport of blood carbon dioxide.

REVIEW QUESTIONS

1. Explain the strategic importance of nitrogen in the food cycle. What is meant by the term "essential amino acids"?
2. Explain the amphoterism of amino acids. Write the equations for the reaction of an amino acid with hydrogen ions and with hydroxyl ions. What type of chemical system reacts the same way?
3. Define the various levels of structural organization of proteins. The complexity of protein structures places them in the class of colloidal substances. List some procedures for separating and purifying proteins that depend on their colloidal properties and explain the principles involved in these procedures.
4. What are the hydrolysis products of simple proteins? Of conjugated proteins?
5. How many amino acid residues are found in a dipeptide? How many peptide bonds? What is a standard chemical test for the peptide bond? For free amino acids?

REFERENCES

Advances in protein chemistry (series). New York, Academic Press.

Annual review of biochemistry, Palo Alto, Calif, Annual Reviews.

Barry, J. M., and Barry, E. M.: An introduction to the structure of biological molecules, Englewood Cliffs, N.J., Prentice-Hall, 1969.

Blackburn, S., editor: Amino acid determination: methods and techniques, New York, 1978, Marcel Dekker, Inc.

Bodansky, M., and Ondetti, M. A.: Peptide synthesis, ed. 2, New York, 1976, John Wiley & Sons, Inc.

Fasman, G. D., editor: Poly-α-mino acids: protein models for conformational studies, New York, 1967, Marcel Dekker, Inc.

Florkin, M., and Stotz, E. H., editors: Comprehensive biochemistry, Vols. 7, 8, Amsterdam, 1963, Elsevier.

Meister, A.: Biochemistry of the amino acids, New York, 1965, Academic Press, Inc.

Needleman, S. B., editor: Advanced methods in protein sequence determination, New York, 1977, Springer-Verlag New York Inc.

Neurath, H., editor: The proteins, ed. 3, Vols. 1-4, New York, 1975-1979, Academic Press, Inc.

Pauling, L.: Chemical bond: a brief introduction to modern structural chemistry, Ithaca, N.Y., 1967, Cornell University Press.

Rich, A., and Davidson, N., editors: Structural chemistry and molecular biology, San Francisco, 1968, W. H. Freeman and Co., Publishers.

Smith, I.: Chromatographic and electrophoretic techniques, ed. 4, Chicago, 1976, Year Book Medical Publishers, Inc.

Chemistry of nucleic acids 17

The nucleic acids, so named because they were first found in cell nuclei, have risen, in the relatively short period of 30 to 40 years, from the status of chemical curiosities to a point of central importance in the efforts of many investigators attempting to unravel some of the secrets of life itself. Now known to occur in all living cells—both in the cytoplasm and in the nucleus—the nucleic acids are considered to have properties that make them the carriers of the information necessary for organisms to produce all the factors necessary for successful living and, ultimately, faithful reproduction. This information is generally referred to as the *genetic code,* or code of life, and will be discussed in more detail in Chapter 23.

CHEMICAL NATURE OF THE NUCLEIC ACIDS

As they occur in cells, the nucleic acids are combined with proteins in the form of nucleoproteins, a class of conjugated proteins. The subsequent discussions are concerned mainly with the nucleic acid portion of the nucleoproteins.

The nucleic acids, the prosthetic groups of the nucleoproteins, are high-molecular weight substances that contain in their structure purine and pyrimidine bases, a pentose, and phosphoric acid combined in polymeric form. The repeating unit of the polymer is the *nucleotide,* which consists of a purine or pyrimidine base attached to a pentose that is also attached to a phosphate group. A purine or pyrimidine base attached only to a pentose is known as a *nucleoside.* These relationships are illustrated here:

Nucleoside: Purine or pyrimidine base—pentose
Nucleotide: Purine or pyrimidine base—pentose—phosphate
Nucleic acid: $[\text{Nucleotide}]_x$
Nucleoprotein: $[\text{Nucleotide}]_x$—protein or nucleic acid—protein

The nucleic acids are subdivided into two major classes on the basis of the type of pentose occurring in the molecule. Those nucleic acids containing ribose are known as ribonucleic acids (RNA) and those containing deoxyribose, as deoxyribonucleic acids (DNA). The major components obtained from RNA and DNA on complete hydrolysis are listed on the next page.

RNA	*DNA*
Adenine	Adenine
Guanine	Guanine
Cytosine	Cytosine
Uracil	Thymine
Ribose	Deoxyribose
Phosphoric acid	Phosphoric acid

From this listing it can be seen that in addition to the different pentoses, DNA contains thymine rather than uracil, as in RNA. Other purine and pyrimidine derivatives have been recognized in some nucleic acids.

It was originally believed that RNA occurred only in the cytoplasm and DNA only in the nucleus. This distinction is no longer valid, RNA having been found in the nucleus and some DNA having been reported to be located in the cytoplasm.

In addition to the genetically important nucleoproteins, there are some free nucleotides of great biologic importance, such as adenosine triphosphate (ATP), nicotinamide adenine dinucleotide (NAD), and their derivatives.

COMPONENTS OF THE NUCLEIC ACIDS

Pentoses

The structural formulas for ribose, found in RNA, and deoxyribose, found in DNA, are shown here:

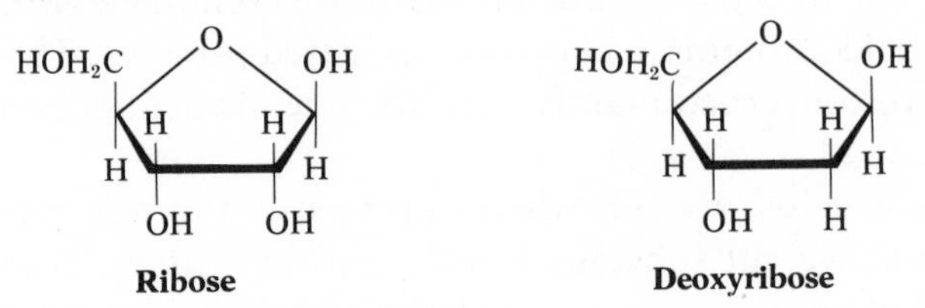

Ribose **Deoxyribose**

Purine bases

The two major purine bases found in all nucleic acids are adenine and guanine:

Adenine **Guanine**

Pyrimidine bases

The major pyrimidine bases found in nucleic acids are cytosine, uracil, and thymine:

Cytosine **Uracil** **Thymine**

As noted, cytosine and uracil appear in RNA, but uracil is replaced by thymine in DNA.

Nucleosides, nucleotides

A purine or pyrimidine base combined with a pentose forms a nucleoside. Addition of a phosphate group on the pentose gives a nucleotide. The 1′-carbon of the pentose is attached to the 9-nitrogen of the purines or the 3-nitrogen of the pyrimidines. The phosphate group on the pentose may be found on the 5′- or 3′-carbon depending on the method of preparation. The structural formulas of the major nucleosides and nucleotides found in DNA are shown here:

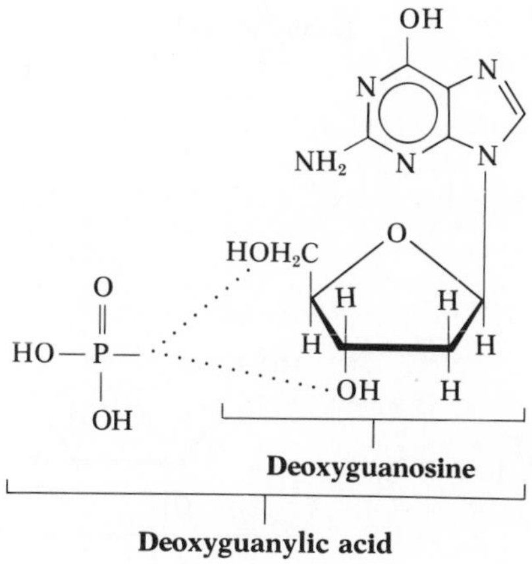

Deoxyguanosine

Deoxyguanylic acid

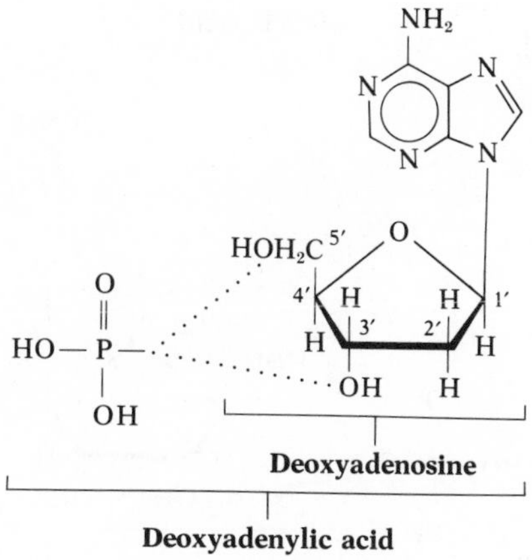

Deoxyadenosine

Deoxyadenylic acid

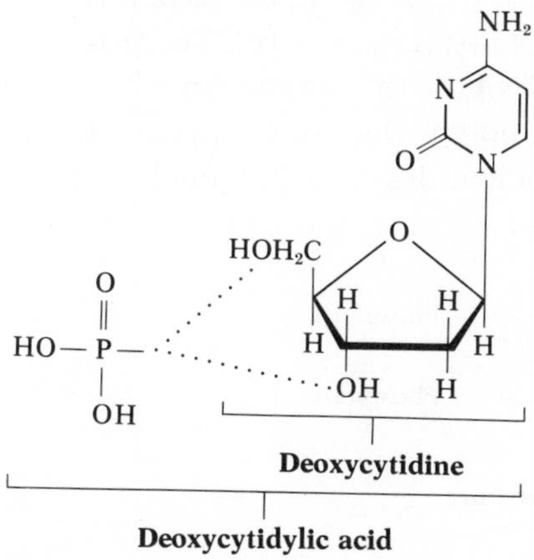

Deoxycytidine

Deoxycytidylic acid

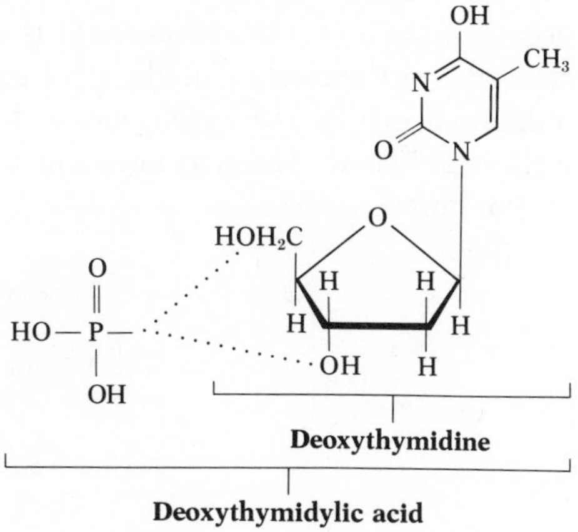

Deoxythymidine

Deoxythymidylic acid

The structural formulas of the major nucleosides and nucleotides found in RNA are shown here:

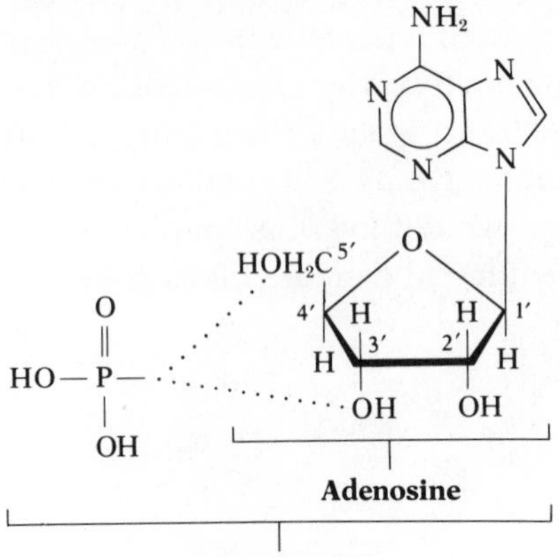

Adenosine

Adenylic acid

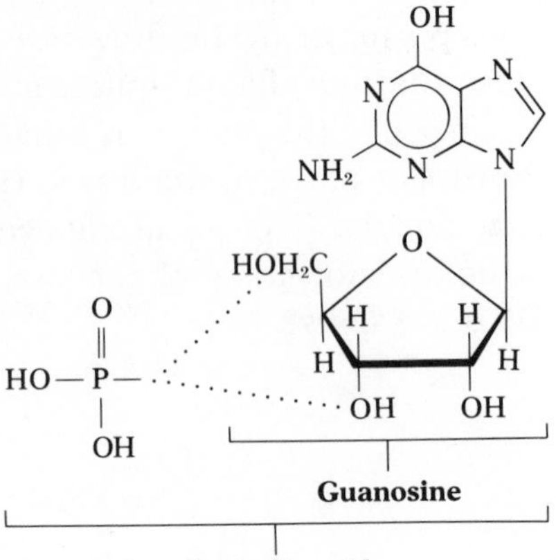

Guanosine

Guanylic acid

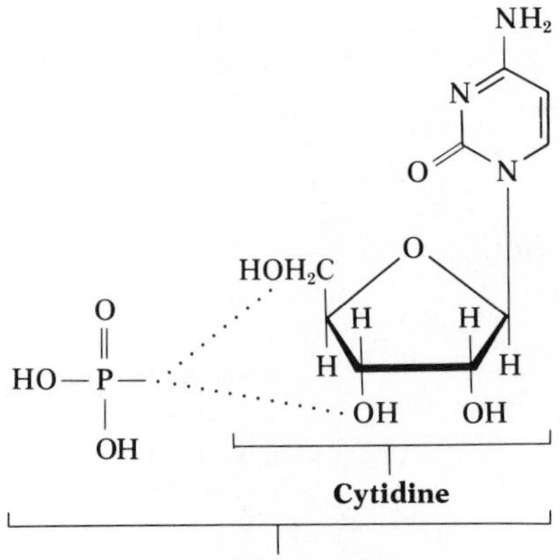

Cytidine

Cytidylic acid

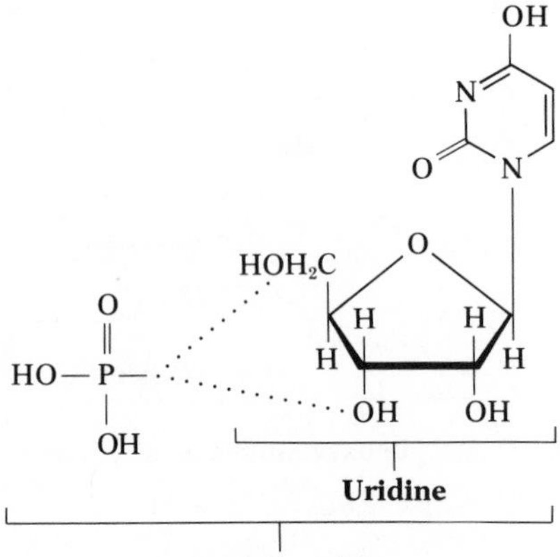

Uridine

Uridylic acid

STRUCTURE OF THE NUCLEIC ACIDS

The structure of the nucleic acids may be discussed in terms of levels of organization as is the structure of the proteins (Chapter 16). The primary structure of the nucleic acids, DNA and RNA, or the manner in which the mononucleotides are combined to form polynucleotides, appears to be well established. The arrangement of nucleotides in a polynucleotide is shown here:

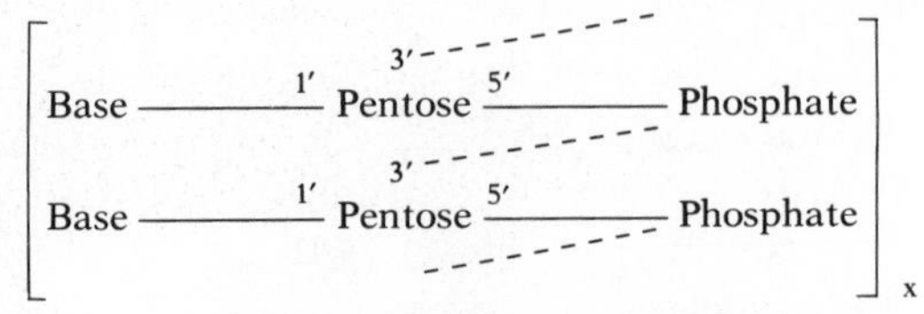

Polynucleotide structure

Examination of this skeletal structure shows a backbone of a pentose-phosphate chain with the bases sticking off to the side. Very little is known at the present time about the specific sequence of bases in polynucleotide molecules.

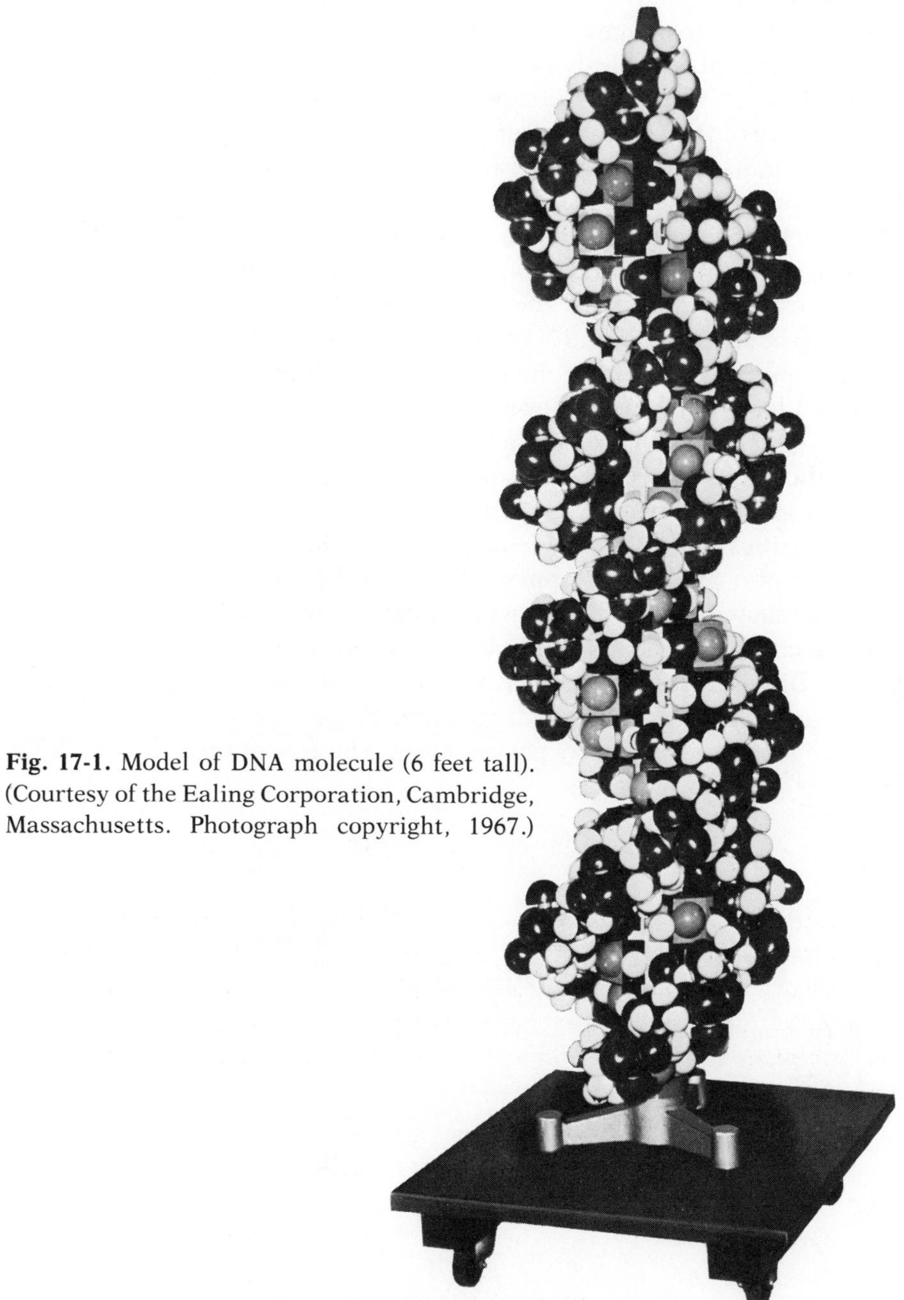

Fig. 17-1. Model of DNA molecule (6 feet tall). (Courtesy of the Ealing Corporation, Cambridge, Massachusetts. Photograph copyright, 1967.)

DNA

All indications are that DNA molecules are long, probably unbranched polymers with high molecular weights ranging from low to high millions. Although little is known about specific base sequences in DNA, there is evidence to indicate the importance of these sequences to the tertiary structure of DNA and to the genetic functions attributed to DNA. These aspects will be discussed more fully later.

Details concerning the secondary and tertiary levels of organization of the structure of DNA are at present best summarized by the structure proposed by James Watson and Francis Crick. In this proposed structure, DNA is considered to be composed of two polynucleotide chains twisted about each other in the form of a double helix. As noted before, the backbone of each polynucleotide consists of a pentose (in this case deoxyribose)-phosphate chain with the purine and pyrimidine bases facing the center of the structure. Fig. 17-1 shows a model of the DNA structure based on these proposals. In fact, the space relationships of this structure are such that for everything to fit properly, the bases must pair off in such a way that a purine always faces a pyrimidine on the plane approximately perpendicular to the central axis of the molecule. This means that a large base (purine) is always paired with a small base (pyrimidine) with the preferred pairings consisting of adenine with thymine and guanine with cytosine: wherever an adenine appears in one of the polynucleotide chains, a thymine must appear in the corresponding position in the second polynucleotide chain. This leads to the complementarity upon which the functions of DNA are based. The DNA structure is stabilized by the formation of hydrogen bonds between hydrogen and oxygen atoms or hydrogen and nitrogen atoms of the closely packed bases. These relationships are shown here:

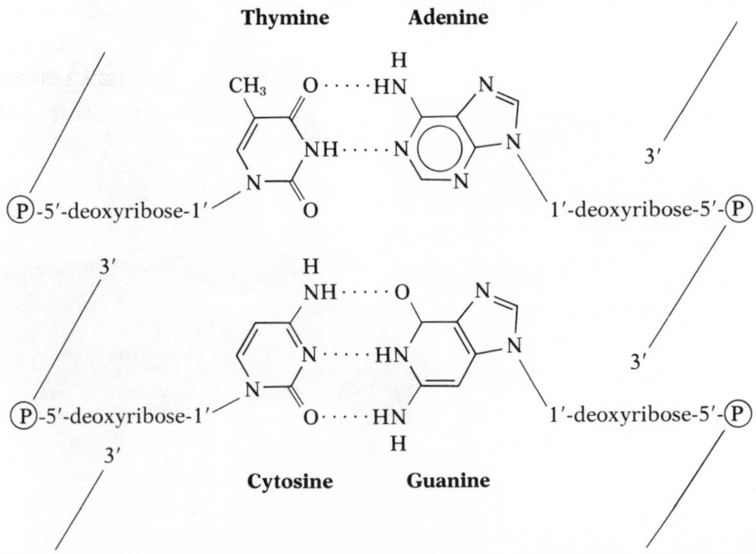

Thymine **Adenine**

Ⓟ-5′-deoxyribose-1′ 1′-deoxyribose-5′-Ⓟ

3′ 3′

Ⓟ-5′-deoxyribose-1′ 1′-deoxyribose-5′-Ⓟ

3′

Cytosine **Guanine**

Much evidence has been accumulated in favor of this structure for DNA proposed by Watson and Crick.

RNA

The secondary and tertiary structures of RNA appear to be less strictly organized than is the case with DNA. The evidence indicates that RNA is usually single-stranded and therefore can form areas of double helix with base-pairing only where the strand folds back on itself—such regions being connected by single-stranded segments of the polynucleotide chain. Where base-pairing does occur in RNA, uracil pairs with adenine, in contrast to DNA in which thymine pairs with adenine.

There are several forms of RNA that have been well established as separate entities. There are ribosomal RNA, transfer RNA, and messenger RNA. Ribosomal RNAs, as the name implies, are found in the ribosome—a small subcellular particle—and have molecular weights of approximately 1 million. Transfer RNAs, sometimes called soluble RNAs, function in protein synthesis as transfer agents for amino acids and have molecular weights in the low range of 25,000 to 30,000. The general struc-

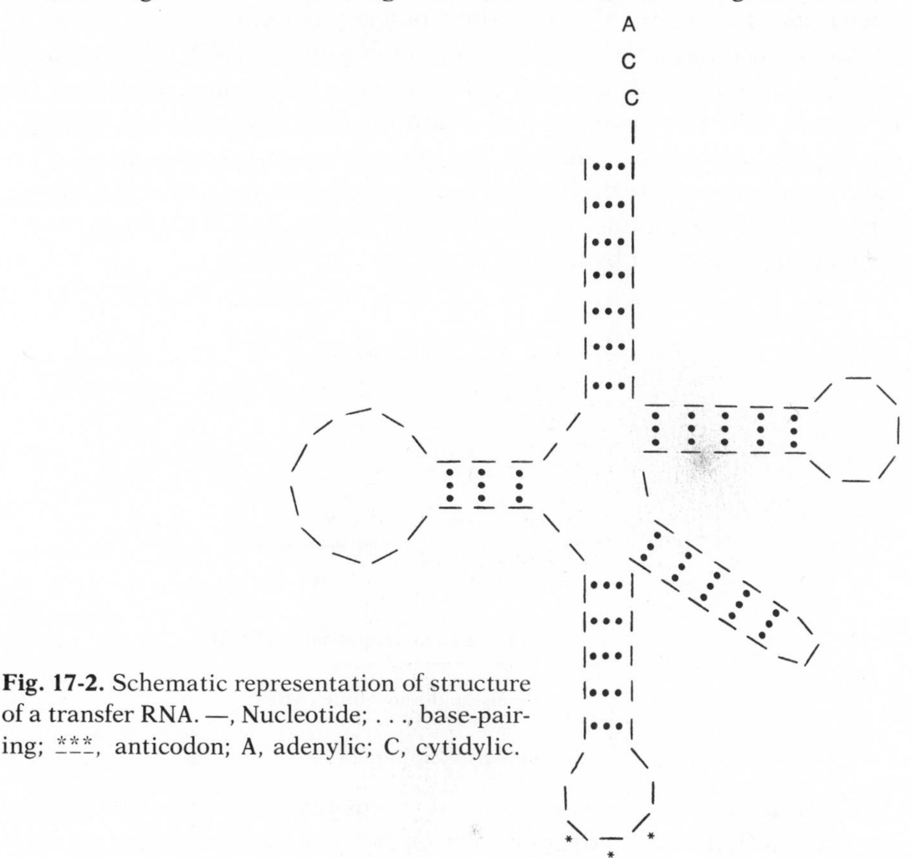

Fig. 17-2. Schematic representation of structure of a transfer RNA. —, Nucleotide; . . ., base-pairing; ***, anticodon; A, adenylic; C, cytidylic.

ture of transfer RNAs is shown in Fig. 17-2. It should be noted that there are areas of base-pairing, single-stranded sections, an A—C—C-nucleotide sequence at one end (p. 355), and a three-nucleotide segment that acts as an anticodon (three-nucleotide segment complementary to a codon) for recognition of the correct position for the transfer RNA to assume on a messenger RNA (p. 356). Messenger RNAs vary in molecular weights from about 30,000 to a few million. More will be said later about the structures and functions of these forms of RNA in Chapters 23 and 24.

Viruses

Viruses are infective agents smaller than bacteria and are composed of a core of DNA or RNA covered by a shell of protein. Some viruses have been found to contain closed circular strands of DNA. In order to survive, viruses must find their way into host cells, which supply the mechanisms and materials required for multiplication of the viruses. This process is discussed in detail in Chapter 23.

Free nucleotides

In addition to the polynucleotides DNA and RNA, there are some free nucleotides that perform important biologic functions.

Adenosine phosphates. Adenosine monophosphate (AMP, adenylic acid, muscle adenylic acid), adenosine diphosphate (ADP), and adenosine triphosphate (ATP) are involved in the capture of energy in the form of high-energy phosphate bonds and transfer of these phosphate groups to supply the energy necessary to drive many metabolic reactions (p. 286). The structures of these compounds are shown here—the position of the high-energy bonds being indicated by a wavy line (~):

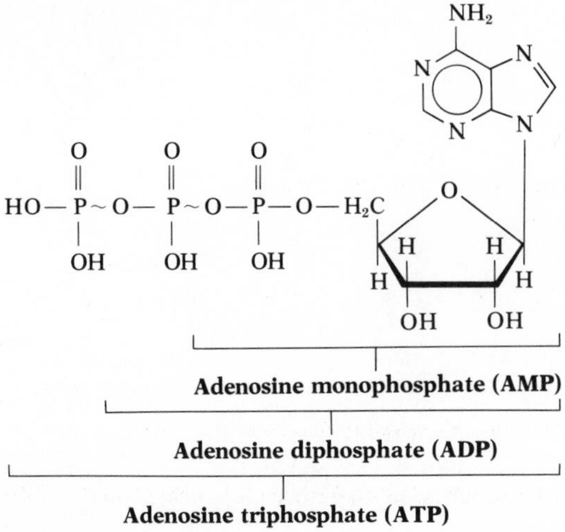

NH_2
N N
N N
O O O
HO—P~O—P~O—P—O—H_2C O
OH OH OH H H
H H
OH OH

Adenosine monophosphate (AMP)

Adenosine diphosphate (ADP)

Adenosine triphosphate (ATP)

Cyclic adenosine monophosphate. In recent years a tremendous surge of research activity has indicated very important functions—particularly with respect to the mechanism of action of hormones (p. 446)—for the free nucleotide, cyclic adenosine 3′,5′-monophosphate (c-AMP), the structure of which is shown here:

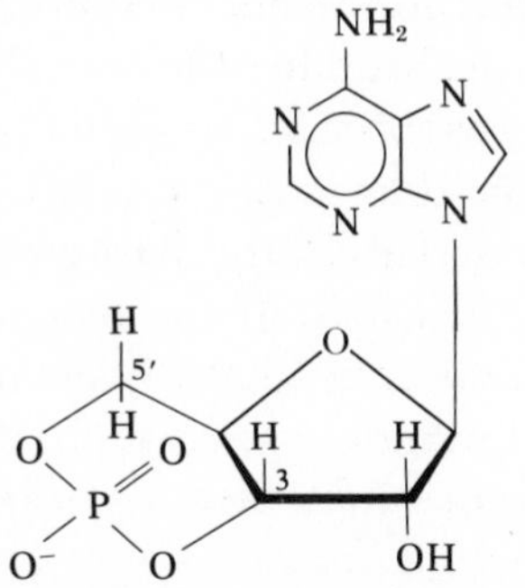

Adenosine 3′,5′-monophosphate (c-AMP)

Nicotinamide adenine dinucleotides. Although the nicotinamide portion of the dinucleotides, nicotinamide adenine dinucleotide (NAD, formerly known as diphosphopyridine nucleotide, DPN) and nicotinamide adenine dinucleotide phosphate (NADP, formerly known as triphosphopyridine nucleotide, TPN), is not a nucleotide by strict definition, NAD and NADP are considered dinucleotides because of their close resemblance to such compounds. NAD and NADP, the structures of which are shown here, act as coenzymes in oxidation-reduction reactions:

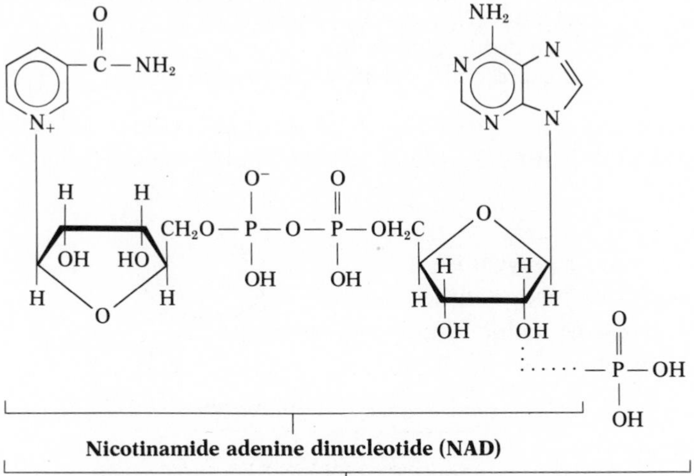

Nicotinamide adenine dinucleotide (NAD)

Nicotinamide adenine dinucleotide phosphate (NADP)

There are other similar compounds that act as coenzymes and that will be discussed later with their metabolic functions.

ANALYSIS OF THE NUCLEIC ACIDS

There are specialized methods available for the isolation, separation, purification, and quantitative determination of nucleic acids. These methods include selective extraction, precipitation, centrifugation, hydrolysis, chromatography, and optical procedures for determining the amount of nucleic acid present in a particular sample.

On complete hydrolysis of nucleic acids—usually by heating with strong acids such as hydrochloric and perchloric acids—the purine and pyrimidine bases can be separated by paper chromatography. The individual bases can then be washed off the paper and quantitatively measured by appropriate optical techniques. Tests for the presence of pentoses include the Bial test and other tests mentioned in Chapter 14. The presence of the phosphate groups may be determined by colorimetric procedures.

SUMMARY

Nucleic acids are found in all living cells—in cytoplasm as well as nucleus—and are believed to be carriers of the genetic code.

CHEMICAL NATURE OF THE NUCLEIC ACIDS

Nucleic acids, the prosthetic groups of the nucleoproteins, are composed of purine and pyrimidine bases, a pentose, and phosphoric acid combined in polymeric form.

Nucleoside: Purine or pyrimidine base—pentose
Nucleotide: Nucleoside—phosphate
Nucleic Acid: [Nucleotide]
Nucleoprotein: Nucleic Acid—protein

RNA contains adenine, guanine, cytosine, uracil, ribose, phosphoric acid. DNA contains thymine in place of uracil and deoxyribose in place of ribose.

COMPONENTS OF THE NUCLEIC ACIDS

Pentoses: Ribose, deoxyribose.
Purine bases: Adenine, guanine.
Pyrimidine bases: Cytosine, uracil, thymine.
Nucleosides, nucleotides:

Base	*Nucleoside*	*Nucleotide*
From RNA:		
Adenine	Adenosine	Adenylic acid
Guanine	Guanosine	Guanylic acid
Cytosine	Cytidine	Cytidylic acid
Uracil	Uridine	Uridylic acid

Base	*Nucleoside*	*Nucleotide*
From DNA*:		
Adenine	d-Adenosine	d-Adenylic acid
Guanine	d-Guanosine	d-Guanylic acid
Cytosine	d-Cytidine	d-Cytidylic acid
Thymine	d-Thymidine	d-Thymidylic acid

STRUCTURE OF THE NUCLEIC ACIDS

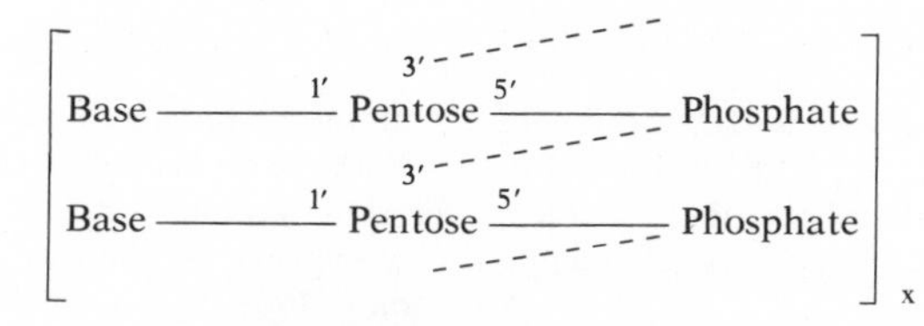

Polynucleotide structure

DNA: According to Watson and Crick, DNA is composed of two long, unbranched, polynucleotide strands wound around each other in the form of a double helix. Each strand has a backbone of pentose-phosphate linkages with the purine and pyrimidine bases facing each other toward the middle of the molecule and forming hydrogen bonds between appropriate groupings. Adenine pairs with thymine; guanine pairs with cytosine. Therefore, the two chains are complementary to each other.

RNA: RNA is less strictly organized than DNA and is usually single-stranded. Where base-pairing does occur, uracil (instead of thymine) pairs with adenine, and guanine pairs with cytosine. Important forms of RNA are ribosomal RNA, transfer RNA, and messenger RNA.

Viruses: Infective particles composed of a core of RNA or DNA covered with a coat of protein.

Free nucleotides:

Adenosine phosphates: Adenosine mono-, di-, and triphosphates are involved in the capture and transfer of energy for use in metabolic reactions.

Nicotinamide adenine dinucleotides: NAD and NADP act as coenzymes in oxidation-reduction reactions.

ANALYSIS OF THE NUCLEIC ACIDS

On complete hydrolysis, the purine and pyrimidine bases can be separated by paper chromatography and quantitatively determined by optical methods. The pentoses can be detected by the Bial test and phosphate by colorimetric procedures.

---

*d = deoxy

REVIEW QUESTIONS

1. What type of ring structure is found in pyrimidines? Purines? Which are larger in size?
2. What are the general types of hydrolysis products of a nucleoside? A nucleotide? What are the specific hydrolysis products of ribonucleotides? Deoxyribonucleotides? What are the major differences in hydrolysis products between ribo- and deoxyribonucleotides?
3. Describe the three-dimensional structure of DNA. What are some of the key considerations that lead to such a structure?
4. What is the relationship between nucleic acids and viruses?
5. Name some important free nucleotides and their functions.

REFERENCES

Adams, R. L. P., Burdon, R. H., Campbell, A. M., and Smellie, R. M. S., editors: Davidson's: the biochemistry of the nucleic acids, ed. 8, New York, 1976, Academic Press, Inc.

Advances in cyclic nucleotide research, New York, Raven Press.

Annual review of biochemistry, Palo Alto, Calif., Annual Reviews.

Barry, J. M., and Barry, E. M.: An introduction to the structure of biological molecules, Englewood Cliffs, N. J., 1969, Prentice-Hall, Inc.

Ciba Foundation Symposium 48 (series): Purine and pyrimidine metabolism, New York, 1977, Elsevier.

Colowick, S. P., and Kaplan, N. O., editors: Methods in enzymology, Vol. 51, New York, 1978, Academic Press, Inc.

Florkin, M., and Stotz, E. H., editors: Comprehensive biochemistry, Vol. 8, Amsterdam, 1963, Elsevier.

Harbers, E.: Introduction to nucleic acids: chemistry, biochemistry and functions, New York, 1968, Van Nostrand, Reinhold Co.

Rich, A., and Davidson, N., editors: Structural chemistry and molecular biology, San Francisco, 1968, W. H. Freeman and Co., Publishers.

Zorbach, W. W., and Tipson, R. S., editors: Synthetic procedures in nucleic acid chemistry, New York, 1973, John Wiley & Sons, Inc.

UNIT FOUR

BIOCHEMISTRY

Enzymes

18

Life in the form we know it depends to a large extent on the properties of a group of remarkable substances known as enzymes. In order to function properly, a living organism must be supplied energy in widely different amounts and rates to meet changing needs and this must be done under mild conditions of pH, temperature, and chemical considerations. Although many of the reactions that occur in the body can be duplicated in a test tube, the drastic laboratory conditions required for most of these reactions to take place and the rates at which they occur are incompatible with the properties of living tissues. Nature has given the enzymes the responsibility for making it possible for reactions to occur in the body at proper rates and under much milder conditions than would otherwise be possible. This means that enzymes act as catalysts.

Thus far, all substances isolated and identified as enzymes have proved to be proteins. Therefore, enzymes react as do other proteins to heat, precipitating agents of many types, and excessive changes in pH by becoming inactivated or denatured. For this reason the body has intricate mechanisms available to keep these important conditions relatively constant within narrow ranges.

As will be seen later, almost every reaction has its own specific enzyme catalyst. With the large number and variety of reactions taking place in all cells, each cell must contain a very large number of different enzymes.

DEFINITIONS

Enzymes may be defined as protein catalysts that facilitate the chemical conversion of various substances. The substance acted on by an enzyme is known as the *substrate.* The substance or substances resulting from enzyme action are the *products* of the reaction.

Some enzymes are associated with relatively small, dialyzable organic compounds when they function as catalysts. These organic compounds are known as *prosthetic groups* when they are firmly attached to the enzyme protein and *coenzymes* when the combination is easily dissociated. These combinations are known as *holoenzymes,* the protein portion being referred to as the *apoenzyme.*

[Enzyme protein—prosthetic group] | (no dissociation)
[Enzyme protein—coenzyme] ⇌ Enzyme protein + coenzyme

$\underbrace{\text{Enzyme protein—}}_{\textbf{Apoenzyme}}$

$\underbrace{\begin{array}{l}\text{[Enzyme protein—prosthetic group]}\\ \text{[Enzyme protein—coenzyme]}\end{array}}_{\textbf{Holoenzyme}}$

Because of the heat stability of prosthetic groups and coenzymes, they may be prepared free of the apoenzyme by heat denaturation of the proteins.

MECHANISM OF ENZYME ACTION

Some reactions, although theoretically possible, do not occur readily unless a "push" (called, in thermodynamic terminology, energy of activation) is administered to start them. The situation is approximately analogous to that of a round rock resting at the edge of a slope. Potentially the rock can roll down the slope to the bottom of the valley under the influence of gravity. All that is required is a little push to get the rock over the edge of the slope. This little push in chemical reactions is often supplied by a catalytic agent, an enzyme in biologic reactions. The catalyst cannot affect the possibility of a reaction occurring, the maximum amount of energy to be derived from a reaction, or the ultimate point to which a reaction may progress (the equilibrium point). A catalyst, or enzyme, only supplies the push that may make the reaction occur at a faster rate and under milder conditions.

Enzyme-substrate complex

Most experimental evidence indicates that in performing its function as a catalyst in the chemical conversion of a substrate, the enzyme combines with the substrate to form an enzyme-substrate complex. As a result of this combination, the appropriate bond in the substrate is somehow strained so that its rupture proceeds at a faster rate than would be possible in the absence of an enzyme. These relationships are illustrated in Fig. 18-1. When the chemical change is completed, the products leave the enzyme, which is then free to act in the conversion by another substrate molecule. The process is summarized by the following:

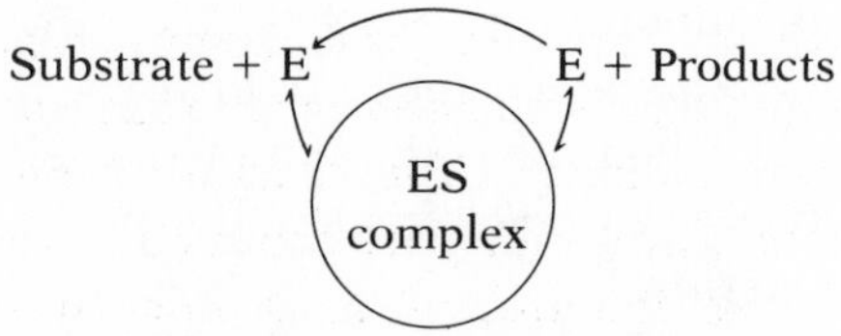

Substrate + E ⟶ ES complex ⟶ E + Products (E returns to start)

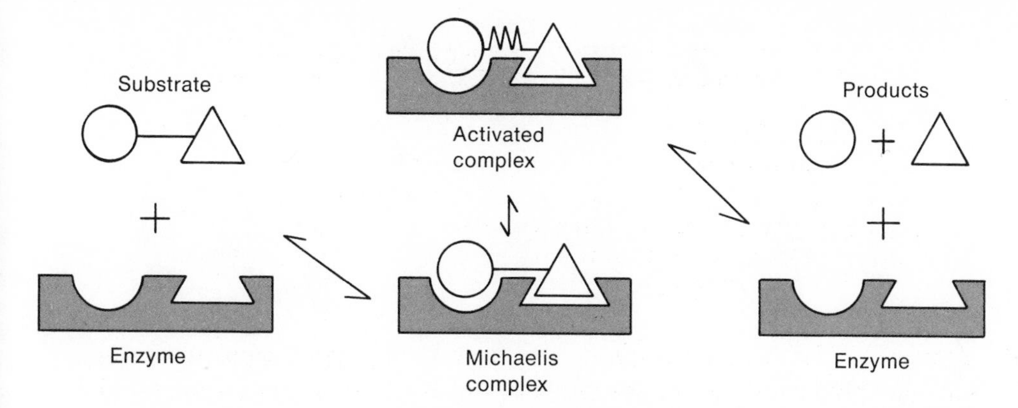

Fig. 18-1. Enzyme action, illustrating the formation and breakdown of an enzyme-substrate complex. (From Cantarow and Schepartz: Biochemistry, ed. 4. Courtesy the W. B. Saunders Co.)

Specificity of enzyme action

As stated, each biologic reaction usually has its own specific enzyme to catalyze it. This specificity of enzyme action can be explained in large part on the basis of the lock and key analogy. Each different substrate-enzyme complex could be represented by a different lock and key combination. Experimental evidence indicates that specificity is caused by the chemical nature of the groups involved in the reaction, the three-dimensional characteristics of these groups (size, shape, location, and the distribution of electrical charges on and around these groups).

Binding site and active site

Enzymes appear to have two types of sites that are crucial to their catalytic activity. The site that holds the enzyme and substrate together is known as the *binding site* and the part of the enzyme that is directly responsible for the chemical conversion of the substrate is termed the *active site* (catalytic site). The two sites need not always be physically separated on the enzyme because the active site must have some substrate binding properties in order to perform its catalytic function.

Investigations of the sequence of amino acids in enzyme molecules at their active sites have shown some regularities between different enzymes. In particular, the amino acids serine and histidine appear to be important in the function of active sites.

Function of prosthetic groups and coenzymes

In enzymatic reactions in which the enzyme operates in partnership with a prosthetic group or a coenzyme, the prosthetic group or coenzyme acts as an acceptor of one of the cleavage products, usually a very small part of the substrate molecule. The remaining portion of the substrate leaves the apoenzyme, while the part attached to the prosthetic group or coenzyme may also be liberated or, as is often the case, pass on along a

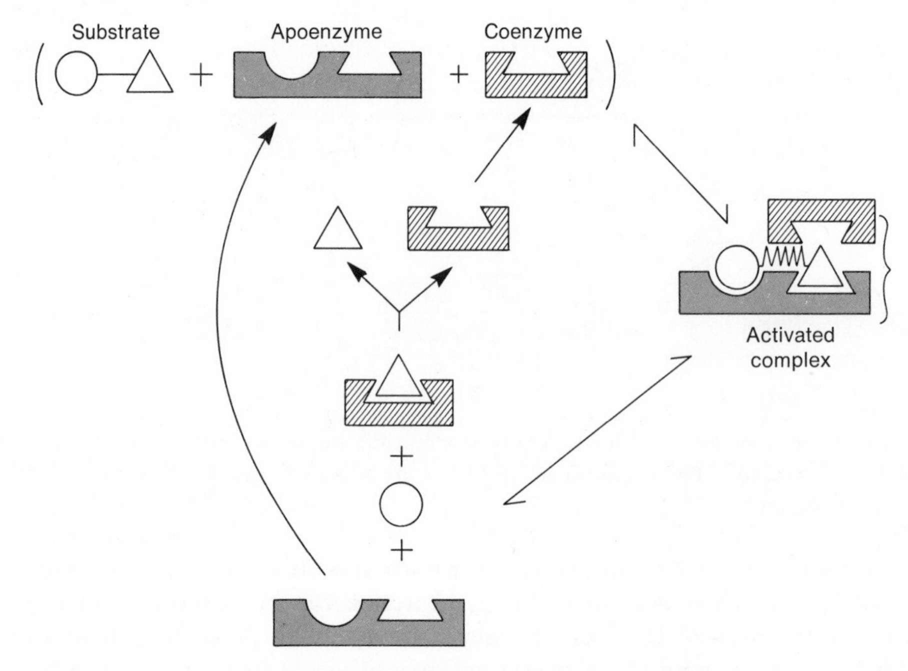

Fig. 18-2. Function of coenzymes. (From Cantarow and Schepartz: Biochemistry, ed. 4. Courtesy the W. B. Saunders Co.)

chain of enzymes for further catalytic conversions. These relationships are illustrated in Fig. 18-2. Note that both apoenzyme and coenzyme (or prosthetic group) are regenerated in the process.

Activators

Activators is the designation generally reserved for specific inorganic ions that are required by some enzymes for their activity. Iron, copper, manganese, magnesium, cobalt, and zinc ions are the more commonly required activators. Ordinarily only one ion functions with a specific enzyme, but in some cases substitutions can be made while retaining satisfactory enzyme activity.

Inhibitors

Any substance that interferes with the activity of an enzyme is an *inhibitor*. Because enzymes are proteins, any substances that denature proteins will automatically become inhibitors. Interference with or conversion of strategic groups on enzymes, prosthetic groups, coenzymes, or ion activators will inhibit enzyme activity.

Although enzymes are usually very specific in their binding with substrates, in some cases the enzyme will bind with another substrate that has the same general structure, size, and functional group relationships that the original substrate has. When an enzyme is tricked into accepting such a substrate, the original process is retarded or completely blocked because the enzyme cannot handle the substitute substrate as well as its specific substrate. The classic example of this type of inhibition, known as *metabolic antagonism,* involves the interference of malonic acid in the conversion of succinic acid to fumaric acid. The similarities in the structures of the acids are shown here:

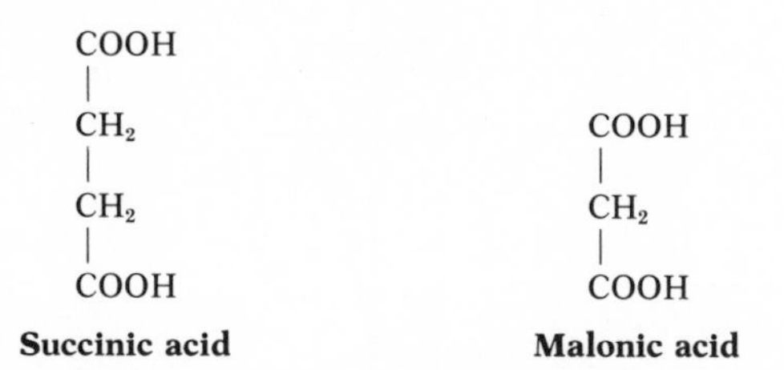

If an increase in the ratio of specific substrate to inhibitor (metabolic antagonist) increases the enzyme activity, the inhibition is termed *competitive inhibition,* and the combination of enzyme with inhibitor is reversible. If the combination of enzyme with inhibitor is irreversible, an increase in the ratio of specific substrate to inhibitor will have no effect on the blocked enzyme activity. This type of inhibition is known as *noncompetitive inhibition.*

The effect of drugs on the metabolism of the body is probably the result of direct or indirect action on enzymes. Some forms of metabolic antagonism achieved by drugs fulfill the aims of Ehrlich's "magic bullet" (p. 5) and others are constantly being sought on the rational basis of selective interference with strategic enzyme activities.

Preenzymes (zymogens, proenzymes)

Some enzymes are in an inactive state when first synthesized. The internationally approved designation for such substances is preenzymes. However, the older names—zymogens and proenzymes—are still widely used. Preenzymes are generally activated by splitting off a portion of the molecule that apparently blocks the enzyme activity. This may be a mechanism that prevents the enzyme from exerting its activity at the wrong time or place. How the activation of the preenzymes is accomplished will be discussed later with the appropriate enzymes.

Isozymes

Lactate dehydrogenase is an example of an enzyme made up of four peptide chains (tetramers). It has been determined that there are five different forms of this enzyme, made up of various ratios of two parent chains, an M chain found predominating in skeletal muscle and an H chain found in heart tissue. Thus there can be the following types of tetramers, known as isozymes: M_4, M_3H, M_2H_2, MH_3, H_4. The various forms can be differentiated on the basis of their electrophoretic, catalytic, and immunologic properties and differences in amino acid composition.

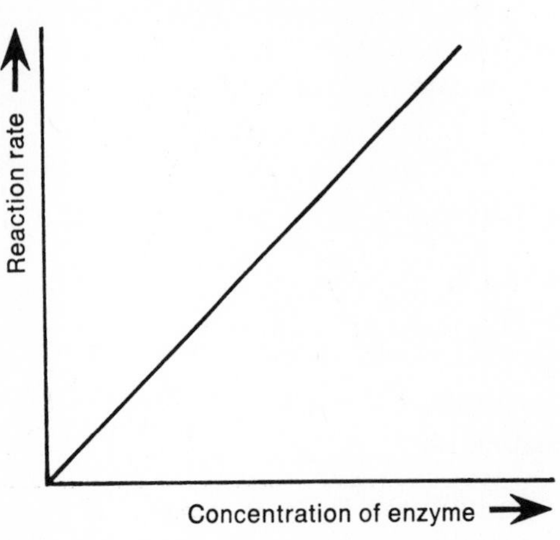

Fig. 18-3. Effect of enzyme concentration on reaction rate.

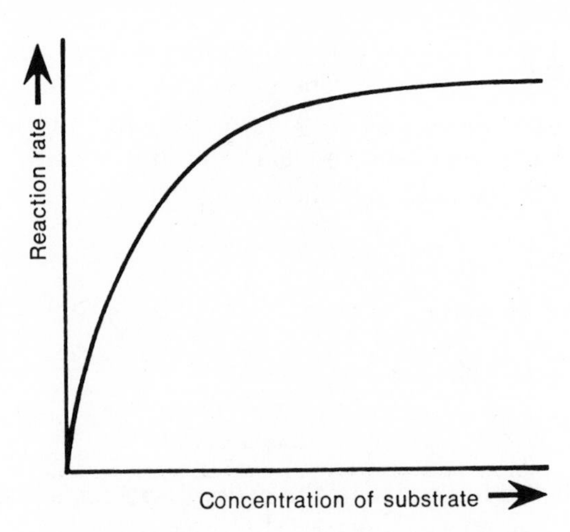

Fig. 18-4. Effect of substrate concentration on reaction rate.

When tissues are injured, the enzymes in the cells tend to leak out and find their way into the bloodstream. Because different tissues have different forms of the lactate dehydrogenase in their cells, a determination of the type of lactate dehydrogenase found in the blood may be of help in making a correct diagnosis of a patient's difficulty.

EFFECT OF CERTAIN FACTORS ON RATE OF ENZYME REACTIONS

Enzyme concentration

Experiments have shown that the rate of an enzyme reaction is directly proportional to the concentration of the enzyme, i.e., the greater the concentration of enzyme, the faster the reaction takes place. This is a logical consequence of the fact that the larger the ratio of enzyme molecules to substrate molecules, the more likely is the possibility that an enzyme molecule will bump into a substrate molecule and cause a reaction. This relationship is illustrated in Fig. 18-3.

Substrate concentration

If the amount of enzyme is kept constant, at first the rate of reaction is directly proportional to the substrate concentration, but then gradually levels off at a maximum value (Fig. 18-4). This maximum value represents the point at which the enzyme molecules are processing substrate molecules as fast as they can. At this point the enzyme is said to be saturated with substrate.

Temperature

It has long been known that the rate of most chemical reactions increases with temperature. The same is true of enzyme reactions, but because enzymes are proteins, after reaching a maximum, the rate decreases as the enzyme is denatured by heating (Fig. 18-5). Therefore, each enzyme

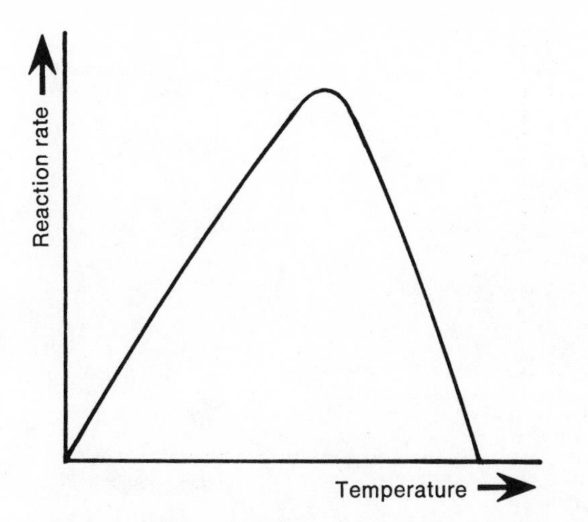

Fig. 18-5. Effect of temperature on reaction rate.

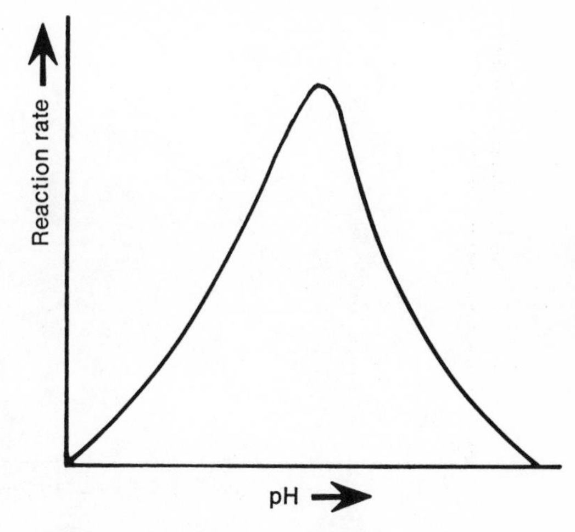

Fig. 18-6. Effect of pH on reaction rate.

has an *optimum temperature* at which it performs best. Most enzymes have an optimum temperature near body temperature.

pH

Again, because of the protein nature of enzymes, the rate of reaction reaches a maximum, the *optimum pH,* as the pH is increased, and then decreases (Fig. 18-6). This is probably caused by the changes in the net charge on proteins, enzymes, and possibly substrates, resulting from changes in pH. The optimum pH represents conditions in which the charges on the enzyme and possibly substrate allow the most efficient catalytic action. Excessive changes in pH, brought on by addition of strong acids or bases, may completely denature and inactivate enzymes.

LOCATION OF ENZYMES IN CELLS

The problem of determining the intracellular location of an enzyme may be approached by two important types of studies. Using the techniques of histochemistry (cytochemistry), which involve special staining or precipitation procedures, the location of specific enzymes in the intact cell can be seen under the microscope.

The other method involves disruption of the cell—differential centrifugation to separate the different particles in the cell (Fig. 18-7)—and experiments to determine which enzymes are associated with which specific particles. The liquid portion of the cell also contains some enzymes and can also be studied by this method. The various metabolic pathways can thus be associated with specific portions of the cell. These relationships will be pointed out later in the discussions on metabolism.

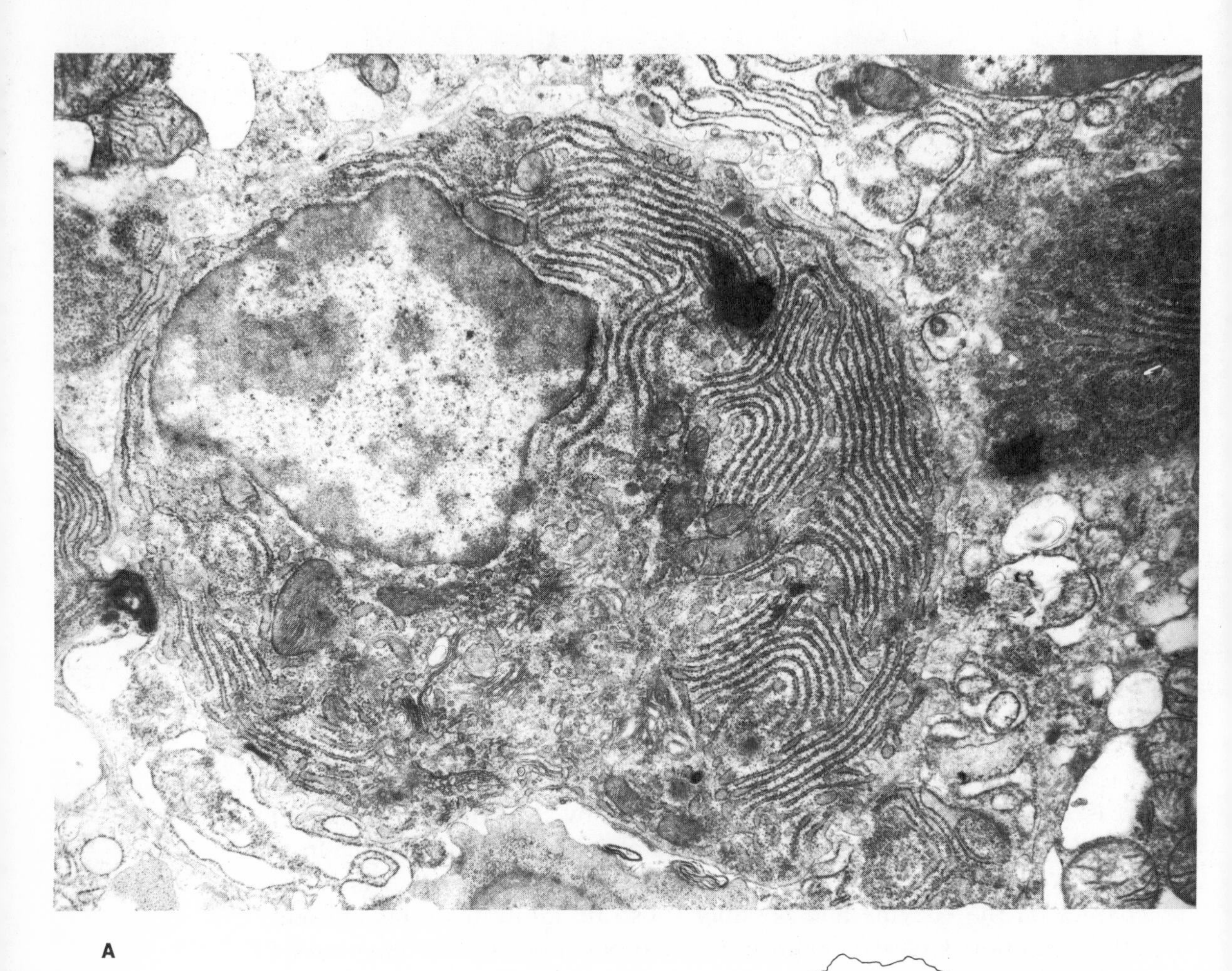

A

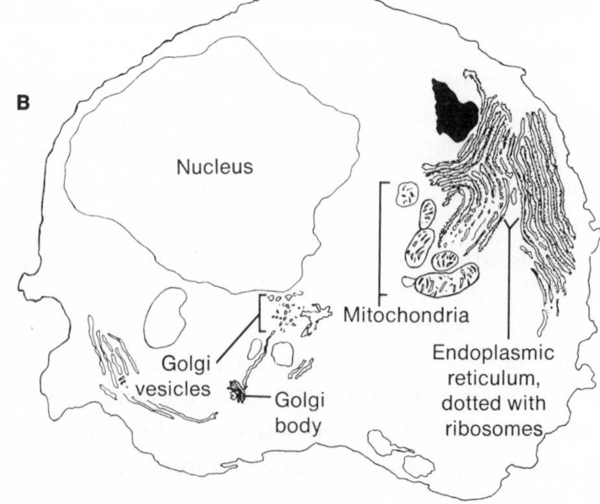

Fig. 18-7. A, Electron microscope view of plasma cell. **B,** The Golgi area (lower center) can be located at lower magnifications, because it does not stain. (×15,000.) (Courtesy J. Barrett, University of Missouri, Columbia, Mo.; from Lane, T. R., editor: Life: the individual, the species, ed. 1, St. Louis, 1976, The C. V. Mosby Co.)

CONTROL OF ENZYMES AND THEIR ACTIVITY

With so many different enzyme functions for each cell to perform, it becomes logical to expect some delicately balanced control or regulatory mechanisms to be in operation. Recent investigations have supplied much evidence in favor of such control systems.

Cellular organization

The organization of the cell into various particles, each with a membrane around it, of itself affords a measure of control on enzymatic processes by limiting the access or exchange of substrate molecules to or between the particles and the liquid portion of the cell. Subtle changes in membrane permeability can have great influence on apparent enzyme activity.

Hormonal influence

That hormones somehow influence enzyme activity is generally accepted. Until relatively recently, however, efforts to explain the manner in which hormones affect enzyme activity had met with little success. At present, most studies are concentrated on developing Sutherland's hypothesis (p. 446) about the mechanism of hormone action, which incorporates previous ideas about the indirect effect of hormones on specific enzymatic reactions and the involvement of hormones with changes in cellular membranes and their functions, which ultimately affect enzyme reactions inside the cell. This mechanism is discussed in Chapter 29.

Genetic influence

Available evidence indicates that the enzyme capabilities of a cell are under genetic control on the basis of the hypothesis that a single gene controls the synthesis of a single enzyme. If the cell has a gene for a certain enzyme, it may produce this enzyme at all times or it may wait for an inducer to stimulate the enzyme's synthesis. If the cell does not have a gene for a certain enzyme, it can never produce this enzyme. That this is a distinct possibility is shown by patients with certain conditions known as *inborn errors of metabolism,* in which an enzyme appears to be missing, resulting in an accumulation of preceding metabolites in large concentrations, often with damaging consequences.

Another genetic disorder for which evidence is accumulating is the production of nonfunctional enzymes or protein products, i.e., enzyme or protein molecules resembling the normal molecules but with enough of an abnormality to make it impossible for these molecules to function normally. Some serious disorders of the blood-clotting system have been traced to such a mechanism.

Negative feedback inhibition

It has been shown that a product toward the end of a long sequence of enzyme reactions may inhibit a reaction near the beginning of the sequence. This phenomenon is known as *negative feedback inhibition* and is a very logical type of control from the standpoint of the economy of the cell, i.e., when sufficient product is available, the sequence of reactions leading to that product is turned off.

NAMING AND CLASSIFICATION OF ENZYMES

At the present time, enzymes are named by using the suffix *-ase* preceded by a term that indicates the substrate or the type of reaction that is taking place. However, many trivial names are still in general use. No single system of nomenclature can take into account all of the possible problems that can and do arise. Listed here is a condensed version of a classification recommended by the Enzyme Commission of the International Union of Biochemistry. In this system, the major classes are divided on the basis of the general type of reaction catalyzed and then subdivided on the basis of the specific group or bond involved in the reaction. A system of four index numbers is used to designate each specific enzyme. Only the first two of these numbers are shown in this list.

Classification of enzymes according to the International Union of Biochemistry

1. Oxidoreductases
2. Transferases, transferring the following groups:
   - 2.1 Carbon
   - 2.2 Aldehyde or ketonic residues
   - 2.3 Acyl
   - 2.4 Glycosyl
   - 2.5 Alkyl or related
3. Hydrolases, acting on following bonds:
   - 3.1 Ester
   - 3.2 Glycosyl
   - 3.3 Ether
   - 3.4 Peptide
   - 3.5 Other carbon—nitrogen bonds
4. Lyases, acting on following bonds:
   - 4.1 Carbon—carbon
   - 4.2 Carbon—oxygen
   - 4.3 Carbon—nitrogen
   - 4.4 Carbon—sulfur
   - 4.5 Carbon—halide
5. Isomerases
   - 5.1 Racemases and epimerases
   - 5.2 Cis-trans isomerases
   - 5.3 Intramolecular oxidoreductases
   - 5.4 Intramolecular transferases
   - 5.5 Intramolecular lyases
6. Ligases, forming following bonds:
   - 6.1 Carbon—oxygen
   - 6.2 Carbon—sulfur
   - 6.3 Carbon—nitrogen
   - 6.4 Carbon—carbon

The oxidoreductases catalyze oxidation-reduction reactions. The transferases are involved in the transfer of various groups that are not free during the transfer process. The hydrolases catalyze the splitting of compounds by the addition of the elements of water across various bonds. The lyases act in the addition or removal of various groups, but in contrast to the transferase reactions, the transferred groups occur free during the reaction. The isomerases are involved in intramolecular rearrangements in which the empirical formula (number of atoms of each type) does not change. The ligases catalyze the linkage of two molecules. Specific examples of some of the various types of enzymes will be encountered in the discussions on metabolism.

SUMMARY

Enzymes make it possible for reactions to occur in the body at appropriate rates and under physiologic conditions of pH, temperature, and chemical concentration. All enzymes are proteins and therefore subject to denaturation by heat, precipitating agents, and excessive changes in pH. Each cell contains many different enzymes.

DEFINITIONS

Substrate: Substance acted upon by an enzyme.
Products: Substance(s) resulting from enzyme action.
Prosthetic group: Small, organic compound firmly attached to enzyme protein and required for enzyme action.
Coenzyme: Small, organic compound loosely attached (easily dissociated) to enzyme protein and required for enzyme action.
Apoenzyme: Protein portion of enzyme.
Holoenzyme: Apoenzyme plus prosthetic group or coenzyme.

MECHANISM OF ENZYME ACTION

Enzymes act as catalysts.

Enzyme-substrate complex: Enzymes combine with substrate to form a complex that causes a strain in an appropriate bond leading to its rupture. After reaction, the complex is broken up, leaving the enzyme free to react with substrate again.

Specificity of enzyme action: Each reaction is catalyzed by its own specific enzyme. Each different enzyme-substrate complex may be considered a different lock and key combination.

Binding site and active site: Binding site is the site that holds enzyme and substrate together. Active site is the site at which catalytic action takes place.

Functions of prosthetic groups and coenzymes: Prosthetic groups and coenzymes act as acceptors of small cleavage products.

Activators: Specific inorganic ions, such as iron and copper, required by some enzymes for activity.

Inhibitors: Substances that interfere with activity of an enzyme.

Metabolic antagonism: Inhibition resulting from similarity in structure of inhibitor and normal substrate.

Competitive inhibition: Inhibition that can be overcome by increasing the ratio of substrate to inhibitor.

Noncompetitive inhibition: Inhibition that cannot be overcome by increasing ratio of substrate to inhibitor.

Preenzymes (zymogens, proenzymes): Inactive form of enzymes that are activated by splitting off portion of protein molecule.

Isozymes: Different forms of an enzyme with similar activity made up of different proportions of parent protein chains. For example, lactate dehydrogenase, a tetramer composed of two parent chains, M and H chains, forms five different isozymes, M_4, M_3H, M_2H_2, MH_3, H_4. Injured tissues may sometimes be identified by the type of isozyme that leaks into the bloodstream.

EFFECT OF CERTAIN FACTORS ON RATE OF ENZYME REACTIONS

Enzyme concentration: The higher the enzyme concentration, the greater the rate of reaction.

Substrate concentration: With increasing substrate concentration (enzyme amount constant), the rate increases to a maximum and levels off at a value representing saturation of the enzyme.

Temperature: Rate rises to a maximum (optimum temperature), then falls off with increasing temperature because of heat denaturation.

pH: Rate rises to a maximum (optimum pH), then falls off with increasing pH.

LOCATION OF ENZYMES IN CELLS

The location of specific enzymes in the intact cell may be determined by the special techniques of histochemistry or by the disruption of the cell—separating the different particles and liquid portion by differential centrifugation, and determining the enzyme activity associated with the various fractions obtained.

CONTROL OF ENZYMES AND THEIR ACTIVITY

Cellular organization: Each cellular particle has a membrane around it and some measure of control is probably associated with changes in membrane permeability.

Hormonal influence: Hormones influence enzymatic activity, but the mechanism(s) are still under investigation. Hormones may affect enzyme activity by way of cyclic adenosine 3′,5′-monophosphate (c-AMP), involving indirect action of hormones and changes in cellular membranes and functions.

Genetic influence: It is believed that a single gene controls the synthesis of a single enzyme. If a gene is missing, as may occur in inborn errors of metabolism, the corresponding enzyme cannot be made. Nonfunctional enzymes may also be produced.

Negative feedback inhibition: A product toward the end of a sequence of enzyme reactions may inhibit a reaction near the beginning of the sequence.

NAMING AND CLASSIFICATION OF ENZYMES

The suffix *-ase* indicates an enzyme; the preceding term indicates the substrate or type of reaction. Many trivial names are still in general use. The major divisions recommended by International Union of Biochemistry are listed here:

1. Oxidoreductases (catalyze oxidation-reduction reactions).
2. Transferases (catalyze transfer of various groups not free during reaction).
3. Hydrolases (catalyze splitting of compounds by addition of water across various bonds).
4. Lyases (catalyze addition or removal of groups that occur free during reaction).
5. Isomerases (catalyze intramolecular rearrangements).
6. Ligases (catalyze the linkage of two molecules).

REVIEW QUESTIONS

1. What are some properties of enzymes that derive from the fact that enzymes are proteins?
2. Explain the effect of temperature on enzyme activity as a result of the fact that enzymes are proteins.
3. Explain the specificity of enzymes on the basis of the lock and key concept. Can competitive and noncompetitive inhibition be explained in similar terms?
4. What are some factors involved in the control of enzyme synthesis and activity? Explain how these factors influence enzyme activity.
5. How may isozyme studies be of diagnostic significance?

REFERENCES

Advances in enzymology: New York, Interscience Publishers.

Baldwin, E.: Dynamic aspects of biochemistry, ed. 5, London, 1967, Cambridge University Press.

Christensen, H. N.: Dissociation, enzyme kinetics, bioenergetics: a learning program for students of the biological and medical sciences, Philadelphia, 1975, W. B. Saunders Co.

Christensen, H. N., and Palmer, G. A.: Enzyme kinetics, ed. 2, Philadelphia, 1974, W. B. Saunders Co.

Cohen, P.: Control of enzyme activity: outline studies in biology, New York, 1976, Halsted Press.

Dickerson, R. E., and Geis, I.: The structure and action of proteins, Menlo Park, Calif., 1969, W. A. Benjamin, Inc.

Fersht, A.: Enzyme structure and mechanism, San Francisco, 1977, W. H. Freeman and Co., Publishers.

Florkin, M., and Stotz, E. H., editors: Comprehensive biochemistry, Vol. 12, Amsterdam, 1964, Elsevier.

Methods in enzymology: New York, Academic Press, Inc.

19 Digestion and absorption

Before the nutrients in food can be made available to the cells of the body, the large molecules must be broken down to a size small enough to be absorbed from the intestinal tract. The total process by which the molecules are prepared for absorption is known as *digestion* and is accomplished by a series of enzymatic changes of a hydrolytic nature (p. 70). These smaller molecules are not available to the body cells until they are transferred from the tube of the digestive tract, across the intestinal walls, and into the bloodstream or the lymphatic system—the process known as absorption. In this sense, the digestive tract may be considered as being external to the body although it is physically surrounded by the body. In other words, the mere fact that food is eaten does not guarantee availability of nutrients to body cells unless the processes of digestion and absorption are occurring normally.

Before food is eaten, it is often processed in one way or another to increase its digestibility. The heat of cooking breaks open starch granules that are covered by a shell of insoluble amylopectin, softens the connective tissues of meat, and makes some proteins (e.g., egg protein) more digestible by causing them to coagulate. Aging of fruits and meats also allows favorable enzymatic changes to increase digestibility of foods. The purpose of any of these means is to increase the accessibility of the food molecules to enzyme action. Some additional benefits of cooking are the possible destruction of microorganisms and harmful enzymes that may be in raw foods.

GENERAL SURVEY OF DIGESTION AND ABSORPTION

The parts of the digestive tract and related organs are shown in Fig. 19-1. The digestive process starts in the mouth, where the food is chewed into small pieces and mixed with saliva. Here digestion of carbohydrates is started. The flow of saliva is stimulated first by psychic and then by mechanical and chemical stimuli. The small portion of food swallowed each time (bolus) then passes relatively rapidly through the pharynx and esophagus and into the stomach. Here the digestion of proteins is started, with little other action. The flow of gastric juice is induced by the action of nervous and hormonal factors originating in the stomach and in the intes-

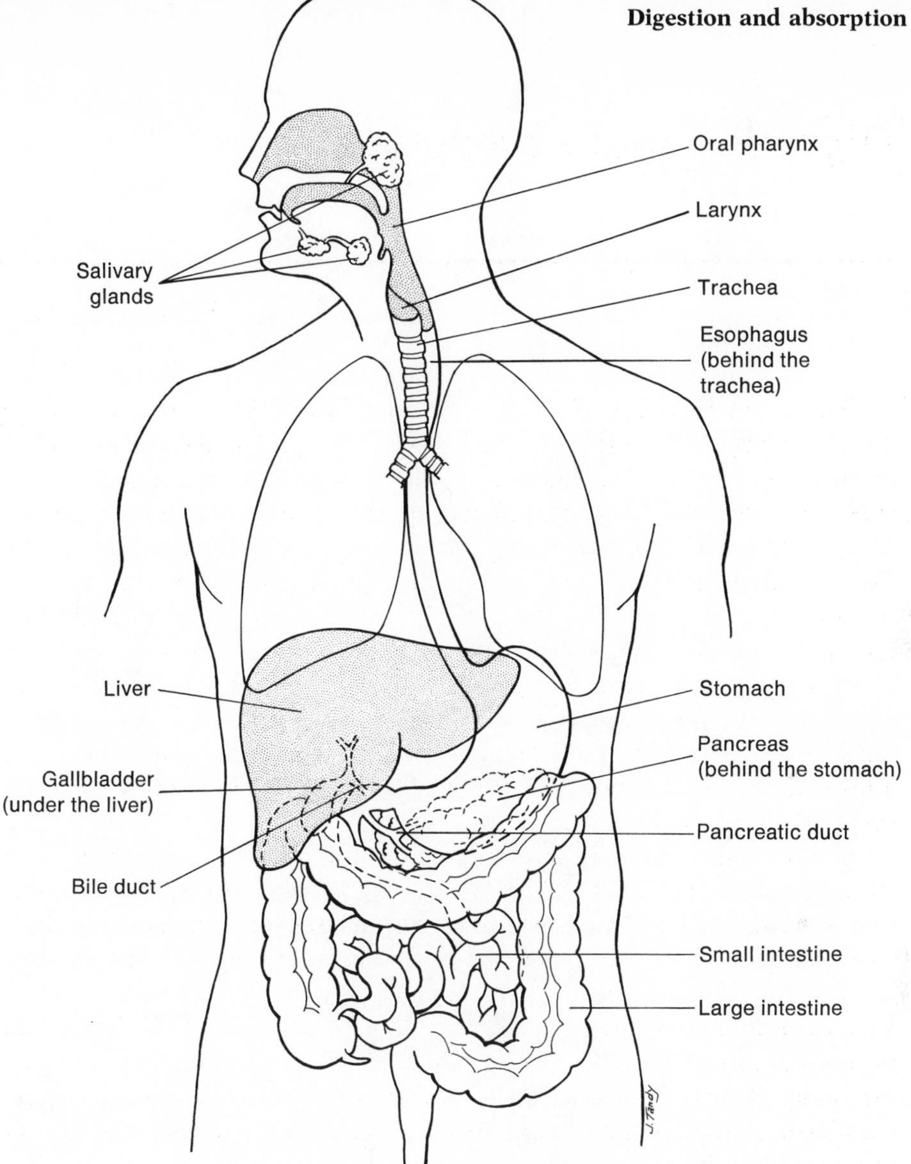

Fig. 19-1. Organs of the digestive system.

tines. After an appropriate period which varies with the type of food eaten (longer with fatty foods), the food is squirted into the duodenum (first section of the small intestine) in small amounts, as controlled by the pyloric sphincter at the junction of the stomach and small intestine.

In the small intestine, where the digestion of carbohydrates and proteins is completed, and the digestion of lipids and nucleic acids is started and completed, digestive enzymes and accessory factors are secreted from several sources. The intestinal mucosa and the pancreas supply the en-

Table 19-1. Digestion

Type of food	Location of action in gastrointestinal tract	Enzyme	Substrate	Products
Carbohydrates	Mouth			
	Saliva	Ptyalin	Polysaccharides	Dextrins, maltose
	Intestine			
	Pancreatic juice	Amylopsin	Polysaccharides, dextrins	Maltose
	Intestinal juice	Maltase	Maltose	Glucose
		Sucrase	Sucrose	Glucose, fructose
		Lactase	Lactose	Glucose, galactose
Proteins	Stomach			
	Gastric juice	Pepsin	Proteins	Smaller polypeptides
		Rennin	Casein	Paracasein curd
	Intestine			
	Pancreatic juice	Trypsin	Proteins, smaller polypeptides	Smaller peptides
		Chymotrypsin	Proteins, smaller polypeptides	Smaller peptides
		Carboxypeptidase	Peptide chains	Smaller peptide chains, amino acids
	Intestinal juice	Aminopeptidase	Peptide chains	Smaller peptide chains, amino acids
		Tri- and dipeptidases	Tri- and dipeptides	Amino acids
Lipids	Stomach			
	Gastric juice	Gastric lipase*	Fats	Fatty acids, glycerol
	Intestine			
	Pancreatic juice	Steapsin	Fats	Fatty acids, glycerol
Nucleic acids	Intestine			
	Pancreatic juice	Ribonuclease	RNA	Oligonucleotides
		Deoxyribonuclease	DNA	Oligonucleotides
	Intestinal juice	Phosphodiesterase	Oligonucleotides	Mononucleotides
		Phosphatases	Mononucleotides	Nucleosides, inorganic phosphate

*Does not operate effectively at the pH of the stomach.

zymes and the gallbladder supplies the bile required for the digestion and absorption of lipids. The flow of pancreatic juice and bile are also stimulated by nervous and hormonal factors.

The products resulting from digestion, and in some cases from bacterial activity, are absorbed mainly in the small intestine. The material remaining after absorption passes into the large intestine (colon), where water is reabsorbed and the feces are concentrated before excretion. (For a summary of enzymes, substrates, and products of digestion see Table 19-1.)

DIGESTION

Mouth—saliva

Food is mixed with saliva in the mouth during the process of chewing. Saliva is a mixture of secretions from the three pairs of salivary glands (the parotid, submaxillary, and sublingual) and the buccal glands. It is a colorless, somewhat viscous fluid of variable composition depending on the type or types of stimulation. The specific gravity is approximately 1.003, indicating a low level of solids. In fact, saliva contains approximately 99.5% water and has a slightly acid pH of about 6.35 to 6.85. The average daily secretion is approximately 1500 ml and its flow is controlled by such stimuli as the thought, sight, odor, or attractiveness of food, as well as the mechanical process of chewing.

The two most important organic constituents in the 0.5% of solids found in saliva are *salivary amylase,* or *ptyalin,* and *mucin,* a glycoprotein. These substances moisten and lubricate the bolus of food for ease in swallowing and movement through the esophagus and start the digestion of carbohydrates. In addition, there are small amounts of other organic substances as well as some inorganic ions such as chloride, bicarbonate, sodium, potassium, and calcium.

Salivary amylase (ptyalin). Starch and glycogen are hydrolyzed to maltose, a disaccharide, by salivary amylase. Chloride, bromide, and calcium ions function as enzyme activators and the optimum pH in sodium chloride solution is 6.9. The enzyme has a wide pH range but is inactivated by a pH below 4 to 5 and by pepsin. Since gastric juice contains pepsin and the stomach contents are at a pH of approximately 1.5 to 2, the action of amylase occurs only until the gastric juice is well mixed with the bolus of food, a period estimated as 15 to 20 min.

Stomach—gastric juice

Normal gastric juice, as saliva, has a specific gravity of about 1.003 and contains only about 0.6% solids, the remainder being water. The major inorganic constituent is hydrochloric acid, with a few salts also present. The most important organic constituents are the enzymes, pepsin and gastric lipase, and mucin (glycoprotein). In infants, a milk-clotting enzyme, rennin, is found in place of pepsin, the rennin gradually disappearing and pepsin appearing in the gastric juice of adults.

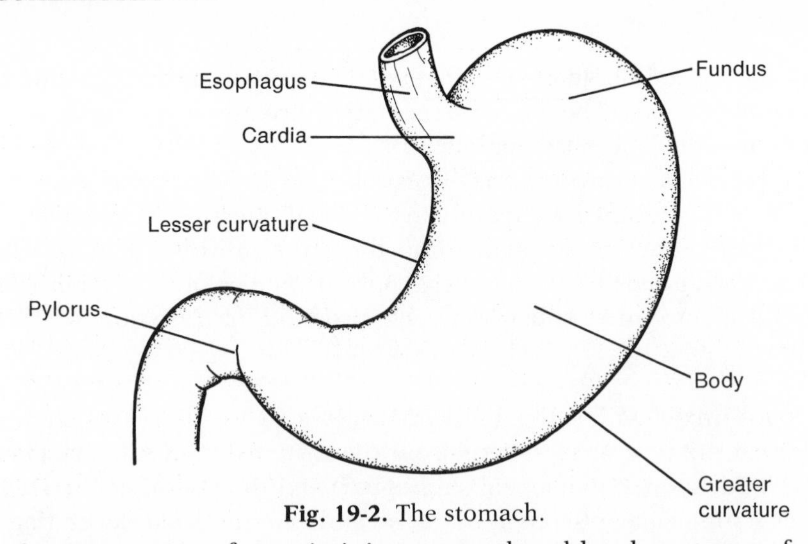

Fig. 19-2. The stomach.

The components of gastric juice are produced by three types of cells: (1) hydrochloric acid by parietal cells; (2) pepsin by chief cells; and (3) mucin by columnar cells. The composition of gastric juice at a particular time depends on the relative secretory activity of the different types of cells, which in turn depends on the type of stimuli being applied, e.g., histamine stimulates the production of hydrochloric acid by parietal cells and the vagus nerve stimulates pepsin production by chief cells. The hormone gastrin, produced by the pyloric region of the stomach (Fig. 19-2) in response to mechanical distention, is secreted into the bloodstream and stimulates secretion of hydrochloric acid. The average daily production of gastric juice may reach as high as 2 or 3 liters.

Hydrochloric acid. The fluid produced by the parietal cells is strongly acidic with a pH of about 0.87. When diluted by the other components, the pH of gastric juice is usually about 1.5 to 2. Among the functions of hydrochloric acid are the maintenance of proper pH for the activity of pepsin, the activation of pepsinogen, stimulation of secretion of secretin in the duodenum, and restriction of the growth of microorganisms in the stomach.

Under certain conditions, as in ulcer patients, more hydrochloric acid is produced than under normal conditions, leading to *hyperacidity*. A decrease in production of hydrochloric acid, or *hypoacidity*, occurs with cancer of the stomach and with advancing age. On occasion an otherwise normal individual has no hydrochloric acid in the gastric juice, a condition referred to as *achlorhydria*.

Pepsin (pepsinogen). *Pepsin* is secreted by the chief cells as the inactive preenzyme (proenzyme, zymogen) *pepsinogen,* which is activated by hydrogen ions. Activation is accomplished by removal of a blocking peptide from pepsinogen. Once some pepsin is formed, it can activate more pepsinogen, a process known as autocatalysis.

$$\text{Pepsinogen} + H^+ \rightarrow \text{Pepsin} + \text{Blocking peptide}$$

The optimum pH range for pepsin is 1.5 to 2.5, a value that coincides with the normal pH range of gastric juice. Pepsin is inactive in alkaline pH and therefore loses its activity when exposed to intestinal juice. Pepsin is an endopeptidase (enzyme that hydrolyzes peptide bonds in the interior of a protein molecule as well as the end peptide bonds) and apparently prefers to act on linkages involving the amino groups of tyrosine or phenylalanine. Smaller peptides resulting from the action of pepsin are known as proteoses and peptones. In the adult, pepsin takes over the milk-clotting function of rennin, found in the gastric juice of infants.

Gastric lipase. Gastric lipase has an optimum pH range of 5.5 to 7.5, depending on the type of lipid acted on, and therefore has little activity in the stomach. It may act when the stomach contents reach the intestine, where the pH becomes alkaline.

Rennin. Rennin is also secreted as a preenzyme that is activated by hydrogen ions. Milk is clotted by rennin to form a curd of insoluble paracasein, which is then digested as are all other proteins.

$$\text{Casein} \xrightarrow[Ca^{++}]{\text{Rennin}} \text{Insoluble paracasein}$$

As noted before, rennin gradually disappears from the gastric juice of infants to be replaced by pepsin in the adult.

Mucus. Gastric *mucus,* containing glycoproteins, is a slippery, slimy substance that is not digested by pepsin. Its function is believed to be to protect the stomach lining from the action of pepsin (self-digestion) and possibly act as a buffer for hydrochloric acid to help control the level of acidity in the stomach. Associated with this fraction is the substance known as *intrinsic factor,* which is required for the intestinal absorption of vitamin B_{12} (p. 434), a vitamin necessary for normal red cell production. In some individuals intrinsic factor is lacking, which leads to a condition known as pernicious anemia resulting from a vitamin B_{12} deficiency.

Small intestine—pancreatic juice, intestinal juice, bile

Pancreatic juice. Pancreatic juice, as secreted by the pancreas, has a specific gravity of about 1.007, is composed of 1.3% solids and 98.7% water, and has an alkaline pH of 7.5 to 8.2 because of its relatively high bicarbonate content. It contains enzymes that act on all the major types of food molecules: proteins, lipids, carbohydrates, and nucleic acids. As with gastric juice, the type of pancreatic juice produced depends on the type of stimulation. Secretin, a hormone produced by the duodenal mucosa as a result of contact with acid from the stomach, causes the secretion of pancreatic juice of high volume and bicarbonate content but low enzyme concentration. Pancreozymin, another hormone produced by the duodenal mucosa, and vagal stimulation increase the secretion of pancreatic juice

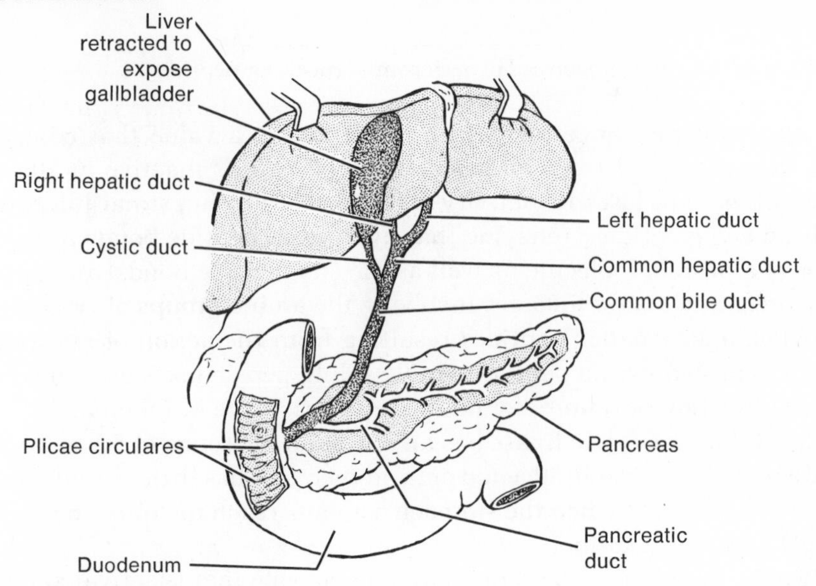

Fig. 19-3. Arrangement of bile ducts and pancreatic duct leading into duodenum.

with high enzyme concentration but low volume. The average daily secretion is about 500 ml and reaches the small intestine by way of the pancreatic duct into the duodenum (Fig. 19-3).

Peptidases (trypsin, chymotrypsin, carboxypeptidase). The proteolytic enzyme *trypsin* is secreted in pancreatic juice as the inactive preenzyme *trypsinogen,* which is converted to active trypsin by the intestinal enzyme enterokinase. Trypsinogen can also be activated by trypsin (autocatalysis):

$$\text{Trypsinogen} \xrightarrow{\text{Enterokinase}} \text{Trypsin}$$

Trypsin is an endopeptidase with an optimum pH range of 8 to 9 and a preference for linkages involving the carboxyl groups of arginine or lysine. Proteins are thus broken down to polypeptides of different sizes, proteoses, peptones, and some amino acids.

Chymotrypsin is also secreted in pancreatic juice as the inactive preenzyme *chymotrypsinogen,* which is activated in the intestine by trypsin:

$$\text{Chymotrypsinogen} \xrightarrow{\text{Trypsin}} \text{Chymotrypsin}$$

Chymotrypsin has proteolytic activity similar to trypsin and operates optimally in the same pH range of 8 to 9, preferably on peptide linkages involving the carboxyl groups of tyrosine or phenylalanine. Also, chymotrypsin has a strong milk-clotting action, which may be important in taking over the function of rennin in the adult.

Carboxypeptidase is an example of an exopeptidase, a proteolytic enzyme that splits off from peptide chains the terminal amino acid with the free carboxyl group. It has an optimum pH of 7.4.

Pancreatic amylase (amylopsin). Pancreatic amylase, with an optimum pH range of 6.5 to 7.2, acts in a manner identical to that of the salivary amylase (ptyalin) in hydrolyzing starch or glycogen to the disaccharide stage—maltose.

Pancreatic lipase (steapsin). Steapsin hydrolyzes fats (glycerides) by splitting off fatty acids, leaving free glycerol if the process goes to completion:

$$\begin{array}{l} H_2C-O-C(=O)-R \\ | \\ HC-O-C(=O)-R \\ | \\ H_2C-O-C(=O)-R \end{array} + 3H_2O \xrightarrow{\text{Steapsin}} \begin{array}{l} H_2C-OH \\ | \\ HC-OH \\ | \\ H_2C-OH \end{array} + 3RCOOH$$

Triglyceride **Glycerol** **Fatty acids**

However, because of the water-insolubility of fats and other factors, conditions in the intestine are not ideal for the complete digestion and absorption of fats and their constituents. These processes are aided by emulsification of fats, particularly by the bile salts in bile (Fig. 19-3), resulting in a great increase in the surface area of the fats available for enzyme action.

Nucleases. Also in pancreatic juice are *ribonuclease* and *deoxyribonuclease,* both endonucleases that break down the corresponding nucleic acids to smaller polymers known as oligonucleotides.

Intestinal juice. Intestinal juice is not as uniform a product as pancreatic juice because the secretory cells in the different areas of the small intestine (duodenum, jejunum, ileum) produce fluids of different composition. Also, varying amounts of white cells and discarded epithelial cells from the mucosal lining of the small intestine appear in the intestinal juice. Since these cells contribute their enzymes to the intestinal fluid, it is sometimes difficult to determine which enzymes are actually secreted into the small intestine and which are there only by accident. Intestinal juice contains about 1.5% solids, half being inorganic salts and the other half organic substances such as enzymes, mucin, and others.

The pH in the small intestine varies according to location, type of food, and extent of bacterial action. It ranges from about 5.5 to 6.0 in the duodenum and from 6.0 to 6.5 in the terminal ileum, becoming more alkaline as the material gets further away from the stomach. The daily secretion of intestinal juice is difficult to estimate but is known to be under nervous system and hormonal control.

Disaccharidases (maltase, sucrase, lactase). Evidence at present indicates that the disaccharidases (carbohydrases) *maltase, sucrase,* and *lactase* are not actually secreted into the intestinal lumen but that the hydrolysis and absorption of the respective disaccharides—maltose, sucrose, and lactose—take place in the brush-border membrane of the epithelial cells lining the small intestine. The presence of disaccharidases in the intestinal juice is probably caused by breakdown of these epithelial cells.

$$\text{Maltose} \xrightarrow{\text{Maltase}} \text{2 Glucose}$$

$$\text{Sucrose} \xrightarrow{\text{Sucrase}} \text{Glucose + Fructose}$$

$$\text{Lactose} \xrightarrow{\text{Lactase}} \text{Glucose + Galactose}$$

Many people have a condition known as *lactase insufficiency,* which may occur because of hereditary deficiency of lactase or by removal of lactase-secreting portions of the small intestine by surgery. With this condition, the lactose, not being hydrolyzed to the monosaccharides, cannot be absorbed. The result is an osmotic pull of water into the intestinal tract, resulting in diarrhea and other intestinal distress symptoms. There have also been reports showing decreased levels of lactase in many ethnic groups after early childhood. This may be one reason why many adults develop a dislike for milk.

Peptidases (aminopeptidase, tri- and dipeptidases). Aminopeptidase is an exopeptidase that hydrolyzes terminal amino acids with a free amino group. The *tri-* and *dipeptidases* split the smaller peptides into constituent amino acids.

Nucleases (phosphodiesterases, phosphatases). Oligonucleotides are split to mononucleotides by the exonuclease *phosphodiesterase,* which removes them one at a time from the end of the nucleotide chain. *Phosphatases* then hydrolyze the mononucleotides to nucleosides and inorganic phosphate:

$$\text{Oligonucleotides} \xrightarrow{\text{Phosphodiesterase}} \text{Mononucleotides}$$

$$\text{Mononucleotides} \xrightarrow{\text{Phosphatases}} \text{Nucleosides + Inorganic phosphate}$$

Nucleosides can be absorbed as such, any further action to break them down to purine and pyrimidine bases and pentoses occurring in individual cells.

Enterokinase. Enterokinase converts trypsinogen to trypsin (see p. 270).

Bile. Bile is secreted by the liver as a golden-yellow fluid with a pH of about 7.0 to 8.5. It is stored in the gallbladder, where it is concentrated by the reabsorption of water and inorganic constituents. At this point, the bladder bile has a pH range from 5.5 to 7.7 and 4% to 17% total solids. The average daily output is about 500 to 700 ml, the solids consisting mostly of organic substances such as bile salts, bile pigments, and cholesterol. The release of bile from the gallbladder is stimulated by the hormone *cholecystokinin,* which is secreted into the bloodstream by the upper intestine under the influence of fat. Bile enters the small intestine in the region of the duodenum (Fig. 19-3).

Bile salts. The bile salts, such as *sodium glycocholate* and *sodium taurocholate* (see Bile acids in Chapter 15), function in the digestion of fats and absorption of the resultant products by causing emulsification because of their surface tension–lowering properties. Such action is also necessary for the proper absorption of cholesterol and the fat-soluble vitamins A, D, and K. The bile salts are absorbed from the intestine into the bloodstream, removed from the blood by the liver, and processed into bile again, leading to an enterohepatic circulation.

Bile pigments. The color of bile is caused mainly by the bile pigment *bilirubin* (p. 380), in the form of its glucuronide. The bile pigments are breakdown products of hemoglobin and will be discussed in greater detail with the metabolism of hemoglobin. An important distinction must be made here between the bile salts, which have an important function in digestion and absorption, and the bile pigments, which have no known function and are in the bile only because nature chose this way of getting these products into the intestine for excretion. Although the bile pigments have no metabolic function, abnormalities in their disposition can have important clinical significance in diseases of the liver and hemolytic conditions, e.g., the condition of *jaundice* (yellow skin) indicates an abnormally high concentration of bile pigments in the skin because of blocked excretion or overproduction, or both, of bile pigments.

Cholesterol. Although the function of cholesterol in the bile is unknown (it was believed to be a means of excretion of blood cholesterol), it is known to cause difficulties by the formation of gallstones, often blocking the flow of bile into the intestine.

ABSORPTION

The process by which molecules are transferred through the walls of the digestive tract into the blood or lymphatic systems is known as *absorption.* The major site of absorption of nutrient molecules from the digestive tract is the small intestine. Alcohol and small amounts of water and other small molecules may be absorbed from the stomach. Water is the major substance absorbed from the colon. In order to present as much absorptive surface as possible, the small intestine is lined with many projections called *villi,* which effectively increase the absorptive surface of the small intestine many times in contrast to a corresponding length of smooth tube. The diagrammatic structure of a villus, including the absorptive surface and blood and lymph systems, is shown in Fig. 19-4.

Substances absorbed by way of the portal blood (generally water-soluble substances) go directly to the liver, where many are used directly and others are converted to new substances before being sent back to the bloodstream and other tissues. Those substances absorbed into the lymphatic system (usually fat-soluble substances) go by way of the thoracic duct into the subclavian vein of the bloodstream, thus bypassing the liver for a time.

Food material remains in the small intestine for several hours depending on the type of food. Fatty foods tend to remain longer than other types. This is probably related to the slower rates and lower efficiencies of digestion and absorption of fats as compared to those of carbohydrates and proteins. Further details about these processes will be discussed with the metabolism of these food substances.

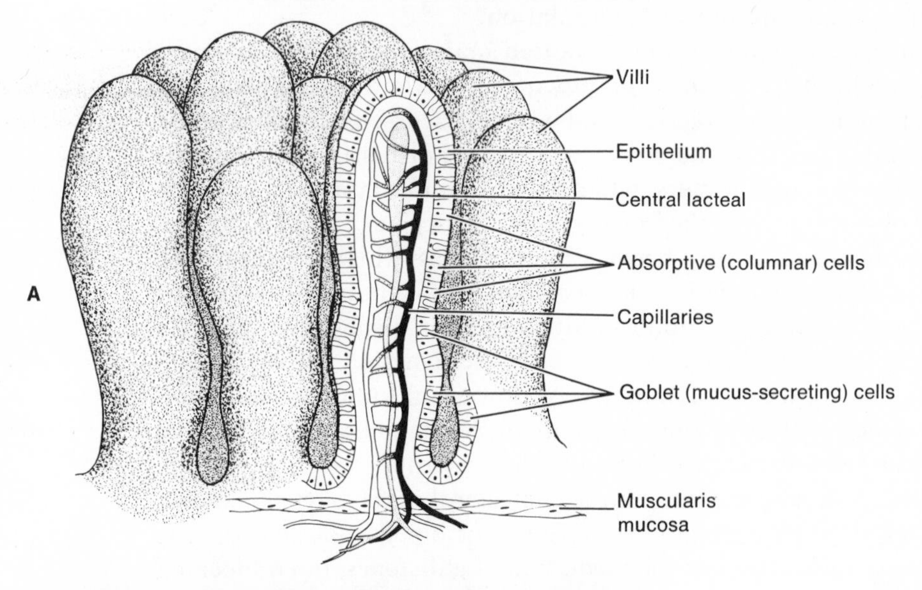

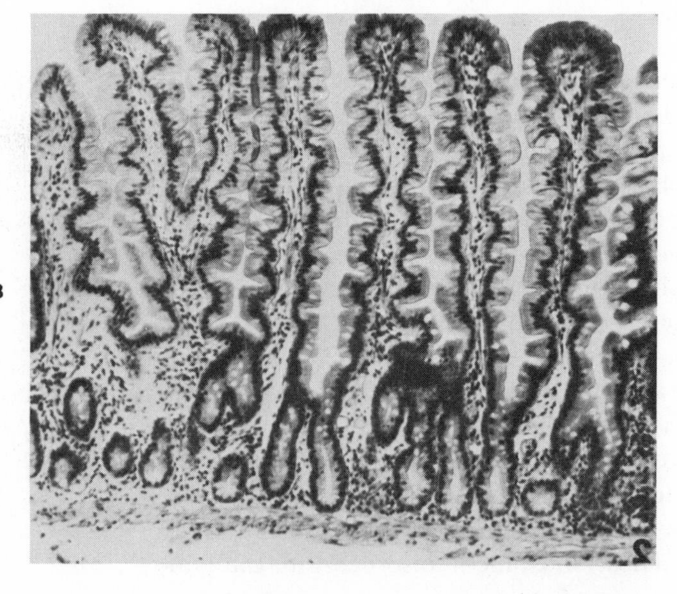

Fig. 19-4. A, Schematic diagram of villi illustrating the organization of the small intestinal mucosa. **B,** Light micrograph of a section of jejunal mucosa. (×100.) (From Trier, J. S.: Reprinted from Federation Proceedings **26:**1391-1404, 1967.)

FECES

On completion of digestion and absorption of foodstuffs on their way through the small intestine, the remaining semiliquid material enters the large intestine, or colon. Included in this material, under normal conditions, may be small amounts of undigested and unabsorbed food, indigestible substances such as cellulose, water, discarded intestinal lining cells, enzymes, large amounts of bacterial cells and their products, bile pigments that provide most of the color of feces, and mucus.

Bacterial action on carbohydrates is known as *fermentation* and results in the production of organic acids such as lactic and butyric, excessive amounts of which may cause diarrhea. Bacterial action on proteins is known as *putrefaction* and results in the production of such substances as *indole* and *skatole,* which are largely responsible for the odor of feces. Gases, such as methane and hydrogen sulfide, also contribute to the odor. Excessive production of these and other gases may lead to the distressing symptoms of distention. Although bacterial action in the intestines produces some potentially useful products, in most instances these substances are too far down the intestinal tract to be absorbed.

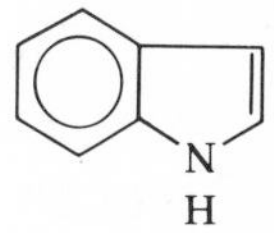

N
H

Indole

CH_3
N
H

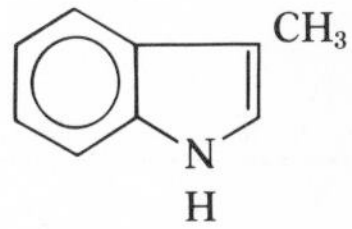

Skatole

The amount of nitrogen (from protein) excreted daily on an average diet ranges from 0.5 to 1.5 g. The amount of lipids ranges from 5% to 25% of the dry weight of the feces. Observing these factors in patients may be of significant help in diagnostic problems.

After reabsorption of most of the water, the feces are excreted from the colon.

SUMMARY

Digestion is a series of enzymatic hydrolytic reactions in the gastrointestinal tract, which converts large food molecules to small ones that can then be absorbed and made available to the cells of the body. Cooking and other types of preparation make foodstuffs more digestible.

DIGESTION

Carbohydrate digestion starts in the mouth and is completed in the small intestine. Lipid digestion takes place mainly in the small intestine. Protein digestion starts in the stomach and is completed in the small intestine. Nucleic acid digestion takes place in the small intestine. Enzyme sources are saliva (mouth), gastric juice (gastric mucosa), pancreatic juice (from pancreas into the small intestine), and intestinal juice (intestinal mucosa). Bile, required for digestion and absorption of lipids, is produced by the liver, stored in the gallbladder, and released into the small intestine. Almost all useful absorption takes place in the small intestine. Remaining material passes into the large intestine, where water is reabsorbed as the feces are concentrated and then excreted. Some enzymes (pepsin, trypsin, chymotrypsin) are secreted in inactive form (preenzyme) and are activated by specific substances. Bile is necessary for the appropriate digestion of fats and absorption of the products. The proper pH must be maintained in the various digestive juices for normal digestion. The flow of juices is controlled by stimuli of nervous system, hormonal, and mechanical origin. For a summary of digestion, see Table 19-1 (p. 266).

ABSORPTION Most useful absorption takes place in the small intestine, where the absorptive surface is increased by many projections called villi. Water-soluble substances are generally absorbed by way of the portal vein, going directly to the liver for use or conversion. Fat-soluble substances are usually absorbed by way of the lymphatic system, bypassing the liver at first. Fatty foods remain in the small intestine longer than other meals.

FECES Unabsorbed material reaches the colon where water is reabsorbed and much of the bacterial action takes place, such as fermentation of carbohydrates (produces organic acids) and putrefaction of proteins (produces indole and skatole). Determination of nitrogen excreted in the feces or amount of lipids excreted may be of diagnostic importance.

REVIEW QUESTIONS

1. What type of chemical reaction predominates in digestive processes? How much energy is required in these reactions? What is the purpose of digestion?
2. What is a preenzyme? How are preenzymes activated? Name some preenzymes and indicate specifically how they are activated.
3. Trace a carbohydrate through the digestive tract (mouth, stomach, small intestine) indicating which enzymes may act on it in each area, the organ of origin of the enzymes, and what the resulting products are. Do the same for a protein, a fat, and a nucleic acid.
4. Examine the following lists and arrange the substances in each list in the order in which they would appear as food travels down the intestinal tract:
   a. Peptones, protein, amino acids, tripeptides.
   b. Starch, glucose, maltose, dextrins.
   c. Nucleotide, nucleoside, nucleic acid.
5. What is the function of the bile and of bile salts? How is this accomplished? What is the pH of the bile? What is the function of the bile pigments? What substance in bile may cause blockage of the bile ducts by formation of gallstones?

REFERENCES

Annual review of physiology: Palo Alto, Calif., Annual Reviews.

Benson, J. A., Jr., and Rampone, A. J.: Gastrointestinal absorption, Ann. Rev. Physiol. **28:**201, 1966.

Davenport, H. W.: Physiology of the digestive tract, ed. 4, Chicago, 1977, Year Book Medical Publishers, Inc.

Grossman, M. I., editor: Gastrointestinal hormones and pathology of the digestive system, New York, 1978, Plenum Publishing Corp.

Johnson, L. R.: Gastrointestinal physiology, St. Louis, 1977, The C. V. Mosby Co.

Physiological reviews: Washington, D.C., American Physiological Society.

Trier, J. S.: Structure of the mucosa of the small intestine as it relates to intestinal function, Fed. Proc. **26:**1391, 1967.

van der Reis, L., and Lazar, H. P.: Human digestive system, Basel, Switzerland, 1972, S. Karger.

White, W. L., Erickson, M. M., and Stevens, S. C.: Chemistry for the clinical laboratory, ed. 4, St. Louis, 1976, The C. V. Mosby Co.

Introduction to intermediary metabolism, biologic oxidation, and bioenergetics

20

In the preceding chapter were discussed the processes whereby food is digested in the gastrointestinal tract and then taken into the body, or absorbed, by way of the small intestine. In subsequent chapters the manner in which the body converts or otherwise disposes of the absorbed food molecules, the biochemical field known as intermediary metabolism, will be discussed in detail. Because of the amount of such detail known, the important general purposes of intermediary metabolism are sometimes overlooked or pushed into the background. In order not to lose sight of some of the major accomplishments or reasons for intermediary metabolism—namely the production, capture, and manipulation of energy resulting from biologic oxidation reactions—these subjects are discussed in this chapter.

INTERMEDIARY METABOLISM

Intermediary metabolism begins after the absorption of food molecules from the small intestine into the blood and lymph systems and ends with the excretion of the resultant waste products. The strategic location of intermediary metabolism in the economy of the organism is shown here:

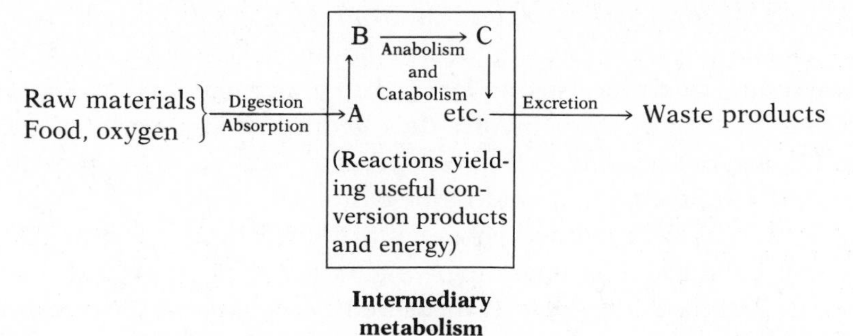

Intermediary metabolism

It should be obvious that the reactions constituting intermediary metabolism are of the greatest importance to the well-being of the organism. It is by studying these reactions that life scientists attempt to define the normal status of an organism in such terms that abnormalities may be recognized quickly and steps taken, if possible, to correct such abnormalities. Before a discussion of specific metabolic pathways in subsequent chapters, a limited examination of the methods used in such studies should be of interest.

Methods of studying intermediary metabolism

In vivo methods. The most obvious way of studying what occurs in a living animal is to study the living animal itself. This is known as the *in vivo* (in life) method. The great advantage of in vivo methods is that, under ideal conditions, studies may be made of reactions as they occur in the natural state.

However, there are several important reasons why it is often not possible to achieve ideal conditions. Because of the complexity of the living organism and the interrelationships of the various reactions taking place, it is generally not a simple task to trace things as they occur. Dangerous or possibly fatal experiments cannot be performed on human beings. The use of other animals in experiments attempting to advance knowledge of the human body has two major drawbacks: (1) there are certain conditions in the human that have not yet been found or duplicated in any other animal, and (2) the fact that something does or does not occur in another animal is incomplete proof that the same is true of the human or any animal other than the one being studied (species differences).

In vitro methods. In order to avoid some of the difficulties inherent in the in vivo methods, various ways were sought to scale down the size and complexities presented by whole organisms. Methods using something less than a whole, living organism are known as *in vitro* (in glass) methods.

One of the in vitro methods that is actually in between in vivo and in vitro, is the perfusion of a whole organ removed from an animal. The organ is supplied with circulating blood or an adequate substitute and it may function for hours after removal from the animal. The effects of various substances can be investigated by adding test substances to the circulating fluid and then looking for resulting products.

Other in vitro methods involve the use of tissue slices or homogenates, purified enzyme systems, cell-free extracts, cell components separated by differential centrifugation, and tissue culture.

A classic in vitro method is known as the Warburg manometric technique. By running reactions involving tissue slices or homogenates in specially designed and carefully measured closed vessels, the consumption of oxygen can be measured very accurately and correlated with the changes observed in the substrates added. More modern automated methods are now available for the performance of experiments of this type.

All of these in vitro methods make it possible to pinpoint the factor under study to a much greater degree than is possible with in vivo methods.

The greatest objection to the in vitro methods is that by disrupting the natural organization of the cells and tissues of an organism, artificial conditions may allow or prevent reactions from taking place that may or may not normally occur in the whole organism. In spite of these and other possible objections, judicious choice of experimental material and methods and appropriate interpretation of results of in vivo and in vitro experiments has led to important advances in the study of the living organism.

Microorganisms. In the same way that certain reactions are common to many species of animals, some reactions occurring in microorganisms also occur in animals. When this is true, the experimenter is sometimes very fortunate because microorganisms make excellent experimental material from the viewpoint of growth rate, numbers, ease of handling, generation time, and space requirements.

Nutritional experiments. Most of the early experiments with living organisms were nutritional, i.e., animals were fed certain substances and various effects were investigated. In many cases the accumulation of a product in the tissues or excreta was looked for as an indicator of a metabolic pathway. This procedure was rather limited until the advent of isotopic methods (discussed next) because massive doses of the starting substance were often required to produce a relatively small increase in the product under investigation. The results were often difficult to interpret because the animals were being exposed to unphysiologically large amounts of materials, and the small effects often observed were actually indirect effects.

Another type of nutritional study that has been used successfully involves the use of prepared diets from which a single substance has been omitted or removed. The effect of such a deficiency on the well-being of the organism can then be determined by observing the animal over a long period, if necessary.

Isotope experiments. The common characteristic of all isotopes is that they differ in mass from other atoms with the same atomic number. Isotopes can then be subdivided into two classes: (1) *stable isotopes,* which can be distinguished only by differences in mass, and (2) *radioactive isotopes,* which, in addition to differences in mass, have unstable nuclei that decompose spontaneously with the emission of radiation, e.g., beta (β) or gamma(γ) rays.

The concentration of both types of isotopes can be measured very accurately in very small amounts, a fact that makes it possible to use what are known as *tracer* doses of test compounds, usually constituting only a small fraction of the amount present under physiologic conditions. Stable isotopes are measured in a device known as a *mass spectrometer,* which separates atoms on the basis of their weights. Radioactive isotopes are measured by electronic means in detector-scaler devices of varying types that can determine not only the amount of isotope present but the type and intensity of the radiation. At the present time, most of the isotopes used in biochemical research are radioactive isotopes, mainly prepared as by-products in atomic piles.

These isotopes are incorporated into suitable compounds by the usual synthetic chemical procedures and are then ready for use in experiments. One important factor that must be taken into consideration is the *half-life* of the isotope. The half-life of a radioactive substance is the time it takes for half the radioactivity to dissipate. The half-lives of some of the more commonly used radioactive isotopes are listed here:

*Isotope**	*Half-life*
^{3}H (tritium)	12.5 yr
^{14}C	5,760 yr
^{32}P	14.3 days
^{131}I	8.1 days
^{55}Fe	4 yr
^{59}Fe	45.1 days
^{36}Cl	440,000 yr
^{38}Cl	37 min
^{24}Na	15.1 hr
^{42}K	12.4 hr

Although at first thought it might appear advantageous to always use the longer-lived isotopes, it may be more strategic, especially in human studies, to use the shorter-lived isotopes—particularly in substances that remain in the body for longer periods (biologic half-life)—thus decreasing the radiation hazard.

In a simple example of the use of such radioactive tracer compounds in a problem some aspects of which were relatively unapproachable by other means, radioactive $^{14}CO_2$ was supplied to a growing green plant in a closed system, and after a time the sugar produced by the plant was isolated and assayed for radioactivity. Radioactive carbon 14 was found in the sugar, indicating that carbon dioxide was a precursor in the production of carbohydrates by green plants. Because the pathway between carbon dioxide and sugar involves many steps, other intermediate compounds can and have been separated in attempts to determine the complete pathway. Another way of determining whether or not a substance is an intermediate on the direct pathway of biosynthesis is to supply the intermediate labeled with a radioactive tracer atom, then isolate a product further down the pathway and determine its radioactivity. If the isolated product has a

*Superscript = atomic mass.

reasonable amount of radioactivity commensurate with the proposed biosynthetic pathway, there is presumptive evidence that the suspected intermediate is on the direct pathway under investigation. As noted before, because of the interrelationships between metabolic pathways, results of isotopic experiments must be interpreted with great care. In spite of these and other problems that must be faced in the use of radioactive isotopes, it is probably fair to say that biochemical research in the last 30 years has made more progress than in the previous 200 years. Much of the work discussed in the following chapters is a direct result of isotopic experiments.

BIOLOGIC OXIDATION

A major function of intermediary metabolism is to extract from the absorbed food molecules the energy required for the maintenance of life. The most important type of energy-producing reaction available to living organisms is the oxidation-reduction reaction (Chapter 6). It should be recalled from inorganic chemistry that when hydrogen is ignited in the presence of oxygen, there is a rapid, explosive oxidation-reduction reaction in which water is formed and a large quantity of energy is produced:

$$2H_2 + O_2 \rightarrow 2H_2O + \text{energy}$$

It should also be recalled that by the expenditure of energy the reaction can be reversed and water can be decomposed into its constituent elements, as is the case in the electrolysis of water:

$$\text{electrical energy} + 2H_2O \rightarrow 2H_2 + O_2$$

The reaction in organic chemistry analogous to the electrolysis of water is photosynthesis:

$$\text{light energy} + 6CO_2 + 6H_2O \rightarrow \underset{\textbf{Sugar}}{C_6H_{12}O_6} + 6O_2$$

The light energy from the sun absorbed in this reaction is, in effect, stored in the products of the reaction, such products ultimately becoming the food molecules of living organisms. It becomes the task of biologic oxidation-reduction reactions, usually referred to simply as biologic oxidation reactions, to extract this energy and make it available to the organism. This is done by oxidizing the food molecules, a process which reverses the photosynthetic reaction:

$$C_6H_{12}O_6 + 6O_2 \xrightarrow{\text{oxidation}} 6H_2O + 6CO_2 + \text{energy}$$

Isotopic studies have shown that the oxygen breathed in by an organism ultimately combines with hydrogen from the food molecules to form water and the oxygen in the carbon dioxide comes mainly from the food molecules.

In considering the inorganic and organic reactions cited, the law of conservation of energy and its implications should be kept in mind. No energy is created or destroyed by these reactions and the reversal of an energy-absorbing reaction cannot produce more energy than was originally absorbed. Also to be kept in mind is that an explosive reaction is unlikely to be of much benefit for a living organism, in which the storage of energy for future use is an important feature rather than immediate use of all energy produced. To this end, nature has devised a system of stepwise reactions in which relatively small packets of energy are handled at any one time. Thus, the overall biologic oxidation reaction generally takes place as a long series of reactions before the hydrogen from a substrate combines with oxygen to form water, rather than as the one-step reaction written for the oxidation of hydrogen directly to water. The maximum energy to be derived is the same whether the reaction takes place in one step or in many steps. It is much easier to walk up ten steps rather than jump in one leap, but the result is the same.

Stepwise process of biologic oxidations

For purposes of this discussion, oxygen is considered as having the highest oxidation potential (strong attraction for electrons). The reverse is true for hydrogen atoms (hydrogen ions plus electrons) of substrate molecules. The task of biologic oxidation is to bring the low oxidation–potential hydrogen atoms together with the high oxidation–potential oxygen to form water, releasing energy along the way. In any reaction between two substances, e.g., A and B, having different oxidation potentials, the substance having the higher potential (A) will become reduced by accepting electrons from the substance having the lower potential (B). In the process of giving up its electrons, substance B becomes oxidized. A shorthand form frequently used to illustrate such reactions is shown here:

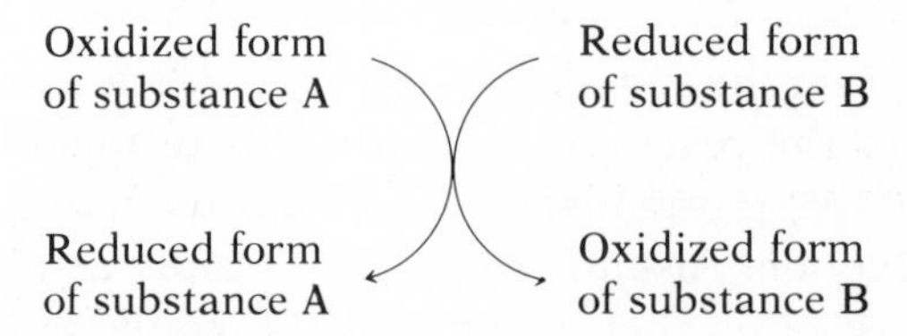

In an actual biologic oxidation, many such steps may be coupled together, forming what is known as an *oxidative chain.* The substances involved in moving the hydrogen ions or electrons, or both, from the substrate to the oxygen are called *carriers.* As the reactions take place, the carriers cycle between oxidized and reduced forms. In an oxidation of the substrate, RH_2, in which the coenzyme nicotinamide adenine dinucleotide (NAD) (Chapter 17) acts as the first carrier, the steps by which the hydrogens and electrons are moved to the oxygen may be written in simplified form as follows (*cyto* indicates cytochromes):

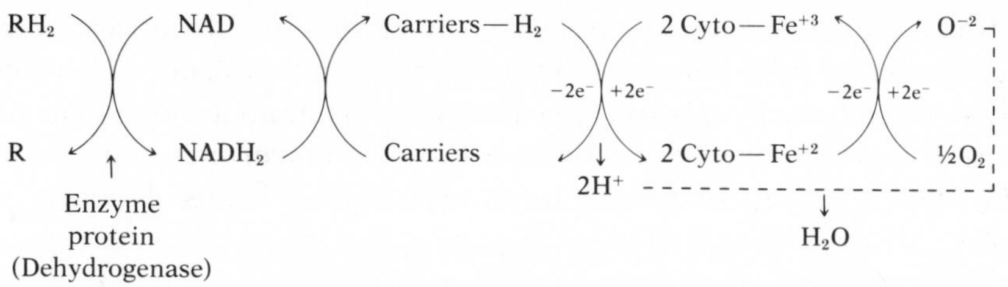

RH_2 → R (Enzyme protein (Dehydrogenase)); NAD → $NADH_2$; $Carriers—H_2$ ⇄ Carriers; −2e⁻ / +2e⁻; 2 Cyto—Fe^{+3} ⇄ 2 Cyto—Fe^{+2}; −2e⁻ / +2e⁻; O^{-2} ⇄ ½O_2; $2H^+$ → H_2O

There are several important points to be noted in the reactions summarized here. The events that take place are started by the removal of two hydrogens (with electrons) from a suitable substrate under the influence of a dehydrogenase enzyme that functions with NAD as the coenzyme. The substrate is oxidized as the NAD is reduced. In general terms, reading from the substrate end to the oxygen end, each succeeding carrier is reduced by reacting with the preceding carrier and as each carrier is reduced, its preceding neighbor is reoxidized to its original state and is ready to react again. Because of the cyclic nature of these reactions, the coenzymes and other carriers involved are required only in low concentrations to oxidize relatively large amounts of substrate. The hydrogen atoms, with their electrons, are transferred together by NAD and following carriers until the cytochrome portion of the oxidative chain is reached. The formula for NAD and the manner in which the hydrogen atoms and their electrons are carried by the nicotinamide portion of NAD are shown here:

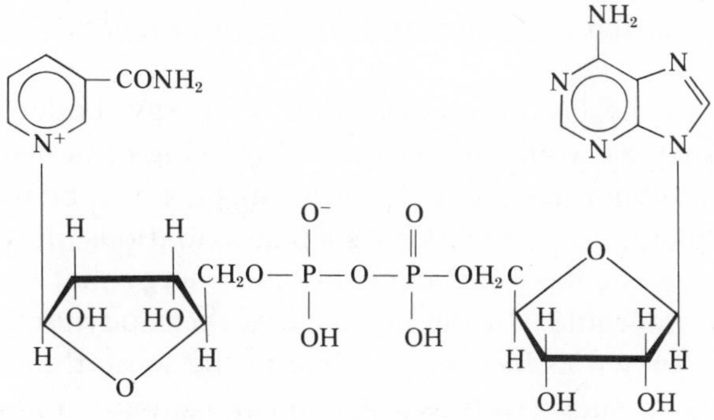

Nicotinamide adenine dinucleotide (NAD)

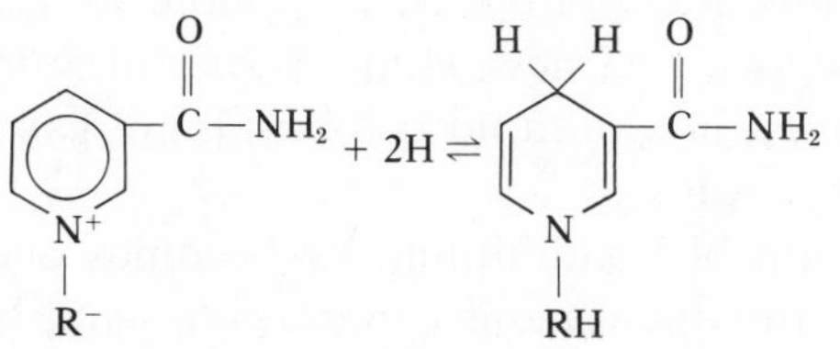

NAD carrying two hydrogen atoms

Since the cytochromes (a series of five different substances, all containing iron in their iron-porphyrin prosthetic groups [p. 382]) transfer only electrons, the hydrogen is released to the medium as ions at the point where the chain meets the cytochromes. The cytochromes transfer electrons by cycling between the reduced and oxidized forms of iron:

$$2e^- + 2\ \text{cyto—Fe}^{+3} \rightleftharpoons 2\ \text{cyto—Fe}^{+2}$$

The iron-porphyrin prosthetic groups are similar to the heme group of hemoglobin, to be discussed later. When the last cytochrome transfers electrons to oxygen to form a reduced oxygen (O^{-2}), reaction takes place immediately with two hydrogen ions to form water.

The oxidative chain discussed is only one of several known to occur in animal tissues. Some vary in the initial coenzymes or carriers and have the cytochrome portions in common. Others may couple almost directly with oxygen, omitting the cytochromes entirely. In each case, the hydrogen atoms and their electrons are transferred by means similar to those already cited. These reactions take place in the subcellular particle known as the *mitochondrion,* popularly known as the "powerhouse" of the cell because of its energy-producing activities.

Studies have shown the points at which various drugs may inhibit components of the oxidative chains, thus interfering with energy production. It has also been shown that cyanide (CN^-) can combine with the last cytochrome and completely inhibit its activity. This stops the functioning of the oxidative chain and death ensues because oxygen can no longer reoxidize the cytochromes which, in turn, reoxidize the initial carriers, as occurs during normal oxidation and energy production.

BIOENERGETICS

According to the law of conservation of energy, under ordinary circumstances, such as would be the case in a living organism, energy can be neither created nor destroyed. Energy can, however, be made available as a result of spontaneous reactions, such as oxidations, in which the conversion of an organic food molecule liberates energy that has been stored in the organic molecule. (In the language of thermodynamics, this liberated energy is known as *free energy*—free in the sense that it is available for useful work of such types as mechanical, osmotic, and electric.) Not all of the important reactions on which life depends are spontaneous. Many of them require the input of energy before they can be made to go. This is quite analogous to a new, shiny, powerful automobile that must have energy, supplied by the rapid combustion of gasoline (an oxidation reaction), to make it run.

Thus the problem of maintaining life becomes one of connecting or coupling reactions in which one is spontaneous, supplying energy to one that requires an input of energy. This is accomplished mainly by extracting energy from oxidized molecules and storing it in the form of what are known as high-energy phosphate compounds (e.g., adenosine triphosphate [ATP]). By another coupling mechanism, this energy in ATP is made available as the driving force in reactions requiring the input of energy. These energy transfers are the subject matter of bioenergetics.

Photosynthesis: ultimate source of energy

As noted before, animals cannot absorb energy from the sun and convert it to carbohydrates as do plants. It is now desirable to examine photosynthesis in a more quantitative sense, since this is the process that ultimately supplies animals with their energy requirements. When low-energy level substances water and carbon dioxide are raised to the high-energy level represented by carbohydrate and oxygen, an input of 688,500 cal of free energy (per mole of glucose, $C_6H_{12}O_6$) from the sun is required to make the reaction take place:

$$6CO_2 + 6H_2O + 688{,}500 \text{ cal} \rightarrow C_6H_{12}O_6 + 6O_2$$

This means that the potential maximum free energy that can be obtained by oxidation of the carbohydrate, the reversal of the photosynthetic reaction, is 688,500 cal. Under physiologic conditions, however, only about 50% of the maximum possible energy is available as free energy because of various inefficiencies. This free energy is available to the organism for

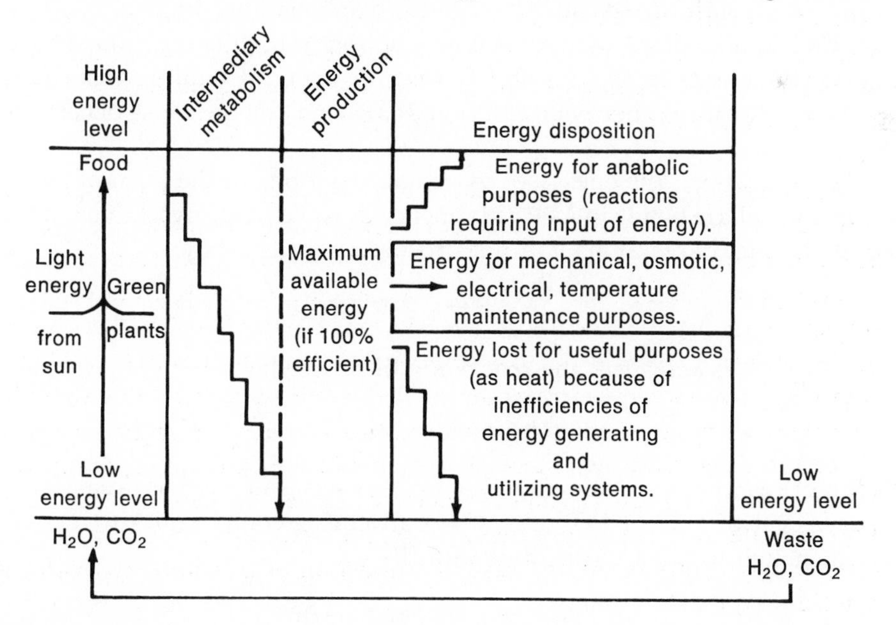

Fig. 20-1. Energy conversion, transfer, and utilization in humans and other animals.

various purposes such as mechanical, osmotic, or electric work, for temperature maintenance, and for any metabolic reactions requiring a source of energy. These relationships are illustrated in Fig. 20-1.

Production and use of high-energy phosphates

As noted before, the energy from oxidation reactions is released in a stepwise manner and is used to produce high-energy adenosine triphosphate bonds (ATP) at different sites in the oxidative chain. Using an abbreviated form of the oxidative chain, the approximate points at which ATP is generated are shown here:

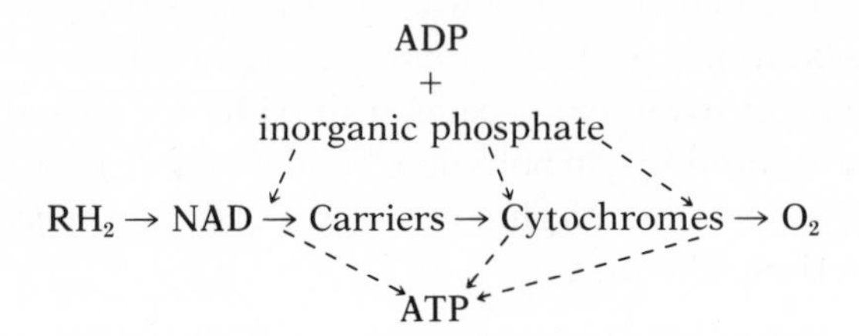

In this process a high-energy phosphate bond is made from inorganic phosphate and transferred to adenosine diphosphate (ADP), forming ATP. Although the exact details of these reactions are not completely established, it is believed that carriers and coupling factors are involved. Of importance here is the phenomenon of *uncoupling* of oxidative phosphorylation, i.e., certain drugs, such as 2,4-dinitrophenol, are known to allow oxidation to take place without the production of high-energy phosphate bonds. This points to the possibility that, in vivo, uncoupling may be a means of controlling energy production. It also may be the cause of certain abnormalities.

Having trapped the energy of oxidation reactions in the form of high-energy phosphate bonds in ATP, this energy must be made available where required. Certain reactions will not proceed spontaneously:

$$A + B \nrightarrow C + D$$

However, if A is *activated*, the reaction will proceed:

$$\text{A-activated} + B \rightarrow C + D$$

By direct reaction with compound A, or by a series of stepwise reactions, the energy available in ATP may be used to bring compound A to a state active enough to react with compound B:

$$A + \text{ATP} \rightarrow \text{A-activated} + \text{ADP}$$

Thus, the energy from ATP is the force that drives the desired reaction and inorganic phosphate is regenerated as a result. Specific examples of such reactions will be discussed later.

Summary of biologic oxidation and bioenergetics

The essential steps and results of oxidation reactions and formation and use of high-energy phosphates are summarized here:

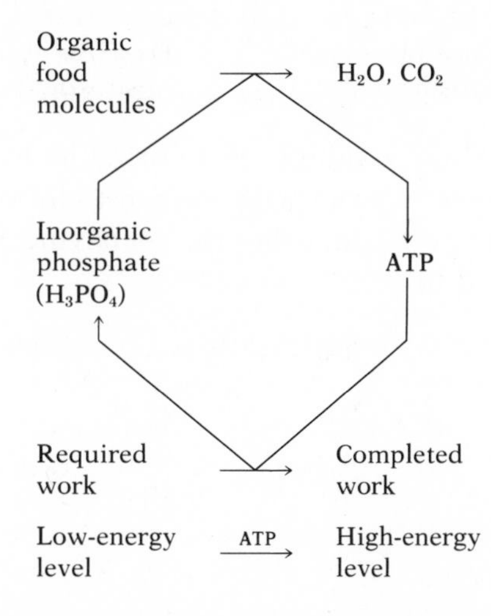

Low-energy level $\xrightarrow{\text{ATP}}$ High-energy level

SUMMARY

INTERMEDIARY METABOLISM

The sum of all chemical reactions occurring in the body—from the absorption of food molecules to the excretion of waste products.

Methods of studying intermediary metabolism:

In vivo methods: Experiments performed on live animals.

In vitro methods: Test-tube experiments, or the equivalent, including perfusion of whole organs, the use of tissue slices, homogenates, purified enzyme systems, cell-free extracts, separated cell components, and tissue cultures.

Microorganisms: Many reactions, more easily studied in microorganisms, also occur in animals.

Nutritional experiments: Studies of the effects of adding or deleting items of diet.

Isotopic experiments: Studies using radioactively labeled tracer compounds to determine pathways of metabolism.

BIOLOGIC OXIDATION

Reactions supplying most of the energy made available in living organisms. Photosynthesis absorbs energy from the sun and stores it in carbohydrates. This energy is then retrieved by oxidation of the carbohydrates in a stepwise series of reactions.

Stepwise process of biologic oxidations:

Each step involves the following process:

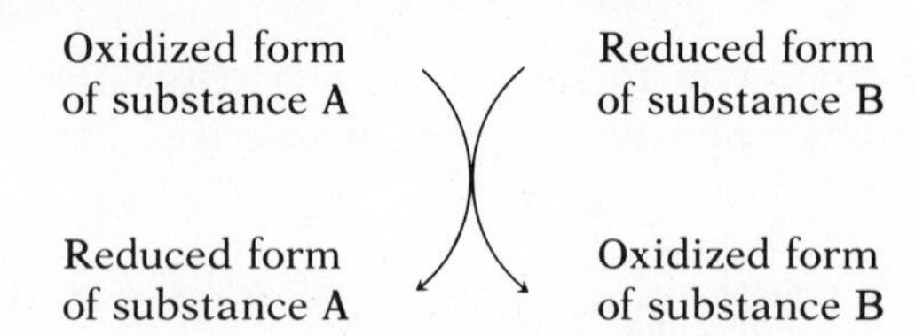

The purpose of biologic oxidation is to bring hydrogen atoms from organic food molecules to react with oxygen to form water and liberate energy. One oxidative chain, using the coenzyme NAD as the first carrier, is summarized here:

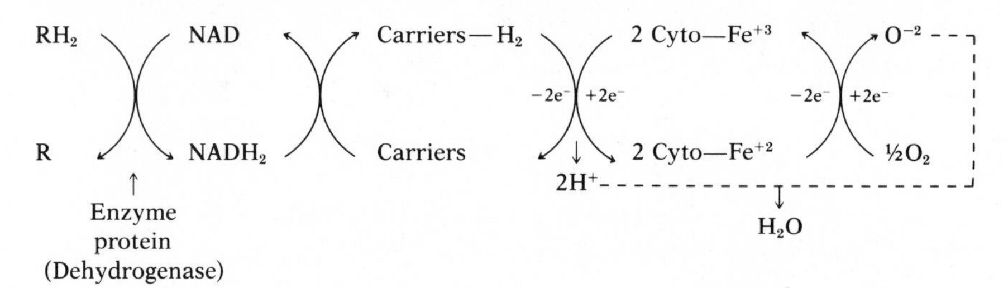

Each carrier is cyclically reduced and then reoxidized, thereby being required only in small concentrations to oxidize relatively large amounts of substrate. These reactions take place in the subcellular particle, the mitochondrion. The process may be interfered with by various drugs that inhibit the different carriers.

BIOENERGETICS Study of the production, capture, and use of free energy in the form of high-energy phosphates.

Photosynthesis: ultimate source of energy:

$$6CO_2 + 6H_2O + 688{,}500 \text{ cal} \rightarrow C_6H_{12}O_6 + 6O_2$$

Of the 688,500 cal stored per mole of carbohydrate by photosynthesis, about 50% is available as free energy under physiologic conditions. This free energy can be used for mechanical, osmotic or electric work, temperature maintenance, and driving reactions requiring an input of energy. The remaining 50% is lost in the form of heat because of inefficiencies.

Production and use of high-energy phosphates: Energy from oxidation reactions is captured in the form of ATP by reactions that couple the oxidative chain to the process of phosphorylation, as summarized here:

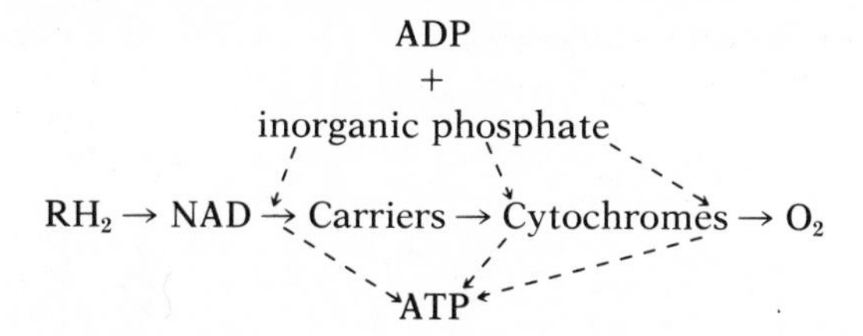

Drugs, such as 2,4-dinitrophenol, may uncouple oxidative phosphorylation, i.e., oxidation takes place but no high-energy phosphate bonds are generated. This may be an energy-control mechanism and also may be the cause of certain abnormalities.

The energy from ATP is made available to drive reactions, which do not occur spontaneously, by direct or indirect reaction with reactants to bring them up to an activated state.

Summary of biologic oxidation and bioenergetics:

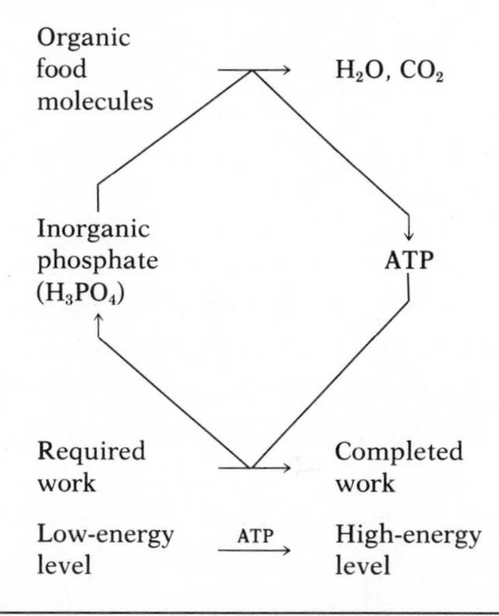

REVIEW QUESTIONS

1. What type of chemical reaction produces most of the energy required by the body? Is it similar to the major type of reaction found in digestive processes?
2. Name some methods of studying intermediary metabolism and the major advantage(s) of each type.
3. Biologic oxidations are carried out in a series of small steps. What types of substances are involved in the oxidative chain? What types of substances are passed along the oxidative chain? What happens to the next succeeding member of the chain as the substances referred to in the preceding question are passed along the chain? As a result, what happens to the preceding substance in the chain?
4. How is the energy released in biologic oxidation reactions trapped for future use? Do biologic oxidation reactions always result in the production of useful energy?
5. What is the ultimate source of all energy in the body? Explain with quantitative considerations.

REFERENCES

Baldwin, E.: Dynamic aspects of biochemistry, ed. 5, London, 1967, Cambridge University Press.

Christensen, H. N.: Dissociation, enzyme kinetics, bioenergetics: a learning program for students of the biological and medical sciences, Philadelphia, 1975, W. B. Saunders Co.

Christensen, H. N., and Cellarius, R. A.: Introduction to bioenergetics: thermodynamics for the biologist, Philadelphia, 1972, W. B. Saunders Co.

Florkin, M., and Stotz, E. H., editors: Comprehensive biochemistry, Vol. 14, Amsterdam, 1966, Elsevier.

Green, D. E., editor: Mechanisms of energy transduction in biological systems, Ann. NY Acad. Sci. **227:**1-680, 1974.

Hoch, F.: Energy transformations in mammals: regulatory mechanisms, Philadelphia, 1971, W. B. Saunders Co.

Klotz, I. M.: Energy changes in biochemical reactions, New York, 1967, Academic Press, Inc.

Lehninger, A. L.: Bioenergetics: the molecular basis of biological energy transformations, ed. 2, Menlo Park, Calif., 1971, W. A. Benjamin Inc.

Lehninger, A. L.: The mitochondrion: molecular basis of structure and function, New York, 1964, W. A. Benjamin Inc.

Racker, E.: A new look at mechanisms in bioenergetics, New York, 1976, Academic Press, Inc.

Wang, C. H., and Willis, D. L.: Radiotracer methodology in the biological environmental and physical sciences, Englewood Cliffs, N.J., 1975, Prentice-Hall, Inc.

Carbohydrate metabolism

21

For teaching purposes, it is a matter of convenience to discuss each metabolic pathway by itself, as much as possible. However, it is well to be aware at the outset that there are many interrelationships between the separate pathways and that, under normal conditions, most of the reaction sequences are taking place at the same time.

Another important consideration is the reversibility or irreversibility of steps in a metabolic pathway. Certain segments of pathways may be able to run in either direction, whereas others may be limited to one direction by a step that is irreversible.

For the scope of this text, although included as reference material, the actual formulas of compounds and names of enzymes involved in metabolic pathways are considered to be of secondary importance. Emphasis will be directed toward the types of changes taking place and the effects of these changes on the economy of the body. With this in mind, a flowchart incorporating these factors will be used to summarize the pathways discussed and it will be added to as further pathways are encountered.

Most of the discussion in this chapter is concerned with the metabolism of glucose, the most important carbohydrate in the body.

ASPECTS OF DIGESTION AND ABSORPTION

As noted previously (p. 267), the polysaccharides starch and glycogen are hydrolyzed to the disaccharide stage, maltose. The disaccharides maltose, sucrose, and lactose are then hydrolyzed to monosaccharides (maltose yields two glucose; sucrose yields glucose and fructose; and lactose yields glucose and galactose) in the brush-border membrane of the intestinal epithelial cells from which they are absorbed.

There are certain observations regarding the rates of absorption of the monosaccharides that indicate that simple diffusion is not the only process involved. It is found that galactose is absorbed faster than glucose, which is absorbed faster than fructose. Also, the rate of absorption of these monosaccharides does not appear to be related to their amounts or concentrations in the intestine, which would be the case if diffusion were the only process of absorption. Therefore, it is believed that in addition to diffusion, a *transport system* functions to some extent in the absorption of glucose and galactose. A transport system involves the concept of carriers that effect the transfer of substances from one side of a membrane to the other with the expenditure of energy.

Under normal conditions of diet and intestinal function, monosaccharides are essentially completely absorbed from the small intestine, the rate of absorption decreasing as the distance from the stomach increases. The monosaccharides are absorbed by way of the portal blood, which leads directly to the liver. In the liver they are either oxidized to provide energy, converted to fatty acids or other substances, stored as glycogen, or released to the systemic circulation for use by other tissues. In other words, the liver exerts a very important influence on the blood sugar level and metabolism of carbohydrates in the body because it has first choice as to the fate of the monosaccharides.

POSSIBLE USES OF GLUCOSE

Storage as glycogen: glycogenesis

Glucose molecules not required for other uses may be converted to glycogen (polysaccharide) and deposited in the liver or other tissues, such as muscle. This process is known as *glycogenesis* (formation of glycogen), and glycogen is thus the animal storage form of glucose, starch being the plant storage form. The amount of glycogen that may be stored in the body is relatively small and definitely limited. Excess amounts of available glucose must therefore be diverted into other pathways.

Oxidation for energy purposes

Glucose can be oxidized completely to carbon dioxide and water in the body to furnish energy as required. This can be accomplished in two different ways. All tissues can oxidize glucose by a pathway known as the glycolytic scheme or the Embden-Meyerhof pathway, which involves the cleavage of the 6-carbon glucose molecule into two 3-carbon halves (pyruvic acid, or pyruvate as it is usually designated in biochemical literature). Under conditions of prolonged or strenuous exercise, the oxidation of glucose in muscle stops at the 3-carbon stage of lactic acid, a reduction product of pyruvic acid. The lactic acid is released into the blood and returned to the liver for reprocessing or further oxidation. Under aerobic conditions, liver and muscle complete the oxidation of pyruvate by way of the tricarboxylic acid cycle.

In addition, liver and adipose tissue can also degrade glucose by a pathway involving pentoses, sometimes called the pentose shunt. This pathway is relatively unimportant in skeletal muscle.

Conversion to other metabolites

Fat. Since the utilization of glucose for storage as glycogen is limited, excess glucose may be converted to fatty acids and glycerol and deposited as triglycerides in adipose tissue, or fat depots. Unfortunately for many individuals, this process is unlimited except on the basis of available glucose. As may be inferred, there are certain molecules common to the metabolic pathways of glucose and fatty acids. Isotopic studies have shown that the conversion of glucose to fatty acids is an irreversible process, as will be explained later. Energy may be obtained from the subsequent oxidation of fatty acids, but fatty acids cannot be used to produce glucose in a manner profitable to the body.

Amino acids. There are certain amino acids, known as nonessential amino acids, which are not required in the diet. In spite of no dietary intake they are found in tissue proteins, leading to the conclusion that they are being synthesized by the body. Isotopic studies have shown that the carbon skeletons of nonessential amino acids may be derived from glucose and these nonessential amino acids can contribute to the carbon skeleton of glucose. The essential amino acids, which are required in the diet,

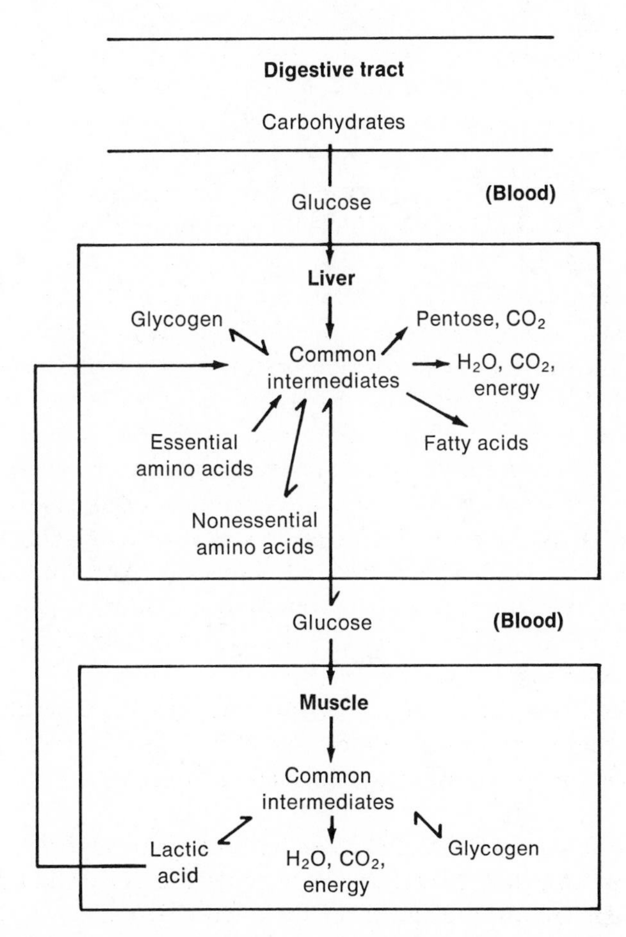

Fig. 21-1. General summary of carbohydrate metabolism.

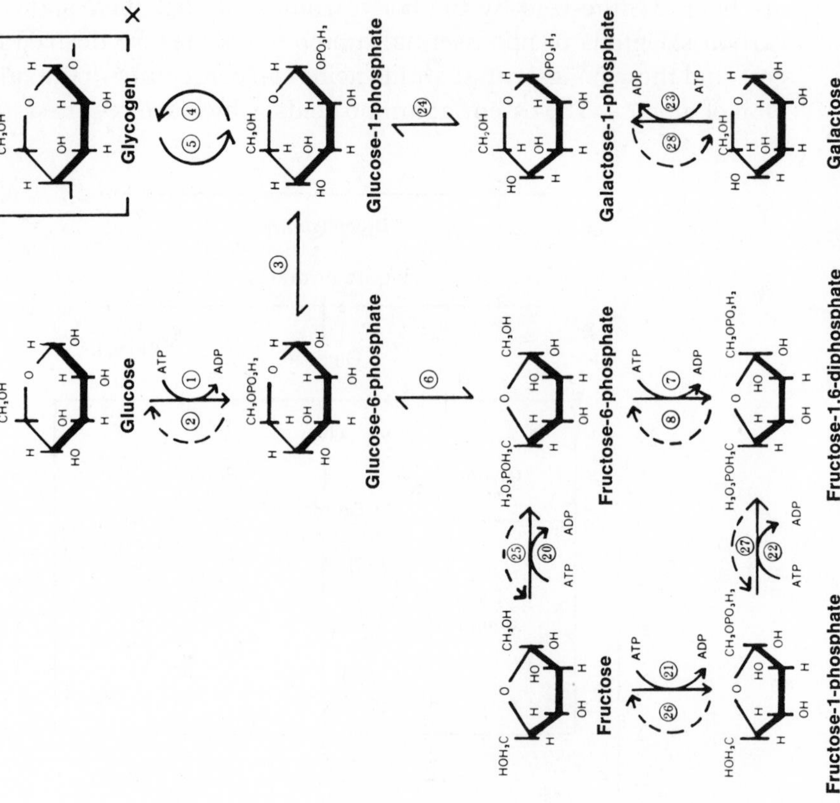
Glycogen
Glucose
Glucose-6-phosphate
Glucose-1-phosphate
Galactose-1-phosphate
Galactose
Fructose-6-phosphate
Fructose-1,6-diphosphate
Fructose
Fructose-1-phosphate
ATP
ADP

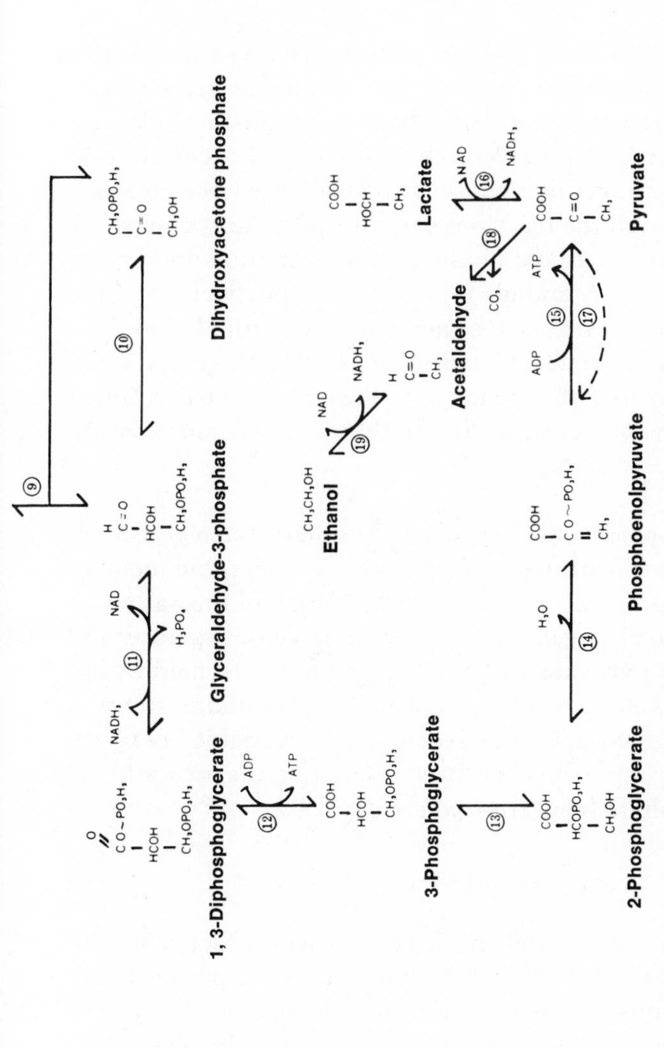

Fig. 21-2. Anaerobic metabolism of glucose (glycolysis) and related hexoses. (~ = high-energy bond.)

can also contribute to the carbon skeleton of glucose, but cannot be formed from glucose or its metabolites.

Other carbohydrates. Some glucose is used for the synthesis of other carbohydrates such as ribose and deoxyribose, components of nucleic acids, and galactose, a component of lactose found in milk.

These general relationships are summarized in Fig. 21-1.

OXIDATION OF GLUCOSE

Discussions of the oxidation of glucose are usually divided into two phases: (1) anaerobic metabolism of glucose (also known as glycolysis or the Embden-Meyerhof pathway), and (2) aerobic metabolism of glucose, involving the Krebs cycle (also known as the tricarboxylic acid cycle or the citric acid cycle). The anaerobic phase is so designated because oxygen is not required, although the reactions can take place in the presence of oxygen. Included in the anaerobic phase are the formation and breakdown of glycogen, glycogenesis and glycogenolysis, respectively, and the interconversions of the hexoses, galactose and fructose, with glucose, this phase ending with the production of pyruvate or lactate and a small amount of energy. The aerobic phase ends with the complete oxidation of glucose to water, carbon dioxide, and the production of a much larger amount of energy.

Anaerobic metabolism of glucose

Formation of glucose-6-phosphate. The sugar-phosphate ester glucose-6-phosphate (G-6-P) occupies a strategic position in the anaerobic metabolism of glucose. As seen in Fig. 21-2, it sits at the junction of the pathways leading from glucose to glycogen, interconversions with fructose and galactose, or formation of pyruvate. Glucose-6-phosphate is formed within cells by an irreversible *kinase* reaction in which a phosphate group is transferred from ATP to glucose (G). The reaction is irreversible because a high-energy phosphate group in ATP has been used up in the formation of a low-energy ester phosphate bond (reaction 1, Fig. 21-2):

$$\text{G} \xrightarrow[\text{hexokinase}]{\text{ATP} \searrow\nearrow \text{ADP}} \text{G-6-P}$$

The activity of hexokinase, found in all cells, is not affected by the concentration of blood glucose. Insulin, which increases the permeability of cells to glucose, can thus indirectly exert an influence on the rate of formation of glucose-6-phosphate in extrahepatic tissues. In liver cells, which are freely permeable to glucose, there is another kinase, glucokinase, the rate of reaction of which is dependent on blood glucose concentration. The synthesis of glucokinase is induced by insulin. Thus insulin controls the rate of formation of glucose-6-phosphate in the liver, to a certain extent, as well as in extrahepatic tissues. Other hormonal controls will be discussed later.

In the liver, glucose-6-phosphate can be converted back to glucose by the action of a specific glucose-6-phosphatase, not by reversal of the hexokinase reaction (reaction 2, Fig. 21-2):

$$\text{G-6-P} \xrightarrow{\text{phosphatase}} \text{G}$$

Muscle does not have such a phosphatase. Therefore, when glucose is phosphorylated in muscle, it cannot be converted back to glucose and returned to the blood as such. It must be stored as glycogen or broken down to pyruvate.

Glycogenesis. The next step in the formation of glycogen is the conversion of glucose-6-phosphate to glucose-1-phosphate (G-1-P) in the presence of phosphoglucomutase. Glucose-1-phosphate reacts with uridine triphosphate (UTP) in a typical activation reaction, forming uridine diphosphate glucose (UDPG) by the action of a pyrophosphorylase. Then, under the influence of UDPG-glycogen glycosyl-transferase (glycogen synthetase, makes 1,4-linkages) and in the presence of a branching enzyme (makes 1,6-linkages) and primer molecules of glycogen, UDPG is converted to glycogen and UDP. UTP is regenerated by reaction of UDP with ATP. These steps are summarized here (reactions 3, 4, Fig. 21-2):

$$\text{G-6-P} \xleftrightarrow{\text{phosphoglucomutase}} \text{G-1-P}$$

$$\text{G-1-P} + \text{UTP} \xrightarrow{\text{pyrophosphorylase}} \text{UDPG} + \text{pyrophosphate}$$

$$\text{UDPG} \xrightarrow[\text{branching enzyme}]{\text{glycogen synthetase}} \text{glycogen} + \text{UDP}$$

$$\text{UDP} + \text{ATP} \xleftrightarrow{\text{phosphokinase}} \text{UTP} + \text{ADP}$$

A form of glycogen storage deficiency resulting from an inherited lack of glycogen synthetase has been described. As a result of not being able to form proper amounts of glycogen, this inborn error of metabolism is characterized by fasting hypoglycemia (less than normal blood glucose levels) with convulsions and mental retardation. Normal individuals store approximately 100 g of glycogen in the liver and 250 g in muscle.

Glycogenolysis. The breakdown of glycogen is accomplished in the presence of a debranching enzyme (breaks 1,6-linkages) and phosphorylase (breaks 1,4-linkages), producing glucose-1-phosphate, which, in turn, is converted to glucose-6-phosphate by the reversible phosphoglucomutase reaction. This brings the molecule back to the busy crossroad of anaerobic glucose metabolism. These reactions are summarized here (reactions 5, 3, Fig. 21-2):

$$\text{Glycogen} \xrightarrow[\text{phosphorylase}]{\text{debranching enzyme}} \text{G-1-P}$$

$$\text{G-1-P} \xleftrightarrow{\text{phosphoglucomutase}} \text{G-6-P}$$

If necessary to maintain the blood sugar level, glucose-6-phosphate can be converted in the liver to glucose and released to the blood (reaction 2, Fig. 21-2):

$$\text{G-6-P} \xrightarrow{\text{glucose-6-phosphatase}} \text{Glucose}$$

The hormone epinephrine stimulates glycogenolysis under conditions requiring increase in blood sugar by its effect on the phosphorylase reaction.

The first recognized genetically caused glycogen storage disease (von Gierke's disease) results from the lack of glucose-6-phosphatase. As a result, hypoglycemia, enlargement of the liver to accommodate increased stores of glycogen, and elevation of blood lactate concentration occur.

Glycolysis: breakdown to pyruvate. Another pathway open to glucose-6-phosphate is the breakdown to pyruvate or lactate, a process known as glycolysis. Glucose-6-phosphate is first converted to fructose-6-phosphate (F-6-P). This reversible reaction is catalyzed by phosphohexose isomerase. In the next step, fructose-6-phosphate is phosphorylated by ATP in a kinase reaction in the presence of phosphofructokinase to form fructose-1,6-diphosphate (F-1,6-P_2). This reaction is irreversible, as is the hexokinase reaction, because a low-energy ester phosphate is formed as a high-energy phosphate bond is used up. However, fructose-6-phosphate can be obtained from fructose-1,6-diphosphate by the action of a specific phosphatase. In effect, there is a biologic reversibility, but not by the same reaction. These reactions are summarized here (reactions 6 to 8, Fig. 21-2):

$$\text{G-6-P} \xleftrightarrow{\text{phosphohexose isomerase}} \text{F-6-P}$$

$$\text{F-6-P} \xrightarrow[\text{phosphofructokinase}]{\text{ATP} \searrow \quad \nearrow \text{ADP}} \text{F-1,6-P}_2$$

$$\text{Inorganic phosphate} + \text{F-6-P} \xleftarrow{\text{phosphatase}} \text{F-1,6-P}_2$$

Phosphofructokinase activity is inhibited by citrate, a component of the Krebs cycle, and this may be a means of controlling carbohydrate breakdown by limiting the amount of pyruvate entering the Krebs cycle.

Fructose-1,6-diphosphate, under the influence of fructose diphosphate aldolase, is split into two triose phosphate molecules, one glyceraldehyde-3-phosphate (glycerald-3-P) and one dihydroxyacetone-phosphate [$(OH)_2$-acetone-P]. The triose phosphates are in equilibrium with each other in the presence of an isomerase, making both halves of the glucose molecule available for the subsequent reactions. In the next step, the conversion of glyceraldehyde-3-phosphate to 1,3-diphosphoglycerate (1,3-P_2-glycerate) is catalyzed by glyceraldehyde-3-phosphate dehydrogenase. This is the first of two reactions in glycolysis that produce high-energy phosphate bonds. The formation of a carboxylic acid group from an aldehyde group is an oxidation reaction that liberates energy, much of which is captured in the form of a high-energy carboxyl phosphate group. This is converted to ATP with the help of phosphoglycerate kinase, as the diphosphoglycerate is converted to 3-phosphoglycerate (3-P-glycerate). Because two trioses are formed per molecule of hexose, this means that at this point two high-energy phosphate bonds are formed per molecule of hexose. The type of phosphorylation that occurs here is termed *substrate level phosphorylation,* resulting from oxidation of substrate rather than operation of the oxidative chain, which does not function under anaerobic conditions. These reactions are summarized here (reactions 9 to 12, Fig. 21-2):

$$\text{F-1,6-P}_2 \xrightarrow[\text{aldolase}]{\text{fructose diphosphate}} \text{Glycerald-3-P} + \text{(OH)}_2\text{-Acetone-P}$$

$$\text{Glycerald-3-P} \xleftrightarrow[\text{isomerase}]{\text{triose phosphate}} \text{(OH)}_2\text{-Acetone-P}$$

$$\text{2 Glycerald-3-P} \xleftrightarrow[\text{H}_3\text{PO}_4]{\text{glycerald-3-P dehydrogenase; NAD} \rightarrow \text{NADH}_2} \text{2(1,3-P}_2\text{ glycerate)}$$

$$\text{2(1,3-P}_2\text{ glycerate)} \xleftrightarrow[\text{phosphoglycerate kinase}]{\text{2ADP} \rightarrow \text{2ATP}} \text{2(3-P glycerate)}$$

In the remaining steps of glycolysis, phosphoglyceromutase converts 3-phosphoglycerate to 2-phosphoglycerate (2-P glycerate) in a reversible reaction. Enolase then catalyzes the dehydration of 2-phosphoglycerate to phosphoenolpyruvate (P-E pyr). The phosphate bond in phosphoenolpyruvate is a high-energy bond (caused largely by the instability of its hydrolysis product, enolpyruvate). The high-energy phosphate is transferred to ADP in the presence of pyruvate kinase, forming pyruvate and two more ATPs per molecule of glucose catabolized. Under anaerobic conditions, lactate dehydrogenase causes the reduction of pyruvate to lactate, as glycolysis is completed. These reactions are summarized here (reactions 13 to 16, Fig. 21-2):

$$\text{2(3-P glycerate)} \xleftrightarrow{\text{phosphoglyceromutase}} \text{2(2-P glycerate)}$$

$$\text{2(2-P glycerate)} \xleftrightarrow[\text{H}_2\text{O}]{\text{enolase}} \text{2(P-E pyr)}$$

$$\text{2 P-E pyr} \xrightarrow[\text{pyruvate kinase}]{\text{2ADP} \rightarrow \text{2ATP}} \text{2 Pyruvate}$$

$$\text{Pyruvate} \xleftrightarrow[\text{lactate dehydrogenase}]{\text{NADH}_2 \rightarrow \text{NAD}} \text{Lactate}$$

Table 21-1. Sugar phosphate esters formed in various tissues*

Tissue	Sugar phosphates formed			
	G-6-P	**F-1-P**	**F-6-P**	**Gal-1-P**
Liver	+	+		+
Brain	+		+	+
Muscle	+	+		

*G = glucose; F = fructose; Gal = galactose; P = phosphate.

Although the kinase reaction converting phosphoenolpyruvate to pyruvate, as ADP forms ATP, is irreversible (reaction 17, Fig. 21-2), there are reactions available to the liver that bypass this obstacle. In conjunction with the phosphatase reactions (reactions 2, 8, Fig. 21-2) and the phosphorylase reaction (reaction 5, Fig. 21-2), the pyruvate kinase bypass contributes to making all of the glycolytic reactions *biologically reversible.* It must be noted, however, that this reversibility is achieved at the price of a loss of energy. This results from the fact that the reactions in the direction of glucose to pyruvate liberate energy. Therefore, in order to reverse this sequence, energy must be expended to drive the reactions backward.

At this point it is interesting to note that the reactions described up to the formation of pyruvate also take place in yeast cells. Then, instead of forming lactate, yeast cells convert pyruvate to carbon dioxide and acetaldehyde by the action of pyruvate carboxylase and alcohol dehydrogenase reduces acetaldehyde to ethyl alcohol, the last step of alcoholic fermentation. These reactions are summarized here (reactions 18, 19, Fig. 21-2):

$$\text{Pyruvate} \xrightarrow[\searrow CO_2]{\text{pyruvate carboxylase}} \text{Acetaldehyde}$$

$$\text{Acetaldehyde} \xleftrightarrow[\text{alcohol dehydrogenase}]{NADH_2 \quad NAD} \text{Ethyl alcohol}$$

Interconvertibility of glucose, fructose, galactose. Various tissues form the different sugar phosphate esters by the action of the appropriate kinases, as listed in Table 21-1.

The different hexoses can contribute to glycolysis or be interconverted by way of the glycolytic reactions.

Fructose-6-phosphate is directly on the glycolytic pathway, as is glucose-6-phosphate. Fructose-1-phosphate is moved onto the pathway by conversion to fructose-1,6-diphosphate by a phosphofructokinase. Galactose-1-phosphate is converted to glucose-1-phosphate in a reaction involving uridine diphosphate glucose and a transferase. Glucose-6-phosphate is then formed from glucose-1-phosphate by the reversible phosphogluco-mutase reaction. These reactions are summarized here (reactions 20 to 24, 3, Fig. 21-2):

$$\text{F} \xrightarrow[\text{kinase}]{\text{ATP} \searrow\nearrow \text{ADP}} \text{F-6-P}$$

$$\text{F} \xrightarrow[\text{kinase}]{\text{ATP} \searrow\nearrow \text{ADP}} \text{F-1-P}$$

$$\text{F-1-P} \xrightarrow[\text{kinase}]{\text{ATP} \searrow\nearrow \text{ADP}} \text{F-1,6-P}_2$$

$$\text{Gal} \xrightarrow[\text{kinase}]{\text{ATP} \searrow\nearrow \text{ADP}} \text{Gal-1-P}$$

$$\text{Gal-1-P} \xleftrightarrow[\text{transferase}]{\text{UDPG} \searrow\nearrow \text{UDP-Gal}} \text{G-1-P}$$

$$\text{G-1-P} \xleftrightarrow{\text{phosphoglucomutase}} \text{G-6-P}$$

Phosphatases are available to make all of the phosphorylation steps biologically reversible (reactions 25 to 28, Fig. 21-2):

$$\text{F-6-P} \xrightarrow{\text{phosphatase}} \text{F}$$

$$\text{F-1-P} \xrightarrow{\text{phosphatase}} \text{F}$$

$$\text{F-1,6-P}_2 \xrightarrow{\text{phosphatase}} \text{F-1-P}$$

$$\text{Gal-1-P} \xrightarrow{\text{phosphatase}} \text{Gal}$$

Energy accounting in glycolysis. Approximately 50,000 cal of free energy are liberated per mole of glucose catabolized in glycolysis. In the process, two high-energy bonds of ATP are used—one to form glucose-6-phosphate from glucose, and one to form fructose-1,6-diphosphate from fructose-6-phosphate (reactions 1, 7, Fig. 21-2). However, two high-energy phosphate bonds in the form of ATP are formed in the oxidative step (reactions 11, 12, Fig. 21-2), and two more as a result of the enolase step (reactions 14, 15, Fig. 21-2). Thus, there is a net gain of two high-energy phosphate bonds in the glycolytic sequence under anaerobic conditions. At approximately 7500 cal per high-energy phosphate, 15,000 of a possible 50,000 calories, or about 30% of the maximum available energy, is recovered in glycolysis.

Location of enzymes. The enzymes, coenzymes, and activating ions required for the functioning of the glycolytic pathway are found in the supernatant fraction of the cytoplasm of disrupted cells after centrifugation. Further details about specific enzymes and their properties may be found in reference texts on enzymology.

Aerobic metabolism of glucose

As noted before, the sequence of glycolytic reactions to the pyruvate stage do not require oxygen but take place in the same way in the presence of oxygen. Under anaerobic conditions, pyruvate is reduced to lactate; but as soon as aerobic conditions are reestablished in the tissues, which is the usual case, lactate is reoxidized to pyruvate. This means that pyruvate is actually the first compound to be considered in the sequence of reactions constituting the aerobic metabolism of glucose or the carbohydrates. Except for one step, all of the aerobic metabolism takes place as a series of reactions called the tricarboxylic acid cycle (citric acid cycle, Krebs cycle). In order to enter the cycle, pyruvate must first be converted to a 2-carbon activated acetyl group. In the subsequent discussion, it must be remembered that for every glucose molecule catabolized, two pyruvate molecules are formed.

Oxidation of pyruvate: formation of acetyl coenzyme A (acetyl-CoA). In a complicated set of reactions involving a system of enzymes, the pyruvate dehydrogenase system, rather than the usual single enzyme, pyruvate is decarboxylated, and the remaining 2-carbon fragment is oxidized and coupled to coenzyme A. Among the substances participating in the reactions are vitamin B_1, coenzyme A, and NAD. The net reaction may be written as follows:

$$\underset{\textbf{Pyruvate}}{CH_3\overset{O}{\overset{\|}{C}}-\overset{O}{\overset{\|}{C}}-OH} + \underset{\textbf{Coenzyme A}}{CoA-SH} \xrightarrow[CO_2]{NAD \quad NADH_2} \underset{\textbf{Acetyl coenzyme A}}{CH_3\overset{O}{\overset{\|}{C}}\sim SCoA} + H_2O$$

It should be noted that the bond between the acetyl group and coenzyme A is a high-energy bond, conserving some of the energy liberated by the oxidation reaction. Also, under aerobic conditions, every $NADH_2$ or $NADPH_2$ formed can be reoxidized via the oxidative chain to generate three high-energy phosphate bonds in the form of ATP. This applies, as well, to the $NADH_2$ formed in the oxidative step in glycolysis (reaction 11, Fig. 21-2).

An important point to keep in mind here is that the metabolism of fatty acids is another source of acetyl-CoA molecules, thereby closely linking the metabolism of carbohydrates and fatty acids. This subject will be taken up in the next chapter.

Tricarboxylic acid cycle (citric acid cycle, Krebs cycle). The structural changes that occur in the tricarboxylic acid cycle are illustrated in Fig. 21-3. The reaction that takes the acetyl-CoA into the tricarboxylic acid cycle is a condensation with oxaloacetate to form citrate and regenerate CoA (reaction 1, Fig. 21-3):

$$\text{Oxaloacetate} \xrightarrow[H_2O]{\text{Acetyl-CoA} \quad \text{CoA}} \text{Citrate}$$

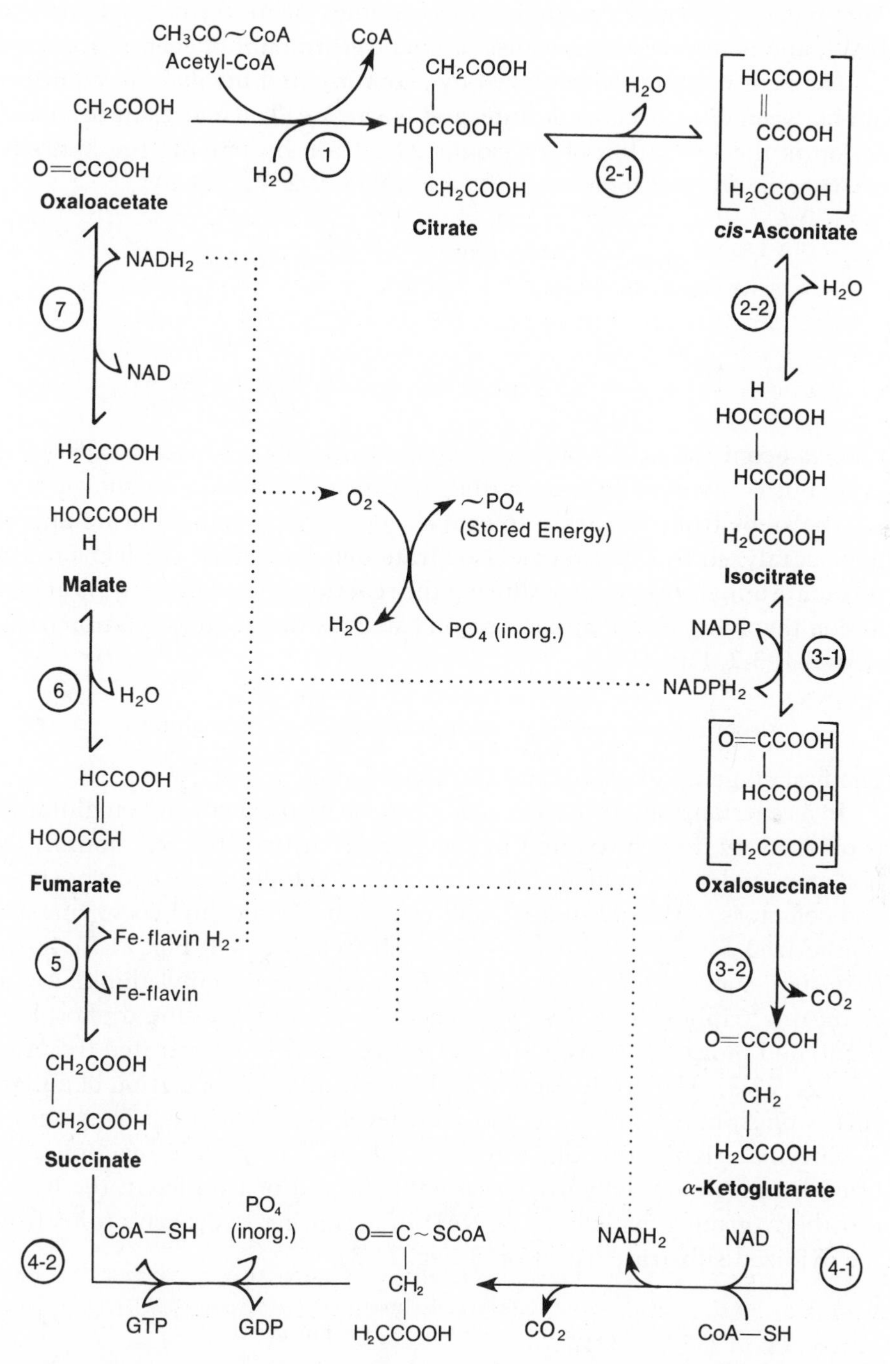

Fig. 21-3. The tricarboxylic acid cycle (also known as the citric acid cycle or the Krebs cycle). (~ = high energy bond.)

This reaction involves a condensing enzyme, citrate synthetase, or synthase, and is irreversible because of the large amount of energy liberated.

The next reaction is catalyzed by aconitase and involves an equilibrium between citrate, *cis*-aconitate, and isocitrate. The net result of this reaction is the formation of an isomer of citrate by moving the hydroxyl group to another carbon atom (reactions 2-1, 2-2, Fig. 21-3):

$$\text{Citrate (—OH on middle C atom)} \xrightarrow{H_2O} \textit{cis}\text{-Aconitate} \xrightarrow{H_2O} \text{Isocitrate (—OH on C atom next to middle C atom)}$$

The *cis*-aconitate is not on the direct pathway of the tricarboxylic acid cycle but is involved by way of the equilibrium reaction, as shown here.

The steps from isocitrate to oxalosuccinate to α-ketoglutarate appear to be catalyzed by one enzyme, isocitrate dehydrogenase, with the oxalosuccinate apparently not free during the reactions. The first stage is an oxidation involving NADP and the second stage is the decarboxylation (reactions 3-1, 3-2, Fig. 21-3):

$$\text{Isocitrate} \xrightarrow{\text{NADP} \quad \text{NADPH}_2} [\text{Oxalosuccinate}] \xrightarrow{CO_2} \alpha\text{-Ketoglutarate}$$

The first stage is reversible but the second stage is not.

In a reaction similar to the one undergone by pyruvate, α-ketoglutarate is oxidatively decarboxylated to succinate in a two-step reaction. In the first step, under the influence of the enzyme α-ketoglutarate decarboxylase and cofactors NAD, vitamin B_1, and coenzyme A, succinyl coenzyme A is synthesized as carbon dioxide is given off (reaction 4-1, Fig. 21-3). In the next step, succinate is produced by the action of succinyl thiokinase as guanosine triphosphate (GTP) is generated from guanosine diphosphate (GDP) and inorganic phosphate, and coenzyme A is regenerated (reaction 4-2, Fig. 21-3). The production of GTP represents the formation of a high-energy phosphate bond at the substrate level. This reaction is irreversible.

Succinate is then oxidized to fumarate in a reversible reaction catalyzed by succinate dehydrogenase with the aid of iron-flavin (Fe-flavin) prosthetic groups, which act as NAD in accepting hydrogen atoms from the oxidized substrate (reaction 5, Fig. 21-3):

$$\text{Succinate} \xrightleftharpoons{\text{Fe-flavin} \quad \text{Fe-flavin } H_2} \text{Fumarate}$$

The reduced iron-flavin groups are reoxidized by the oxidative chain by coupling with the chain at the cytochrome region. This means that only two high-energy phosphate bonds are formed per molecule of iron-flavin oxidized, rather than three as with $NADH_2$:

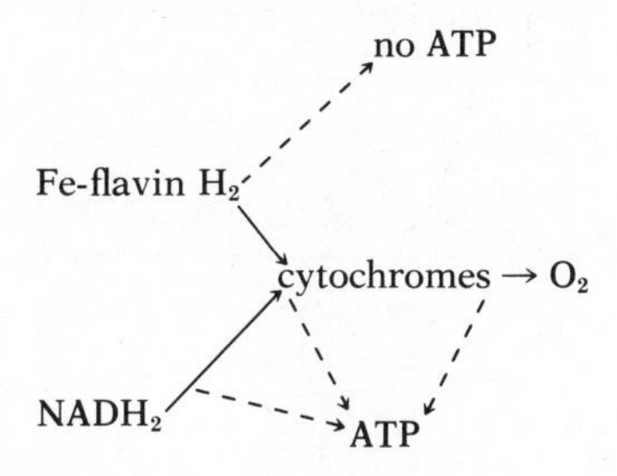

The reversible hydration of fumarate to malate is catalyzed by fumarase (reaction 6, Fig. 21-3):

$$\text{Fumarate} \xrightleftharpoons{H_2O} \text{Malate}$$

In the last step of the tricarboxylic acid cycle, malate is dehydrogenated in a reversible reaction involving malate dehydrogenase and NAD. This reaction produces oxaloacetate that is now ready to go around the cycle again (reaction 7, Fig. 21-3):

$$\text{Malate} \xrightleftharpoons{\text{NAD} \quad \text{NADH}_2} \text{Oxaloacetate}$$

The net oxidation reaction of pyruvic acid may be written as follows:

$$CH_3COCOOH + 5(O) \rightarrow 2H_2O + 3CO_2$$

The three carbon atoms end up in carbon dioxide as a result of the decarboxylations of pyruvate before the tricarboxylic acid cycle, and oxalosuccinate and α-ketoglutarate in the cycle (reactions 3-2, 4, Fig. 21-3). Five water molecules are actually formed but three are used in hydration reactions (reactions 1, 4, 6, Fig. 21-3), leaving a net production of two molecules of water as in the equation.

Energy accounting in aerobic oxidation. An accounting of the net energy produced by the aerobic oxidation of glucose can be found in Table 21-2. Of a maximum available free energy of 688,500 cal per mole of glucose, a net of 38 ATP bonds are formed. At approximately 7500 cal per ATP, 285,000 cal are conserved, or about 41%. Thus, in efficiency and absolute total, the aerobic oxidation of glucose is of greater use to the body than anaerobic oxidation.

Attention has been called to some of the simpler mechanisms for controlling the production of energy to fit the requirements of the body. Many other intricate mechanisms beyond the scope of this text are also in opera-

Table 21-2. Energy accounting in aerobic oxidation of glucose

Reaction	Text reference*	High-energy phosphate			
		Used	Generated		Net
			Substrate level	Oxidative chain level	
Hexokinase	1, Fig. 21-2	−1			−1
Phosphohexokinase	7, Fig. 21-2	−1			−1
X2 { Oxidative step	11, Fig. 21-2		+2	+6	+8
X2 { Enolase step	14, Fig. 21-2		+2		+2
X2 { Pyruvate → citrate	p. 302	−2	+2	+6	+6
X2 { Isocitrate → oxalosuccinate	3-1, Fig. 21-3			+6	+6
X2 { Ketoglutarate → succinate	4, Fig. 21-3		+2	+6	+8
X2 { Succinate → fumarate	5, Fig. 21-3			+4	+4
X2 { Malate → oxaloacetate	7, Fig. 21-3			+6	+6
				TOTAL	+38

*Text reference gives reaction number in figure cited.

tion and others are still under intensive investigation. There can be no doubt of the great importance of such controls.

Location of enzymes. The enzymes and cofactors required for the oxidation of pyruvate and the functioning of the tricarboxylic acid cycle and oxidative chain are found almost exclusively in the mitochondria. This is in contrast to the enzymes of the glycolytic pathway, which are found in the cell supernatant—a situation which, because of the spatial separation of the two systems, may provide a measure of control.

Pentose shunt

For muscle tissue, the glycolytic scheme and the tricarboxylic acid cycle discussed represent the most important pathways for carbohydrate metabolism. In liver and adipose tissue there is another possible way of degrading glucose known as the pentose shunt. As may be inferred from the name, a pentose, ribose-5-phosphate, is one of the possible products of this pathway (the aldehyde carbon, C-1, of glucose is lost in a decarboxylation reaction). Isotopic studies indicate that the ribose in RNA, and ultimately the deoxyribose in DNA, arise from this pathway. Another important product of this pathway is $NADPH_2$, which is used for reducing purposes as required in other reactions such as the synthesis of fatty acids.

It would appear that the primary purpose of the pentose shunt is not energy production. When it runs in cyclical fashion, one carbon dioxide molecule is produced per turn of the cycle, and hexose is regenerated. This means that it requires six cycles to degrade the equivalent of one molecule of glucose. Therefore, if six glucose molecules start down the pathway, after six cycles six molecules of carbon dioxide are formed and five molecules of glucose can be regenerated. Glucose-6-phosphate and fructose-6-phosphate are involved in the cycle, which must therefore be in some kind of equilibrium with the glycolytic scheme in tissues where both pathways are operating.

MUSCLE CONTRACTION

Much of the energy generated from the oxidation of carbohydrate is used up in muscle contraction. In spite of great research efforts in this field, the exact details of the process of muscle contraction are matters of continuing disagreement. However, it is generally agreed that the act of contraction involves an interaction between actomyosin, muscle protein, and ATP, causing the muscle fiber to shorten and converting ATP to ADP and inorganic phosphate.

Since muscle frequently performs under anaerobic conditions and ATP cannot be regenerated by the oxidative chain under such conditions, there must be a storage form of high-energy phosphate bonds for such periods. This function is filled by phosphocreatine, which reacts with ADP to regenerate ATP when required during periods of exercise.

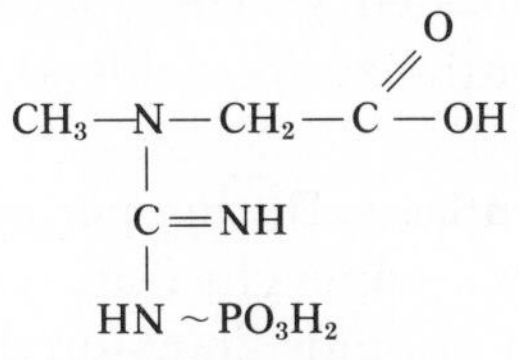

$$\begin{array}{l} CH_3-\underset{\displaystyle \begin{array}{c} | \\ C=NH \\ | \\ HN\sim PO_3H_2 \end{array}}{N}-CH_2-\overset{\displaystyle O \atop //}{C}-OH \end{array}$$

Phosphocreatine

When muscle is at rest, aerobic conditions are reestablished, the oxidative chain produces more ATP, and the ATP rebuilds the supply of phosphocreatine. These relationships are summarized here:

Resting muscle (Aerobic conditions)

Oxidative chain $\rightarrow$ ATP
Creatine + ATP $\rightleftharpoons$ Creatine $\sim PO_4$ + ADP

Contracting muscle (Anaerobic conditions)

ATP + Actomyosin $\rightarrow$ ADP + Muscle contraction
Creatine $\sim PO_4$ + ADP $\rightleftharpoons$ Creatine + ATP

BLOOD GLUCOSE CONCENTRATION

As implied in the previous discussions, the body attempts to maintain the blood sugar (glucose) at a relatively constant level. The normal fasting venous blood sugar concentration averages about 80 mg per 100 ml blood with a range from 65 to 110 mg. The blood glucose level reflects an equilibrium between factors supplying glucose to the blood and those removing it from the blood.

Factors contributing to blood glucose concentration

Intestinal absorption. After meals, there may be a temporary increase in blood glucose (hyperglycemia) because of the rapid absorption of glucose from the small intestine.

Glycogenolysis. When the blood glucose falls to a low level (hypoglycemia), the rate of glycogenolysis in the liver increases in an attempt to put more glucose into the blood.

Gluconeogenesis. During periods of restricted or absent dietary intake of carbohydrates, the process of converting amino acids to glucose, *gluconeogenesis,* increases in an attempt to supply the required blood glucose and glycogen. Gluconeogenesis will be discussed in greater detail with amino acid metabolism.

Factors removing glucose from blood

Liver, muscle, other tissues. After absorption into the portal blood, glucose goes first to the liver, where some is stored as glycogen, some is oxidized for energy purposes, some is converted to other necessary metabolites, and some is released to the systemic blood for use by the rest of the body. Muscle also removes glucose from the blood for storage as glycogen and for oxidation for energy production. The remaining tissues of the body also use some glucose for energy production.

Amino acid synthesis. As needed, glucose may be used to furnish the carbon skeleton of the nonessential amino acids. Details of this synthesis will be discussed later.

Fatty acid and glycerol synthesis. During periods of prolonged excess dietary intake of carbohydrates, some glucose is diverted to the synthesis of the fatty acids and glycerol of lipids and stored in adipose tissue.

Kidney (renal threshold). Should the blood glucose concentration rise above approximately 160 mg per 100 ml, the excess begins to spill out into the urine. In a sense, the kidney acts as a dam and when the capacity of the dam is exceeded, the excess spills over the top. This occurs in uncontrolled diabetes and other conditions.

Hormonal effects in carbohydrate metabolism

Insulin. The hormone insulin, produced by the pancreas (beta cells of of the islands of Langerhans), acts to decrease blood glucose by favoring glycogenesis in the liver and increased use of glucose by other tissues. Thus, as the blood glucose rises above normal levels, the production of insulin is stimulated, which in turn stimulates glycogenesis in the liver and increased utilization of glucose in the tissues, bringing about a decrease in blood glucose. The reverse occurs in the face of a decreasing blood glucose concentration.

Epinephrine (adrenaline). Epinephrine is produced by the adrenal medulla and acts generally in a manner directly opposite to that of insulin. It favors glycogenolysis in liver and muscles by stimulating phosphorylase activity and appears to decrease the uptake of glucose by tissue cells.

Glucagon. Glucagon is formed in the pancreas (alpha cells of the islands of Langerhans) and stimulates glycogenolysis in the liver, leading to an increase in blood glucose.

Glucocorticoids. Certain hormones produced by the adrenal cortex act to stimulate gluconeogenesis and in general in a manner opposite to that of insulin, thereby having a hyperglycemic effect.

Thyroid hormone. Thyroxine, the thyroid hormone, stimulates hepatic glycogenolysis, leading to an increase in blood glucose.

Adenohypophyseal (anterior pituitary) hormones. Adrenocorticotropic and thyrotropic hormones generally produce the same effects as the glucocorticoids and thyroxine because of their actions in stimulating the production of the latter hormones. Other factors from the anterior pituitary also have hyperglycemic activity.

With so many enzymes involved in the metabolic pathways of carbohydrate metabolism and so many factors exerting an influence on these processes, ultimate control of the blood glucose concentration depends on a delicate balance between the various factors. The multiplicity of factors, each with somewhat different activities, allows for rather fine adjustments.

Glucose tolerance: diabetes mellitus

Under ordinary conditions, in a normal individual ingesting a relatively large dose of glucose at one time (100 g), there is a sharp rise in glucose concentration from the normal fasting level of 80 mg per 100 ml of blood (average) to about 120 mg per 100 ml (average), the peak being reached usually within 30 to 60 min. Thereafter there is a sharp decrease that

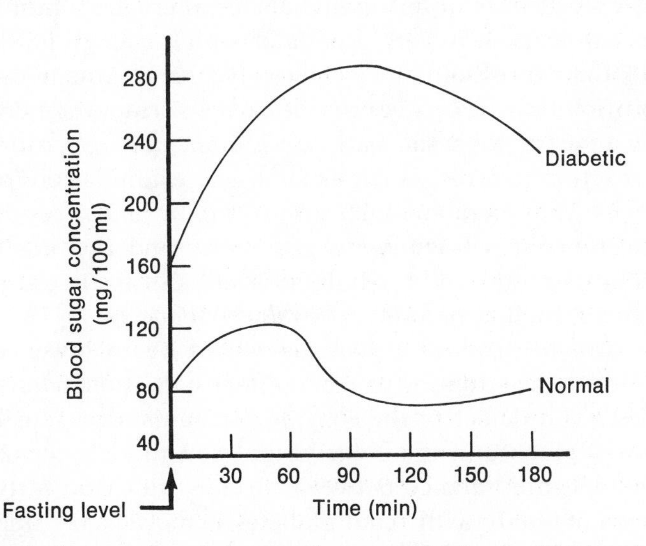

Fig. 21-4. Typical normal and abnormal curves obtained with the glucose tolerance test.

usually drops the blood glucose concentration somewhat below the normal fasting level in approximately 2 hr. The normal fasting level is usually regained after 3 hr. This course of events is graphed in Fig. 21-4.

In patients with diabetes who have a partial or total lack of insulin, the fasting level is usually well above the normal figure and glucose may be found in the urine—a condition known as *glycosuria.* This results from impaired storage and catabolism of glucose in the tricarboxylic acid cycle. Another result is the buildup of an excess of acidic ketone bodies (ketonemia, ketonuria, to be discussed with lipid metabolism), which may lead progressively to acidosis (decreased pH in blood), diabetic coma, and death unless controlled with insulin. The glucose tolerance test in such an individual shows a slower but higher rise to a peak usually well above the renal threshold value (greater than 180 mg per 100 ml blood) and a slower return toward the patient's fasting level, usually requiring more than 3 hr (Fig. 21-4). The difference between the normal and diabetic glucose tolerance test results appears to be related mainly to the lack of insulin response in the diabetic individual. In view of these large differences, the glucose tolerance test has been used effectively as a diagnostic procedure.

Investigations have indicated that individuals subjected to partial or total gastrectomy show changes from normal in their glucose tolerance curves that tend to resemble, in some aspects, the curves of diabetic individuals. The reasons for this are not yet known.

Nondiabetic glycosurias. As noted before, the hyperglycemia that occurs in diabetes may cause glucose to be spilled into the urine if the renal threshold is exceeded. In addition to diabetes, there are a number of conditions that can cause a hyperglycemia of a high enough level to exceed the normal kidney threshold and produce glycosuria. Among these are the rapid absorption of a large quantity of carbohydrate, temporarily overloading the glucose disposal pathways—*alimentary glycosuria;* strong emotional reaction to stress, such as facing an examination (on the subject of carbohydrate metabolism?), severe pain or anger, resulting in the secretion of epinephrine—*emotional glycosuria;* and chemically induced forms of diabetes produced by administration of drugs in animals for research purposes, such as *phlorhizin* or *alloxan diabetes.*

Another way to produce glycosuria would be to lower the kidney threshold. This is the situation in a hereditary condition, *renal glycosuria,* in which there is a defect in the enzyme mechanism responsible for the reabsorption of glucose in the kidney tubules. Although normal in other respects, the delicate balance between carbohydrate and fatty acid metabolism in individuals with renal diabetes is more easily upset than in normal people.

It is also possible for other sugars, such as pentoses, fructose, galactose, or lactose (during lactation), to appear in the urine for hereditary or other reasons. These may be mistaken for glucose—*false glycosuria*—unless more specific tests than the Benedict test are used for identification of the sugar in the urine.

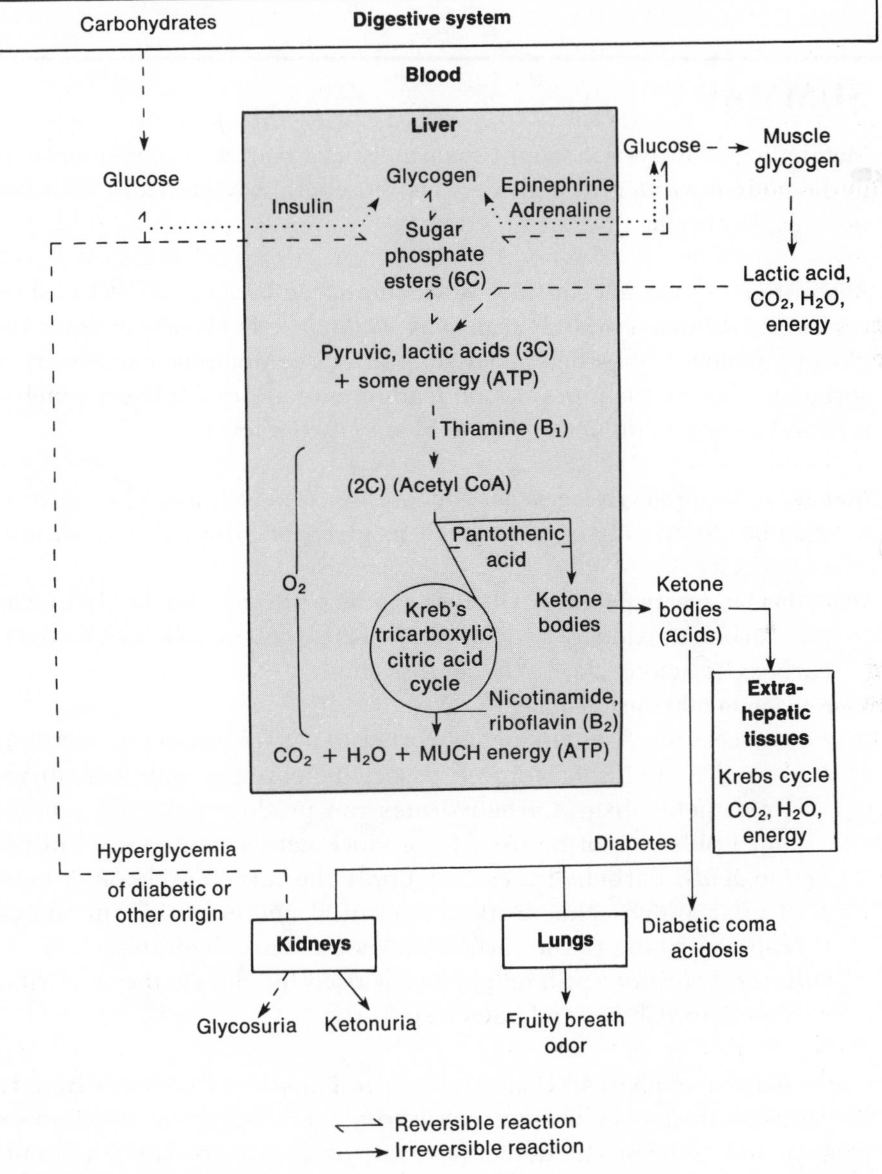

Fig. 21-5. Flow chart of carbohydrate metabolism. (Pathways: - - - - = carbohydrate; —— = common.)

SUMMARY OF CARBOHYDRATE METABOLISM AND ITS RELATIONSHIP TO PROTEIN AND LIPID METABOLISM

Fig. 21-5 is a summary of carbohydrate metabolism in simplified form. Also shown in this flowchart are some of the points of relationship with protein and lipid metabolism. These subjects will be discussed in greater detail in the following chapters and will be added to the diagram.

SUMMARY

Metabolic pathways are taught separately as a matter of convenience, but in the body the different pathways are closely interrelated and their reactions are occurring at the same time.

ASPECTS OF DIGESTION AND ABSORPTION

Absorption of monosaccharides is accomplished by a combination of processes: (1) diffusion and (2) transport. Galactose is absorbed faster than glucose, which is absorbed faster than fructose. Monosaccharides are absorbed by way of the portal blood leading directly to the liver, which exerts an important influence on the blood glucose level.

POSSIBLE USES OF GLUCOSE

Storage as glycogen: glycogenesis: Glucose not required for other purposes may be stored in liver and muscle as glycogen. This process is limited in capacity.

Oxidation for energy purposes: Glucose can be oxidized completely to water and carbon dioxide by two pathways: (1) glycolysis followed by the tricarboxylic acid cycle, or (2) pentose shunt.

Conversion to other metabolites:

Fat: Excess glucose may be converted to fatty acids and glycerol and stored as lipids in adipose tissue, indicating a relationship with lipid metabolism. Carbohydrates can produce net lipid synthesis but lipids cannot be used to produce net carbohydrate synthesis.

Amino acids: Carbohydrates can supply the carbon skeleton of nonessential amino acids. Both nonessential and essential amino acids can contribute to the carbon skeleton of carbohydrates.

Other carbohydrates: Some glucose is used for the synthesis of ribose and deoxyribose, and galactose.

OXIDATION OF GLUCOSE

Divided into two phases: (1) anaerobic metabolism (glycolysis or Embden-Meyerhof pathway)—does not require oxygen but can take place in oxygen, ends with the production of pyruvate or lactate and small amount of energy; (2) aerobic metabolism—requires oxygen and involves the tricarboxylic acid cycle (citric acid cycle, Krebs cycle), ends with production of water, carbon dioxide, and large amount of energy.

Anaerobic metabolism of glucose: See Fig. 21-2.

Formation of glucose-6-phosphate: Reaction 1.

Glycogenesis: Reactions 3, 4.

Glycogenolysis: Reactions 5, 3, 2.

Glycolysis: breakdown to pyruvate: Reactions 6, 7, 9-16. Yeast fermentation—reactions 18, 19.

Interconvertibility of glucose, fructose, galactose: Reactions 20-24, 3, 25-28.

Energy accounting in glycolysis: Net gain of two ATPs at 7500 cal per ATP is 15,000 cal out of a maximum of 50,000 cal per mole of glucose, or 30% efficiency.

Location of enzymes: Glycolytic enzymes found in supernatant fraction of cytoplasm after centrifugation.

Aerobic metabolism of glucose: Same as anaerobic up to pyruvate stage.

Oxidation of pyruvate: formation of acetyl-CoA:

$$\text{Pyruvate} + \text{CoA} \xrightarrow[CO_2]{NAD \;\; NADH_2} \text{Acetyl-CoA}$$

Tricarboxylic acid cycle (citric acid cycle, Krebs cycle): See Fig. 21-3. Net result is the breakdown of one 2-carbon acetyl group to carbon dioxide and water per turn of the cycle.

Energy accounting in aerobic oxidation: Net gain of 38 ATPs is 285,000 cal of a maximum of 688,500 cal per mole of glucose, or 41% efficiency.

Location of enzymes: Aerobic oxidation and oxidative chain enzymes and cofactors are found in the mitochondria.

Pentose shunt: Occurs in liver and adipose tissue. Glucose can be oxidized by this pathway but it appears to function more for the purpose of synthesizing ribose and $NADPH_2$ than producing energy.

MUSCLE CONTRACTION

Muscle contraction appears to result from interaction of actomyosin (muscle protein) with ATP. High-energy phosphate is stored in muscle as phosphocreatine. The following reactions take place:

$$\text{ATP} + \text{Actomyosin} \rightarrow \text{ADP} + \text{muscle contraction}$$

$$\text{Creatine} \sim PO_4 + \text{ADP} \rightleftharpoons \text{Creatine} + \text{ATP}$$

$$\text{Oxidative chain} \rightarrow \text{ATP}$$

BLOOD GLUCOSE CONCENTRATION

Normally about 80 mg per 100 ml blood and closely regulated.

Factors contributing to blood glucose concentration:

Intestinal absorption: May cause temporary hyperglycemia.

Glycogenolysis: Stimulated by low blood glucose concentration.

Gluconeogenesis: Conversion of amino acids to glucose also stimulated by low blood glucose levels.

Factors removing glucose from blood:

Liver, muscle, other tissues: For storage as glycogen, oxidation for energy, conversion to other metabolites.

Amino acid synthesis: Glucose may be converted to nonessential amino acids.

Fatty acid and glycerol synthesis: Excessive amounts of glucose may be converted and stored as lipids.

Kidney (renal threshold): At blood glucose concentrations above 160 mg per 100 ml, the kidney spills glucose into the urine (glycosuria).

Hormonal effects in carbohydrate metabolism:

Insulin: Acts to decrease blood glucose by favoring hepatic glycogenesis and increased use of glucose by other tissues. Production of insulin is stimulated by high blood glucose levels.

Epinephrine (adrenaline): Favors glycogenolysis in liver and muscle and acts opposite to insulin.

Glucagon: Stimulates hepatic glycogenolysis.

Glucocorticoids: Stimulate gluconeogenesis and have hyperglycemic effect.

Thyroid hormone: Stimulates hepatic glycogenolysis.

Anterior pituitary hormones: Have hyperglycemic activity. Control of the blood glucose concentration depends on a delicate balance between the various hormones and other mechanisms involved.

Glucose tolerance: diabetes: See Fig. 21-4 for comparison of normal and diabetic glucose tolerance curves. This technique is used as a diagnostic procedure.

Nondiabetic glycosurias: Any situation leading to hyperglycemia above the renal threshold or lowering the renal threshold itself may result in glycosuria, e.g., alimentary glycosuria, emotional glycosuria, phlorhizin or alloxan diabetes, renal glycosuria, false glycosuria (sugars other than glucose giving positive Benedict's test).

SUMMARY OF CARBOHYDRATE METABOLISM AND ITS RELATIONSHIP TO PROTEIN AND LIPID METABOLISM

For a summary of carbohydrate metabolism see Fig. 21-5.

REVIEW QUESTIONS

1. The oxidation of glucose is divided into how many phases? What are the names generally applied to these phases? Which require(s) oxygen? Which can operate without oxygen? Which provides most of the energy produced by the oxidation of glucose? How much of the total theoretical energy in glucose is made available to the body by its oxidation?
2. Besides oxidation for energy purposes, what other uses are possible for glucose in the body? What are some of the considerations involved in controlling the use to which glucose may be put?
3. Name some factors that may lead to an increase in blood glucose; to a decrease in blood glucose.
4. What is the effect of insulin on carbohydrate metabolism? Which hormones act to oppose some effects of insulin? What is gluconeogenesis and what class of hormones stimulates this process?
5. What hormonal problem is involved in diabetes? How is the kidney involved in diabetes? Name some conditions other than diabetes in which glycosuria may occur. What are the differences in tolerance to a large dose of glucose between normal people and people who have diabetes?

REFERENCES

Advances in carbohydrate chemistry (series), New York, Academic Press, Inc.

Annual review of biochemistry, Palo Alto, Calif., Annual Reviews.

Annual review of physiology, Palo Alto, Calif., Annual Reviews.

Baldwin, E.: Dynamic aspects of biochemistry, ed. 5, London, 1967, Cambridge University Press.

Dickens, F., Randle, P. J., and Whelan, W. J., editors: Carbohydrate metabolism and its disorders, New York, 1968, Academic Press, Inc.

Gregory, R. P. F.: Biochemistry of photosynthesis, ed. 2, New York, 1977, John Wiley & Sons, Inc.

Hers, H. G.: The control of glycogen metabolism in the liver, Ann. Rev. Biochem. **45:** 167, 1976.

Krebs, H. A.: The history of the tricarboxylic acid cycle, Perspect. Biol. Med. **14:**154, 1970.

Litwack, G., editor: Biological actions of hormones, 6 vols., New York, 1970-1979, Academic Press, Inc.

Lowenstein, J. M., editor: Methods in enzymology, Vol. 13, New York, 1969, Academic Press, Inc.

Stanbury, J. B., Wyngaarden, J. B., and Frederickson, D. S., editors: The metabolic basis of inherited disease, ed. 4, New York, 1978, McGraw-Hill Book Co.

White, W. L., Erickson, M. M., and Stevens, S. C.: Chemistry for the clinical laboratory, ed. 4, St. Louis, 1976, The C. V. Mosby Co.

22 Lipid metabolism

Because of the diversity of types of substances classified as lipids (Chapter 15), it is not possible to present as unified a picture of the metabolism of the lipids as was possible in the case of the carbohydrates with their closely interrelated metabolic pathways. However, an important aspect of lipid metabolism is the close relationship between the fats (triglycerides) and carbohydrates. As noted in Chapter 21, the fatty acid component of the fats can furnish acetyl coenzyme A (acetyl-CoA), which can be degraded by way of the tricarboxylic acid cycle. The metabolism of the fatty acids will be discussed in detail in this chapter.

Both fats and carbohydrates may be used by the body as a source of energy by way of the tricarboxylic acid cycle, the fats furnishing more than twice as much energy as the carbohydrates (approximately 9 kcal per gram fat; approximately 4 kcal per gram carbohydrate or protein). In spite of this numerical advantage in favor of the fats, the body prefers to use carbohydrate as its primary fuel. Excessive use of fats for energy purposes may lead to serious difficulties.

The metabolism of other lipids, such as the phospholipids and sterols, are concerned mainly with functions other than energy production.

ASPECTS OF DIGESTION AND ABSORPTION

At best, the digestion and absorption of lipids take place under difficult conditions. The enzymes must function at nonoptimum pH and the water-solubility of the enzymes makes it difficult for them to reach the fat-soluble substrates. This makes the bile salts, emulsifying agents, important aids in the proper digestion and absorption of lipids. Monoglycerides and soaps of the fatty acids also have emulsifying activity.

For triglycerides, investigations indicate that complete hydrolysis is not necessary for absorption, i.e., it is believed that mono- and diglycerides are absorbed as well as fatty acids and glycerol. The longer-chain fatty acids (11 or more carbons) are absorbed by way of the lymphatic system and appear in the lymph mainly in the form of triglycerides. This means that resynthesis of the triglycerides takes place in the intestinal mucosal cells. The shorter-chain fatty acids (10 or less carbons) may be absorbed into the portal blood leading to the liver. During the process of digestion, very small lipid particles known as *micelles* are formed. These particles are composed of a mixture of fatty acids, monoglycerides, and bile salts. The mechanism of absorption of these substances is still debated, the two leading theories being simple diffusion and a process called *pinocytosis*, an engulfing of lipid material by the wall of the absorbing cell. In the lymph, the absorbed triglycerides appear as *chylomicrons*, very small particles composed of a core of triglyceride, with an outer film of phospholipid, cholesterol, and protein. This increases the solubility of these substances for transport purposes.

Phospholipids appear to be absorbed as is or as products of complete hydrolysis. Because of their emulsifying activity, phospholipids may be of importance in the absorption and transport of triglycerides, as noted before, and cholesterol.

Although cholesterol is easily absorbed, mainly in the free form after hydrolysis of its esters, most plant sterols, such as ergosterol (a precursor of vitamin D_2), are poorly absorbed. Fortunately, after irradiation of ergosterol by sunlight, the resulting product is easily absorbed. Cholesterol is absorbed by the lymphatic system, appearing in the lymph mainly as the reesterified product.

The optimum absorption of fat-soluble vitamins and precursors such as vitamin A and the carotenes (precursors of vitamin A), vitamin D and its precursors, and vitamin K all depend on the presence of bile salts and good conditions for fat absorption. Deficiencies of one or more of these vitamins may occur as a result of obstruction of the bile duct. Other conditions in which the digestion or absorption of fat is deranged for any of a number of possible reasons, leading to an increase in the fat content of the feces (steatorrhea), also may lead to such vitamin deficiencies.

DISTRIBUTION OF LIPIDS IN TISSUES

Triglycerides (neutral fat) are found mainly in adipose tissue (fat depots) and normally comprise almost 10% of the body weight. The liver also contains a small amount of triglycerides, as well as some of all the other classes of lipids, because of its important role in the metabolism of these substances.

When needed for energy purposes, free fatty acids can be rapidly mobilized from adipose tissue, released to the blood, in which they are found bound to plasma albumin, and transported in this way to the liver and other tissues for oxidation. The old idea of the inertness of adipose tissue gave way in the face of isotopic experiments to a picture of dynamic equilibrium depending on the state of health and nutrition.

In most tissues other than adipose, phospholipids are present in greater concentrations than other lipids. This is probably related to the many functions attributed to the phospholipids. These functions will be discussed later in this chapter.

Sphingomyelin and glycolipids are noted for their presence in brain and nerve tissue.

Cholesterol, which is found free in low concentration in most tissues, is also found in high concentration in brain tissue. In plasma and liver, cholesterol occurs free and esterified. Esterified cholesterol is present in relatively high concentration in the adrenal cortex, probably because of its role as precursor in the synthesis of other steroids.

Intracellularly, the triglycerides are found in the soluble portion of the cytoplasm. The phospholipids occur as lipoprotein complexes and are found in the nucleus, microsomes, and mitochondria and may be especially important as a component of the membranes (p. 199) around these structures.

One of the manifestations of certain diseases, lipidoses, is the abnormal accumulation of specific lipids, e.g., cerebrosides in spleen, liver, and bones in Gaucher's disease. In obesity, however, in the absence of any known metabolic abnormality, the accumulation of fat seems to be caused only by the ingestion of calories in excess of requirements. Thus far, no better explanation has received wide acceptance, although many have been proposed.

FATTY ACID METABOLISM

Both the synthesis and catabolism of fatty acids involve acetyl-CoA as an intermediate. This fact is the key to the close relationship of fatty acid metabolism with carbohydrate metabolism, in which acetyl-CoA is also an important intermediate.

Synthesis of fatty acids

The major pathway for the synthesis of fatty acids is cytoplasmic, i.e., extramitochondrial, and is summarized in Fig. 22-1. In the first reaction, under the influence of the enzyme acetyl-CoA carboxylase and in the presence of ATP, carbon dioxide, and biotin (which transfers the carbon dioxide group), acetyl-CoA is converted to malonyl-CoA (reaction 1, Fig. 22-1):

$$\text{Acetyl-CoA} \xrightarrow[\text{Biotin}]{\text{ATP},\ CO_2 \quad \text{ADP}} \text{Malonyl-CoA}$$

Acetyl-CoA also reacts with a protein molecule known as the acyl carrier protein, abbreviated ACP-SH to indicate that it has an -SH group in its structure. The enzyme catalyzing this reaction is acetyl-CoA-ACP transacylase (reaction 2, Fig. 22-1):

$$\text{Acetyl-CoA} + \text{HS-ACP} \rightarrow \text{Acetyl-ACP} + \text{CoA-SH}$$

ACP-SH is part of a multienzyme complex known as fatty acid synthetase. The function of ACP-SH is to move the growing acyl group, attached by way of the -SH, along the synthetic cycle from one enzyme to the next.

The acetyl-ACP then forms a complex with the condensing enzyme (E_c-SH) β-ketoacyl-ACP synthetase. In the process, ACP-SH is regenerated (reaction 3, Fig. 22-1):

$$\text{Acetyl-ACP} + E_c\text{-SH} \rightarrow \text{Acetyl-}E_c + \text{ACP-SH}$$

In the next reaction, malonyl-CoA, produced in reaction 1, reacts with ACP-SH to form malonyl-ACP and CoA-SH (reaction 4, Fig. 22-1):

$$\text{Malonyl-CoA} + \text{ACP-SH} \rightarrow \text{Malonyl-ACP} + \text{CoA-SH}$$

The enzyme in this reaction is malonyl-CoA-ACP-transacylase.

At this point, the acetyl-E_c complex from reaction 3 reacts with the malonyl-ACP from reaction 4 to produce acetoacetyl-ACP, also giving off carbon dioxide and regenerating the condensing enzyme, E_c-SH (reaction 5, Fig. 22-1):

$$\text{Acetyl-}E_c + \text{Malonyl-ACP} \rightarrow \text{Acetoacetyl-ACP} + E_c\text{-SH} + CO_2$$

The enzyme involved is β-ketoacyl-ACP synthetase. It should be noted that the carboxyl carbon atom of the malonyl-CoA is lost as carbon dioxide in the reaction as the remaining two carbon atoms become incorporated as the ones which correspond to the carboxyl end of the growing fatty acid. This leaves the two carbon atoms from the acetyl-E_c complex to become the ones corresponding to the end of the fatty acid farthest away from the carboxyl group.

In the next reaction, the acetoacetyl-ACP is reduced to β-hydroxybutyryl-ACP with the help of β-ketoacyl-ACP reductase and $NADPH_2$ (reduced NADP) (reaction 6, Fig. 22-1):

$$\text{Acetoacetyl-ACP} \xrightarrow{NADPH_2 \searrow \ \nearrow NADP} \beta\text{-Hydroxybutyryl-ACP}$$

β-Hydroxybutyryl-ACP is then dehydrated by β-hydroxyacyl dehydrase to form the α, β-unsaturated butyryl-ACP (reaction 7, Fig. 22-1):

$$\beta\text{-Hydroxybutyryl-ACP} \xrightarrow{\nearrow H_2O} \alpha,\beta\text{-Unsaturated butyryl-ACP}$$

In the last reaction of the first cycle, the α,β-unsaturated butyryl-ACP is reduced by enoyl-ACP reductase and $NADPH_2$ to form butyryl-ACP (reaction 8, Fig. 22-1):

$$\alpha,\beta\text{-Unsaturated butyryl-ACP} \xrightarrow{NADPH_2 \searrow \ \nearrow NADP} \text{Butyryl-ACP}$$

To start the second cycle, the butyryl-ACP reacts with E_c-SH to form butyryl-E_c, which then reacts with another molecule of malonyl-ACP to form a six carbon diketo-ACP (dotted line, Fig. 22-1). Thus, at each turn of the cycle, two carbon atoms are added to the growing fatty acid at the carboxyl end of the molecule. After 7 cycles, palmitic acid is released from the enzyme complex by palmityl deacylase (reaction 9, Fig. 22-1):

$$\text{Palmityl-ACP} \rightarrow \text{Palmitic acid} + \text{ACP-SH}$$

The high requirement for $NADPH_2$ in this process (fourteen $NADPH_2$ per molecule of palmitic acid produced) may be met by the production of $NADPH_2$ in the pentose shunt (p. 306), which takes place to a quanti-

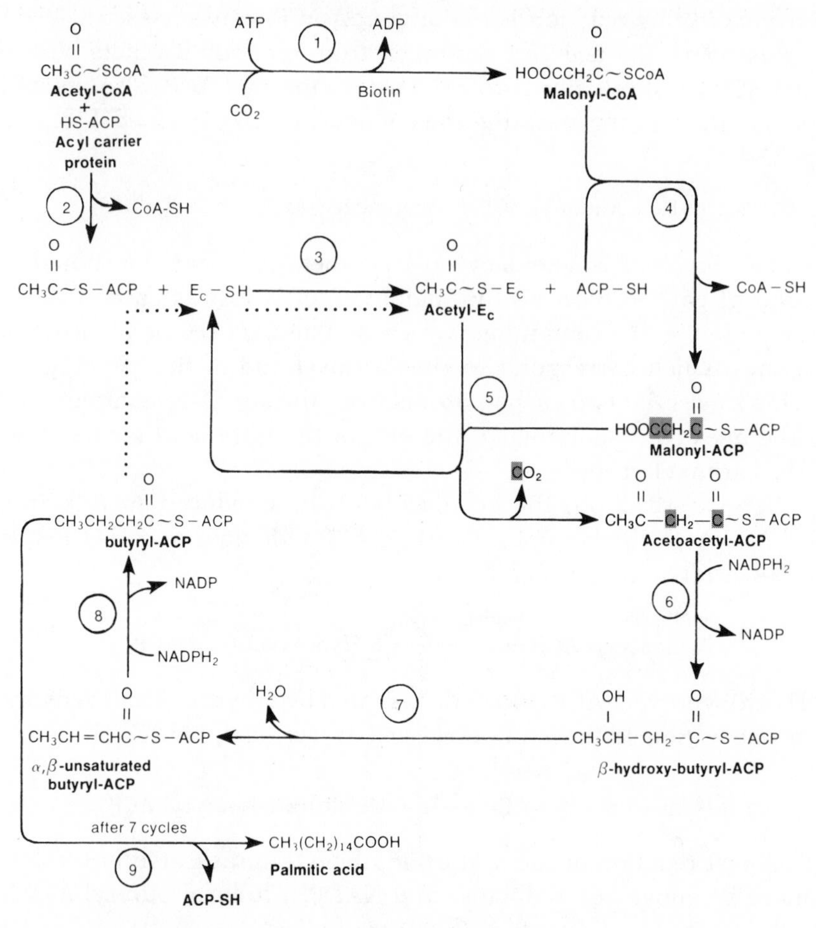

Fig. 22-1. Extramitochondrial fatty acid synthesis. Note the positions of the newly incorporated atoms from the malonyl-ACP. The condensing enzyme (E_c-SH) is β-ketoacyl-ACP synthetase.

tatively significant extent in liver and adipose tissue. This is another point of close connection between fatty acid and carbohydrate metabolism.

Insulin acts to stimulate fatty acid synthesis by increasing the supply of acetyl-CoA as a result of increasing glucose utilization. Citrate stimulates the formation of malonyl-CoA, whereas the end products, such as palmityl-CoA, inhibit the pentose shunt and other key steps. It appears that these control mechanisms are attempting to coordinate the rate of synthesis with the rate of utilization of fatty acids and with the state of carbohydrate metabolism.

The reactions in the mitochondrial pathway of fatty acid synthesis, summarized in Fig. 22-2, are similar to those just described for extramitochondrial synthesis, i.e., condensation, reduction, dehydration, reduction. In contrast to the extramitochondrial pathway, malonyl-CoA and ACP are not involved, and one of the reduction steps requires $NADH_2$ instead of $NADPH_2$. Evidence indicates that the mitochondrial pathway serves mainly for lengthening existing fatty acid chains rather than complete synthesis.

The synthesis of fatty acids requires an expenditure of energy that must be supplied by the breakdown of carbohydrates. Isotopic studies

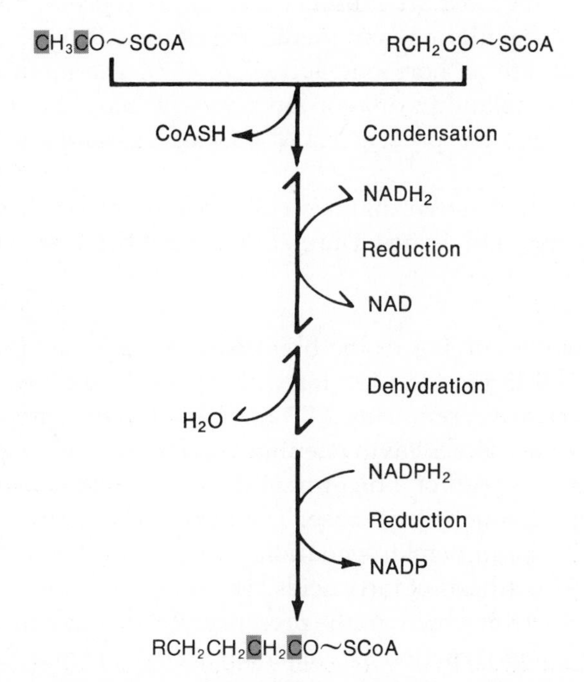

Fig. 22-2. Mitochondrial fatty acid synthesis. Note the positions of the newly incorporated carbon atoms.

have shown that the rate of turnover or replacement of fatty acids in the various tissues is high in liver and adipose tissue and intestinal mucosa, and low in muscle, skin, and nerve tissue.

Interconversions

Using reactions such as the ones just discussed, interconversions can be made between fatty acids by addition or removal of 2-carbon units. Enzymes are also available for the formation of some types of unsaturated fatty acids. The unsaturated fatty acids that cannot be synthesized directly are linoleic, linolenic, and arachidonic (p. 202), and are known as *essential fatty acids*. However, if sufficient linoleic acid only is obtained in the diet, linolenic and arachidonic acids can be synthesized by further desaturation and elongation of the linoleic acid.

Synthesis of prostaglandins, thromboxanes. As noted before (p. 202), it has been determined that some essential fatty acids are converted to prostaglandins in the body. Because many important functions have been attributed to the prostaglandins, it is now possible to explain why these fatty acids are essential and must be obtained in the diet. The biosynthesis of some important derivatives of the essential fatty acid, arachidonic acid, is illustrated in Fig. 22-3.

Among the more important functions of prostaglandins for which evidence has been presented are muscle contraction, reproduction, lipid metabolism, blood coagulation, and mediation of the effects of cyclic adenosine monophosphate on hormone action (p. 447). The intense interest in the effects of prostaglandins on reproductive physiology has led to clinical trials ranging from the prevention of miscarriages to the induction of labor.

It has also been reported that aspirin prevents the synthesis of certain prostaglandins by body tissues. The significance of this observation will be discussed later (p. 373).

Breakdown of fatty acids

The reactions occurring in the breakdown of fatty acids are summarized in Fig. 22-4. In reaction 1, a fatty acid is activated by conversion to an acyl-CoA derivative, requiring ATP and CoA. The next step (reaction 2), a dehydrogenation, uses a flavin adenine dinucleotide (FAD, p. 346) as the prosthetic group instead of a nicotinamide adenine dinucleotide. The enzyme is an acyl-CoA dehydrogenase. The next two reactions, 3 and 4, involve a hydration and dehydrogenation that are similar to the reactions occurring in the synthesis of fatty acids. The resulting compound is a β-ketoacyl-CoA derivative, which, in the presence of CoA and a thiolase, breaks up into two portions (reaction 5), one being acetyl-CoA and the remainder an acyl-CoA product containing two less carbon atoms than the fatty acid that started through the series of reactions. This series of reactions can then be repeated until the fatty acid is completely chopped into 2-carbon acetyl-CoA units. This type of fatty acid breakdown was first proposed by Knoop as the beta-oxidation theory.

The acetyl-CoA formed can then funnel into the tricarboxylic acid cycle for further oxidation and production of more energy. Because the tricarboxylic acid cycle produces CoA in the reaction that forms citrate, the rate of operation of the cycle can have an influence on the supply of CoA

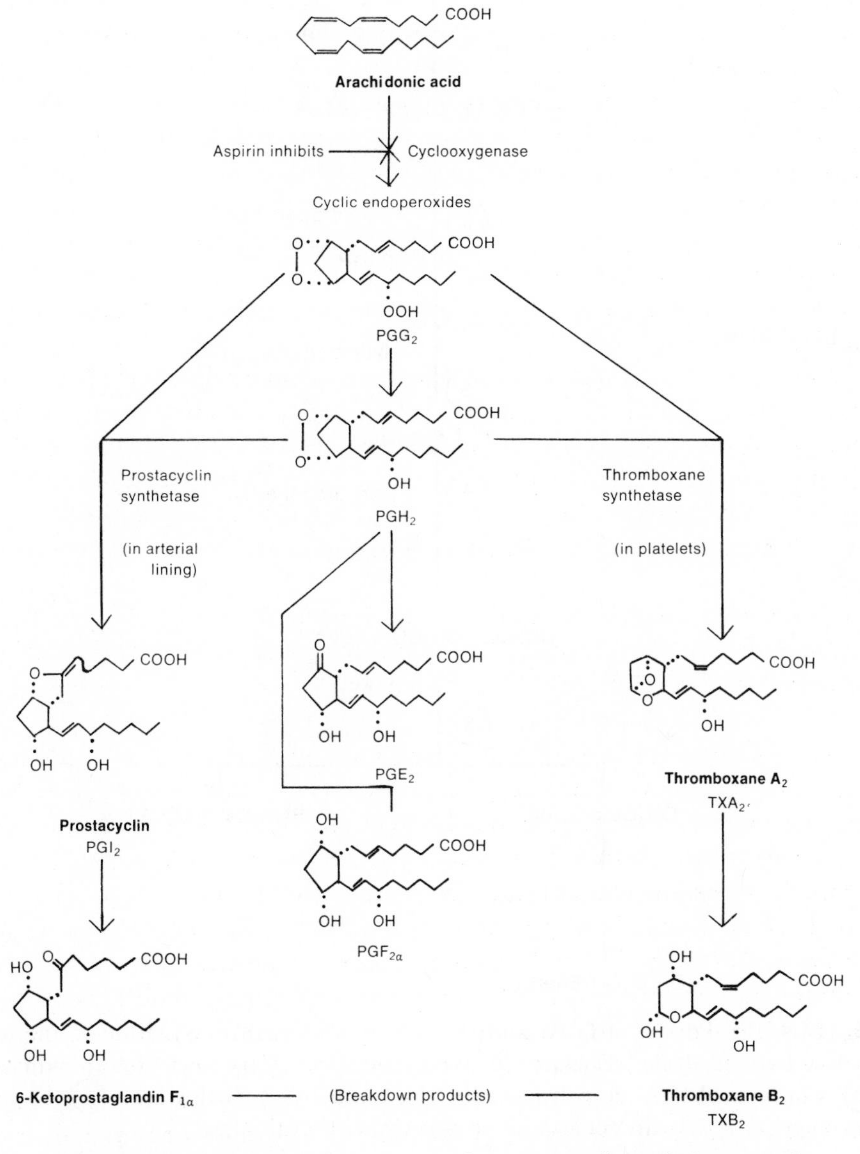

Fig. 22-3. Biosynthesis of important derivatives of arachidonic acid.

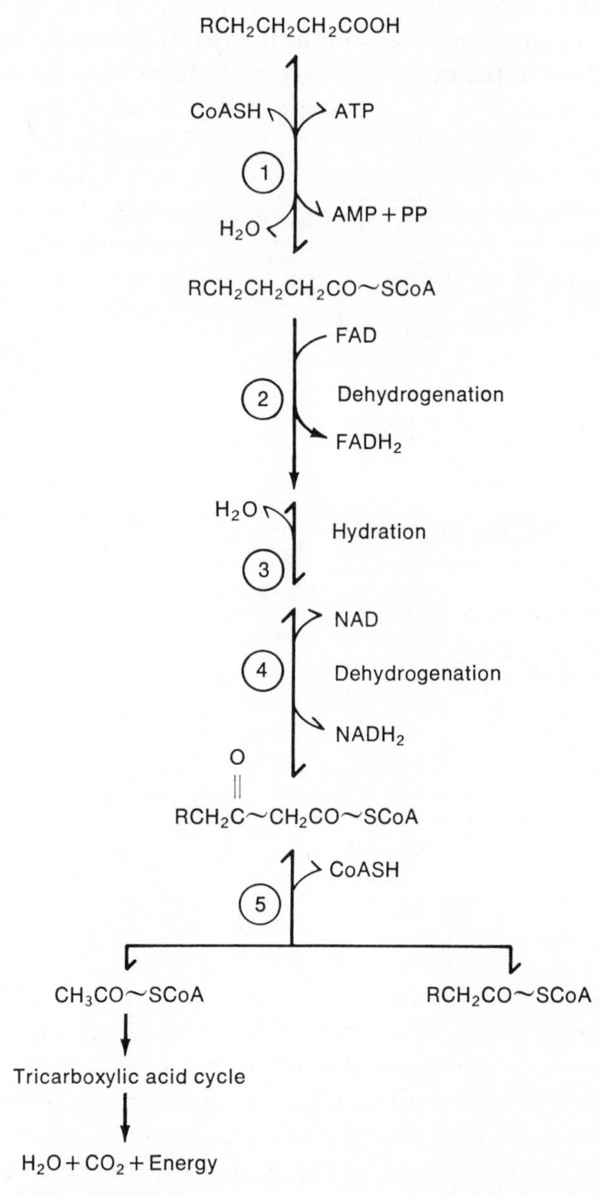

Fig. 22-4. Breakdown of fatty acids to acetate and further oxidation in the tricarboxylic acid cycle. (Enzymes for reaction: (1) = fatty acyl-CoA thiokinase; (2) = fatty acyl-CoA dehydrogenase; (3) = enoyl-CoA hydratase; (4) = β-hydroxyacyl-CoA dehydrogenase; (5) = β-ketoacyl-CoA thiolase.)

(reaction 1, Fig. 22-4) and, thereby, an influence on whether synthesis or breakdown of fatty acids may gain the upper hand. This is another delicate connection between fatty acid and carbohydrate metabolism and is discussed later.

It should also be noted by comparing the pathways of synthesis and breakdown of the fatty acids that some of the reactions are similar in all the pathways. Each pathway is also essentially concerned with the manipulation, by condensation or breakdown, of a 2-carbon unit.

Ketone body formation (ketogenesis). There is an important side reaction that occurs as a result of the normal accumulation of small amounts of acetyl-CoA. This involves the condensation of two molecules of acetyl-CoA to form acetoacetyl-CoA and CoA:

$$2CH_3CO \sim CoA \rightleftharpoons CH_3COCH_2CO \sim CoA + CoA$$

Acetyl-CoA **Acetoacetyl-CoA**

In the liver, a thiolesterase catalyzes the hydrolysis of acetoacetyl-CoA to acetoacetate, a reaction that is essentially irreversible in the liver but not in extrahepatic tissues:

$$CH_3COCH_2CO \sim CoA \rightarrow CH_3COCH_2COOH + CoA$$

Acetoacetyl-CoA **Acetoacetate**

Most of the acetoacetate formed in the liver is reduced to β-hydroxybutyrate and some is converted to acetone by a spontaneous decarboxylation:

$$CH_3COCH_2COOH$$

Acetoacetate

β-Hydroxybutyrate dehydrogenase ($NADH_2$ → NAD) → $CH_3CHOHCH_2COOH$ **β-Hydroxybutyrate**

Spontaneous decarboxylation (CO_2) → CH_3COCH_3 **Acetone**

The three compounds shown, acetoacetate, β-hydroxybutyrate, and acetone, are known as *ketone bodies*. The liver releases these compounds to the blood, which carries them to extrahepatic tissues, where they can be put back on the tricarboxylic acid cycle pathway for complete oxidation. This occurs under normal circumstances and the concentration of ketone bodies in the blood remains quite low. However, in situations in which fat metabolism increases as carbohydrate metabolism decreases, as is the case in diabetes mellitus, the concentration of ketone bodies in the blood rises to abnormal levels (ketonemia), some ketones appear in the urine (ketonuria), and a clinical condition of ketosis develops.

Energy accounting in fatty acid oxidation

As an example, the complete oxidation of palmitate produces approximately 2,340,000 cal per mole, of which 975,000 cal are recovered in the form of high-energy phosphate—an efficiency of about 42%, a figure that is approximately the same for glucose.

FAT (TRIGLYCERIDE) METABOLISM

Synthesis and catabolism

The synthesis and metabolism of the fatty acid component of fats have already been described. The glycerol portion is produced from an intermediate of glycolysis:

$$\text{Dihydroxyacetone-phosphate} \xrightarrow[\text{Glycerol-phosphate dehydrogenase}]{NADH_2 \quad NAD} \text{L-Glycerolphosphate}$$

The two components are put together to form triglycerides in the series of reactions shown in Fig. 22-5. Although this is the major pathway of synthesis, a number of tissues have alternate possibilities.

Triglycerides, or neutral fat, are found stored mainly in adipose tissue and liver. It has been found that the type of triglyceride formed for deposition may be altered to some extent by the type of fat supplied in the diet.

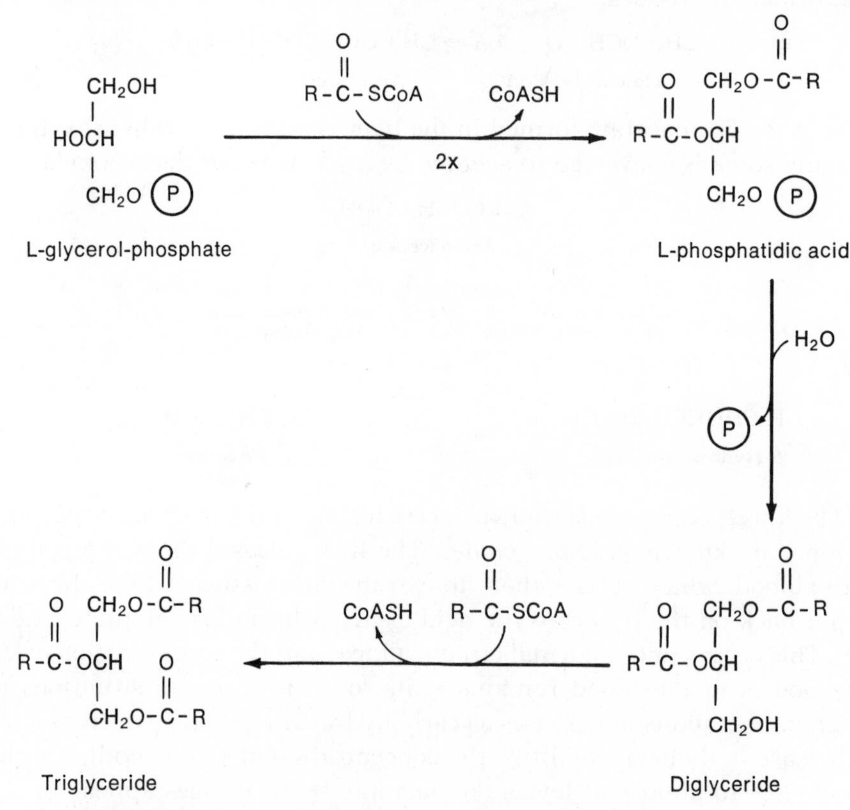

Fig. 22-5. Synthesis of triglycerides. ((P) = phosphate.)

Therefore, pigs being fattened for the production of bacon and lard, for which a high melting point of the fat is desirable, are fed a diet high in saturated fats (p. 201).

Fats must be hydrolyzed by intracellular lipases before they can be oxidized. Glycerol enters the glycolytic scheme (p. 298) and the fatty acids are oxidized as already described (p. 322).

PHOSPHOLIPID METABOLISM

Synthesis and catabolism

Among the simpler phospholipids are the lecithins (contain the nitrogenous base choline) and the cephalins (contain the nitrogenous base ethanolamine) (p. 205). These compounds may be synthesized from a diglyceride intermediate, as in triglyceride synthesis, by reaction with the appropriate base in the activated form of a cytidine-diphosphate base, as summarized here:

$$\text{Choline or ethanolamine base} \xrightarrow{\text{ATP} \quad \text{ADP}} \text{Base-phosphate} \xrightarrow{\text{CTP} \quad \text{PP}} \text{CDP-choline or CDP-ethanolamine}$$

$$\underset{\textbf{Diglyceride}}{\begin{array}{l} \qquad\qquad\quad \text{O} \\ \qquad\qquad\quad \| \\ \text{O} \quad \text{CH}_2\text{O}-\text{C}-\text{R} \\ \| \qquad | \\ \text{R}-\text{C}-\text{OCH} \\ \qquad\quad | \\ \qquad\quad \text{CH}_2\text{OH} \end{array}} \xrightarrow{\text{CDP-Base} \quad \text{CMP}} \underset{\textbf{Lecithin or cephalin}}{\begin{array}{l} \qquad\qquad\qquad\quad \text{O} \\ \qquad\qquad\qquad\quad \| \\ \text{O} \qquad\quad \text{CH}_2\text{O}-\text{C}-\text{R} \\ \| \qquad\qquad | \\ \text{R}-\text{C}-\text{O}-\text{CH} \\ \qquad\qquad\quad | \\ \qquad\qquad\quad \text{CH}_2-\text{O}-\text{P}-\text{Base} \end{array}}$$

All tissues can synthesize phospholipids for their own use, but the liver synthesizes large amounts for export as plasma phospholipids.

Catabolically, the phospholipids are hydrolyzed completely by a series of reactions, the components being degraded by pathways already described, except for the bases. Details concerning the intermediary metabolism of the more complicated phospholipids are left for more advanced texts.

Functions of phospholipids

The phospholipids are believed to perform many functions, some of which have not been definitely established but are still being investigated. Some of these functions are listed here.

As a structural component. Phospholipids have received much attention as components of cellular membranes (p. 199), including membranes around such particles as mitochondria. The functioning of the oxidative chain and of oxidative chain phosphorylation in mitochondria is inactivated by removal of the phospholipids, which may be controlling or participating in the transport of metabolites from one side of the membrane to the other.

In blood coagulation. Phospholipids containing the base ethanolamine or the amino acid serine function in the complicated process of blood coagulation (p. 372).

In absorption and transport of lipids. Phospholipids can act as emulsifying agents during digestion and absorption of lipids and are considered to be an important component of the coating of chylomicrons (p. 317) in the form of lipoproteins. Phospholipids are apparently also involved in the transport of lipids in the blood.

STEROL METABOLISM—CHOLESTEROL

The most important sterol to animal metabolism is cholesterol. Isotopic studies have shown that cholesterol is synthesized from small molecules in a long series of steps involving condensations, transformations, and ring closures. The first important step is the condensation of acetyl-CoA with acetoacetyl-CoA, establishing a relationship with carbohydrate and fatty acid metabolism. Cholesterol is probably synthesized in all tissues, but the liver is the major site.

Much of the interest in recent years in the metabolism of cholesterol has centered on its relationship to atherosclerosis, abnormalities in the circulatory system, and heart attacks. Some observers believe that a high concentration of cholesterol in the blood leads to deposition of cholesterol in the arteries, hardening of the arteries, and eventually heart failure. At present however, there are no unequivocal data available to prove cholesterol is the direct causative agent for these conditions. There are some long-range large group studies under way that may provide definitive evidence.

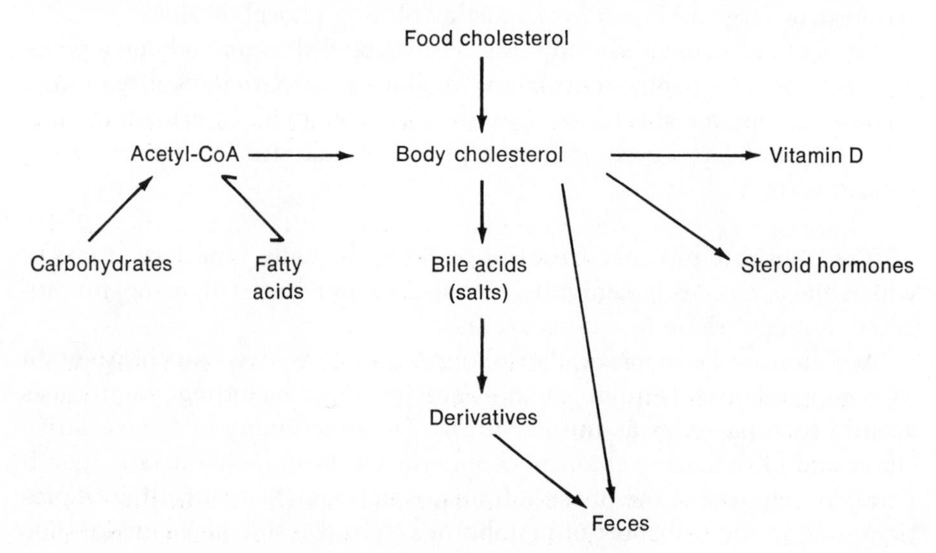

Fig. 22-6. Cholesterol metabolism.

Attempts have been made to lower the blood cholesterol level with drugs and diet restrictions. These attempts have not been very successful because the rate of synthesis of cholesterol by the liver is inversely related to the dietary intake and, in the presence of low blood cholesterol, the liver tends to produce more.

In addition, cholesterol is a key intermediate in the synthesis of vitamin D (p. 209) and bile acids (p. 209), and has been found to be incorporated into the sex hormones, estrogens (p. 452) and androgens (p. 451), and adrenocortical hormones (p. 456).

Bile salts are reabsorbed from the intestine and returned to the liver in the blood by an enterohepatic circulation. In this way, the rate of synthesis of the bile salts can be controlled by the concentration in the blood. Some cholesterol and derivatives of the bile salts are lost in the feces. The metabolism of cholesterol is summarized in simplified form in Fig. 22-6.

HORMONAL EFFECTS IN LIPID METABOLISM

Insulin

As a result of the increased utilization of glucose, stimulated by insulin, including increased production of $NADPH_2$ by way of the pentose shunt, there is an increased synthesis of fatty acids (lipogenesis). Insulin also increases the synthesis of triglycerides in liver and adipose tissue and decreases the rate of release of fatty acids from adipose tissue. These actions decrease the rate of ketogenesis.

In insulin deficiency the reverse situations occur, including an increased synthesis of cholesterol because of the increased availability of 2-carbon fragments.

Adrenocortical hormones

The glucocorticoids tend to increase the rate of release of fatty acids from adipose tissue which, in the absence of adequate insulin or carbohydrate metabolism, leads to ketogenesis and increased synthesis of cholesterol and fatty acid–containing lipids.

Anterior pituitary hormones

Adrenocorticotropic hormone (ACTH), in addition to stimulating the production of adrenocortical hormones, increases the release of fatty acids from adipose tissue, thus leading to ketogenesis.

Epinephrine

In the presence of adrenocortical and thyroid hormones, epinephrine also increases mobilization of fatty acids from adipose tissue and subsequently ketogenesis.

Thyroid hormone

Thyroid hormone tends to decrease plasma cholesterol and other lipids but, in conjunction with inadequate insulin and carbohydrate metabolism, increases fatty acid release from adipose tissue and ketogenesis.

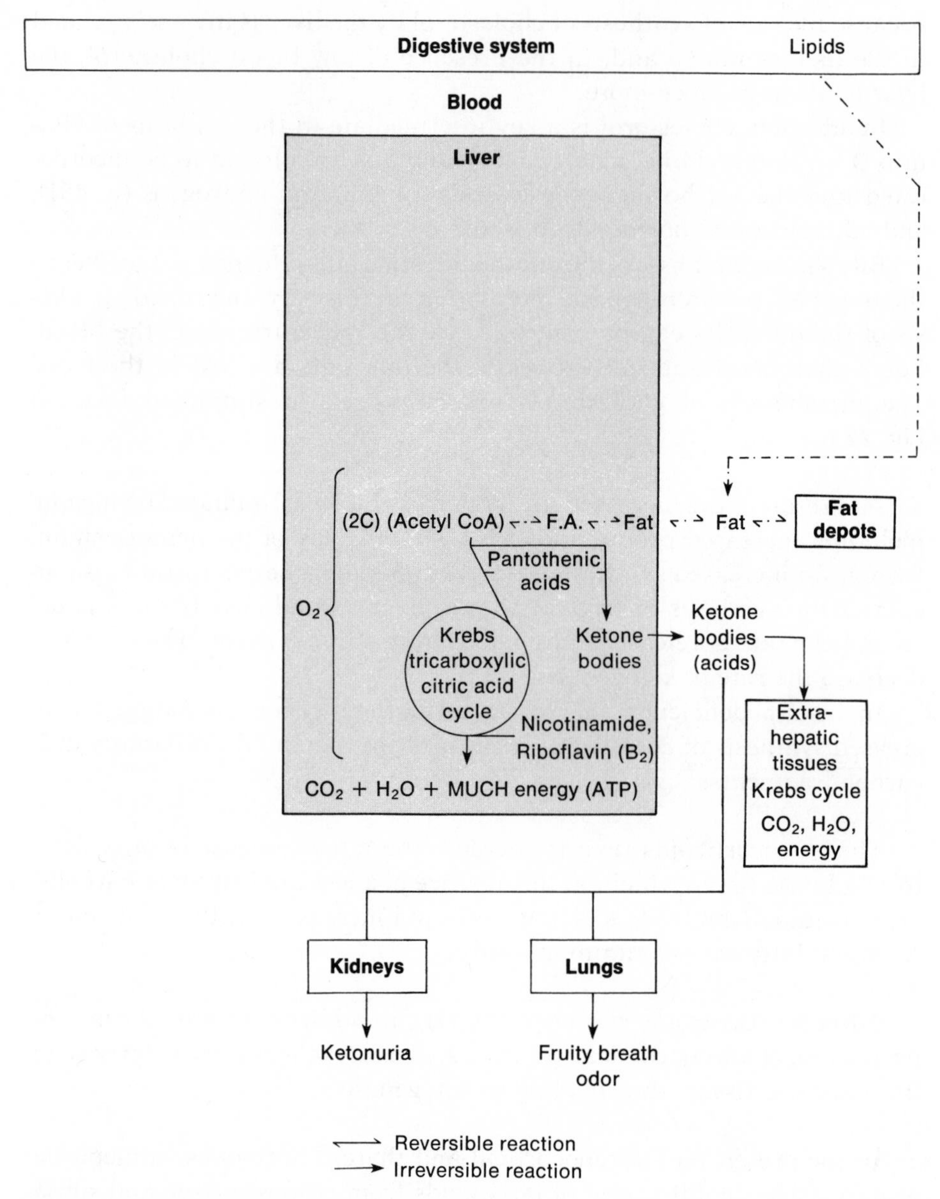

Fig. 22-7. Flowchart of lipid metabolism. (Pathways: —-—-— = lipid; ———— = common.)

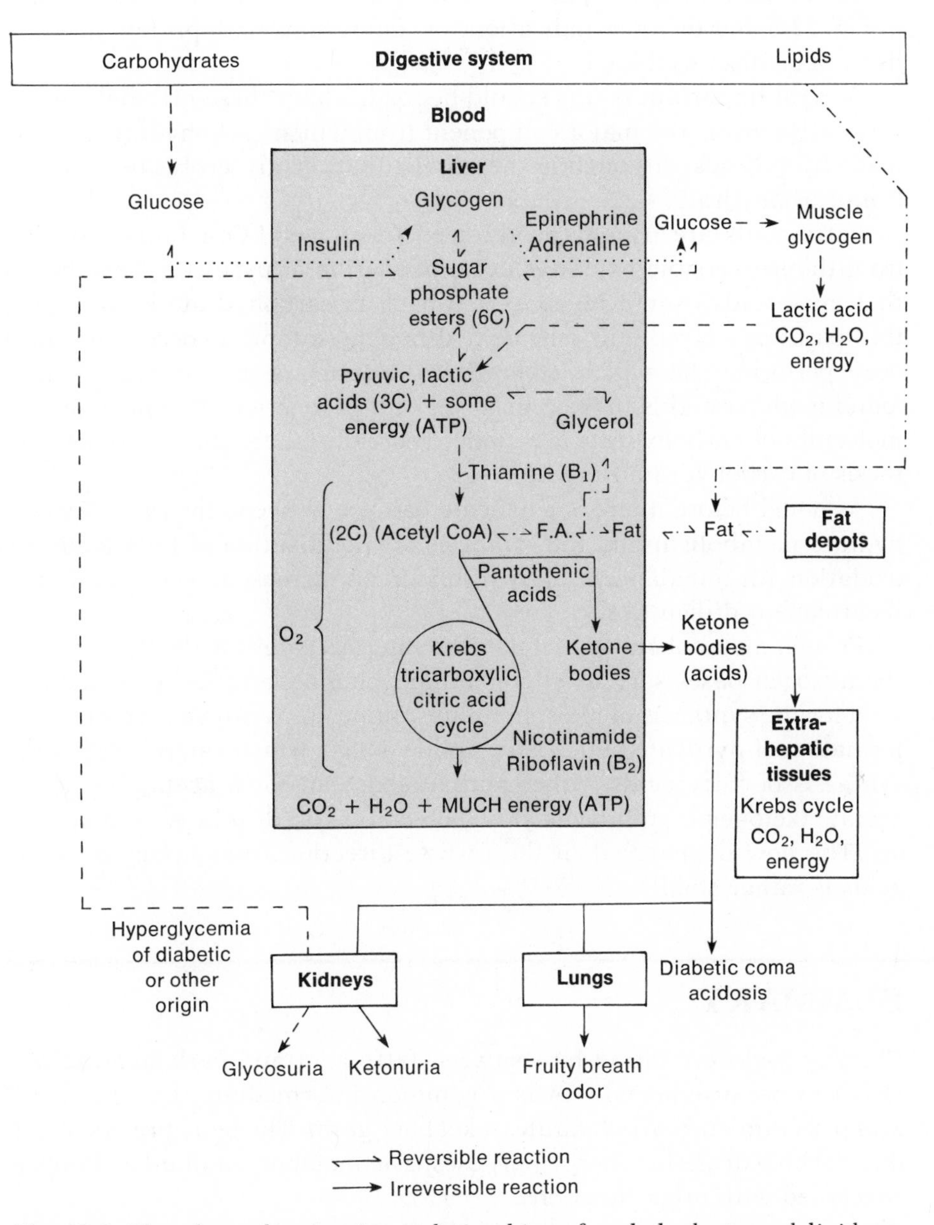

Fig. 22-8. Flowchart showing interrelationships of carbohydrate and lipid metabolism. (Pathways: - - - - = carbohydrates; —-—-— = lipids; ———— = common.)

SUMMARY OF LIPID METABOLISM AND ITS RELATIONSHIP TO CARBOHYDRATE AND PROTEIN METABOLISM

A simplified summary of lipid metabolism is shown in Fig. 22-7. The detailed relationship of lipid to carbohydrate metabolism is summarized in Fig. 22-8 and includes indications of connections with protein metabolism, to be discussed later.

Several important points should be made about these metabolic pathways. Glycerol is the major component from which carbohydrate may be made from lipids, although in the usual situation, glycerol is being made from carbohydrate.

Fatty acids are usually synthesized from acetyl-CoA units resulting from the irreversible oxidative decarboxylation of pyruvate. This means that fatty acids cannot be converted back to carbohydrate by reversing this pathway (glycolytic scheme). Although isotopic experiments have been performed in which atoms from acetyl groups are subsequently found in glucose, this may occur as a result of reactions in which no new molecules of carbohydrate are made. Therefore, there can be no net synthesis of carbohydrate from lipid.

As noted before, there is a delicate balance between the rate of carbohydrate metabolism and the synthesis or mobilization of fatty acids for oxidation. An imbalance usually results in an increase in ketogenesis and its attendant difficulties.

Protein metabolism contributes the intermediates for the formation of the nitrogen bases such as ethanolamine, choline, and serine, which are used for the synthesis of phospholipids. Amino acids can contribute to the formation of pyruvate (glucogenic amino acids), which can be used in the synthesis of fatty acids. Other amino acids can form acetate or acetoacetate (ketogenic amino acids), which can be used in fatty acid synthesis. The flow of material in the reverse direction from lipids to amino acids is rather small.

SUMMARY

There is a close relationship between fatty acid and carbohydrate metabolism because acetyl-CoA is a common intermediate. Fats furnish 9 kcal per gram and carbohydrates 4 kcal per gram. The body prefers to oxidize carbohydrate for energy purposes. The metabolism of other lipids is concerned with other functions.

ASPECTS OF DIGESTION AND ABSORPTION

Good emulsification (bile salts) required for optimum digestion and absorption of lipids and fat-soluble vitamins. Triglycerides are hydrolyzed to lower glycerides, absorbed, resynthesized to triglycerides, which appear in the lymph. Absorption of lipid particles, micelles, may occur by diffusion or pinocytosis. In lymph, triglycerides appear as chylomicrons (triglyceride core, outer film of phospholipid, cholesterol, and protein). Phospholipids are absorbed as is or as hydrolysis products. Cholesterol is absorbed free and appears esterified in lymph.

DISTRIBUTION OF LIPIDS IN TISSUES

Triglycerides are found mainly in adipose tissue, some in liver, which contains some of all other lipids. Fatty acids may be mobilized from adipose tissue and transported in blood bound to albumin. Phospholipids are the lipids of greatest concentration in most tissues other than adipose. Sphingomyelin and glycolipids are found in brain and nerve tissue. Cholesterol is also found in high concentration in brain, in low concentration in most other tissues, and in high concentration as an ester in adrenal cortex. The triglycerides are found in the soluble cytoplasm and the phospholipids appear to be important as a component of cellular and subcellular membranes.

FATTY ACID METABOLISM

Synthesis: Fatty acids are synthesized by condensation of acetyl units in mitochondrial and extramitochondrial pathways. See Fig. 22-1 for extramitochondrial pathway involving acetyl-CoA, malonyl-CoA, ACP, and $NADPH_2$ (supplied by pentose shunt). Insulin and citrate stimulate and palmityl-CoA inhibits synthesis of fatty acids. See Fig. 22-2 for mitochondrial pathway, which adds acetyl units onto existing, smaller-chain, fatty acids.

Interconversions: Most of the fatty acids are interconvertible, and unsaturated acids can also be synthesized except for the essential fatty acids, linoleic, linolenic, and arachidonic, which must be obtained in the diet.

Synthesis of prostaglandins, thromboxanes: Essential fatty acids act as precursors in the biosynthesis of prostaglandins, e.g., arachidonic acid is converted to the prostaglandins, prostacyclin and thromboxane (Fig. 22-3). The prostaglandins are believed to have a wide variety of functions, including smooth muscle contraction, reproduction (conception to initiating labor), lipid metabolism, blood coagulation, mediating the effects of cyclic adenosine monophosphate, and others.

Breakdown of fatty acids: Also occurs by way of acetyl units (Knoop beta-oxidation scheme) as shown in Fig. 22-4. A delicate balance exists between production of acetyl-CoA and its utilization in carbohydrate and fatty acid metabolism.

Ketone body formation: Acetoacetate, β-hydroxybutyrate, and acetone are known as ketone bodies, which are synthesized normally in small amounts and in larger amounts when acetyl-CoA accumulates. The liver cannot metabolize ketone bodies, which are released into the blood for transport to other tissues for complete oxidation. In ketonemia abnormal amounts of ketone bodies are found in blood.

Energy accounting in fatty acid oxidation: Of 2,340,000 cal, 975,000 cal (42%) are recovered as high-energy phosphate in the complete oxidation of 1 mole of palmitate.

FAT (TRIGLYCERIDE) METABOLISM

Synthesis and catabolism: Fatty acids are synthesized as already described. Glycerol is synthesized from the glycolytic intermediate, dihydroxyacetone-phosphate. See Fig. 22-5 for the synthesis of triglycerides. Fats are hydrolyzed before oxidation, glycerol entering the glycolytic scheme and fatty acids being oxidized as already described.

PHOSPHOLIPID METABOLISM

Synthesis and catabolism: Diglyceride reacting with an appropriate base in form of CDP-base produces phospholipid. Phospholipids are completely broken down by hydrolytic reactions. All tissues synthesize phospholipids; liver makes large amounts as plasma phospholipids.

Functions of phospholipids:

As a structural component: In cellular and subcellular membranes. Oxidative chain and oxidative phosphorylation inactivated by removal of phospholipid.

In blood coagulation: Phospholipids containing ethanolamine and serine function in blood coagulation.

In absorption and transport of lipids: Phospholipids have emulsifying activity and appear in the outer layer of chylomicrons.

STEROL METABOLISM—CHOLESTEROL

Synthesized from small molecules such as acetate in long series of reactions. Implicated in circulatory abnormalities and heart attacks. Cholesterol is a key intermediate in synthesis of vitamin D, bile acids, steroid hormones (Fig. 22-6).

HORMONAL EFFECTS IN LIPID METABOLISM

Insulin: Increases lipogenesis, triglyceride synthesis; decreases ketogenesis.

Adrenocortical hormones: Increases ketogenesis in absence of adequate insulin or carbohydrate metabolism.

Anterior pituitary hormones: Increases ketogenesis.

Epinephrine: Increases ketogenesis in presence of adrenocortical and thyroid hormones.

Thyroid hormone: Decreases plasma cholesterol, increases ketogenesis if insulin and carbohydrate metabolism are inadequate.

SUMMARY OF LIPID METABOLISM AND ITS RELATIONSHIP TO CARBOHYDRATE AND PROTEIN METABOLISM

See Figs. 22-7 and 22-8. Glycerol may form carbohydrate or lipid. Fatty acids are synthesized from acetyl-CoA arising from the irreversible decarboxylation of pyruvate. Therefore, there can be no net synthesis of carbohydrate from fatty acids. There is a delicate balance between the rate of carbohydrate metabolism and the synthesis or mobilization of fatty acids for oxidation.

REVIEW QUESTIONS

1. What are some important considerations in the proper digestion and absorption of lipids? What is the distribution of lipids in the tissues and what are their established or proposed functions?
2. What is the key unit in the biosynthesis and breakdown of fatty acids? What is the significance of the essential fatty acids? What important substances are produced biosynthetically from essential fatty acids?
3. What is the key unit in the biosynthesis of cholesterol? What is another source of body cholesterol? Is there any relationship between these sources? For what important substances is cholesterol a key intermediate? What health conditions may be related to cholesterol metabolism? Has the relationship been definitely established?
4. What are the important points at which fatty acid metabolism and carbohydrate metabolism are related? Can there be a net synthesis of carbohydrate from lipid? As the oxidation of fat increases, what substance may be produced in excessive amounts? How is this dangerous? How is diabetes also related to the excessive production of this substance?
5. What is the effect of insulin on lipid metabolism? Is this related to its effect on carbohydrate metabolism? Explain. What are the effects of some other important hormones on lipid metabolism?

REFERENCES

Advances in lipid research, New York, Academic Press, Inc.

Annual review of biochemistry, Palo Alto, Calif., Annual Reviews.

Annual review of physiology, Palo Alto, Calif., Annual Reviews.

Baldwin, E.: Dynamic aspects of biochemistry, ed. 5, London, 1967, Cambridge University Press.

Florkin, M., and Stotz, E. H., editors: Comprehensive biochemistry, Vol. 18, New York, 1970, Elsevier.

Gurr, A. I., and James, A. T.: Lipid biochemistry: an introduction, ed. 2, New York, 1976, Halsted Press.

Kharasch, N., and Fried, J., editors: Biochemical aspects of prostaglandins and thromboxanes, New York, 1977, Academic Press, Inc.

Litwack, G., editor: Biological actions of hormones, New York, 1970-1979, Academic Press, Inc.

Lowenstein, J. M., editor: Methods in enzymology, Vol. 14, New York, 1969, Academic Press, Inc.

Snyder, F., editor: Lipid metabolism in mammals, Vol. 1, New York, 1977, Plenum Press.

Stanbury, J. B., Wyngaarden, J. B., and Fredrickson, D. S., editors: The metabolic basis of inherited disease, ed. 4, New York, 1978, McGraw-Hill Book Co.

White, W. L., Erickson, M. M., and Stevens, S. C.: Chemistry for the clinical laboratory, ed. 4, St. Louis, 1976, The C. V. Mosby Co.

23 Nucleic acid metabolism

With the determination and establishment of the various functions and mechanisms of action of the nucleic acids, their importance in life processes has become quite clear. As mono-, di-, and trinucleotides, the nucleic acids function as coenzymes, prosthetic groups, and transfer agents in many metabolic reactions. As the polynucleotides RNA and DNA, the nucleic acids are involved in the processes of reproduction of the organism and directing, controlling, and supplying key components for all metabolic reactions.

The chemistry of the nucleic acids and their components has already been discussed in Chapter 17. The metabolism of these substances will be discussed in this chapter, as well as the strategic significance of the polynucleotide forms, DNA and RNA. Although the nucleic acids exist largely as nucleoproteins, only the nucleic acid portion will be emphasized here.

ASPECTS OF DIGESTION AND ABSORPTION

The major problem in this area is the determination of how far the digestion of nucleic acids goes in the lumen of the small intestine (pp. 266, 271). There is evidence to indicate that some breakdown of nucleic acids in the lumen beyond the stage of mononucleotides occurs as a result of the release of intracellular enzymes from broken mucosal cells. Therefore, the degradation of mononucleotides may normally occur in the mucosal cells, as does the final breakdown of disaccharides.

Mononucleotides are absorbed easily from the small intestine, further breakdown can occur in mucosal cells, and the products are absorbed mainly by way of the portal system. As was the case with the carbohydrates, these products reach the liver before the other tissues.

METABOLISM OF NUCLEIC ACID COMPONENTS

Pentose metabolism

Synthesis and catabolism. Ribose is synthesized from glucose by way of the pentose shunt (p. 306). Its catabolism to carbon dioxide may occur by a cycling of the pentose shunt pathway reactions or by way of the glycolytic scheme as a result of the production of triose and hexose intermediates in the pentose shunt reactions.

Prevailing evidence indicates that deoxyribose is synthesized from ribose at the ribonucleotide (nucleoside diphosphate) stage.

Incorporation into nucleosides, formation of nucleotides. Ribose becomes attached to the precursors of the purine ring before the ring actually forms. In pyrimidine synthesis, the ribose is added on toward the end of the pathway, after the ring has been formed. Nucleosides may also be formed directly with preformed purines and pyrimidines. The nucleosides are converted to nucleotides with ATP in kinase reactions.

Purine metabolism

Synthesis and catabolism. The purines (p. 238) are synthesized from small molecules and fragments. The origins of the atoms in the purine ring structure are shown in Fig. 23-1. As with many other substances, the liver is the most important source of purines in the body. Only a small amount of preformed purines is used for the synthesis of nucleic acids, which means that the body must synthesize the remainder. The desired end-products are adenylate and guanylate, the ribose-5′-phosphate being added to the purine precursors early in the synthetic pathway before the ring structure is formed.

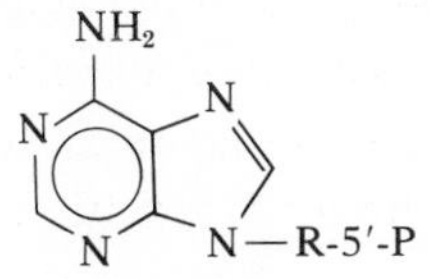

NH_2 N N N N—R-5′-P

Adenylate

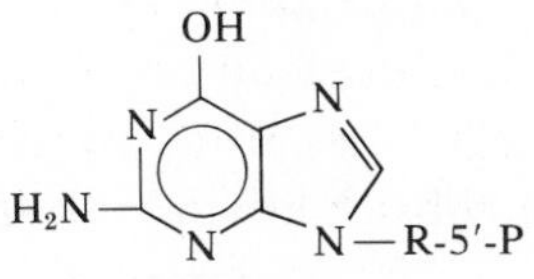

OH N N H_2N— N N—R-5′-P

Guanylate

Adenylate may be converted to guanylate in the body. Introduction of the 1-carbon fragments at C-2 and C-8 involves the folic acid (a B-group vitamin) coenzymes and may be inhibited by metabolic antagonists, such as aminopterin, which have been used in the treatment of leukemias.

In preparation for their incorporation into nucleic acids, adenylate, guanylate, and their deoxy equivalents are converted to triphosphates by kinase reactions.

After complete hydrolysis of the nucleotides, adenine and guanine are converted, in humans, to urate (uric acid) mainly in the liver, and the urate is subsequently excreted in the urine.

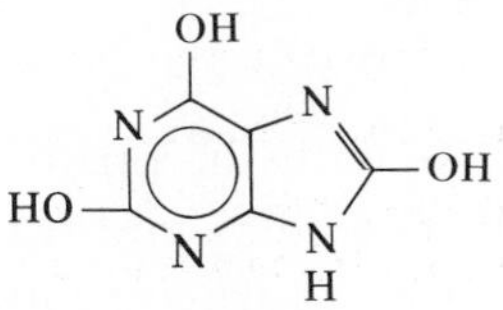

OH N N —OH HO— N N H

Urate

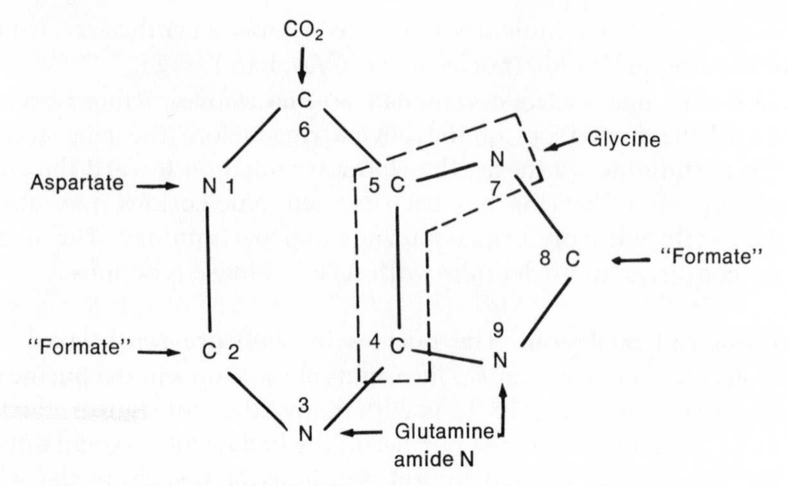

Fig. 23-1. Purine ring structure showing origin of atoms. ("Formate" = 1-carbon fragment at oxidation level of formic acid HCOOH; glutamine amide N = ⓃH_2—C(=O)—$CH_2CH_2CH(NH_2)$—C(=O)—OH.)

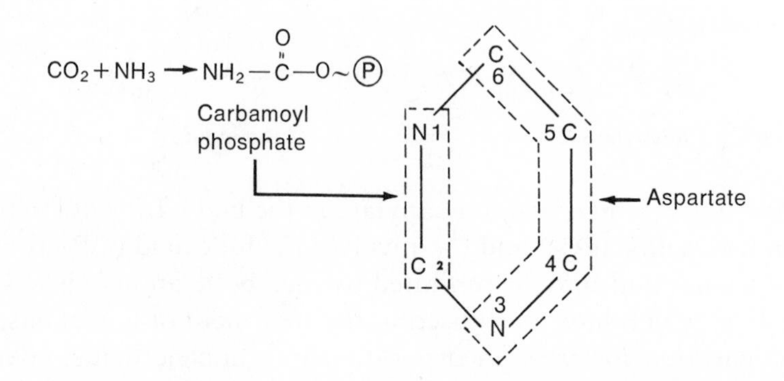

Fig. 23-2. Pyrimidine ring structure showing origins of atoms.

Dietary purines, not used for nucleic acid synthesis, are also catabolized to urate. In the condition known as gout, which is an inborn error of purine metabolism, there appears to be an overproduction as well as an under-excretion of urate.

Pyrimidine metabolism

Synthesis and catabolism. As with the purines, the pyrimidines (p. 238) are synthesized from small molecules and fragments. The origins of the atoms of the pyrimidine ring structure are shown in Fig. 23-2. In contrast to the purine synthetic pathway, the ribose-5′-phosphate is added to the

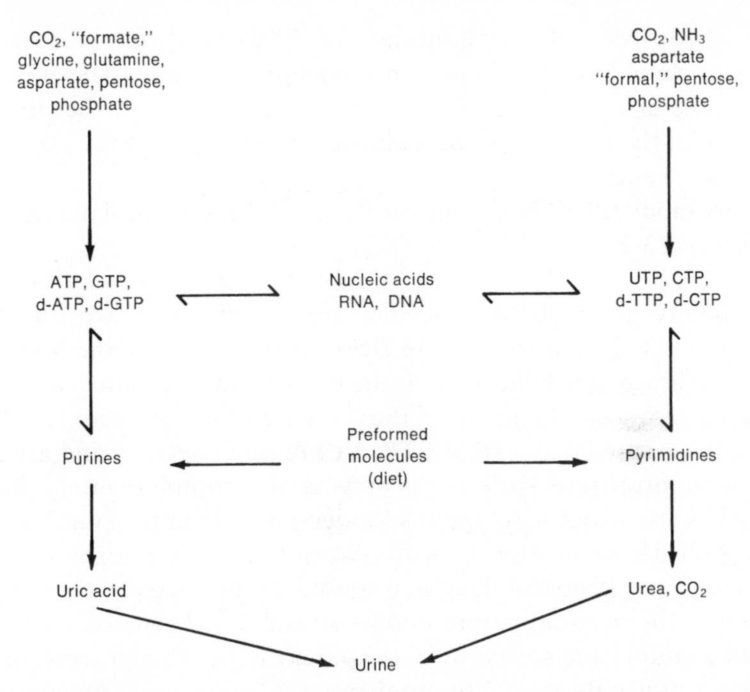

Fig. 23-3. General summary of purine and pyrimidine metabolism. 1-Carbon fragment at oxidation level of formic acid, HCOOH = "formate" at oxidation level of formaldehyde, HCHO = "formal.")

pyrimidine precursors near the end of the pathway after the pyrimidine ring has been formed. The liver is the site of most of the pyrimidine synthesis but more preformed pyrimidine nucleotide precursors may be used for the synthesis of nucleic acids than with the purines. Uridylate, which is the first pyrimidine product, is converted to cytidylate and thymidylate, the corresponding deoxy derivatives and triphosphates, which are used in RNA and DNA synthesis.

NH_2 / R-5′-P	OH / R-5′-P	OH, CH_3 / R-5′-P
Cytidylate	**Uridylate**	**Thymidylate**

Thymidylate is synthesized by the insertion of a methyl group in position 5 of the pyrimidine ring of deoxyuridylate by a folic acid coenzyme transferring a 1-carbon fragment at the formaldehyde (HCHO) level of oxidation. This reaction is also inhibited by folic acid antagonists.

After hydrolysis, the pyrimidines are catabolized in a series of reactions to carbon dioxide, acetate, propionate, and urea. The urea is excreted in the urine, the acetate joins the acetyl pool for further metabolic changes, and the propionate, as a glucogenic substance, may also join the acetyl pool eventually.

The metabolism of the purines and pyrimidines is summarized in brief form in Fig. 23-3.

NUCLEIC ACID METABOLISM

DNA synthesis (replication)

As already noted, DNA molecules are composed of two intertwined strands of nucleic acid (p. 242). In DNA synthesis, under the influence of DNA polymerase, it is believed that the two strands separate—at least in a small area at any one moment. In the presence of all four deoxynucleoside triphosphates (d-ATP, d-TTP, d-GTP, d-CTP), each open strand attracts to itself the appropriate molecules carrying the complementary bases as specified by the structure of the DNA (adenine to thymine, guanine to cytosine, p. 242). These relationships are shown in Fig. 23-4. On completion of the process, two identical daughter double strands are produced that are identical to the original parent double strand. Furthermore, each daughter DNA contains one strand from the parent DNA. This process, in which the parent molecule is exactly duplicated, is known as *semiconservative replication* and it takes place in the nucleus (except, possibly, for the small amount of mitochondrial DNA).

The characteristics of this process make DNA a most appropriate substance for the preservation of genetic information because each existing strand of DNA specifies exactly the composition of its complementary strand. Before a cell divides, the concentration of DNA doubles, one-half going to each daughter cell.

Drugs, antibiotics, nucleic acid analogues (compounds resembling the normal nucleic acids), and radiation may interfere with DNA replication in two important ways:

1. By inhibiting the replication of normal DNA
2. By causing the formation of DNA that is not an exact copy of its parent DNA, possibly leading to a mutation or a nonfunctional DNA

The body produces enzymes able to repair the various types of damage that may occur, but some significant damage may persist if the repair mechanisms are overwhelmed or if they are not functioning at normal effectiveness.

RNA synthesis (transcription)

As in DNA synthesis, in RNA synthesis there is at least some separation of the double strands of DNA. In the presence of RNA polymerase and the four required ribonucleoside triphosphates (ATP, UTP, GTP, CTP), one strand attracts to itself the necessary molecules with complementary

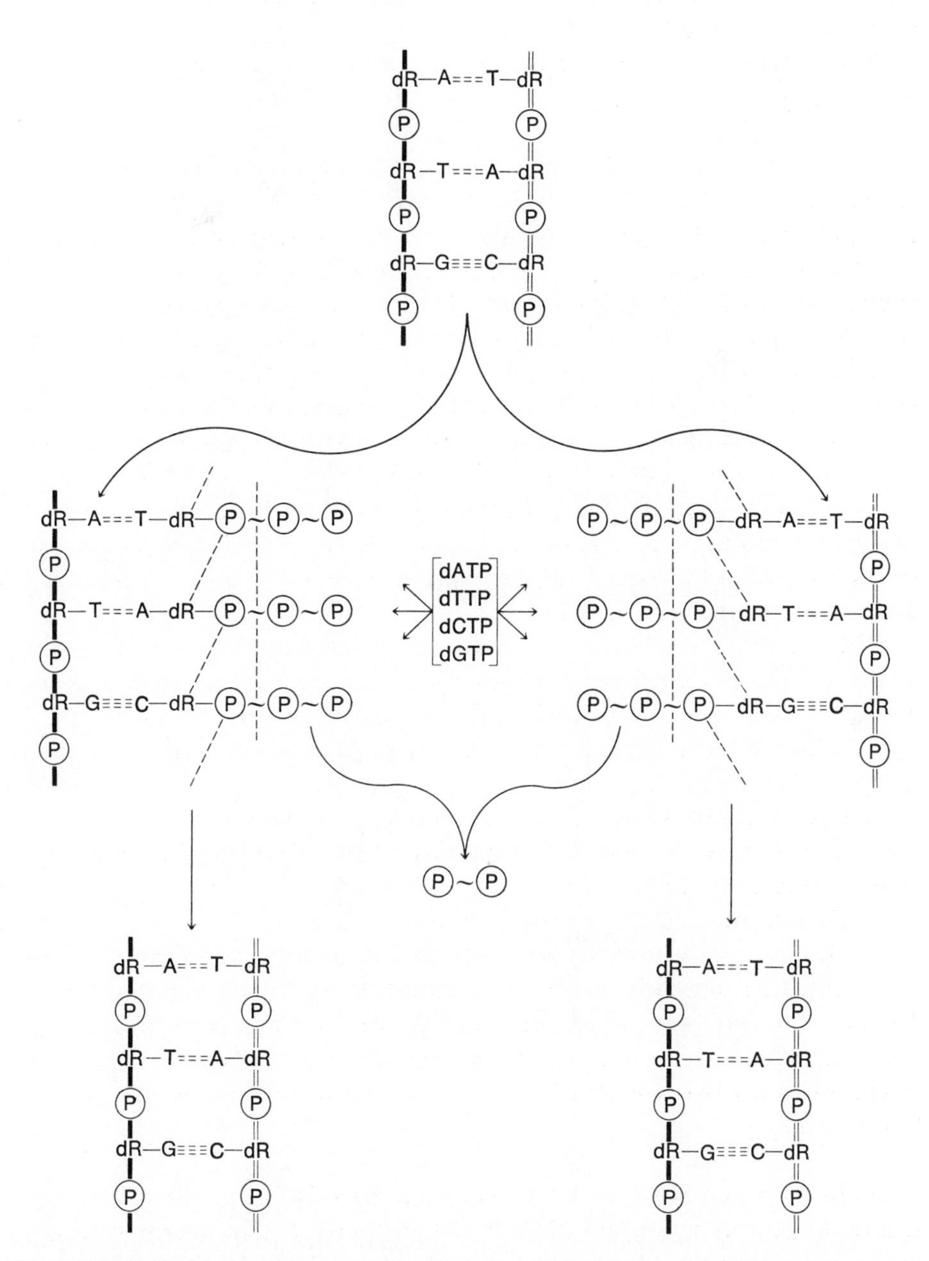

Fig. 23-4. Probable mechanism of replication of DNA. (Redrawn from Cantarow, A., and Schepartz, B.: Biochemistry, ed. 4, Philadelphia, 1967, W. B. Saunders Co.)

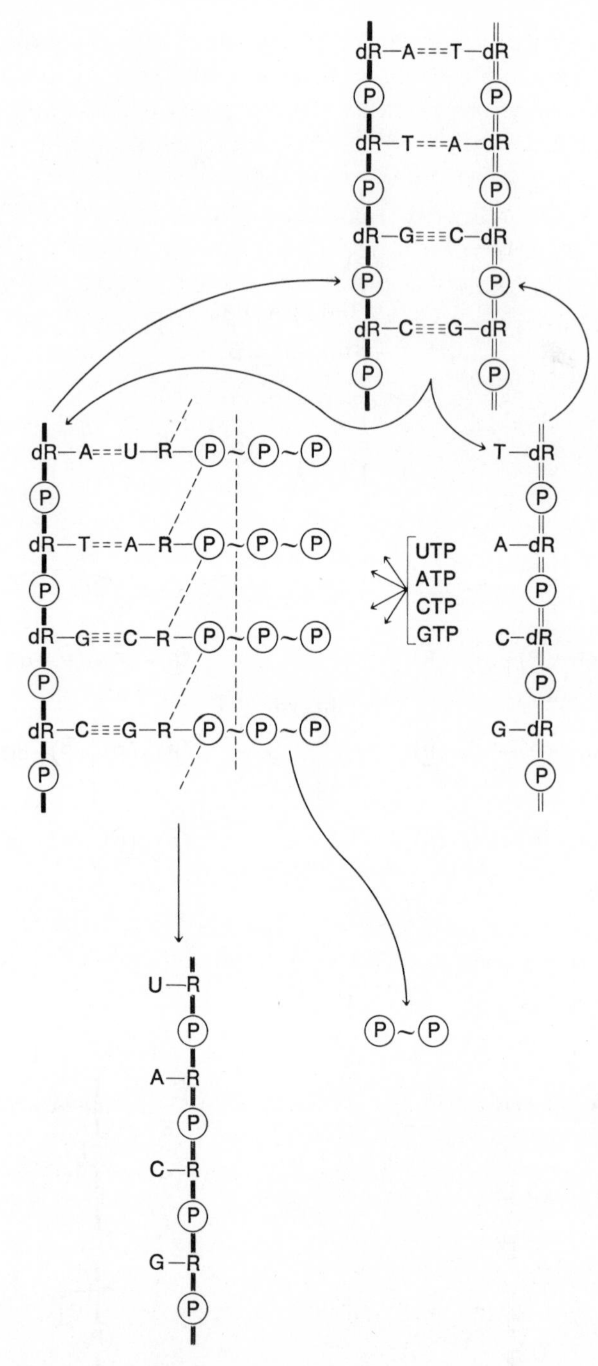

Fig. 23-5. Probable mechanism of transcription of DNA to RNA. (Redrawn from Cantarow, A., and Schepartz, B.: Biochemistry, ed. 4, Philadelphia, 1967, W. B. Saunders Co.)

bases (adenine to uracil, guanine to cytosine) to form a single strand of RNA, the DNA strands closing up to assume the normal structure after the strand of RNA has been released. This process is illustrated in Fig. 23-5.

Since the information from DNA has been transferred to RNA, the process is known as *transcription.* Again it takes place mainly in the nucleus (except for mitochondrial RNA), although most of the RNA synthesized is transported to the cytoplasm. Unlike the relative uniformity of DNA molecules, there are several types of RNA—messenger or mRNA, ribosomal or rRNA, and transfer or tRNA, which, in the past, was sometimes also referred to as soluble or sRNA—that vary in composition and quantity, depending on the state of the cell. Another major difference between DNA and RNA synthesis is the use of both strands in DNA replication but only one strand in RNA transcription. It should also be noted that in RNA synthesis uracil replaces thymine as the complementary base of adenine. RNA synthesis is also interfered with by various drugs, analogues, antibiotics, and radiation. The replacement or turnover rate of RNA is greater than that of DNA, as would be expected on the basis of their functions, discussed later.

Reverse transcription. The central dogma of the Watson and Crick theory of transfer of genetic information showed this transfer to be unidirectional, as follows:

DNA → DNA (replication)

DNA → RNA (transcription)

RNA → protein (translation, p. 356)

Now it has been accepted that it is also possible to produce DNA on the basis of information supplied by RNA, a phenomenon referred to as *reverse transcription:*

RNA → DNA (reverse transcription)

This process occurs under the influence of an enzyme known as RNA-dependent DNA polymerase, or reverse transcriptase, which thus far has been related mainly to tumor-causing (oncogenic) viruses (p. 344), but which has also been found in some normal tissues.

Catabolism of nucleic acids

After complete hydrolysis of the nucleic acids, the components are catabolized by the separate pathways already described.

Functions of nucleic acids

In genetics. DNA appears to be the substance in which resides all the genetic information required by all animals, plants, bacteria, and viruses, except for some viruses that apparently use RNA for this function. As already noted, this is related to the concept of DNA replication, in which the principle of complementarity ensures the exact replication of molecules, and to the transcription of RNA from DNA. The full implications of these processes will be better understood after the discussion of protein synthesis in the next chapter.

There is much evidence to indicate the involvement of the nucleic acids in genetics. One of the earliest and most famous observations to indicate such involvement was the transfer of a hereditary characteristic from a culture of bacteria possessing such a characteristic to a culture not possessing it. This was accomplished by growing the latter culture in the presence of a DNA extract of the culture of bacteria that did possess the hereditary characteristic. The process is known as *transformation* and the DNA extract is called the *transforming factor.* This type of process may ultimately serve as the basis of therapeutic manipulation of deficient or abnormal human genetic material. Research into the feasibility and possible applications of altering the composition of existing DNA molecules (recombinant-DNA studies) and of producing synthetic DNA molecules has been the subject of intense interest not only for scientists but also for the general public. Serious controversies over many ethical issues have arisen as a result of such research activities.

In protein synthesis. It is currently believed that DNA not only specifies the types of proteins to be synthesized by each cell but also controls the synthesis of all the necessary enzymes (which are also proteins), processes that also involve RNA. Experiments have indicated that with increased protein synthesis, the RNA content of cells increases. Involved in protein synthesis are the types of RNA—mRNA, tRNA, and rRNA (p. 343). One very important concept in protein synthesis is the one gene–one enzyme hypothesis, i.e., for every specific protein or enzyme to be produced, there must be a unique piece of DNA to specify its structure. More complete details about protein synthesis are in the next chapter.

In viruses: relation to carcinogenesis. Viruses resemble genetic material, or genes, in that they can reproduce themselves exactly, as does DNA, and they are also subject to mutation. Some plant viruses have been crystallized as nucleoproteins and animal viruses also appear to be nucleoproteins. There are both RNA- and DNA-type viruses, consisting of a core of nucleic acid covered by a cylindrical coat of protein.

Viruses appear to bridge the gap between living and nonliving matter. By themselves, viruses have no enzymatic activity and cannot reproduce. However, if they manage to enter a living cell, they can reproduce by causing the host cell to change its metabolic pattern in order to produce new viral particles. Viral infection of animal cells stimulates the production of substances known as *interferons* (proteins) that inhibit the synthesis of viral nucleic acids. Experimental attempts are being made to minimize or eradicate virus infections by the administration of substances that may start the production of interferons.

For many years viruses have been implicated as the causative agent of tumor formation in some animals, but intensive efforts to obtain completely unequivocal evidence that viruses also cause tumor formation in humans have thus far been unsuccessful. The discovery of reverse transcriptase (p. 343), particularly in relationship to virus-infected tissues, increased the tempo of work in this area. An important part of the hypothesis that viruses may be causing cancer is the proposal that after the virus causes a cell to produce DNA necessary for the reproduction of the virus, this DNA is somehow incorporated into the host cell DNA and the machinery for growth of the tissue according to viral directions rather than normal cell requirements is established. This could lead to the uncontrolled growth characteristic of cancer and other tumors.

In mutations: correlation with carcinogenesis. The same substances and procedures, e.g., chemicals and radiation, that produce mutations are also known to alter the structure of DNA. This change in structure is believed to be the basis for the mutation and carcinogenesis is believed by some to be caused by such mutations, which can result in drastically different cellular growth. This idea draws support from the observations that some of the mutagenic agents have been found to be carcinogenic and that some types of malignant tumors, at least in lower animals, appear to be caused by viruses.

In cell differentiation. Among other possible functions of the nucleic acids, it has been proposed that they may be involved in the timing mechanism required for proper cell differentiation. Research in this area is continuing but has not yet yielded definitive results.

In memory process. Experiments have been conducted in attempts to transfer learned activities from one organism to another by the direct transfer of RNA from a trained organism to an untrained organism. The results are still controversial.

Intracellular locations of nucleic acids

DNA is found only in the nucleus, except for the very small amount of mitochondrial DNA. RNA is found in all subdivisions of the cell, nucleus and cytoplasm, in particles and supernatant fluid.

Free nucleotides

Although DNA and RNA generally receive more attention, cells also contain some important nucleotides of much simpler composition, which are sometimes referred to as *free nucleotides*.

Table 23-1. Nucleotide coenzymes (free nucleotides)

Name	Structure	Function
Flavin mononucleotide (FMN)	Flavin-ribitol-P (vitamin B_2)	Oxidation-reduction
Flavin adenine dinucleotide (FAD)	Flavin-ribitol-P | Adenine-ribose-P	Oxidation-reduction
Nicotinamide adenine dinucleotide (NAD)	Nicotinamide-ribose-P (B-vitamin) | Adenine-ribose-P	Oxidation-reduction
Coenzyme A (CoA)	Adenine-ribose-P | P—P | Thiolethylamine-pantothenate (HS-R) (B-vitamin)	Transacylation

Nucleotide coenzymes. There are a number of free nucleotides, some with vitamins in their structure, that act as coenzymes in metabolic reactions. The components of the structures of a few of these nucleotides and their functions are listed in Table 23-1.

In addition to their roles as intermediates in the synthesis of nucleic acids, some of the mono-, di-, and triphosphates of the purine and pyrimidine nucleosides found in DNA and RNA may also serve as coenzymes in important reactions.

Cyclic adenosine 3′,5′-monophosphate (c-AMP). c-AMP (p. 245) is synthesized in cells from ATP under the influence of a membrane-bound enzyme, adenyl cyclase:

$$\text{ATP} \xrightarrow[\text{(membrane)}]{\text{adenyl cyclase}} \text{c-AMP} + \underset{\textbf{Pyrophosphate}}{\text{PP}}$$

The manner in which c-AMP is involved in the actions of hormones is discussed later (p. 447). Another enzyme inside cells, a cyclic nucleotide phosphodiesterase, can act on c-AMP and convert it to 5′-adenosine monophosphate (5′-AMP):

$$\text{c-AMP} \xrightarrow{\text{phosphodiesterase}} 5'\text{-AMP}$$

This reaction may be one means of controlling the cellular concentration of c-AMP.

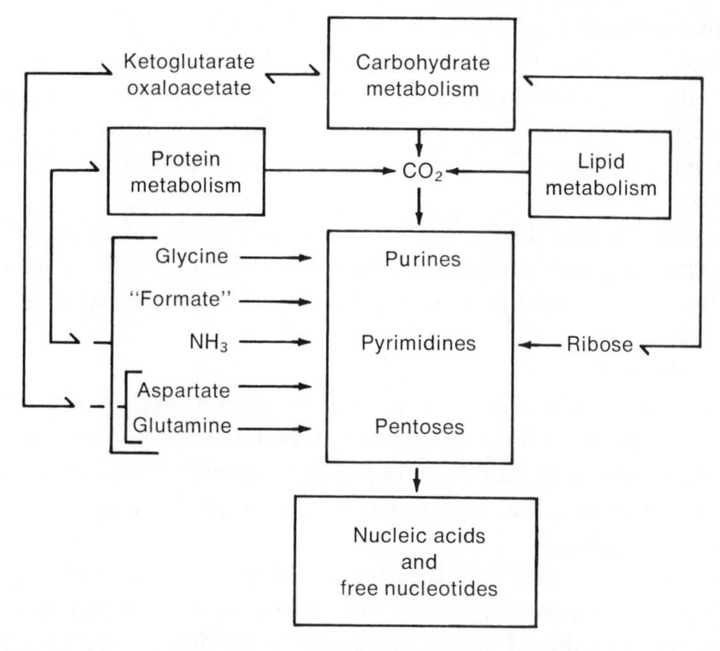

Fig. 23-6. Relationship of nucleic acid metabolism to protein, carbohydrate, and lipid metabolism.

SUMMARY OF NUCLEIC ACID METABOLISM AND ITS RELATIONSHIP TO PROTEIN, CARBOHYDRATE, AND LIPID METABOLISM

The relationships of nucleic acid metabolism to protein, carbohydrate, and lipid metabolism are summarized in Fig. 23-6. It should be noted that protein metabolism provides most of the raw material required for nucleic acid synthesis, and that lipid metabolism contributes only carbon dioxide. Not shown on the diagram is the fact that the nucleic acids and free nucleotides control and catalyze all intermediary metabolism by determining the enzymes to be produced by a cell, participating in their synthesis, and acting as coenzymes in many reactions.

SUMMARY

Nucleotides function as coenzymes, prosthetic groups, and transfer agents. The polynucleotides RNA and DNA are involved in reproduction and in directing, controlling, and supplying intermediates for all metabolic reactions.

ASPECTS OF DIGESTION AND ABSORPTION

The degradation of mononucleotides may actually occur inside the mucosal cells, not in the intestinal lumen. The products of nucleic acid digestion are absorbed by way of the portal system.

METABOLISM OF NUCLEIC ACID COMPONENTS

Pentose metabolism:

Synthesis and catabolism: Ribose is synthesized from glucose by way of the pentose shunt and is catabolized by the shunt pathway or by returning to the glycolytic pathway. Deoxyribose is synthesized by conversion of ribose in the nucleoside diphosphate stage.

Incorporation into nucleosides, formation of nucleotides: Pentose is incorporated into nucleosides during formation of purine and pyrimidine rings. Nucleosides are converted to nucleotides by ATP in kinase reactions.

Purine metabolism:

Synthesis and catabolism: Synthesis is summarized in Fig. 23-1. The liver is the most important site of synthesis. Adenylate may be converted to guanylate. Catabolically, adenylate and guanylate are converted to urate, which is excreted by the kidney.

Pyrimidine metabolism:

Synthesis and catabolism: Synthesis is summarized in Fig. 23-2. The liver is the most important site of synthesis. The pyrimidines are interconvertible. Catabolically, the pyrimidines are converted to carbon dioxide, acetate, propionate, and urea. Urea is excreted in urine; acetate and propionate may join the acetyl pool.

Fig. 23-3 is a diagrammatic summary of purine and pyrimidine metabolism.

NUCLEIC ACID METABOLISM

DNA synthesis (replication): The probable mechanism of replication of DNA is shown in Fig. 23-4. The result of DNA replication is the production of two identical daughter double strands, each containing one strand from the parent double strand, and both identical to the parent double strand. In this way DNA preserves its genetic information. The concentration of DNA doubles before cell division. Drugs, antibiotics, analogues, and radiation may inhibit DNA replication or induce mutations.

RNA synthesis (transcription): The probable mechanism of transcription of DNA to RNA is shown in Fig. 23-5. A single strand of RNA is formed, complementary to one strand of the DNA from which it is transcribed. There are several types of RNA—messenger RNA (mRNA), ribosomal RNA (rRNA), and transfer RNA (tRNA), sometimes also called soluble RNA (sRNA). RNA synthesis is also disturbed by various chemical agents and radiation.

Catabolism of nucleic acids: See catabolism of the components of nucleic acids.

Functions of nucleic acids:

In genetics: Preserve and transmit genetic information and determine the metabolic capability of the organism and its cells.

In protein synthesis: DNA specifies the types of proteins a cell may synthesize and, in conjunction with RNA, controls the synthesis of all the necessary enzymes. As protein synthesis increases, RNA synthesis increases.

In viruses: Viruses resemble genetic material in being able to reproduce themselves and in being subject to mutation. On infection, viruses convert the metabolism of host cells to the manufacture of new viral particles, the viral nucleic acid acting as mRNA.

In mutations, correlation with carcinogenesis: Mutations are believed to result from changes in structure of DNA, such as those caused by chemical agents and radiation, which are also carcinogenic. It is believed by some that carcinogenesis results from DNA mutation.

In cell differentiation: It has been proposed that nucleic acids control the timing in cell differentiation.

In memory process: Experiments are being performed in attempts to demonstrate a function for nucleic acids in the memory process.

Intracellular location of nucleic acids: DNA is found in the nucleus, except for very small amount of mitochondrial DNA. RNA is found in all parts of the cell, in particles as well as supernatant fluids.

Free nucleotides:

Nucleotide coenzymes: Simpler nucleic acids act as coenzymes and transfer agents, e.g., FMN, FAD, NAD, CoA. The mono-, di-, and triphosphates of the purine and pyrimidine nucleosides of RNA and DNA also act as coenzymes and carriers.

Cyclic AMP: c-AMP is synthesized in cells from ATP by membrane-bound enzyme, adenyl cyclase, and can be converted to 5′-AMP by a cyclic nucleotide phosphodiesterase. c-AMP is involved in hormone action.

SUMMARY OF NUCLEIC ACID METABOLISM AND ITS RELATIONSHIP TO PROTEIN, CARBOHYDRATE, AND LIPID METABOLISM

See Fig. 23-6. Protein metabolism provides most of the raw materials for nucleic acid synthesis; carbohydrate, mainly the pentose; lipid metabolism, only carbon dioxide. The nucleic acids and free nucleotides then control all intermediary metabolism.

REVIEW QUESTIONS

1. List, with examples, some of the various functions of nucleotides and nucleic acids. Where are nucleic acids found in cells?
2. Which organ is mainly responsible for the biosynthesis of purines and pyrimidines? At what stage is pentose incorporated into nucleotides? How are the deoxyribose compounds synthesized? Where does the ribose originate? Where do other portions of the purines and pyrimidines come from?
3. In the replication of DNA, two molecules are produced from an existing one. How are these two molecules related to each other and to the original molecule? What significance does this have with respect to the accepted function of DNA?
4. How many molecules are produced per unit transcription (synthesis) of RNA? What acts as the template for this process? What is the relationship between the RNA and its template? Name some of the various types of RNA.
5. What are the relationships between viruses and RNA and DNA? What are some factors which can change the structure of DNA? What may result from such changes? What is the significance of reverse transcriptase?

REFERENCES

Adams, R. L. P., Burdon, R. H., Campbell, A. M., and Smellie, R. M. S., editors: Davidson's: The biochemistry of the nucleic acids, ed. 8, New York, 1976, Academic Press, Inc.

Advances in cyclic nucleotide research, New York, Raven Press.

Annual review of biochemistry, Palo Alto, Calif., Annual Reviews.

Annual review of physiology, Palo Alto, Calif., Annual Reviews.

Ciba Foundation Symposium 48 (new series): Purine and pyrimidine metabolism, New York, 1977, Elsevier.

Colowick, S. P., and Kaplan, N. O., editors: Methods in enzymology, Vol. 51, New York, 1978, Academic Press, Inc.

Kornberg, A.: DNA synthesis, San Francisco, 1974, W. H. Freeman and Co., Publishers.

Stanbury, J. B., Wyngaarden, J. B., and Fredrickson, D. S., editors: The metabolic basis of inherited disease, ed. 4, New York, 1978, McGraw-Hill Book Co.

Watson, J. D.: The double helix, New York, 1968, Atheneum Publishers.

Watson, J. D.: Molecular biology of the gene, ed. 3, New York, 1976, W. A. Benjamin, Inc.

Protein metabolism

24

It was in connection with investigations on protein metabolism that observations first led to the important concept of the dynamic nature of intermediary metabolism. With the availability of stable and radioactive isotopes, it became possible to trace the course of metabolites through their various reactions in the body and it quickly became obvious that not only were body constituents being replaced constantly and relatively rapidly but that molecules used in synthetic processes came from metabolic pools and not necessarily from the immediate diet. In effect, an equilibrium exists between dietary intake, anabolic or synthetic processes, and catabolic or degradative processes, as shown here for the amino acid pool:

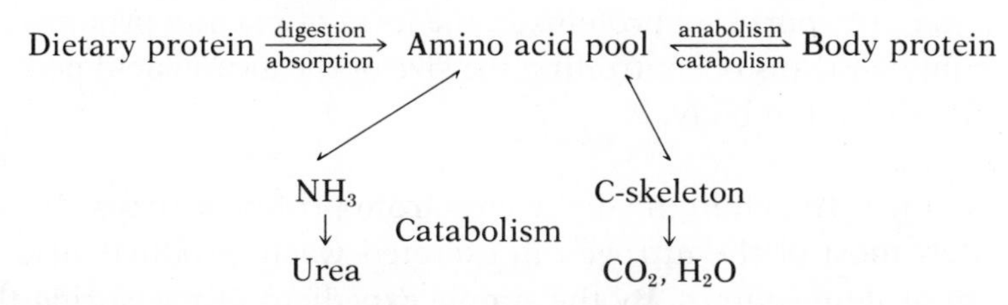

As noted in the previous chapters on metabolism, these concepts were quickly extended to all of intermediary metabolism.

The metabolism of proteins is somewhat unique in that in order to synthesize or degrade proteins, other proteins (enzymes) are required. Enzymes are also required for the metabolism of the many nonprotein substances. This situation makes the synthesis of proteins of prime importance to the life process. Progress in the area of protein synthesis continues at a rapid pace.

The discussion of the plasma proteins will be included in the chapter on blood. Analytic studies of these important proteins have been found of great value as an aid in the diagnosis of some abnormal clinical conditions.

The metabolism of the amino acids, the basic units of the proteins, is usually included with protein metabolism. Only the reactions common to all amino acids will be discussed in this chapter. Discussions of metabolic pathways unique to each individual amino acid must be left for more advanced texts.

ASPECTS OF DIGESTION AND ABSORPTION

The main points with respect to digestion and absorption have already been discussed (pp. 266, 273). In summary, proteins are generally hydrolyzed completely to amino acids before absorption by way of the portal blood. There is some evidence to indicate that very small peptides may also be absorbed without harm.

It is believed that under unusual circumstances, very small amounts of larger peptides may be absorbed from the intestinal tract. When this occurs, a chain of events is set off that may have very serious effects on the individual. An intact foreign protein (antigen) in the circulation is known to cause the body to produce antibodies (protein molecules that combine specifically with molecules of antigen) in an effort to dispose of the foreign protein. This sensitizes the individual to that particular foreign protein. Should this same protein be absorbed intact at a later time, the reactions may be anything from a mild response to death, in severe cases. The absorption of intact protein molecules is believed to be the basis for the development of allergies.

The rate and efficiency of absorption of amino acids are generally high. Because amino acids are absorbed mainly by the portal blood, the liver has first choice regarding their fate. This is probably due to the role of the liver as a major exporter of proteins in the form of plasma proteins and is also probably a means of controlling the size of the amino acid pool available to the rest of the body.

NUTRITIONAL ASPECTS OF PROTEIN METABOLISM

Nitrogen balance

Most of the nitrogen in the diet comes from protein sources. As will be shown later, most of the nitrogen in excreted waste products also originates from protein sources. By the simple expedient of measuring the intake of nitrogen (food) and the output (mainly in urine and feces)—in effect the *nitrogen balance*—much can be learned about the status of the individual's protein metabolism.

At nitrogen equilibrium, intake and output of nitrogen are equal. This situation may occur in a normal adult on a diet adequate in protein, carbohydrates, lipids, vitamins, and other required nutrients. A positive nitrogen balance—when intake exceeds output—generally occurs in young, growing children, in pregnancy, and in convalescence after severe diseases—all conditions in which more new protein is being synthesized than required for replacement. In negative nitrogen balance, the output must exceed the intake. This means that the excess over intake must come from body proteins, a situation that occurs in cases of fever, surgical procedures, and severe diseases.

Essential amino acids

From nutritional experiments, mainly in animals other than humans, it has been found that certain of the naturally occurring amino acids must be supplied in the diet in order to maintain adequate growth in the young and nitrogen equilibrium in the adult. This means that the animal either cannot synthesize these amino acids at all or does so in inadequate amounts. In any case, these amino acids are known as *essential* amino acids from the point of view of being required in the diet. It is generally agreed that there are at least eight essential amino acids required by adult human beings. These are:

Isoleucine	Phenylalanine
Leucine	Threonine
Lysine	Tryptophan
Methionine	Valine

Two other amino acids, histidine and arginine, are also believed to be essential for humans by some investigators. One of the difficulties in determining the essentiality of a dietary constituent for humans is the fact that such experiments usually can be carried out for only short periods, whereas with other animals, such as the rat, the experiments can continue throughout their entire life span, or a significant part of it.

Optimal ratios for dietary amino acids

The composition of the mixture of dietary amino acids that is utilized most efficiently by an animal is similar to the composition of the body proteins of the animal. Any large imbalance caused by the increased ratio of a single amino acid, or its complete lack in the diet, especially of the essential amino acids, may result in toxic reactions or inefficient absorption or utilization. Furthermore, if a single essential amino acid is fed a few hours after an otherwise complete mixture, the benefit to the animal is drastically reduced, because by the time the missing amino acid is absorbed, the liver has changed the composition of the amino acid pool by oxidation of amino acids which could not be used for synthetic purposes (a complete mixture is required for protein synthesis).

Biologic value of proteins

The ultimate biologic value of a protein is not determined solely by consulting a nutritional table. The protein must be eaten, digested, absorbed, and utilized—all at adequate levels—before it has any biologic value. An abnormality in any one or a combination of these processes automatically decreases the biologic value of any protein. Also, there are certain proteins that are completely lacking or are low in one or more essential amino acids (e.g., gelatin has no tryptophan, and zein, corn protein, has no lysine and is low in tryptophan). Reliance on such proteins as a sole source will not maintain an animal in nitrogen balance. Mixing such proteins in the diet will increase the biologic value of the total mixture as one protein makes up for the lack of an amino acid in another protein.

Lack of storage form of protein

There is no storage form of protein comparable to glycogen for carbohydrates or adipose tissue for lipids. This is reflected by the rapid adjustment of nitrogen output to any changes in intake. In protein deprivation, proteins required for vital functions, e.g., the plasma proteins, must be maintained at the expense of such momentarily less strategic proteins as muscle protein. Such a situation cannot continue indefinitely without dire results.

Protein-sparing effect

Since protein, in times of stress, may be used for oxidation for energy purposes by way of carbohydrate pathways (gluconeogenesis), it follows that if the diet contains adequate amounts of carbohydrate and lipids to satisfy the caloric requirements of the body, less protein will be required to fulfill its strategic functions. This phenomenon is known as the *protein-sparing effect.*

SYNTHESIS AND CATABOLISM OF PROTEINS

Protein synthesis

The key reaction in the synthesis of proteins is the formation of the peptide bond (p. 220):

$$R_1-\underset{|}{\overset{NH_2}{CH}}-COOH + R_2-\underset{|}{\overset{NH_2}{CH}}-COOH \rightarrow R_1-\overset{NH_2}{\overset{|}{CH}}-\underbrace{\overset{O}{\overset{\|}{C}}-NH}_{\text{Peptide bond}}-\overset{R_2}{\overset{|}{CH}}-COOH + H_2O$$

This reaction requires an input of energy to make it proceed in the direction of protein synthesis. This is accomplished, as in previous instances, by prior activation reactions, as will be shown. The probable mechanism of protein synthesis by the cytoplasmic ribosomal system is illustrated in Fig. 24-1, which shows the addition of glycine to a growing peptide chain. (For diagrammatic simplicity, the actual structure of DNA, a double helix, and the currently accepted four-leaf clover type structure of tRNA are not used in Fig. 24-1.) Protein synthesis also occurs to a small extent in the nucleus and mitochondria.

Amino acid activation. Since the formation of the peptide bond requires the input of energy, activation of the amino acids may be considered as a first step in protein synthesis. The activating enzymes, located in the cytoplasm, are known as aminoacyl-tRNA synthetases, each enzyme being specific for a single amino acid. The activation reaction involves a specific synthetase, the specific amino acid, and ATP and results in the formation of an aminoacyl-adenylate-synthetase complex (referred to, for simplicity, as an activated amino acid) and pyrophosphate (Fig. 24-1). Amino acid analogues such as ethionine (analogue of methionine) may interfere with protein synthesis at this point by being activated and ultimately incorporated into protein as an abnormal constituent.

Transfer of activated amino acids to transfer RNA (tRNA). Also located in the cytoplasm are nucleic acids known as transfer RNA (tRNA). These are relatively small molecules (molecular weight, 20,000 to 30,000), each different one containing a terminal nucleotide sequence of -cytidylic-cytidylic-adenylic (-C-C-A). Although the tRNA molecules are specific for a

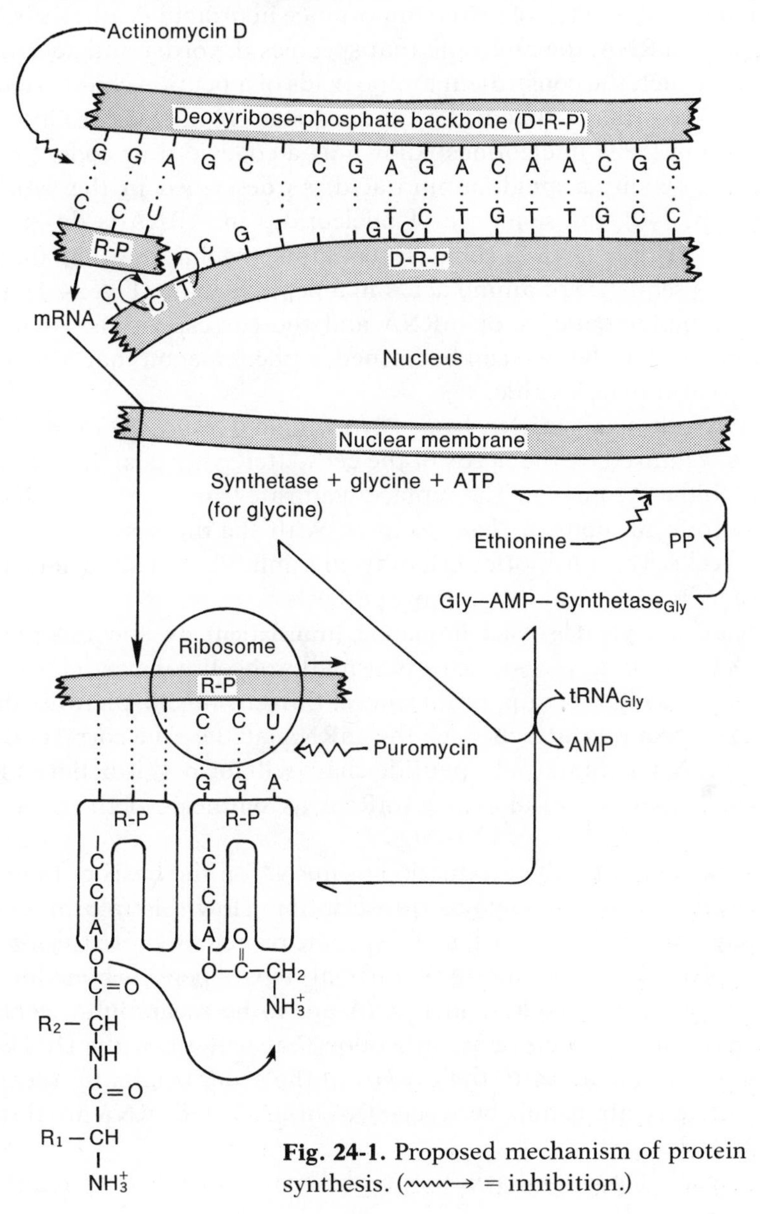

Fig. 24-1. Proposed mechanism of protein synthesis. (⟿ = inhibition.)

given amino acid, it appears that there may be more than one tRNA capable of carrying a particular amino acid. The aminoacyl portion of the complex formed in the first step (already described) is transferred to the appropriate tRNA on the -C-C-A end (Fig. 24-1 and p. 243). In this reaction, the synthetase must be able to recognize both its specific amino acid and the tRNA that will accept that amino acid.

Synthesis of messenger RNA (mRNA), template for protein synthesis (transcription). Another step of prime importance in protein synthesis is the production of mRNA, the molecule that specifies the order and acts as a template on which the constituent amino acids of a peptide chain are to be arranged. The order is specified by the sequence of nucleotides in mRNA, a sequence of three nucleotides, known as a codeword or codon, being required to specify a single amino acid. As described in the synthesis of mRNA (p. 340), the sequence of nucleotides in mRNA is designated by DNA. Therefore, DNA ultimately contains the information that determines the sequence of amino acids in a peptide chain (Fig. 24-1). Because of the complementarity of mRNA and the corresponding DNA, hybrid combinations of the two can be formed, a phenomenon that has proved to be of great research value.

The synthesis of mRNA from DNA is called *transcription* and it takes place as required by the needs of the cell. After synthesis, the mRNA is released to the cytoplasm. The replacement rate (turnover) of mRNA varies widely from an hour, or less, to days, with the different types found in animal cells. The antibiotic actinomycin D inhibits transcription of mRNA and thereby also inhibits protein synthesis.

Ribosomes; peptide bond formation (translation). In the cytoplasm the mRNA forms an aggregate with the small, subcellular particles known as ribosomes, which form polyribosomes. Current theory pictures the ribosome as becoming attached to the mRNA at the end corresponding to where the N-terminus of the peptide chain will form. It is believed that the ribosomes then travel along the mRNA as synthesis of the protein continues.

As stated before, the production of mRNA on the basis of information contained in DNA is known as transcription. How this information is ultimately used in the formation of a specific protein is known as *translation* and involves tRNA and mRNA in the final steps of peptide bond formation. If the trinucleotide codons in mRNA are to be meaningful, each tRNA must have in its structure an anticodon that will allow the tRNAs to deliver their amino acids to the mRNA in the order originally specified by DNA. These relationships between DNA, mRNA, and tRNA are illustrated in Fig. 24-2.

The final step is the formation of the peptide bonds between the adja-

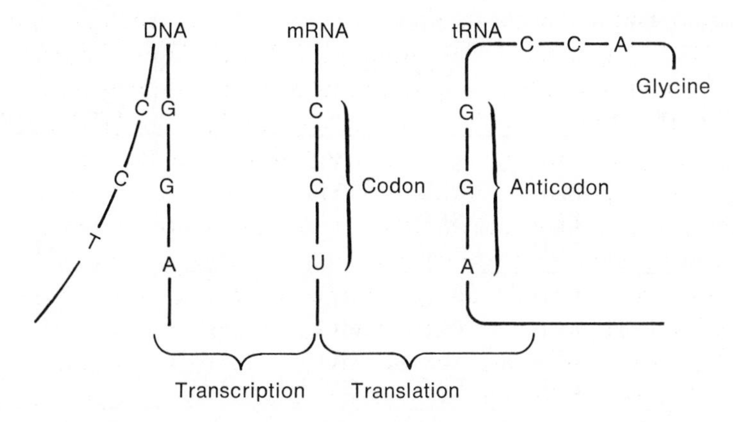

Fig. 24-2. Relationship between DNA, mRNA, and tRNA.

cent properly aligned amino acids. This is believed to start with ribosomes at the end of mRNA where the free amino end of the peptide will form. The tRNA carrying the end amino acid becomes attached to the mRNA by codon-anticodon attraction (hydrogen bonding between complementary nucleotides). The same occurs with the tRNA carrying the second amino acid. Under the influence of two aminoacyl-transferases and other factors, the amino group of the second amino acid forms a peptide bond with the carboxyl group of the first amino acid. As a result, the tRNA carrying the first amino acid is freed from the mRNA and the process of forming peptide bonds continues along the mRNA (Fig. 24-1). In animals, it is believed that each mRNA codes for a single, specific peptide chain.

Antibiotics and drugs such as puromycin, streptomycin, tetracyclines, and chloramphenicol inhibit or affect synthesis in various ways.

Specification of protein conformation. It was previously believed that information in addition to that supplied by mRNA was necessary to specify the three-dimensional shape to be assumed by a protein. However, it is now accepted that interactions among properly grouped amino acids are sufficient for this purpose, i.e., it is the sequence of amino acids (primary level of organization, p. 223), that actually determines the final shape of the protein molecule. One of the best examples of this is the total synthesis of ribonuclease (an enzyme that hydrolyzes ribonucleic acids), from amino acids with no experimental procedure used to force the resulting molecule into a definite conformation. That the synthetic molecule folds itself into its required conformation is attested to by the fact that its physical and chemical properties, particularly its enzymatic activity, very closely resemble those of the natural enzyme.

The genetic code. By various ingenious experiments, some involving syn-

Table 24-1. Codons of the genetic code

First nucleotide	Second nucleotide				Third nucleotide
	U	**C**	**A**	**G**	
U	PHE	SER	TYR	CYS	U
	PHE	SER	TYR	CYS	C
	LEU	SER			A
	LEU	SER		TRP	G
C	LEU	PRO	HIS	ARG	U
	LEU	PRO	HIS	ARG	C
	LEU	PRO	GLUN	ARG	A
	LEU	PRO	GLUN	ARG	G
A	ILEU	THR	ASPN	SER	U
	ILEU	THR	ASPN	SER	C
	ILEU	THR	LYS	ARG	A
	MET	THR	LYS	ARG	G
G	VAL	ALA	ASP	GLY	U
	VAL	ALA	ASP	GLY	C
	VAL	ALA	GLU	GLY	A
	VAL	ALA	GLU	GLY	G

U = uridylate; C = cytidylate; A = adenylate; G = guanylate; PHE = phenylalanine; LEU = leucine; ILEU = isoleucine; MET = methionine; VAL = valine, SER = serine; PRO = proline; THR = threonine; ALA = alanine; TYR = tyrosine; HIS = histidine; GLUN = glutamine; ASPN = asparagine; LYS = lysine; ASP = aspartic acid; GLU = glutamic acid; CYS = cysteine; TRP = tryptophan; ARG = arginine; GLY = glycine.

thetic mRNAs, it has been possible to assign probable codons for each amino acid. As many as six codons have been assigned to some amino acids. A list of codons is found in Table 24-1. There are several interesting points to be noted. Because this is a triplet code based on four different nucleotides, a total of 64 codons are possible. Because each amino acid (except possibly methionine) has more than one codon assigned to it, the code is said to be *degenerate.* The degeneracy is mainly confined to the third nucleotide, the first two generally remaining invariant, as seen in Table 24-1.

Tissue protein synthesis and turnover. All tissues synthesize their own specific proteins. As noted previously (p. 352), the liver not only produces its own protein but is also responsible for the production of most of the plasma proteins (p. 373). Other examples of proteins produced for extracellular functions are enzymes, such as those synthesized in the pancreas for use in the small intestine, and protein hormones, such as insulin, which affect most of the cells of the body.

An interesting corollary of the dynamic state of protein metabolism is

Table 24-2. Protein turnover rates in humans

Protein	Half-life in days
Whole body	80
Liver	10
Serum albumin	15-25
Muscle	180

the relatively high rate of turnover shown by some tissue proteins. An example of a very high rate of turnover is that of a rat liver enzyme, tryptophan pyrrolase, which has a half-life of 2 to 4 hr, whereas rat liver protein has a half-life of 6 days. Some protein turnover rates in humans are listed in Table 24-2.

High rates of turnover are usually associated with tissues of high metabolic activity, such as intestinal mucosa, liver, and kidney, whereas low rates of turnover are seen in muscle, brain, and skin.

Catabolism of proteins

Intracellular breakdown of proteins is believed to be the function of enzymes known as *cathepsins*. The cathepsins are found in a subcellular structure known as a *lysosome*, which settles with the mitochondrial material after centrifugation of disrupted cells. The mechanism of action of the cathepsins and the conditions under which they are released from the lysosomes are still being investigated.

Because of the dynamic nature of protein metabolism there is a constant breakdown of some tissue proteins as new molecules are synthesized. In the adult, the two processes are approximately at equilibrium under normal conditions. The ultimate catabolic fate of the various parts of the amino acid molecules is discussed later.

HORMONAL EFFECTS IN PROTEIN METABOLISM

Growth hormone (somatotropin)

Growth hormone has a protein anabolic effect, i.e., it promotes a positive nitrogen balance resulting in increased deposition of proteins in the tissues.

Testosterone

Testosterone also has a protein anabolic effect.

Insulin

Insulin apparently increases the rate of entry of amino acids into cells and incorporation into proteins by increasing the synthesis of mRNA. It is also required for the protein anabolic effect of growth hormone.

Glucocorticoids

As a result of their action in increasing gluconeogenesis, the glucocorticoids tend to cause a negative nitrogen balance.

Thyroid hormone

Physiologic doses of thyroid hormone cause a protein anabolic effect, but large doses cause a protein catabolic effect.

GENERAL METABOLISM OF AMINO ACIDS

In this section will be discussed the general metabolic reactions in which most of the amino acids participate. The catabolism of amino acids usually starts with the removal of the amino group and is then followed by the separate disposal of the nitrogen of the amino group and the remaining carbon skeleton.

Removal of amino group

The amino group is removed from amino acids by two general reactions—oxidative deamination and transamination.

Oxidative deamination. The enzyme L-amino acid oxidase catalyzes the oxidative deamination of most L-amino acids:

$$R-\underset{\displaystyle NH_2}{CH}-COOH \xrightarrow{\text{Oxidase}} R-\overset{\displaystyle O}{\overset{\|}{C}}-COOH + NH_3$$

This type of reaction produces an α-keto acid and ammonia, which continue along separate pathways. The same type of reaction is carried out by L-glutamate dehydrogenase, which acts specifically on L-glutamic acid.

Transamination. The reaction in which an amino group is transferred from one amino acid to an α-keto acid, forming a new amino acid and a new α-keto acid, is known as transamination. In this reaction, in which pyridoxal phosphate, a B-group vitamin, acts as coenzyme, the ammonia does not appear free at any time, and the reaction is reversible.

$$R_1-\underset{\displaystyle NH_2}{CH}-COOH + R_2-\overset{\displaystyle O}{\overset{\|}{C}}-COOH \underset{\text{Pyridoxal phosphate}}{\overset{\text{Transaminase}}{\rightleftharpoons}} R_1-\overset{\displaystyle O}{\overset{\|}{C}}-COOH + R_2-\underset{\displaystyle NH_2}{CH}-COOH$$

Glutamic acid is involved in some of the most active transaminations and elevated levels of serum glutamate-oxaloacetate and glutamate-pyruvate transaminases are found to be of diagnostic aid in instances of possible damage to heart and liver tissues. When such damage occurs, some heart muscle or liver cells die and the transaminases normally present inside the cells (different types and concentrations in different cells) leak out into the bloodstream. Analysis of the blood for the presence and amounts of the various transaminases can be very helpful in determining the type and extent of damage.

Disposition of the nitrogen

Synthesis of purines, pyrimidines, other compounds. As already described, the ammonia produced in deamination of amino acids may be used in the synthesis of purines (p. 337) and pyrimidines (p. 338). It may also be used in the reductive amination of α-keto acids, from carbohydrate metabolism, to form new amino acids:

$$R-\overset{\overset{\large O}{\|}}{C}-COOH \xrightarrow{NH_3} R-\overset{\overset{\large NH_2}{|}}{CH}-COOH$$

Glutamine formation: detoxication. Free ammonia is a toxic substance in cells and may lead to coma unless removed or detoxified. In a reaction involving glutamic acid and ATP, glutamine is formed in the extrarenal tissues:

$$HOOCCH_2CH_2-\overset{\overset{\large NH_2}{|}}{CH}-COOH \xrightarrow[NH_3 \quad H_3PO_4]{ATP \quad ADP} NH_2-\overset{\overset{\large O}{\|}}{C}-CH_2CH_2-\overset{\overset{\large NH_2}{|}}{CH}-COOH$$

In this form, the ammonia is no longer toxic and is carried in the blood to the kidneys, where it is hydrolyzed and the regenerated ammonia is excreted:

$$NH_2-\overset{\overset{\large O}{\|}}{C}-CH_2CH_2-\overset{\overset{\large NH_2}{|}}{CH}-COOH \xrightarrow[H_2O]{Glutaminase} HOOCCH_2CH_2-\overset{\overset{\large NH_2}{|}}{CH}-COOH + NH_3$$

Much of the glutamine formation takes place in the liver.

Excretion as ammonia. If deamination of amino acids occurs in the kidneys, as it may, the ammonia may be excreted directly into the urine.

Urea formation: ornithine cycle. The ornithine cycle, by which urea is formed from ammonia and carbon dioxide, is an interesting example of the way nature attains important goals by the use of a series of little steps. As seen in Fig. 24-3, all of the changes that take place occur on the carbon atom furthest from the carboxyl group of ornithine. The carbon chain itself is untouched. The synthesis of urea starts with the condensation of carbamoyl phosphate (formed from ammonia and carbon dioxide by the enzyme carbamoyl phosphate synthetase, p. 338) and ornithine to form citrulline with the aid of the enzyme transcarbamylase. The citrulline condenses with aspartate to form argininosuccinate by the action of the enzyme argininosuccinate synthetase. Then under the influence of the enzyme argininosuccinase, the argininosuccinate is cleaved to form arginine and fumarate. Finally, in the presence of the enzyme arginase, urea is split off from arginine as ornithine, one of the starting substances in this cycle, is regenerated. The urea, which is the major nitrogenous waste product of protein catabolism, is carried in the blood to the kidneys for excretion. Evidence indicates that the liver is the sole site of formation of urea.

Excretion as creatinine. Creatine, which, as phosphocreatine, acts in

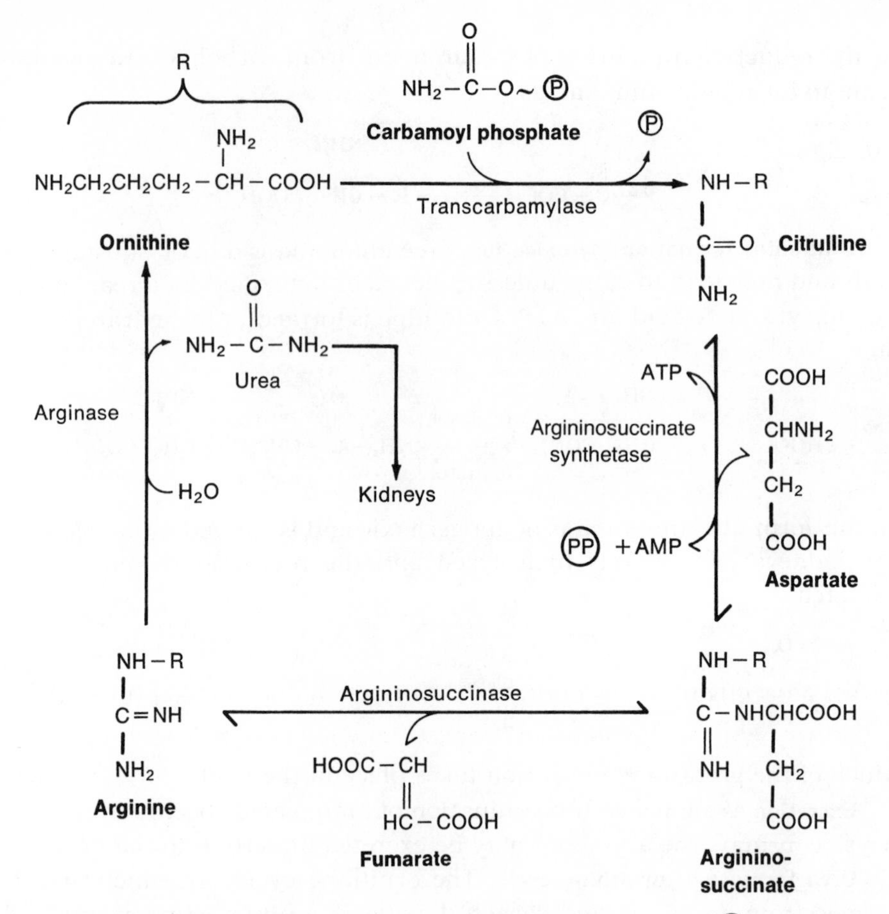

Fig. 24-3. Production of urea by way of the ornithine cycle. ((P) = phosphate; (PP) = pyrophosphate.)

muscle contraction (p. 307), is synthesized in a series of reactions from the three amino acids, glycine, arginine, and methionine. One of the end products of muscle metabolism is creatinine, the anhydride of creatine. Creatinine is excreted in a rather constant daily amount by a given individual and this appears to be directly related to the muscle mass. The amount of creatinine found in a 24-hour urine collection is used to verify the completeness of the collection. The relationships of these compounds to each other are shown in Fig. 24-4.

Disposition of the carbon skeleton

Resynthesis of amino acids. As noted before (p. 361), the α-keto acids formed in deamination reactions may be reductively reaminated to produce the original amino acids. Obviously, the net direction these reactions take depends on the requirements of the body at any particular time.

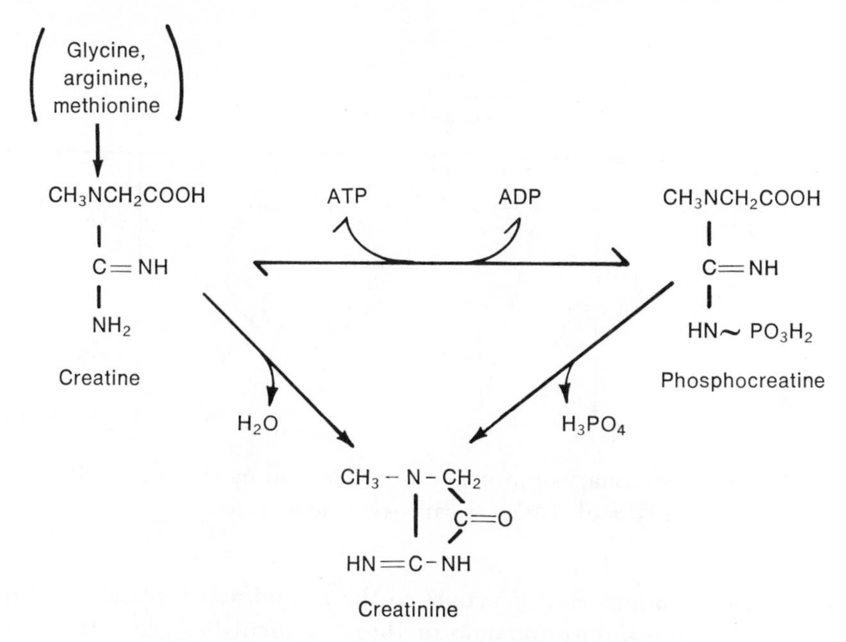

Fig. 24-4. Creatine and related compounds.

Formation of carbohydrate intermediates: gluconeogenesis. The α-keto acids formed from many amino acids by transamination reactions may enter carbohydrate pathways (glucogenic amino acids). A few, which become involved in carbohydrate metabolism directly, are shown here:

$$\begin{array}{ccccccc} \text{Glucose} \rightleftharpoons & \text{Pyruvate} & \rightleftharpoons & \text{Oxaloacetate} & \rightleftharpoons & \alpha\text{-Ketoglutarate} \\ & \downarrow\uparrow & & \downarrow\uparrow & & \downarrow\uparrow \\ & \text{Alanine} & & \text{Aspartate} & & \text{Glutamate} \end{array}$$

The process in the direction of carbohydrate metabolism is termed gluconeogenesis and requires the presence of pyridoxal phosphate, a B-group vitamin, as coenzyme. Gluconeogenesis is reversible for the nonessential but not for the essential amino acids. The adrenal glucocorticoid hormones stimulate gluconeogenesis.

Formation of ketone bodies. A few amino acids—leucine, isoleucine, phenylalanine, tyrosine—give rise to ketone bodies and are known as ketogenic amino acids. Only leucine is exclusively ketogenic. The others are also glucogenic.

Energy accounting for amino acid oxidation. Assuming that the nitrogen is excreted as urea, and that the carbon skeleton is oxidized to carbon dioxide and water by way of the tricarboxylic acid cycle, 1 mole of glutamic acid would give a maximum of 490,000 cal. Of this total 191,000 cal (39%) is conserved in the form of ATP. This is approximately the same efficiency achieved in carbohydrate and fatty acid oxidations (pp. 306, 326).

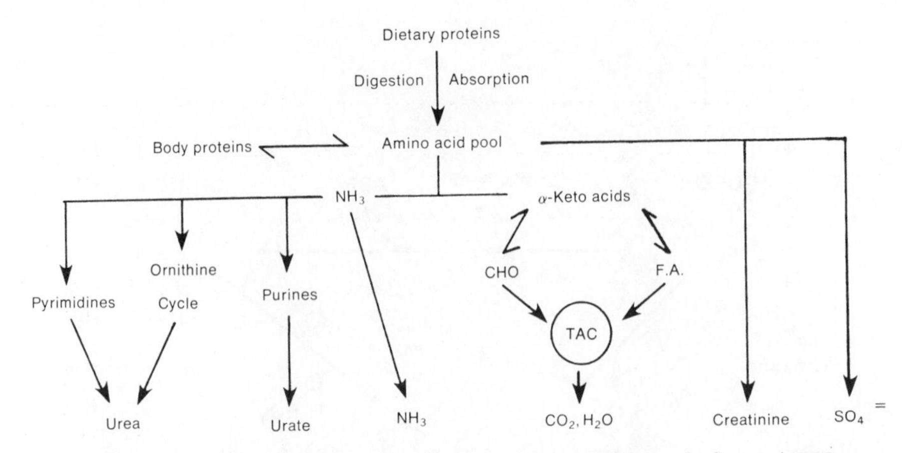

Fig. 24-5. General summary of protein and amino acid metabolism. (CHO = carbohydrate; FA = fatty acid; TAC = tricarboxylic acid cycle.)

Disposition of sulfur

Formation of sulfate. Sulfur occurs in the amino acids cystine and methionine. The essential amino acid methionine furnishes the sulfur in the synthesis of cystine, a nonessential amino acid. Cystine appears in relatively large concentration in hair and fingernails. Most of the sulfur of these amino acids is eventually oxidized to sulfate, which is excreted in the urine mainly as inorganic sulfate with a small fraction as organic or ethereal sulfate. In a certain inborn error of metabolism, the ability of the kidney to reabsorb cystine is decreased and large amounts of cystine may be excreted in the urine (cystinuria).

SUMMARY OF PROTEIN METABOLISM AND ITS RELATIONSHIP TO CARBOHYDRATE AND LIPID METABOLISM

A simplified summary of protein and amino acid metabolism is shown in Fig. 24-5. The relationship of protein metabolism to carbohydrate and lipid metabolism is found in Fig. 24-6. Note that pyruvate appears to be the most important transfer point.

At this point it is almost superfluous to emphasize the dynamic nature of intermediary metabolism and the close relationships between carbohydrate, lipid, and protein metabolism. Yet in spite of these interrelationships, for survival an individual must consume a varied diet in order to supply the required essential amino acids, essential fatty acids, and vitamins. All of the reactions shown may take place simultaneously, the net direction of metabolism being determined by the state and requirements of the body at a particular time. Under normal conditions, the proper flow of these reactions results from an exquisite system of coordination and controls. It should be understood that in times of stress, small deviations from normal may result in massive dislocation of the metabolic

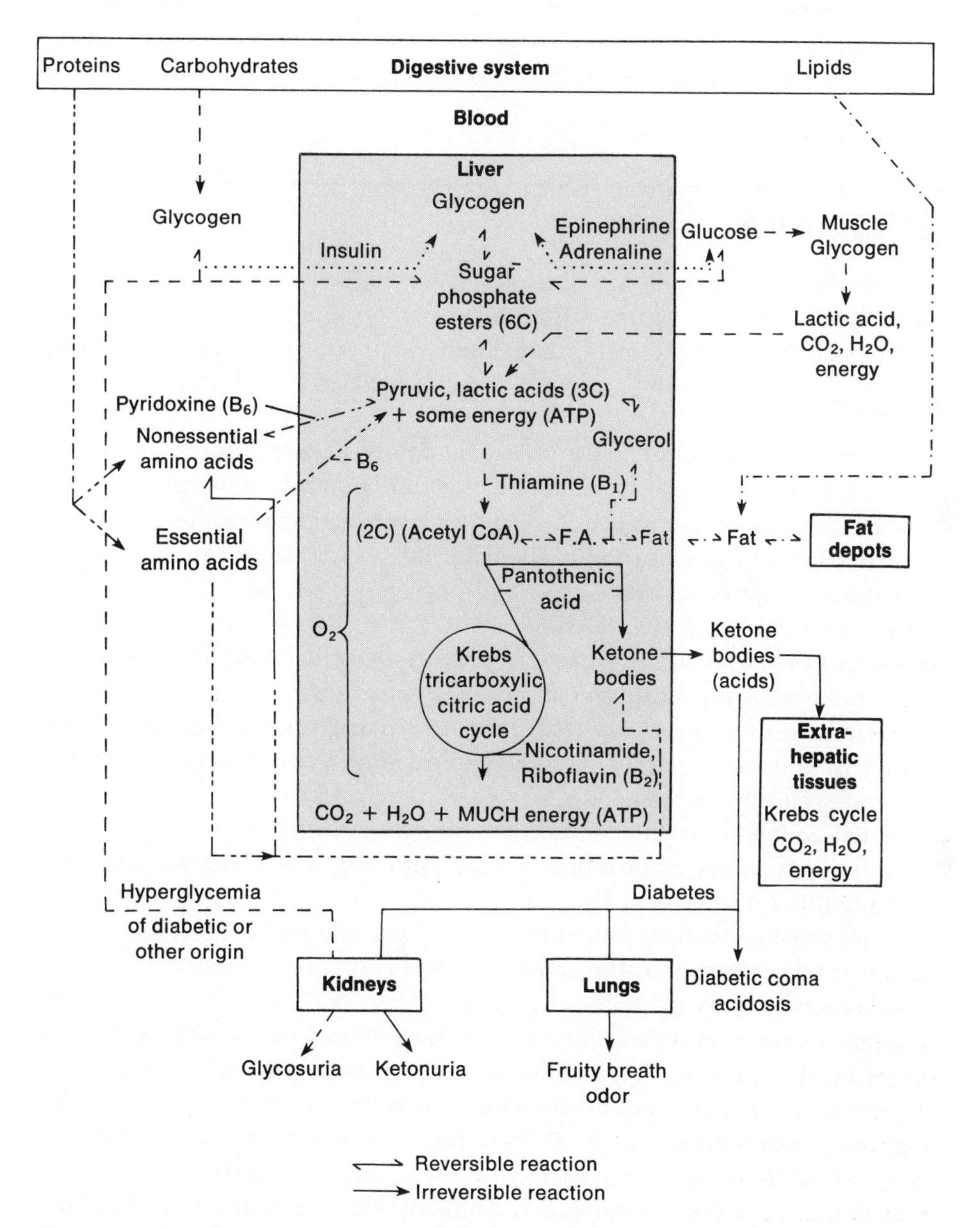

Fig. 24-6. Flowchart summarizing interrelationships of carbohydrate, lipid, and protein metabolism. (Pathways: — — — — = carbohydrate; — · — · = lipid; — - - — - - = protein; ——— = common.)

machinery. Much effort is being expended in biochemical research in attempts to correlate various disease states and abnormalities with the specific metabolic reactions that have gone awry. These efforts have been successful in many instances in recent years but much more remains to be done.

SUMMARY

The concept of the dynamic nature of intermediary metabolism was first developed as a result of investigations on protein metabolism. An equilibrium exists between dietary intake, anabolic processes, and catabolic processes.

ASPECTS OF DIGESTION AND ABSORPTION

Proteins are hydrolyzed to amino acids before absorption by way of the portal blood. In unusual circumstances, small amounts of peptides may be absorbed, leading to formation of antibodies and making the individual sensitive to the foreign protein. The liver has first choice regarding the fate of absorbed amino acids.

NUTRITIONAL ASPECTS OF PROTEIN METABOLISM

Nitrogen balance: Positive nitrogen balance—output less than intake—occurs in growing children or convalescent adults. Negative nitrogen balance—output greater than intake—occurs in fever, surgical procedures, severe diseases. Nitrogen equilibrium—output same as intake—occurs in normal adult.

Essential amino acids: Isoleucine, leucine, lysine, methionine, phenylalanine, threonine, tryptophan, valine, and perhaps histidine and arginine must be supplied by diet.

Optimal ratios for dietary amino acids: For best efficiency, all amino acids must be consumed at the same time in a ratio approximating the composition of body protein.

Biologic value of proteins: Depends on the amount and types of protein eaten, then normal digestion, absorption, and utilization. Some proteins lack one or more essential amino acids.

Lack of storage form of protein: Protein has no storage form comparable to glycogen for carbohydrates or adipose tissue for lipids.

Protein-sparing effect: If caloric requirements are met with carbohydrates and lipids, less protein is required in the diet.

SYNTHESIS AND CATABOLISM OF PROTEINS

Protein synthesis: Key reaction is the formation of the peptide bond, requiring an input of energy by way of activation reactions. The cytoplasmic ribosomal system of protein synthesis is illustrated in Fig. 24-1.

Amino acid activation: Amino acid + ATP + specific synthetase → aminoacyl-adenylate-synthetase complex (activated amino acid).

Transfer of activated amino acids to transfer RNA: See Fig. 24-1.

Synthesis of messenger RNA, template for protein synthesis: mRNA is transcribed from DNA. Each sequence of three nucleotides on mRNA acts as a codon (codeword), specifying the order of amino acids to be assembled in the peptide. This order was originally specified by DNA.

Ribosomes: mRNA functions in the cytoplasm in aggregation with ribosomes that attach to mRNA at the end corresponding to the free amino end of the growing peptide chain and move along the mRNA as synthesis continues. The tRNAs, with amino acids attached, line up on the mRNA on the basis of complementary anticodons. The amino group of the second amino acid reacts with the carboxyl group of the first amino acid to form a peptide bond. This is repeated down the line as the peptide grows.

Specification of protein conformation: The three-dimensional shape of proteins appears to be specified by the order in which the amino acids are lined up in a peptide (primary structure).

The genetic code: This is a triplet code based on four different nucleotides with sixty-four possible codons. This means that each amino acid (20+) may have, and does have, more than one codon. For this reason the code is termed degenerate.

Tissue protein synthesis and turnover: Turnover rates may vary from hours to days. High turnover rates are associated with tissues of high metabolic activity such as liver, intestinal mucosa, and kidney.

Catabolism of proteins: Cathepsins are responsible for intracellular breakdown of proteins.

HORMONAL EFFECTS IN PROTEIN METABOLISM

Growth hormone (somatotropin): Has protein anabolic effect.

Testosterone: Has protein anabolic effect.

Insulin: Increases rate of entry of amino acids into cells and increases protein synthesis by stimulating synthesis of mRNA.

Glucocorticoids: Increases gluconeogenesis, causing negative nitrogen balance.

Thyroid hormone: In physiologic doses, has protein anabolic effect; in large doses, has catabolic effect.

GENERAL METABOLISM OF AMINO ACIDS

Removal of amino group:

Oxidative deamination: Produces α-keto acids:

$$R-\underset{}{\overset{NH_2}{\overset{|}{CH}}}-COOH \xrightarrow{\text{oxidase}} R-\overset{O}{\overset{\|}{C}}-COOH + NH_3$$

Transamination: Transfer of amino group from an amino acid to a keto acid forming a new amino acid and a new keto acid with no free ammonia appearing in the reaction. Requires pyridoxal phosphate (B-group vitamin) as coenzyme:

$$R_1-\overset{\overset{NH_2}{|}}{CH}-COOH + R_2-\overset{\overset{O}{\|}}{C}-COOH \xrightleftharpoons[\text{Pyridoxal phosphate}]{\text{Transaminase}} R_1-\overset{\overset{O}{\|}}{C}-COOH + R_2-\overset{\overset{NH_2}{|}}{CH}-COOH$$

Disposition of the nitrogen:

Synthesis of purines, pyrimidines, other compounds: Also in reductive amination of α-keto acids to form amino acids:

$$R-\overset{\overset{O}{\|}}{C}-COOH \xrightarrow{NH_3} R-\overset{\overset{NH_2}{|}}{CH}-COOH$$

Glutamine formation: detoxication: Toxic ammonia is made harmless by formation of glutamine from glutamic acid. Glutamine is carried to the kidneys, hydrolyzed, and the regenerated ammonia is excreted.

Excretion as ammonia: Ammonia produced in kidney by deamination of amino acids may be excreted directly.

Urea formation: ornithine cycle: See Fig. 24-3. Urea is synthesized in the liver from ammonia and carbon dioxide and is the major nitrogenous waste product of protein metabolism.

Excretion as creatinine: Creatinine excretion is directly related to the muscle mass (Fig. 24-4).

Disposition of the carbon skeleton:

Resynthesis of amino acids: By reductive amination of the α-keto acid to produce the original amino acid.

Formation of carbohydrate intermediates: gluconeogenesis: For example, alanine $\rightleftharpoons$ pyruvate, aspartate $\rightleftharpoons$ oxaloacetate, glutamate $\rightleftharpoons$ α-ketoglutarate.

Formation of ketone bodies: Some amino acids are ketogenic.

Energy accounting for amino acid oxidation: 191,000 of 490,000 cal (39%) is conserved as ATP.

Disposition of sulfur:

Formation of sulfate: Sulfur, from cystine and methionine, is eventually oxidized to sulfate, mainly inorganic with some organic (ethereal) sulfate, and excreted in the urine.

SUMMARY OF PROTEIN METABOLISM AND ITS RELATIONSHIP TO CARBOHYDRATE AND LIPID METABOLISM

See Fig. 24-5 for a summary of protein and amino acid metabolism.

See Fig. 24-6 for flow chart of carbohydrate, lipid, and protein metabolism.

Pyruvate appears to be the most important transfer point. In spite of the intricate interrelationships, a varied diet is required to supply essential amino acids, essential fatty acids, vitamins, and adequate calories.

REVIEW QUESTIONS

1. At what stage of protein hydrolysis are the products ordinarily absorbed? Explain what is meant by the phenomenon referred to as the dynamic nature of protein metabolism. How is this related to the concept of nitrogen balance? What is the significance of each of the three stages of nitrogen balance? What is an essential amino acid?
2. How much protein can be stored in the body? Explain what is meant by the biologic value of a protein. What is the biologic value of a protein such as gelatin? What may be done to alter its biologic value?
3. List the steps involved in protein synthesis indicating what occurs at each step. Show the relationship between DNA, mRNA transcribed from a specified portion of DNA, and the anticodon region of the corresponding tRNA. How is this related to the genetic code?
4. How is protein metabolism related to carbohydrate and lipid metabolism? What reactions may be involved in converting amino acids to non–amino acid compounds? Can glycogen be produced starting with amino acids? Can essential fatty acids be produced from amino acids?
5. Growth hormone has a protein anabolic effect. How is this related to the function of growth hormone? Why do the glucocorticoids cause a negative nitrogen balance?

REFERENCES

Advances in protein chemistry, New York, Academic Press, Inc.

Annual review of biochemistry, Palo Alto, Annual Reviews.

Annual review of physiology, Palo Alto, Annual Reviews.

Baldwin, E.: Dynamic aspects of biochemistry, ed. 5, London, 1967, Cambridge University Press.

Fox, J. L., editor: Protein structure and evolution, New York, 1976, Marcel Dekker, Inc.

Litwack, G., editor: Biological actions of hormones, 6 vols., New York, 1970-1979, Academic Press, Inc.

Munro, H. N., and Allison, J. B., editors: Mammalian protein metabolism, 4 vols, New York, 1964-1970, Academic Press, Inc.

Needleman, S. B., editor: Advanced methods in protein sequence determination, New York, 1977, Springer-Verlag New York Inc.

Neurath, H., and Hill, R. L., editors: The proteins, ed. 3, vols. I-IV, New York, 1975-1979, Academic Press, Inc.

Smith, A. E.: Protein biosynthesis, New York, 1976, Halsted Press.

Stanbury, J. B., Wyngaarden, J. B., and Fredrickson, D. S., editors: The metabolic basis of inherited diseases, ed. 4, New York, 1978, McGraw-Hill Book Co.

White, W. L., Erickson, M. M., and Stevens, S. C.: Chemistry for the clinical laboratory, ed. 4, St. Louis, 1976, The C. V. Mosby Co.

25 Blood, respiration

Because of its all-important function in the transport of oxygen and nutrient molecules to the tissues and carbon dioxide and waste products away from the tissues for excretion, the blood must be in effective equilibrium with the major metabolic pathways of the body. As a result, the composition of the blood is a reflection of the state of health, nutrition, and metabolism of the body and, as such, analytic determinations of various constituents of the blood are almost invariably used as an aid to diagnosis of abnormal conditions. Taking blood samples during medical examinations of one type or another is very common because of the ease with which samples can be obtained and because the blood is free of extraneous materials or complicating factors.

For purposes of discussion, the major components of the blood are usually divided into the formed elements, i.e., the cells in the blood, and the plasma, or liquid portion of the blood. In one of the formed elements, the red cell, is hemoglobin, a red pigment of strategic importance in respiration. In the plasma are the plasma proteins, important substances of many functions.

BLOOD

Volume

The normal volume of blood in an individual is about 5 liters. This volume will vary, of course, with the size and weight of the individual, and women generally have a somewhat smaller volume than men. Small losses of blood, whether by accident, functional bleeding, or blood donation, generally go unnoticed, one important reason being that the rate of replacement is very rapid normally. The greatest danger to maintenance of blood volume is massive hemorrhage. Excessive loss of fluid and electrolytes as a result of diarrhea or vomiting may also tend to decrease the blood volume. Such changes may be very critical in young babies.

Formed elements

Hematocrit. On centrifugation of normal blood, it is found that the formed elements, red cells, white cells, and platelets, constitute about 45% of the total volume of the sample. The ratio of the volume of cells to the total volume is termed the *hematocrit,* which in a normal individual is about 0.45 or 45%. Since most of the cell volume is made up of red cells, determination of the hematocrit of a blood sample may supply a quick evaluation of the red cell status of an individual.

Red blood cells (erythrocytes). The red blood cell count in the normal male is approximately 5 million per cubic millimeter and is approximately 4.5 million in the normal female. Red cells are formed in the bone marrow, where a normal complement of hemoglobin (p. 376) is incorporated into the red cell as it matures. When released into the circulation, the red cell, a smooth, round, biconcave cell, normally has no nucleus and the hemoglobin remains inside the cell for the life of the cell. Isotopic and other types of measurements have shown that the normal average lifetime of the red cell is approximately 120 days. Determinations of red cell count, characteristics of shape and color, and average lifetime are of importance in the diagnosis and response to treatment of the various anemias.

The function of the red cell is to transport oxygen from the lungs to tissues and carbon dioxide from tissues to the lungs. The red cell membrane is very selective in its permeability. Chloride ions are freely permeable through this membrane, a factor of great importance in the gas transport function of the red cell (p. 385). Energy requirements are met by the glycolytic and pentose shunt pathways. The NADH and $NADPH_2$ produced by these pathways maintain the iron in hemoglobin in the functional ferrous (Fe^{+2}) state. The red cell has no tricarboxylic acid cycle system.

White blood cells (leukocytes). In contrast to the red cell, the various types of white blood cells have nuclei, are much larger, and are more irregularly shaped. Normal white cell counts range from about 4000 to 10,000, or higher, per cubic millimeter of blood. This number may increase rapidly and extensively in response to bacterial infections, making white cell counts of potential importance in the differential diagnosis of acute infections.

Evidence indicates that the γ-globulins (antibodies) are formed in plasma cells. Lymphocytes are found to contain relatively large amounts of γ-globulin, possibly acting as a storage site for these important infection-fighting proteins. Heparin, an anticoagulant (p. 372), is synthesized in mast cells.

Platelets (thrombocytes). As the name implies, platelets are relatively small particles, believed to be formed by splitting of large megakaryocytes. Normal platelet counts range from about 250,000 to 500,000 per cubic millimeter of blood. Platelets contain proteins and relatively high concentrations of phospholipids (mainly cephalin), which are liberated on lysis and function in the mechanism of blood clotting (see next). Among the proteins in platelets is the enzyme thromboxane synthetase, which is involved in the synthesis of thromboxane A_2 (p. 322). Evidence has been reported indicating that thromboxane A_2 promotes platelet clumping, an important step in blood clotting.

Blood clotting. A detailed analysis of the very complicated mechanism of blood clotting (coagulation) is beyond the scope of this text. For purposes of simplification, the mechanism may be divided into two major steps. In one step, prothrombin, a glycoprotein produced by the liver and found in the α_2-globulin fraction of the plasma proteins (p. 373), is converted to thrombin in the presence of many plasma factors, platelets, and calcium ions. In effect, the interaction of these many factors, estimated as ten or more, is to produce a prothrombinase activity, as a result of which prothrombin, which may be considered a pre-enzyme, is converted to the enzyme thrombin:

$$\begin{array}{c}\text{Many plasma factors, platelets, calcium ions}\\ \downarrow \\ \text{Prothrombin} \xrightarrow{\text{Prothrombinase}} \text{Thrombin}\end{array}$$

Vitamin K is required for the synthesis of prothrombin (pp. 429, 440) and may also be required for the synthesis of one of the factors involved in the conversion of prothrombin to thrombin.

In the second major step, fibrinogen, a soluble plasma protein (p. 373), is converted to insoluble fibrin under the influence of thrombin, and a clot is formed from the fibrin:

$$\text{Fibrinogen} \xrightarrow{\text{Thrombin}} \text{Fibrin} \rightarrow \text{Fibrin clot}$$

Anticoagulants. Since calcium ions are required for the conversion of prothrombin to thrombin, coagulation may be prevented by removing them from the scene of action. Among the agents that can effectively remove or tie up calcium ions in an inactive form are oxalate and citrate, the calcium salts of which do not ionize, and binding agents of one kind or another, such as ethylenediaminetetraacetate (EDTA).

The blood system itself has a built-in anticoagulant system consisting of heparin and antithrombins. Heparin, which is made by mast cells lining the walls of blood vessels, is rapidly destroyed by the enzyme heparinase and is normally not detected in blood. However, in conjunction with a cofactor, believed to be an antithrombin, heparin acts as an antiprothrombin and an antithrombin, interfering with both major steps in the clotting mechanism. It is believed that this system limits the formation of a clot to the immediate area of injury. At times, however, clots do form in the blood vessels, sometimes breaking loose and finally lodging in a strategic blood vessel of the heart or brain, leading to serious difficulty or death.

Prostaglandins also are involved in the fine control of the clotting system. Thromboxane A_2 stimulates platelet clumping, whereas prostaglandin I_2 (prostacyclin, p. 323), synthesized from prostaglandins G_2 and H_2 by the enzyme prostacyclin synthetase, which is found in the arterial lining, inhibits platelet clumping. Aspirin at one time was believed to be useful in the prevention of heart attacks resulting from blood clotting because of its ability to inhibit the synthesis of some prostaglandins such as thromboxane A_2. At present, however, it appears that aspirin also inhibits the formation of prostaglandin I_2, which prevents platelet clumping, leaving the net result of aspirin treatment in doubt until further research can define the quantitative effects on the various processes involved in blood clotting.

Dicumarol, a naturally occurring metabolic antagonist of vitamin K (p. 429), is used sometimes in cases where it is desired to increase the coagulation time, i.e., to delay clotting. It acts by interfering with the activity of vitamin K in the synthesis of prothrombin and another clotting factor.

PLASMA

Distinction between plasma and serum

The formed elements of the blood may be separated from the remainder by centrifugation in the presence of an anticoagulant. The cells are forced to the bottom of a centrifuge tube and the supernatant liquid may be drawn off the top. This liquid is known as *plasma,* and it contains all of the plasma proteins, as well as salts and other substances. As noted before, the plasma constitutes approximately 55% of normal blood.

Normal blood, in the absence of an anticoagulant, will clot soon after being drawn. The supernatant liquid resulting from centrifugation of this mixture is called *serum,* which differs from plasma in that fibrinogen, one of the plasma proteins, has been removed by formation of a clot. Serum is probably used more often than plasma in clinical determinations in order to avoid interference from the fibrinogen fraction.

Plasma proteins

The plasma proteins are made up of a large number of different types of proteins, with wide ranges in size, composition, and properties. Attention will be directed in this discussion to the major fractions listed in Table 25-1. Together these proteins appear in normal plasma at a concentration of approximately 7% (7g/100 ml).

Largely on the basis of isotope experiments with perfused livers and hepatectomized animals, it has been determined that the liver is responsible for the synthesis of all the albumin, most of the globulins except γ-globulin, and all the fibrinogen. Under normal conditions, an individual replaces about 15 to 20 g of plasma protein per day. The plasma proteins are in dynamic equilibrium with tissue proteins but it is found that at times of protein deprivation, the plasma proteins are maintained as close as possible to normal concentrations at the expense of tissue proteins.

Table 25-1. Major plasma protein fractions

Fraction	Relative percent of total plasma proteins
Albumin	60
α_1-Globulin	5
α_2-Globulin	9
β-Globulin	12
Fibrinogen	4
γ-Globulin	10

As noted before, γ-globulins are made in the plasma cells of bone marrow, spleen, and lymph nodes (reticuloendothelial system).

A great amount of research, both basic and clinical, has been performed with the plasma proteins because of the ease with which samples are obtained in relatively pure form from blood, and because of the interest in the close correspondence between the clinical state of an individual and the status of his or her plasma proteins. In most instances, the plasma or serum proteins must be fractionated before they can be studied. Separation methods already described for such purposes for proteins in general (p. 225) are also applicable to the plasma proteins. In particular, paper electrophoresis and its many variations have proved to be exceedingly useful for all types of work on the plasma proteins. A simple separation of serum protein components illustrated in Fig. 16-4 produced five major fractions. The relative percent of each fraction to the total serum protein may be determined from the electropherogram by several methods, some largely automated. The data obtained from an electropherogram in the form of a curve and relative percentages are shown in Fig. 25-1. Other electrophoretic procedures with simple variations may produce as many as 20 to 30 fractions, if necessary.

A normal serum protein pattern shows approximately the following relative percentages of the major fractions: albumin, 60%; α_1-globulin, 4%; α_2-globulin, 8%; β-globulin, 12%; γ-globulin, 16%. In any serious illness, the relative percent of albumin almost invariably decreases as the relative percent of one or more of the globulins increases. Because of the simplicity in procedure, speed, low cost of materials, and small sample size requirement, paper electrophoresis can be used on a routine basis as an aid in clinical diagnosis. In a few instances, the serum protein patterns are so distinctive (multiple myeloma, nephrosis, agamma- or hypogammaglobulinemia) that further evaluation of the patient is almost unnecessary.

Functions of the plasma proteins. Albumin has the largest responsibility for most of the functions listed for the plasma proteins. This is so because of the quantitative predominance of albumin (60% of the plasma protein) and because of its low molecular weight, it is also present in the largest number of molecules. The functions of the plasma proteins include:

1. Nutritive. Body proteins are in equilibrium with plasma proteins. Studies have shown that albumin is efficiently used for this purpose.

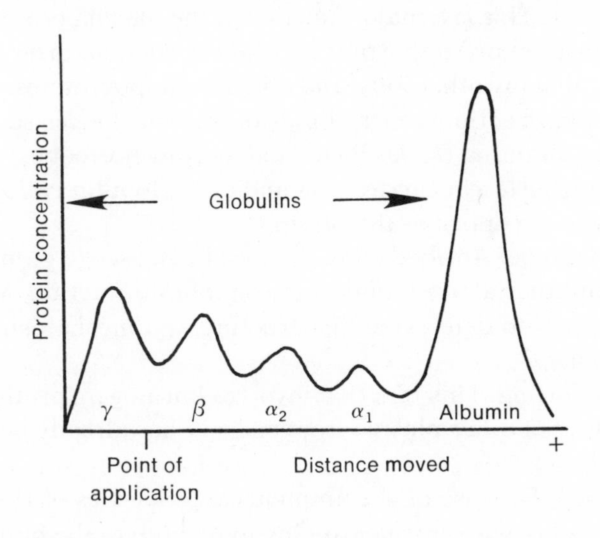

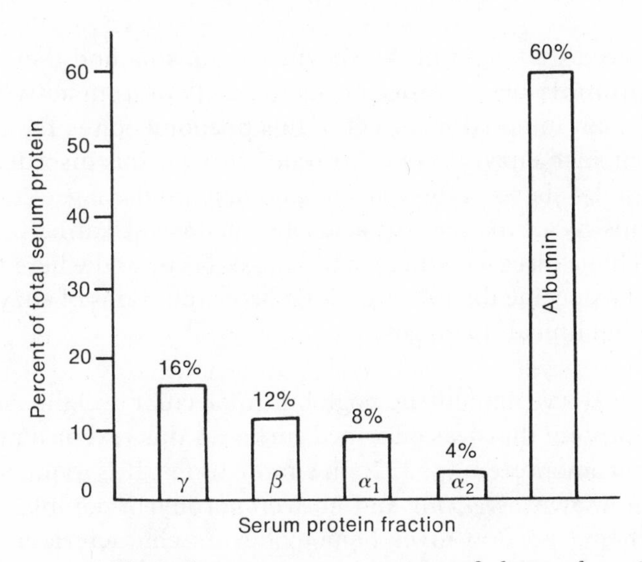

Fig. 25-1. Curve and block diagram representations of electrophoretic pattern of normal human blood serum.

2. Osmotic. The colloid osmotic pressure of the plasma proteins, which functions to maintain the proper distribution of water in blood and tissues, is mainly created by albumin, which supplies the largest number of molecules. Should there be a decrease in the albumin fraction, for any reason, less water would be brought into the blood from the tissues, leaving more water behind, and a condition known as *edema* would result.
3. Transport. This is a major function of the plasma proteins. Albumin transports many important substances such as drugs, fatty acids, calcium, many other ions, and bilirubin. Lipoproteins, in the α- and β-globulin fractions, carry lipid-soluble substances such as the fat-soluble vitamins (A, D, E, K) and steroid hormones. A β-globulin named transferrin carries iron and an α-globulin called ceruloplasmin carries copper in the plasma.
4. Immunologic. Antibodies to bacterial diseases (e.g., mumps, measles, diphtheria) are found in the γ-globulin fraction. Much work is in progress to determine the structure and mechanism of action of antibodies.
5. Blood clotting. This function involves fibrinogen, prothrombin, and a number of other plasma fractions and has already been discussed (p. 372).
6. Buffering. Because of the amphoteric properties of the constituent amino acids, the plasma proteins contribute to the buffering capacity of the blood (p. 219).

Enzymes in plasma

It is believed that most of the enzymes in plasma find their way there by leakage from tissues or broken cells rather than from active secretion. Nevertheless, an interesting aspect of this phenomenon is the correlation of changes in such enzyme concentration with various disorders. Certain transaminase levels rise after cardiac and hepatic damage. Cancer of the prostate leads to an increase in acid phosphatase. Alkaline phosphatase increases in bone disease. Other such changes occur and where the change is unique for a specific disorder, analytic determinations of enzyme activity are used in clinical diagnosis.

HEMOGLOBIN

The respiratory pigment hemoglobin (molecular weight, 68,000) is a conjugated protein that was classified earlier in this text as a porphyrinoprotein or chromoprotein (p. 222). In simple terms, it is made up of a specific protein known as *globin* and an iron-porphyrin complex known as *heme.* The heme portion gives hemoglobin its characteristic red color, which is why it is classed as a chromoprotein. In actual structure, as a

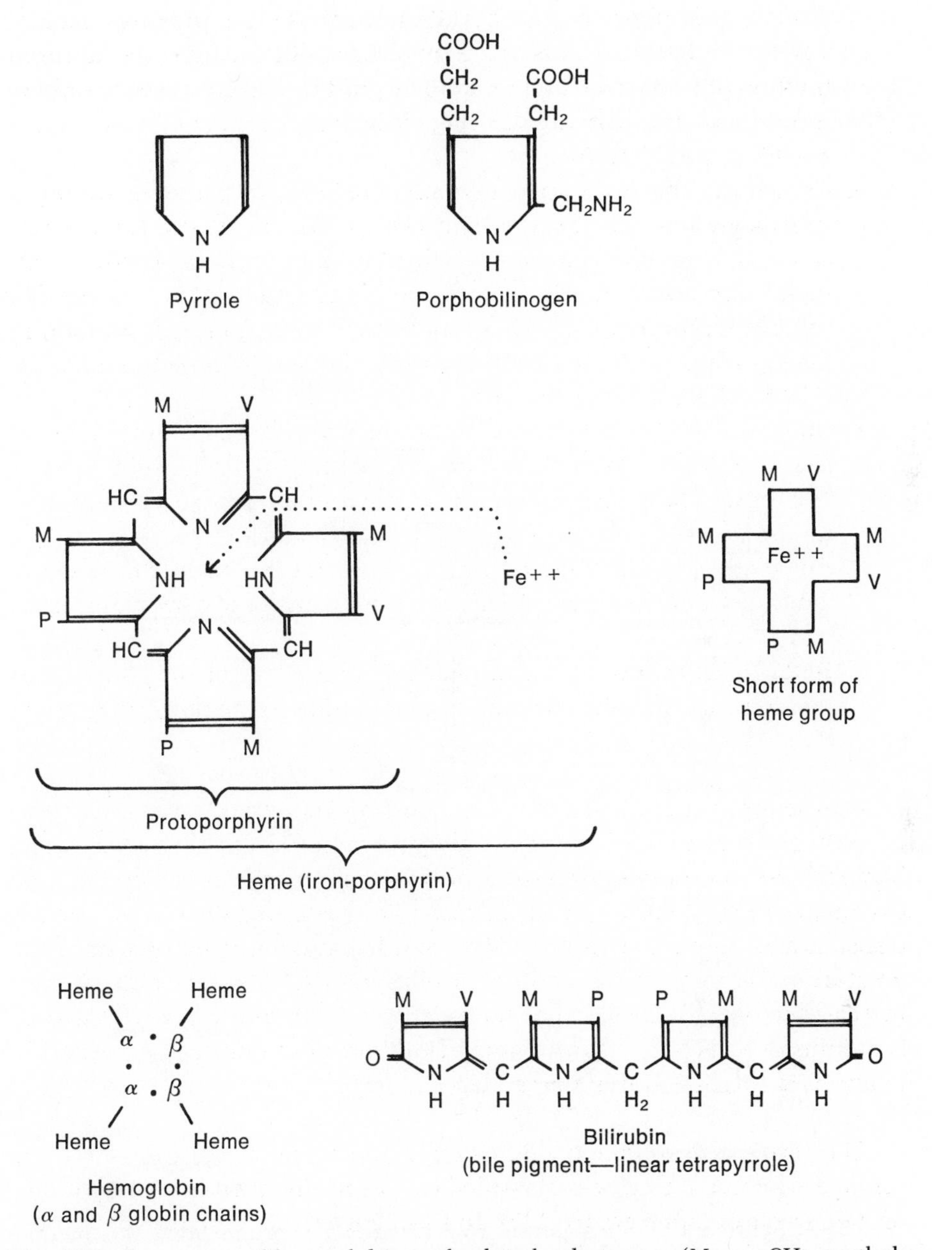

Fig. 25-2. Structures of hemoglobin and related substances. (M = $—CH_3$, methyl; V = $—CH{=}CH_2$, vinyl; P = $—CH_2CH_2COOH$, propionate; α,β = globin [protein] chains.)

functioning molecular unit, hemoglobin consists of four peptide chains grouped together—two α and two β chains—with a heme group attached to each of the four peptide chains. The molecule is close to being round in shape, with the heme groups lying on the outside surface and approximately uniformly spaced from each other (p. 224). These structures and related substances are illustrated in Fig. 25-2.

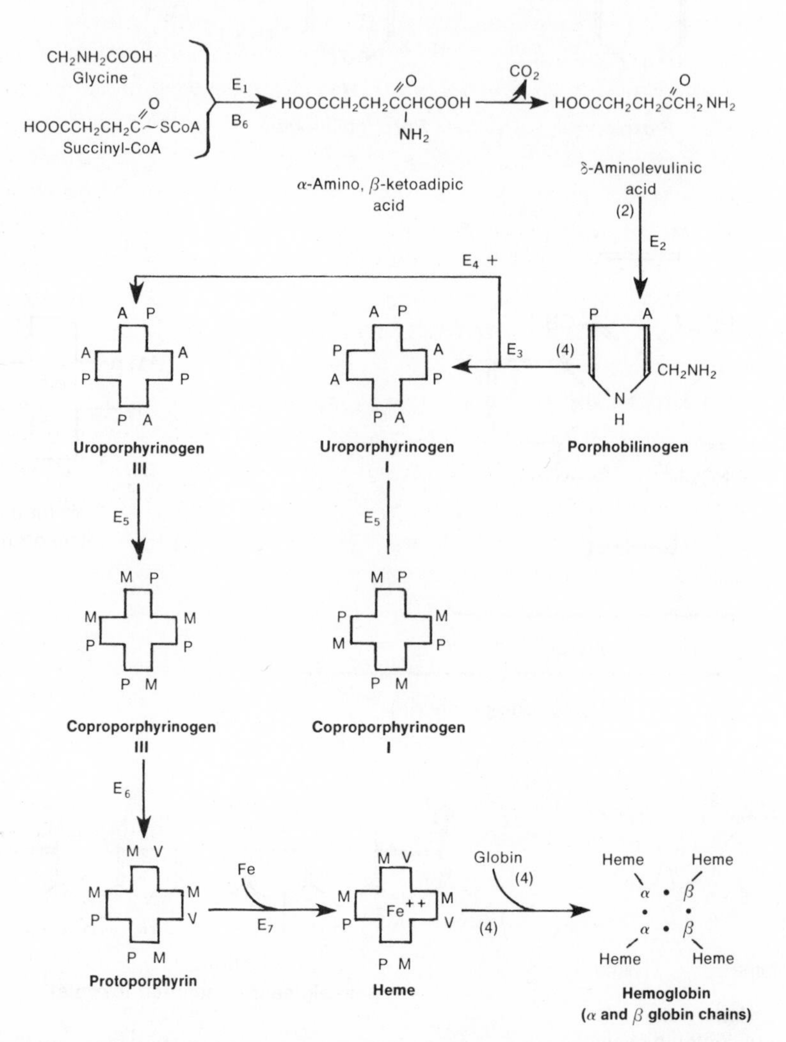

Fig. 25-3. Synthesis of hemoglobin. (A = acetate; P = propionate; M = methyl; V = vinyl; B_6 = pyridoxal phosphate; E_1 = δ aminolevulinic acid synthetase; E_2 = δ aminolevulinic acid dehydratase; E_3 = uroporphyrinogen I synthetase; E_4 = uroporphyrinogen III cosynthetase; E_5 = uroporphyrinogen decarboxylase; E_6 = coproporphyrinogen oxidative decarboxylase; E_7 = heme synthetase.)

Synthesis of hemoglobin

The pathway by which hemoglobin is synthesized is illustrated in Fig. 25-3. In the first step, the amino acid glycine and succinyl-CoA combine in the presence of the enzyme δ-aminolevulinic acid (ALA) synthetase and pyridoxal phosphate (B-group vitamin) to form α-amino, β-ketoadipic acid, which decarboxylates spontaneously to form ALA. Two molecules of ALA then condense under the influence of the enzyme ALA dehydratase to form the pyrrole ring derivative, porphobilinogen (PBG). Under the concerted action of two enzymes, uroporphyrinogen I synthetase and uroporphyrinogen III cosynthetase, four molecules of PBG condense to form uroporphyrinogen III, the main product, and normally, a very small amount of uroporphyrinogen I (note that in Fig. 25-3 the short form of the porphyrin ring structure is used wherever possible, as in Fig. 25-2). Following a series of decarboxylations in the presence of the enzyme uroporphyrinogen decarboxylase, uroporphyrinogen III and I are converted to coproporphyrinogen III and I. The I series does not continue any further on the synthetic pathway. Coproporphyrinogen III is converted to protoporphyrin by the enzyme system coproporphyrinogen oxidative decarboxylase. At this point, under the influence of the enzyme heme synthetase, iron as Fe^{+2}, is introduced into the center of the porphyrin structure to form heme. In the final stage, two α-globin and two β-globin protein chains combine with four heme groups (1 heme group per globin chain) to form the functioning molecule of hemoglobin. Portions of this process take place in the mitochondrion and other portions in the cytoplasm, as shown in the simplified version of the biosynthesis of hemoglobin presented in Fig. 25-4.

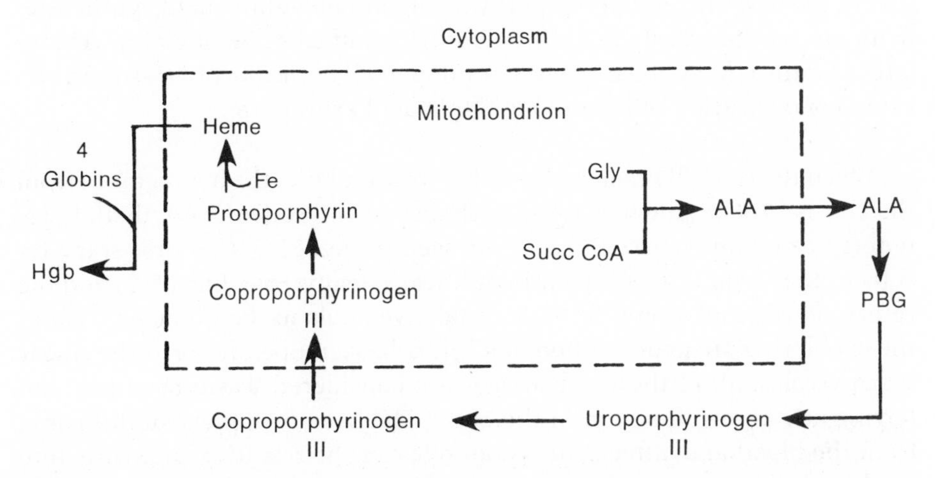

Fig. 25-4. Simplified version of biosynthesis of hemoglobin, showing stages occurring in mitochondrion and in cytoplasm. (Gly = glycine; Succ CoA = Succinyl CoA; ALA = δ-aminolevulinic acid; PBG = porphobilinogen; Hgb = hemoglobin.)

The synthesis of hemoglobin takes place in the red bone marrow and occurs in the red cell as it matures. In addition to pyridoxal phosphate, other B-vitamins required for blood formation are folic acid, vitamin B_{12}, and pantothenic acid. Because of the vitamin B_{12} requirement, intrinsic factor is also required to ensure the proper absorption of the vitamin from the intestinal tract (p. 434). A deficiency of vitamin B_{12} leads to pernicious anemia. In addition to iron, a trace of copper is also required. Vitamin C, because of its antioxidant properties, also appears to be beneficial in this process.

Under normal conditions approximately 8 g of hemoglobin are replaced every day. The average hemoglobin concentration is about 15 g per 100 ml blood. Hemoglobin synthesis is controlled by feedback inhibition (p. 259), i.e., when sufficient hemoglobin is produced, steps near the beginning of the synthetic pathway are slowed down. The hormone erythropoietin, found in the blood of anemic patients, stimulates the production of red cells. As noted before, the red cell has an average lifetime of 120 days.

Decreased hemoglobin synthesis may occur as a result of nutritional deficiencies, inhibition of various types (e.g., poisoning) of one or more enzymes in the synthetic pathway, or disorders in the control system. The cause of a particular anemia is not too easily tracked down in many cases.

A number of genetic conditions have been described, known as *porphyrias*, in which there is an accumulation of one or more of the precursors of hemoglobin in individuals as a result of a defect in one or another of the enzymes along the synthetic pathway. This induces an increased production of the first enzyme in the pathway, δ-aminolevulinic acid synthetase, in an attempt to supply the individual with sufficient hemoglobin. Therefore, a feature that appears to be common to all of the porphyrias is an elevated concentration of δ-aminolevulinic acid synthetase.

Catabolism of hemoglobin

When the red cell reaches the end of its useful life, the hemoglobin from the cell is broken down mainly in tissues of the reticuloendothelial system (spleen, bone marrow, liver). As seen in Fig. 25-5, this first stage results in the formation of bilirubin (a linear tetrapyrrole, Fig. 25-2) and the return of iron and globin to their respective pools in the body. As soon as the cyclic ring structure of hemoglobin is broken open to form the linear tetrapyrroles, all of these substances are considered waste products, not having any known function in the body. The liver then picks up bilirubin from the blood and, after processing, releases the resulting pigments into the bile. In the intestines a series of reductions takes place, forming bilinogens that are excreted and converted to bilins on exposure to oxygen.

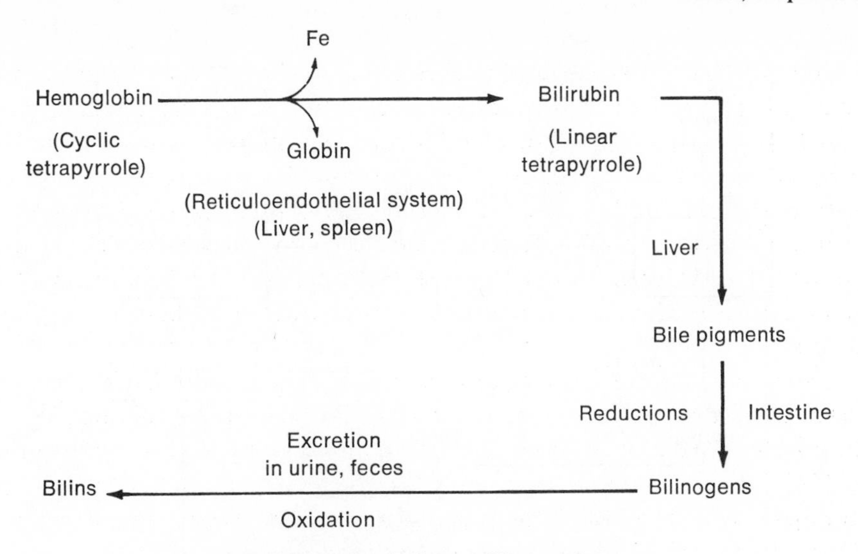

Fig. 25-5. Catabolism of hemoglobin.

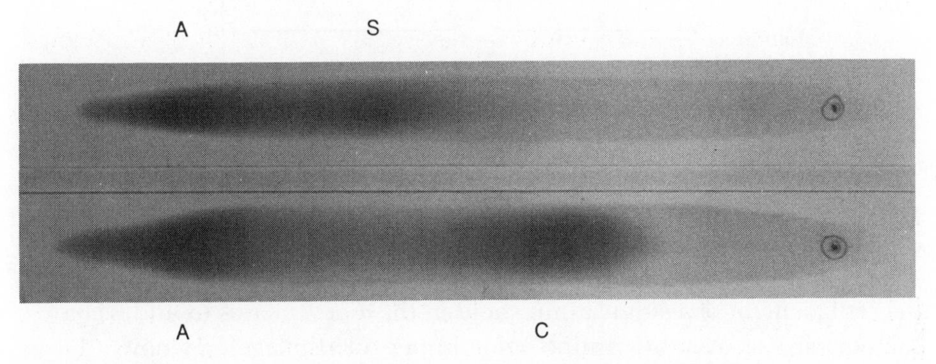

Fig. 25-6. Electrophoretic patterns showing the relative mobilities of hemoglobins. (A = normal adult; S = sickle cell; and C.) (Courtesy of J. D. Toporek.)

Bilirubin is toxic in high concentrations, especially in infants, in whom brain damage may result. The blood concentration of bilirubin may rise as a result of obstruction in the bile duct or an increased rate of hemoglobin catabolism.

Abnormal hemoglobins

With paper electrophoresis, it became possible to examine the hemoglobin of patients with various blood disorders. It was found that in some cases the hemoglobin was different than hemoglobin from normal individuals. An electrophoretic comparison of two such hemoglobins (S and C) with normal hemoglobin (A) is illustrated in Fig. 25-6. Further research showed that the abnormal hemoglobins varied from normal on the basis of

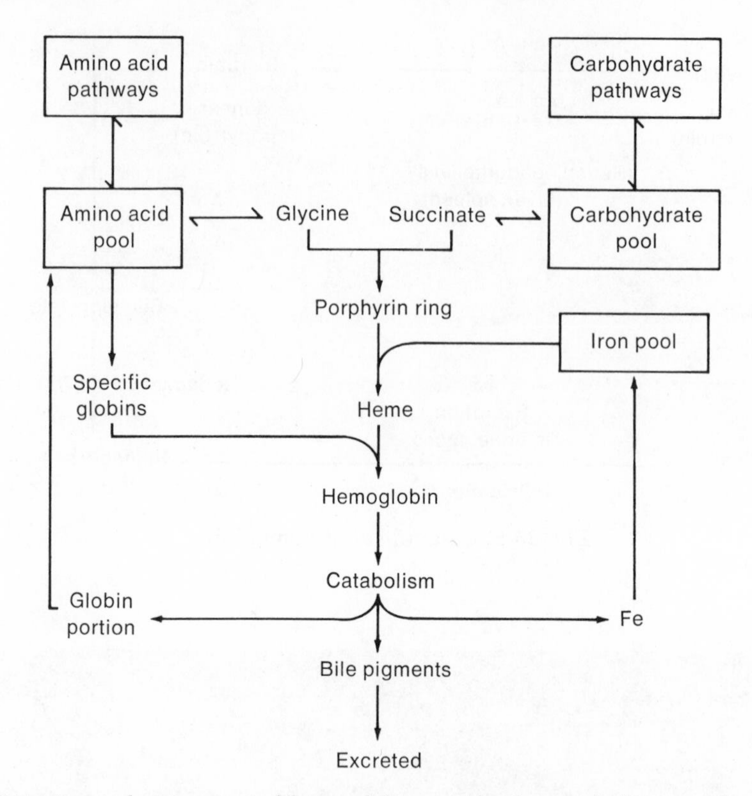

Fig. 25-7. General summary of hemoglobin synthesis, catabolism, and relationships to other metabolic pathways.

the replacement of a single amino acid in the α or β chains (α and β chains each consist of over 140 amino acids in peptide linkage). By now a large number of such abnormal hemoglobins have been reported. Although a few cause serious difficulties, most are not harmful.

Summary of hemoglobin metabolism

A simplified summary of hemoglobin synthesis and catabolism and the relationships with other metabolic processes is presented in Fig. 25-7. Although the emphasis has been on hemoglobin, other porphyrinoproteins of importance, believed to be synthesized and catabolized in a similar manner, are myoglobin in muscle, the cytochromes (p. 284), and the enzyme catalase, which breaks down hydrogen peroxide (H_2O_2) as it forms in the tissues.

RESPIRATION

Transport of oxygen: respiratory cycle

The importance of hemoglobin in respiration can be illustrated with the following facts. In the resting state, an individual requires about 250 ml of oxygen per minute. A liter of blood can carry about 2 ml of oxygen in physical solution and the maximum blood flow of 25 liters per minute would deliver about 50 ml of oxygen to the tissues, a figure far short of the minimum requirement. In combination with hemoglobin, a liter of blood can carry about 200 ml of oxygen, more than enough to satisfy the usual requirements. The reversible reaction between hemoglobin and oxygen thus constitutes a very important function of hemoglobin (Hb):

$$Hb + O_2 \rightleftharpoons HbO_2$$

Of strategic value in this connection is the interesting behavior of hemoglobin as the relative concentration (partial pressure) of oxygen in a gas mixture changes. At the higher partial pressures, as is the case with air in the lungs, the affinity of hemoglobin for oxygen increases sharply, i.e., hemoglobin combines more readily with oxygen to form HbO_2. At the lower partial pressures of oxygen, as is the situation in the tissues, the affinity of hemoglobin for oxygen decreases sharply, shifting the equilibrium of the reaction toward the breakdown of HbO_2. This makes hemoglobin ideally suited for its task of picking up oxygen in the lungs and releasing it to the tissues.

In contrast to the structure of the normal, functioning hemoglobin molecule, which is made up of four heme groups, two α globin chains and two β globin chains, myoglobin, the muscle porphyrinoprotein, is composed of only a single heme group and a single globin chain (p. 224). Myoglobin has a much greater affinity for oxygen (i.e., holds on to it much more tightly) than does hemoglobin but lacks the property of hemoglobin of a changing affinity for oxygen as the partial pressure of oxygen changes. The more complicated quaternary structure of hemoglobin (p. 224) is required for such an effect. The oxygen held by myoglobin may represent a small reserve supply of oxygen for a contracting muscle in which the blood supply may be hindered or from which the hemoglobin oxygen supply has already been removed.

Carbon monoxide also combines with hemoglobin to form carboxyhemoglobin. In fact, the combination with carbon monoxide is much stronger and more stable than with oxygen. In effect, this means that any hemoglobin that combines with carbon monoxide can no longer be used to transport oxygen. If sufficient hemoglobin is thus removed from its oxygen transport function, death results. For this reason and because it is odorless, carbon monoxide, found in heating gas and automobile exhaust fumes, is a dangerous gas.

To complete the respiratory cycle, carbon dioxide must also be transported from the tissues, released in the lungs, and breathed out into the air. About 200 ml of carbon dioxide are produced per minute by the tissues of an individual at rest. As with oxygen, very little of the carbon dioxide can be carried in the blood in physical solution. Hemoglobin, the red cells,

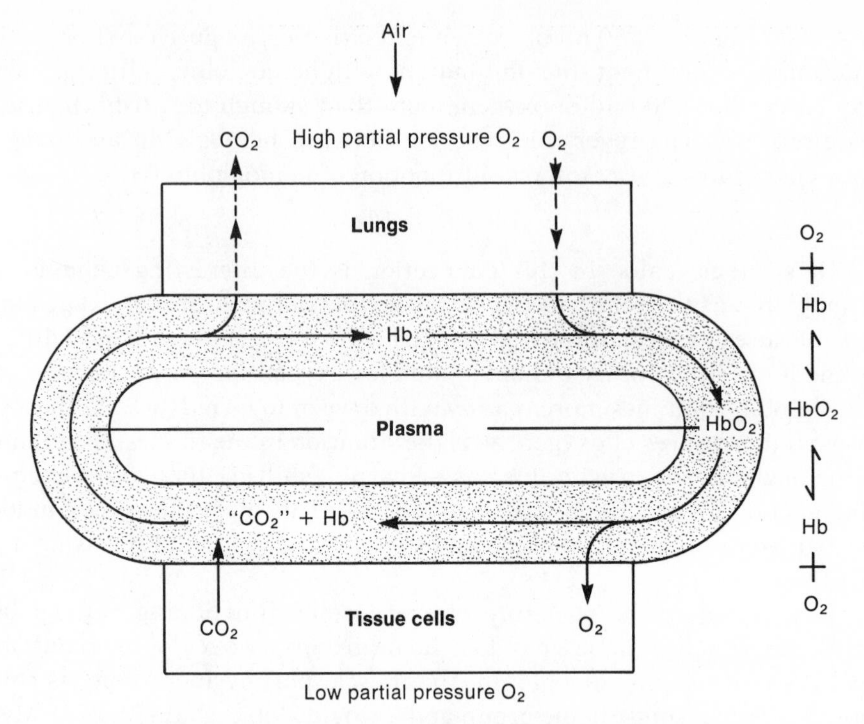

Fig. 25-8. Simplified version of respiratory cycle.

and the plasma carry the carbon dioxide to the lungs for excretion with very little change in pH. A simplified version of the respiratory cycle is diagrammed in Fig. 25-8.

Transport of carbon dioxide

Some of the details of how the carbon dioxide is transported from the tissues to the lungs are presented here.

Considered separately, the details concerning the transport of carbon dioxide in the blood are not particularly difficult. However, since a number of processes are occurring at the same time, it sometimes becomes confusing to try to understand the total situation. The separate processes that must be taken into account may be listed as follows (as they take place in the tissues):

1. The reaction of carbon dioxide with water occurs very slowly in plasma but occurs rapidly in red cells, where it is catalyzed by the enzyme carbonic anhydrase:

$$CO_2 + H_2O \rightleftharpoons H_2CO_3 \rightleftharpoons H^+ + HCO_3^-$$

2. The hydrogen ions that would accumulate in the red cells from the preceding reaction must be buffered to prevent a decrease in pH (increase in acidity). This is accomplished by the buffering action of de-

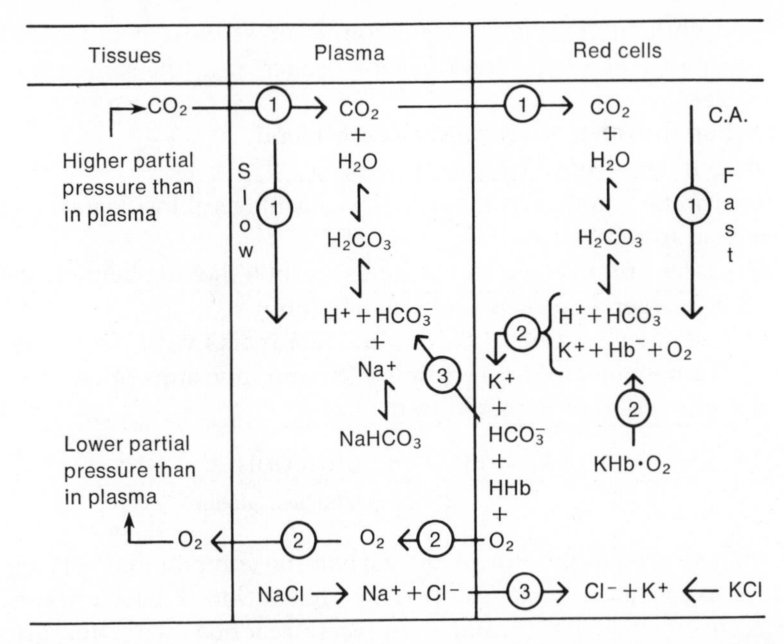

Fig. 25-9. Transport of carbon dioxide in blood. (C.A. = carbonic anhydrase.)

oxygenated hemoglobin that is formed as oxygenated hemoglobin breaks down to release oxygen to the tissues:

$$Hb \cdot O_2^- \rightleftharpoons Hb^- + O_2$$

The deoxygenated hemoglobin is less acid than the oxygenated form and can bind more hydrogen ions:

$$Hb^- + H^+ \rightleftharpoons HHb$$

This keeps the pH from changing and is known as the *isohydric transport* of carbon dioxide.

3. As the bicarbonate ion (HCO_3^-) concentration increases rapidly in the red cells, bicarbonate ions diffuse out into the plasma. In order to maintain electrical neutrality, chloride ions (Cl^-) from the plasma diffuse into the red cells. This process is known as the *chloride-bicarbonate shift*.
4. Inside the red cell, negative ions such as bicarbonate, chloride, and oxygenated hemoglobin are balanced by potassium ions (K^+). In the plasma, negative ions such as bicarbonate and chloride are balanced by sodium ions (Na^+).

These processes and their relationships to each other are illustrated in Fig. 25-9. In this diagram, the separate phases, 1 to 3, are identified to correspond with the previous discussion. It must be remembered, however, that in vivo, all of the reactions are occurring at the same time. The net results are:

1. Carbon dioxide goes from tissues to blood.
2. Oxygen goes from blood to tissues.
3. Bicarbonate ions go from red cell to plasma as chloride ions go from plasma to red cell.

These processes are reversed in the lungs, carbon dioxide being lost from the blood as oxygen is picked up by hemoglobin.

A much smaller portion of carbon dioxide reacts with hemoglobin to form carbaminohemoglobin (reaction with amino groups of amino acids in the globin portion of hemoglobin):

$$CO_2 + HbNH_2 \rightleftharpoons \underset{\textbf{Carbaminohemoglobin}}{HbNHCOOH}$$

Deoxygenated hemoglobin forms the carbamino compounds more readily than does oxygenated hemoglobin. Thus, formation of carbaminohemoglobin occurs in the tissues and the reverse reaction, with liberation of carbon dioxide, takes place in the lungs.

SUMMARY

The composition of the blood is a reflection of the state of health, nutrition, and metabolism of the body. Blood is composed of formed elements (cells) and a liquid portion (plasma).

BLOOD

Volume: In human beings normal volume is about 5 liters. Hemorrhage, diarrhea, vomiting may decrease blood volume.

Formed elements:

Hematocrit: Ratio of cell volume to total volume. Normal hematocrit is about 45% and is used as a measure of red cell status.

Red blood cells (erythrocytes): Normal male red cell count—5 million per cubic millimeter; female—4.5 million. Red cells are formed in bone marrow where hemoglobin is made before red cells are released. Average life of red cells is 120 days. Functions to transport oxygen from lungs to tissues and carbon dioxide from tissues to lungs.

White blood cells (leukocytes): Normal white cell count—4000 to 10,000 per cubic millimeter. White cells increase in response to bacterial infections. Plasma cells make γ-globulins, which may be stored in lymphocytes. Heparin is made in mast cells.

Platelets (thrombocytes): Normal platelet count—250,000 to 500,000 per cubic millimeter. Function in blood clotting mechanism.

Blood clotting: Two major stages:

$$\text{Many plasma factors, platelets, calcium ions}$$
$$\downarrow$$
$$\text{Prothrombin} \xrightarrow{\text{Prothrombinase}} \text{Thrombin}$$

$$\text{Fibrinogen} \xrightarrow{\text{Thrombin}} \text{Fibrin} \rightarrow \text{Fibrin clot}$$

Vitamin K is required for synthesis of prothrombin.

Anticoagulants: Removal of calcium ions by oxalate, citrate, or EDTA prevents coagulation. Heparin and dicumarol also act as anticoagulants. Clots do form sometimes in blood system and block strategic blood vessels, possibly resulting in death. Prostaglandins are involved in the control of the coagulation process: thromboxane A_2 promotes platelet clumping, whereas prostacyclin (PGI_2) inhibits platelet clumping. Net effect of aspirin on these processes is uncertain at present.

PLASMA

Distinction between plasma and serum: Plasma—total liquid portion of blood. Serum—plasma from which fibrinogen has been removed after clotting.

Plasma proteins: Major fractions are albumin, α_1- and α_2-globulins, β-globulin, fibrinogen, and γ-globulin. Normal concentration in plasma is 7 g/100 ml. Liver makes all of the albumin, fibrinogen, and most of the globulins, except γ-globulins. Plasma proteins are replaced at the rate of 15 to 20 g per day. Paper electrophoresis is used often to separate the plasma proteins. Serious illness is usually reflected by a change in the normal plasma protein pattern; the relative percent albumin decreases as one or more globulins increase.

Functions of the plasma proteins:

Nutritive: Body proteins are in equilibrium with the plasma proteins. Albumin is used efficiently.

Osmotic: Albumin supplies the largest number of molecules for development of the colloid osmotic pressure of blood which maintains normal distribution of water in body. Edema may result from decreased albumin concentration.

Transport: Albumin and other fractions must transport many important substances.

Immunologic: Bacterial antibodies are found in γ-globulin fraction.

Blood clotting: Fibrinogen, prothrombin, and other factors found in plasma proteins.

Buffering: By way of amphoteric properties of constituent amino acids.

Enzymes in plasma: Reflect damage to tissues with rise in enzyme concentration, e.g., certain transaminases increase after cardiac or hepatic injury.

HEMOGLOBIN A conjugated protein (molecular weight, 68,000) made up of four peptide chains, two α, two β chains, grouped together, with a heme group attached to each peptide chain.

Synthesis of hemoglobin: Requires glycine, succinyl-CoA, and iron as raw materials. See Fig. 25-3 for details. Synthesis takes place in red bone marrow and requires the following B-group vitamins: pyridoxal phosphate, folic acid, vitamin B_{12}, and pantothenic acid. Also requires intrinsic factor for absorption of vitamin B_{12}, deficiency of which leads to pernicious anemia. Average hemoglobin concentration is about 15 g per 100 ml blood and replacement rate is 8 g per day. Synthesis is controlled by feedback inhibition and production of red cells is stimulated by the hormone erythropoietin.

Catabolism of hemoglobin: Fig. 25-5 summarizes the catabolism of hemoglobin. High concentrations of bilirubin are toxic, especially for infants.

Abnormal hemoglobins: Many abnormal hemoglobins have been reported, most differing from normal by the replacement of a single amino acid in the α or β chain.

Summary of hemoglobin metabolism: Fig. 25-7 summarizes hemoglobin metabolism.

RESPIRATION **Transport of oxygen: respiratory cycle:** Hemoglobin reacts reversibly with oxygen, picking up oxygen in the lungs, releasing it in the tissues:

$$Hb + O_2 \rightleftharpoons HbO_2$$

Carbon dioxide is carried by the blood from the tissues to the lungs. See Fig. 25-8 for respiratory cycle.

Transport of carbon dioxide: Fig. 25-9 summarizes the transport of carbon dioxide from tissues to lungs. Carbon dioxide is transported with no change in pH (isohydric transport). In tissues, bicarbonate goes from cells to plasma as chloride goes from plasma to cells (chloride-bicarbonate shift). Reactions that occur in tissues reverse when blood reaches lungs.

REVIEW QUESTIONS

1. Why are blood samples so commonly taken from hospital patients? What are some of the items usually looked for and the significance attached to these items?
2. What are some of the important factors involved in blood clotting? How do they operate? Under what circumstances may it be useful to prevent the clotting of blood? How may this be done and what is the principle(s) involved?
3. What are the major plasma proteins and their functions? Where are they made? What happens to the plasma protein pattern in disease states?
4. In simple terms, describe the structure of the hemoglobin molecule and its functions. What are some of the substances required for its synthesis? What are the breakdown products of hemoglobin? Which are excreted and which are reutilized?
5. How is carbon dioxide carried in the blood from the tissues to the lungs, with little, if any, change in pH (isohydric transport)? What is the chloride-bicarbonate shift?

REFERENCES

Bianchi, R., Mariani, G., and McFarlane, A. S., editors: Plasma protein turnover, Baltimore, 1976, University Park Press.

Bissell, D. M.: Formation and elimination of bilirubin, Gastroenterology **69:**519, 1975.

Brewer, G. J., editor: International conference on red cell metabolism and function, New York, 1972, Plenum Press.

Bunn, H. F., Forget, B. G., and Ranney, H. M.: Hemoglobinopathies, Philadelphia, 1977, W. B. Saunders Co.

Davenport, H. W.: The ABC of acid-base chemistry: the elements of physiological blood-gas chemistry for medical students and physicians, ed. 6, Chicago, 1974, University of Chicago Press.

Gidari, A. S., and Levere, R. D.: Enzymatic formation and cellular regulation of heme synthesis, Semin. Hematol. **14:**145, 1977.

Hardisty, R. M., and Weatherall, D. J., editors: Blood and its disorders, Philadelphia, 1974, J. B. Lippincott Co.

Meyer, U. A., and Schmid, R.: Hereditary hepatic porphyrias, Fed. Proc. **32:**1649, 1973.

Putnam, F. W., editor: The plasma proteins, ed. 2, 3 vols., New York, 1975-1977, Academic Press, Inc.

Seegers, W. H., editor: Blood clotting enzymology, New York, 1967, Academic Press, Inc.

Surgenor, D. M., editor: The red blood cell, ed. 2, 2 vols., New York, 1974, 1975, Academic Press, Inc.

White, W. L., Erickson, M. M., and Stevens, S. C.: Chemistry for the clinical laboratory, ed. 4, St. Louis, 1976, The C. V. Mosby Co.

With, T. K.: Bile pigments; chemical, biological and clinical aspects, New York, 1968, Academic Press, Inc.

26 Urine: a product of the kidney

As seen in the preceding chapter, an important function of the lungs is the disposal of carbon dioxide, a volatile waste product of metabolism. The task of eliminating some of the solid, nonvolatile waste products of metabolism is an important function of the kidneys. In the process of carrying out this function, the kidney is also involved in water, electrolyte, and acid-base balance (Chapter 27). As a result, the kidney exerts a large measure of control in maintaining the normal and relatively constant composition of the blood and related body fluids. It does so by regulating the reabsorption of many of the substances originally filtered from the blood and in some instances adds more of these substances to the fluid that ultimately becomes the urine.

It was previously noted that the composition of the blood may reflect the nutritional and health status of an individual. Since urine is made largely of materials brought to the kidneys by the blood, the composition of the urine may also provide information of importance in diagnosing various disorders.

One of the most important properties of the kidney is its ability to concentrate the substances being excreted. This becomes obvious from an examination of Fig. 26-1, in which the composition of urine is compared with that of blood plasma from which it is formed. Especially noteworthy is the tremendous concentration of urea in urine as compared to that in blood plasma. The concentration of solids, and therefore the specific gravity, in the urine varies during the day depending on the amount of water ingested and produced as a result of metabolism. Normally, when more water is available the urine volume is larger and the specific gravity is lower than when less water is available. These observations provide the basis for a procedure, known as a *clearance test,* for determining the functional status of the kidney.

THE KIDNEY IN URINE FORMATION

The nephron

The functional unit of the kidney in which urine is produced is the *nephron* (Fig. 26-2). It has been estimated that each human kidney contains more than one million such units. The nephron consists of a structure known as Bowman's capsule, which funnels into a narrow, tubular section, which in turn leads into the collecting tubules. Inside the Bowman's capsule lies a jumble of capillaries known as the glomerulus, the inner wall of the capsule lying very close to the capillaries. It is here that the formation of urine starts with a glomerular filtrate of the blood. The tubule near Bowman's capsule is known as the proximal portion and it is separated from the distal portion by the loop of Henle. The capillaries emerge from the capsule and twist around the proximal and distal tubules, where reabsorption and adjustment of the glomerular filtrate take place, resulting in the formation of urine.

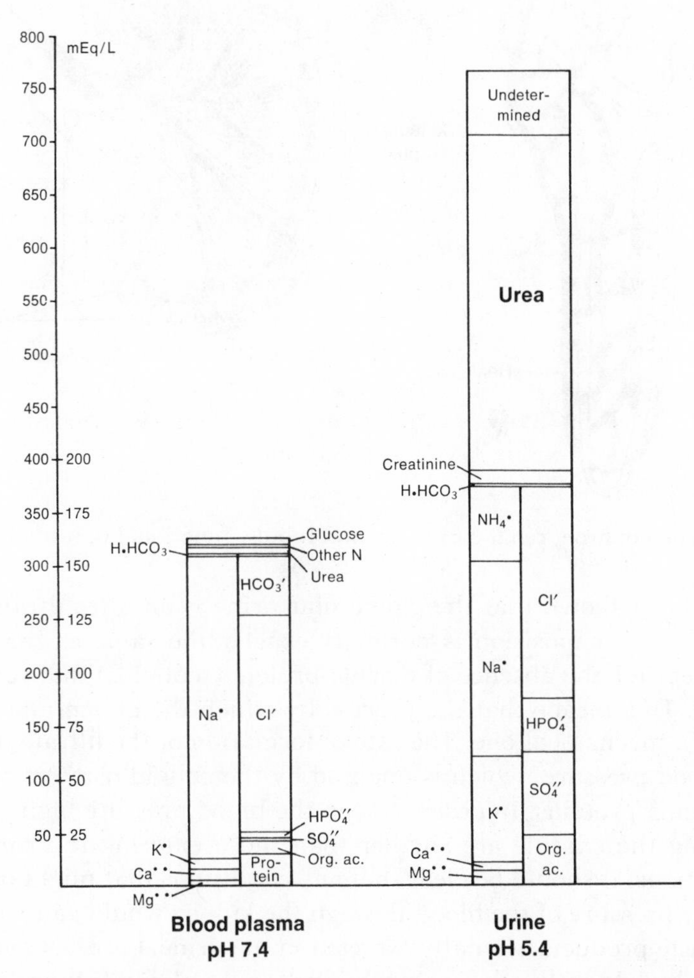

Fig. 26-1. Comparison of composition of blood plasma and urine. (Adapted from Gamble, J. L.: Chemical anatomy, physiology, and pathology of extracellular fluid, Cambridge, Mass., 1942, Harvard University Press. Copyright 1942, by J. L. Gamble; 1947, 1954, by the President and Fellows of Harvard College.)

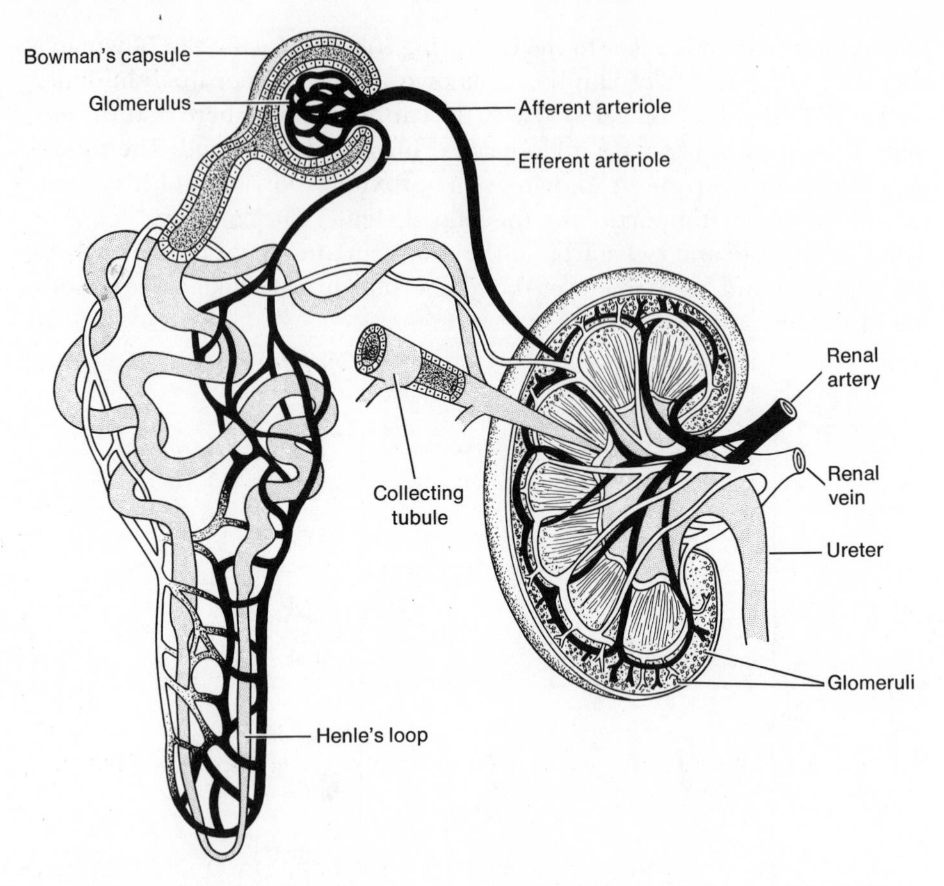

Fig. 26-2. The nephron, renal corpuscle, and tubule, functional unit of the kidney.

Formation of glomerular filtrate

It has been shown that the glomerular urine is an ultrafiltrate of the blood, i.e., its composition is normally exactly the same as that of the blood except for the absence of plasma proteins (colloids) that remain in the blood. This means that the process by which the glomerular fluid is formed is a mechanical one. The rate of formation of the filtrate depends on the blood pressure, which is opposed by the colloid osmotic pressure of the plasma proteins. In order to keep the blood pressure high, the vessels leaving the capsule are smaller than those entering it. From these considerations, it should be clear that any conditions that might decrease the flow or pressure of the blood through the kidney would cause a retention of waste products normally excreted in the urine. Cardiac failure, dehydration, and hemoconcentration are among the clinical states in which kidney function is impaired because of lowered blood pressure.

Glomerular filtrate is produced normally at the rate of approximately 125 ml per minute or 180,000 ml per day, a rather astonishing figure when

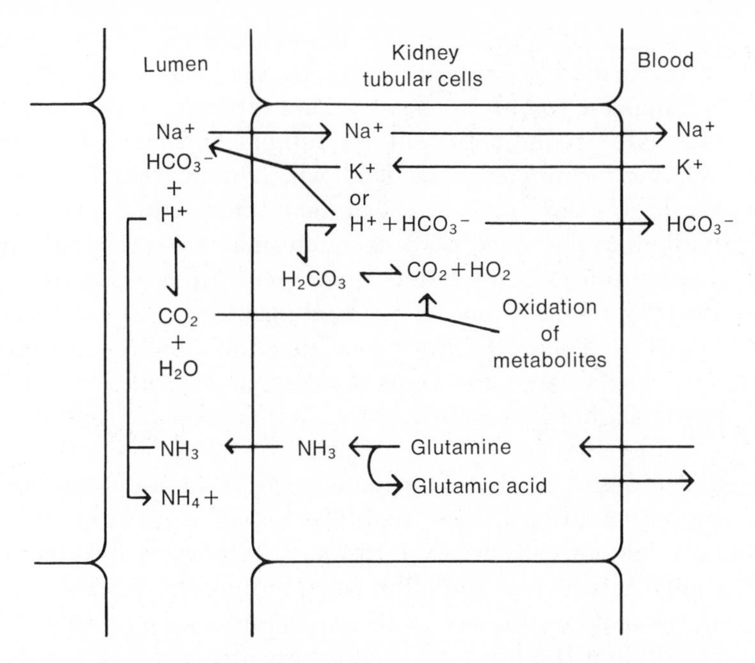

Fig. 26-3. Mechanisms involved in the acidification of urine.

compared to the average daily urine output of about 1000 to 2000 ml, about 1% of the glomerular filtrate.

Tubular reabsorption and acidification of urine

A comparison of the rates at which glomerular filtrate and urine are formed and the observation that glomerular filtrate has glucose in it, whereas urine normally does not, are direct indications that as the filtrate passes through the tubules, large-scale reabsorption of water and other substances takes place. Among the substances normally reabsorbed are the following (percentage reabsorbed in parentheses): water (99%), sodium chloride (99%), bicarbonate ions (100%), potassium ions (reabsorption inversely related to amount of sodium ion reabsorbed), urea (40%), glucose (100%). In passing through the glomerulus, the blood plasma loses about 20% of its water but no protein. This results in an increased colloid osmotic pressure, an important factor in the large task of reabsorption by the tubules.

Reabsorption of sodium ion is stimulated by the hormone aldosterone (p. 457), which results in the excretion of more potassium ion. The antidiuretic hormone (p. 462) regulates the absorption of water by the tubules. If absent, as in the condition known as diabetes insipidus, the volume of urine increases greatly (possibly up to 20 liters per day), and the specific gravity decreases correspondingly (to as low as 1.001).

In addition to reabsorption, the tubules function in the process of acidification of the urine as required. Hydrogen ions may enter the tubular fluid in exchange for sodium ions, which are returned to the blood as sodium bicarbonate. Hydrogen ions in the tubular fluid may combine with bicarbonate ions, forming carbonic acid, which breaks down to water and carbon dioxide. The carbon dioxide can then diffuse into the tubular cells and eventually into the blood. Hydrogen ions may also combine with ammonia, formed from glutamine and amino acids (p. 361), in the tubular fluid to form ammonium ions. Some hydrogen ions also combine with monohydrogen phosphate (HPO_4^{-2}) ions to form dihydrogen phosphate ions ($H_2PO_4^-$). The various reactions and exchanges taking place in the tubules during acidification of urine are summarized in Fig. 26-3.

Excretion by tubular cells

It has been observed that some substances, such as potassium ions, not present in glomerular filtrate are ultimately found in urine. This indicates that the tubular cells are excreting such substances into the tubular urine. Certain drugs such as penicillin and p-aminosalicylic acid are eliminated from the body in this way. Other substances such as phenolsulfonphthalein (PSP) and Diodrast are used to test this aspect of kidney function.

Certain abnormalities in the transport processes of the tubular cells, which move substances into or out of these cells, result from inborn errors of metabolism (inherited defect), e.g., renal glycosuria, aminoaciduria. Similar disorders may result sometimes as a secondary effect of infectious diseases.

Clearance tests

Since the kidney represents the only means of excreting certain waste products, it may be of great importance in times of illness to determine the state of kidney function. By using a substance that is excreted in the urine but not metabolized by the body and comparing the amount found in the urine (cleared by the kidney from the blood) with the amount remaining in the blood, much valuable information about kidney function can be obtained pertinent to the nephron and the blood flow. Such procedures are known as clearance tests. Inulin is a polysaccharide that is not metabolized by the body and is also not reabsorbed by the tubules after it is filtered from the blood into the glomerular fluid. In other words, it is cleared exclusively by glomerular filtration after intravenous administration. Inulin clearance can thus be used to determine the rate of formation of glomerular filtrate, normally about 125 ml per minute. Comparing the rate of inulin clearance with the rate of clearance of a substance that is partially reabsorbed or excreted by the tubules furnishes enough data to calculate the rate of reabsorption or tubular excretion of that substance.

Diodrast (which is excreted by glomerular filtration and tubular excretion), urea (which is partially reabsorbed), and phenolsulfonphthalein (which is cleared mainly by tubular excretion) are also used in kidney function tests.

URINE

Normal characteristics

Color. The first property of urine usually noted is its color, which normally varies from light yellow with large volumes to darker shades approaching brown or red as the urine becomes more concentrated. The color is caused mainly by two substances—urochrome, a sulfur-containing compound of unknown composition, and urobilin, a breakdown product of hemoglobin (p. 380). The color may be changed by certain drugs or foods, e.g., methylene blue produces a greenish color, phenols, a brown or black color, and rhubarb, a reddish color.

Normal, fresh urine, slightly acid, is water-clear. Turbidity is found in alkaline urine or develops on standing as a result of ammoniacal fermentation, phosphates precipitating out at the higher pH. If present in high enough concentrations, uric acid, urates, or calcium oxalate precipitate at low pH. Turbidity may also be caused by the presence of large numbers of leukocytes and epithelial cells in cases of urinary tract infections.

Odor. Fresh, normal urine has a slight characteristic odor. On standing, it develops an odor of ammonia resulting from the bacterial conversion of urea to ammonia. The typical odor associated with eating asparagus can be detected in the urine within a very short time after ingestion, indicating the rapidity and efficiency with which digestion and elimination of waste products take place.

Volume. Normal daily excretion of urine varies widely from about 800 to 2000 ml, depending to a large extent on the eating and drinking habits of the individual. The minimum amount necessary to excrete solid waste products is approximately 500 ml. Most urine is produced during the daytime active period, less during the night.

As noted before, an abnormally large volume of urine (polyuria) is excreted in the condition of diabetes insipidus. A less extreme polyuria also occurs in diabetes mellitus. Drugs that cause polyuria are known as diuretics.

At the other extreme, conditions causing extensive damage to kidneys, such as immunologic reactions or heavy metal poisoning, may result in the complete cessation of urine formation or anuria. Other conditions in which water is lost excessively before the kidney can excrete it, as in vomiting, diarrhea, or edema, lead to a decreased production of urine or oliguria.

Specific gravity. The specific gravity of the urine is a measure of the concentration of solids in the urine. Usually, the higher the volume, the lower the specific gravity. Normally, the specific gravity varies from about 1.010 to 1.030. This is a reflection of the ability of the kidneys to concentrate or dilute the urine as required by the state of the body.

pH. The pH of normal urine varies over a wide range from about 4.8 to 8.0 with an average of about 6.0. Urinary pH is a reflection of dietary intake. A high protein diet, the metabolism of which produces acids, results in an acid urine. When metabolized, most fruits and vegetables produce bicarbonate, which leads to a less acid or alkaline urine. The temporary decrease in urinary acidity that occurs shortly after eating is known as the *alkaline tide.* An increase in ketone bodies, which may occur in starvation, carbohydrate deprivation, or diabetes mellitus, causes an increase in urinary acidity.

Inorganic constituents

As seen in Fig. 26-1, the major inorganic ions excreted in the urine are sodium, potassium, ammonium, chloride, phosphate, and sulfate.

Sodium, potassium. As noted before, the kidney, which regulates the excretion of sodium, the main cation (positive ion) of extracellular fluid, and potassium, the main cation of intracellular fluid, is of great importance in the maintenance of water, electrolyte, and acid-base balance (p. 404). Normal daily excretion of sodium is about 3 to 5 g and of potassium is about 2 to 4 g, the ratio being about 5 sodium to 3 potassium. Aldosterone increases the reabsorption of sodium, thereby causing an increase in the excretion of potassium.

Ammonia. Ammonia, formed in the tubular epithelium from glutamine and other amino acids (p. 361), also plays an important part in acid-base balance. By combining with hydrogen ions in the tubules, the acidity is lowered, enabling more hydrogen ions to come from the tubule cells in exchange for sodium ions, which are reabsorbed (p. 394). The normal daily excretion of ammonium nitrogen is 0.3 to 1.2 g, with an average of about 0.7 g. The output of ammonia increases with increasing urinary acidity resulting from high protein diets, diabetic acidoses (ketone bodies), or ingestion of acid-forming substances such as ammonium salts.

Chloride. As seen in Fig. 26-1, chloride ions are excreted in larger amounts than any of the other inorganic ions. Of all the solid constituents, only urea is excreted in larger quantity. The normal daily output of chloride ions ranges from about 5 to 9 g. As the plasma concentration of chloride ions decreases, the excretion of chloride ions in the urine decreases as a result of an increase in tubular reabsorption.

Phosphate. Phosphorus is excreted in the urine mainly in the form of the phosphate ions HPO_4^{-2} and $H_2PO_4^-$, the ratio varying with and controlling the urinary acidity and the amounts varying with the diet. The normal daily excretion of phosphorus ranges from 0.7 to 1.5 g, with an average of 1.1 g. The parathyroid hormone causes an increased excretion of phosphate by the kidney as a result of decreased tubular reabsorption. Turbidity in alkaline urines is often caused by precipitated phosphates.

Sulfate. Sulfur is found in the urine in the form of neutral sulfur (5% to 15% of total sulfur, in such compounds as the amino acids methionine and cystine, thiosulfate, and urochrome, p. 395) and sulfate (85% to 95% of total sulfur), both inorganic (SO_4^{-2}) and organic (ethereal sulfate). The normal daily excretion of sulfur ranges from 0.7 to 1.5 g, with an average of about 1.0 g and varies with the dietary protein, because most of the sulfur originates in the amino acids methionine and cystine.

One of the ethereal sulfate products arises from indole, which is formed as a result of putrefaction of protein in the large intestine. Some indole goes to the liver, where it is converted to indoxyl and conjugated with sulfate (detoxified), ultimately appearing in the urine as sodium or potassium indoxyl sulfate (ethereal sulfate), known as indican. The amount of indican appearing in the urine can be used to estimate the extent of intestinal putrefaction, which is normally small.

Bicarbonate. Bicarbonate ions (HCO_3^-) are not found in the final urine of normal acidity, most of the bicarbonate ions of the glomerular filtrate being reabsorbed in the proximal tubules and the remainder in the distal tubules. However, bicarbonate ions play an important role in the acidification of urine (pp. 394, 413).

Bicarbonate ions may appear in the urine if an excessive concentration exceeding the limits of reabsorption is reached in the glomerular filtrate. As a result, the urine will become more alkaline.

Organic constituents

Among the more important organic constituents of urine are urea, uric acid, creatinine, amino acids, and ketone bodies.

Urea. Urea, which is the major end product of amino acid catabolism (p. 361), is the urinary solid excreted to the largest extent (about 50% of total urinary solids, 80% to 90% of the nonprotein nitrogen, NPN). The normal daily excretion of urea is about 25 to 30 g. The amount of urea excreted varies directly with the protein intake.

An abnormal level of urea in the blood indicates kidney dysfunction and is a very serious situation because the kidney represents the only means of excretion of urea, which is toxic in higher concentrations. This has led to the development of artificial "kidneys"—machines in which the blood is dialyzed outside the body in an attempt to eliminate such waste products as urea. Although these machines are still quite expensive to produce and maintain, they have been used successfully to prolong the lives of

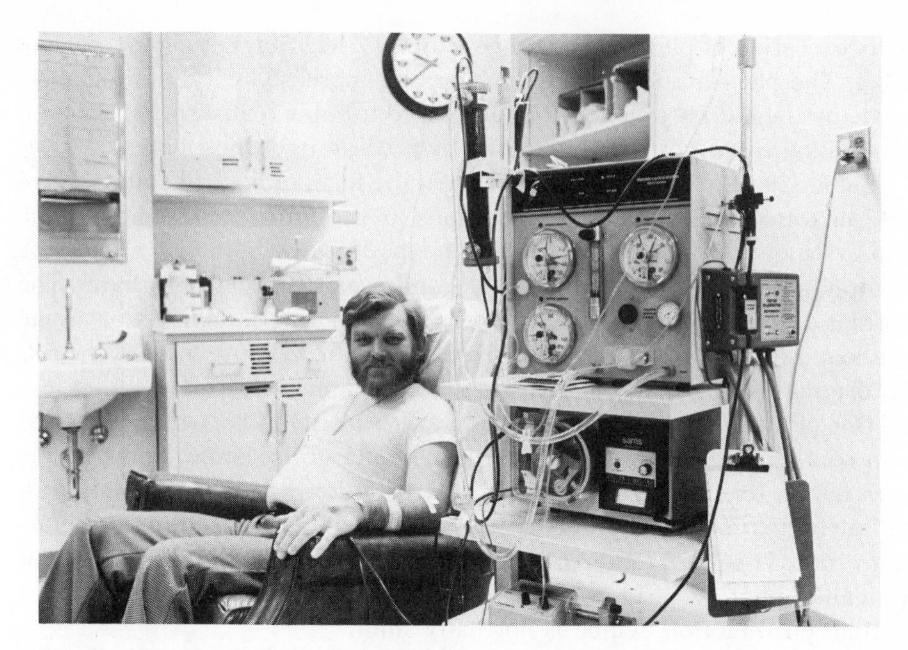

Fig. 26-4. Experimental artificial kidney apparatus being tested on a volunteer patient.

patients suffering from acute or chronic kidney disorders (Fig. 26-4). More recent developments are bringing closer the realization of portable artificial kidney devices.

Uric acid. Uric acid is the major waste product of purine catabolism (p. 337). The normal daily excretion of uric acid is about 0.7 g. It is present in normal urine of pH 6 as a soluble urate of sodium or potassium but in urines of lower pH (higher acidity), it may precipitate out as the insoluble, free acid. The excretion of uric acid increases on high purine diets or conditions in which tissue nucleoprotein breakdown increases, as in leukemia. In gout, there appears to be an increased production of uric acid, raising the concentration in blood, and leading to the deposition of urates in the joints, which can be a very painful condition.

Creatinine and creatine. The origin and function of creatine have already been discussed (p. 361). Creatinine, which is excreted in the urine, is formed from creatine in constant amounts, which are directly related to the muscle mass. In fact, the excretion of creatinine for any individual is so constant that analytic determinations of creatinine in 24-hr urine collections are used to verify the completeness of the collection. The normal daily excretion of creatinine ranges from about 0.7 to 1.8 g.

A figure known as the *creatinine coefficient,* which is equal to the number of milligrams of creatinine excreted daily per kilogram of body weight, is used as an indication of muscular development. The values for men range from 20 to 26, for women, from 14 to 22, and lower for children, depending on their development.

Creatine is not excreted in the urine under normal circumstances because at normal blood levels it is completely reabsorbed by the kidney tubules. Creatinuria may occur in muscle diseases, such as muscular dystrophy, where muscle tissue is being broken down at an accelerated rate.

Amino acids. A small amount of amino acid nitrogen is excreted in the urine, the normal daily excretion reaching about 0.5 to 1.0 g, the remainder being reabsorbed by the tubules. Aminoacidurias may occur because of increased blood concentrations of amino acids or disorders in tubular reabsorption by the kidney. Both types of conditions may result from acquired or inherited causes. Severe liver damage may result in aminoaciduria because of impaired urea formation, which leaves more unconverted amino acids in the blood. Cystinuria is an example of aminoaciduria resulting from an inherited abnormality (inborn error of metabolism) in reabsorption.

Ketone bodies. A small amount of ketone bodies (acetoacetic acid, β-hydroxybutyric acid, and acetone [p. 325]) is excreted normally in the urine. The daily output is about 100 mg. With high-fat diets, during carbohydrate restriction, and in the clinical condition of diabetes mellitus, the excretion of ketone bodies may increase greatly.

Pathologic constituents

Glucose (carbohydrate). Normally there is very little excretion of carbohydrates in the urine. Glucose and lactose may appear during pregnancy and lactation. Pentoses may sometimes appear as a result of dietary intake or inherited defect. Some form of the Benedict test or glucose oxidase test (p. 193) is usually used to screen urine for the presence of carbohydrate.

The various types of glycosuria were discussed with the carbohydrates (p. 310). Renal glycosuria is the only type resulting from an inherited defect in tubular reabsorption.

Protein. Protein in amounts detectable by ordinary tests is not found in urine under normal circumstances, most of the protein remaining in the capillaries as the blood is filtered in the glomerulus, while the remainder is reabsorbed in the tubules. In various kidney diseases, such as nephritis and nephrosis, the permeability of the glomerular membrane is changed or the reabsorption of protein is decreased and protein does pass into the urine. Urinary tract disorders below the kidney may also cause proteinuria. Proteinuria may be detected by heat coagulation or more precise tests as described previously (p. 230).

Ketone bodies. As noted before, excess amounts of ketone bodies in the urine may occur in diabetes mellitus, carbohydrate deprivation, starvation, or on high-fat diets. Since these acid substances must be neutralized in the formation of urine, it is possible, in the presence of large quantities of ketone bodies, to deplete the body of sodium and potassium ions, leading to an acidosis. These ions are spared by the formation of ammonia by the kidney (p. 394), forming ammonium ions by combination with hydrogen ions in the urine and returning the sodium or potassium ions to the blood.

Blood. Blood in the urine, *hematuria,* results from hemorrhage in the kidneys or the rest of the urinary tract. Excessive hemolysis of red blood cells may result in the excretion of some hemoglobin in the urine or *hemoglobinuria.*

Large amounts of blood or hemoglobin can be noted by the red color of the urine. The benzidine test (blue color) is used to detect smaller amounts.

Bile. In cases of obstructive jaundice (blocking of the bile ducts), bile may be detected in the urine by the color of the bile pigments (green) or by chemical tests.

SUMMARY

The kidney excretes solid, nonvolatile waste products of metabolism and is involved in water, electrolyte, and acid-base balance. The composition of urine reflects the state of metabolism in the body. An important function of the kidney is its ability to concentrate materials (Fig. 26-1).

THE KIDNEY IN URINE FORMATION

See Fig. 26-2 for a diagram of the nephron, the functional unit of the kidney. Important features are the glomerulus (capillaries), Bowman's capsule, and the tubules.

Formation of glomerular filtrate: Glomerular urine is an ultrafiltrate of the blood (only protein is absent). The production of glomerular filtrate amounts to 180,000 ml per day but only 1000 to 2000 ml (about 1%) ends up as urine.

Tubular reabsorption and acidification of urine: Large-scale reabsorption of water (99%) and other substances takes place in the tubules. The tubules function in acidification of urine, hydrogen ions being exchanged for sodium ions and ammonia combining with hydrogen ions to form ammonium ions (NH_4^+) if necessary to decrease urinary acidity (Fig. 26-3).

Excretion by tubular cells: Some substances not found in glomerular filtrate are added to or excreted into the tubular fluid by the tubular cells.

Clearance tests: Information on the state of kidney function, glomerular filtration, tubular reabsorption may be obtained from clearance tests with substances such as inulin and Diodrast, which are not metabolized in the body.

URINE

Normal characteristics:

Color: Normally light yellow and varies with concentration. Color is caused by urochrome and urobilin. Alkaline urine may be turbid (precipitated phosphates). Strongly acid urine may also be turbid (precipitated urates and calcium oxalate).

Odor: Slight, characteristic odor. Develops ammoniacal odor on standing (bacterial breakdown of urea).

Volume: Normal daily volume—800 to 2000 ml. Polyuria—abnormally large volume, as in diabetes insipidus and mellitus. Drugs that cause polyuria are called diuretics. Anuria—complete cessation of urine production. Oliguria—decreased production of urine resulting from loss of water in other ways (diarrhea, vomiting).

Specific gravity: A measure of the concentration of solids in urine. Normally 1.010 to 1.030.

pH: Varies from 4.8 to 8.0, average 6.0.

Inorganic constituents:

Sodium, potassium: Excretion of sodium ions and potassium ions are controlled by kidney. Normal daily excretion of sodium, 3 to 5 g; potassium, 2 to 4 g.

Ammonia: Formed in tubules from glutamine and other amino acids and spares sodium or potassium ions by combining with hydrogen ions. Normal daily excretion, 0.3 to 1.2 g ammonium nitrogen (average 0.7 g).

Chloride: Normal daily excretion, 5 to 9 g chloride ions.

Phosphate: Excreted as monohydrogen phosphate (HPO_4^{-2}) or dihydrogen phosphate ($H_2PO_4^-$) ions, the ratio varying with and determining the urinary pH. Normal daily excretion of phosphorus, 0.7 to 1.5 g; average, 1.1 g.

Sulfate: Neutral sulfur, 5% to 15% of the total sulfur (methionine, cystine, urochrome); inorganic and organic sulfate, 85% to 95% of the total sulfur. Normal daily excretion of sulfur, 0.7 to 1.5 g, average 1.0 g.

Bicarbonate: Bicarbonate ions are not found in normal urine but play important role in acidification of urine.

Organic constituents:

Urea: Major waste product of amino acid catabolism making up about 50% of total urinary solids. Normal daily excretion, 25 to 30 g. Abnormal amounts are toxic. Artificial kidneys have been used to keep urea levels down.

Uric acid: Major waste product of purine catabolism. Normal daily excretion, 0.7 g. Precipitates in strongly acid urines. Uric acid production increases in gout, with resulting deposition of urates in the joints.

Creatinine and creatine: Creatinine is produced in constant amounts, directly related to muscle mass. Creatinine content is used to verify completeness of 24-hr urine collections. Creatinine coefficient is the number of milligrams of creatinine excreted per day per kilogram of body weight. Normal daily excretion, 0.7 to 1.8 g. Creatine is not normally excreted in urine but may appear as a result of muscle disease.

Amino acids: Small, normal daily excretion, 0.5 to 1.0 g. Aminoacidurias may occur because of increased blood concentrations of amino acids or defective tubular reabsorption, resulting from acquired diseases or inherited disorders, or as a result of severe liver damage (decreased conversion of amino acids to urea).

Ketone bodies: Small, normal daily excretion, 100 mg. Increases on high fat diets, carbohydrate restriction, and in diabetes mellitus.

Pathologic constituents:

Glucose (carbohydrate): Normally not in urine. Glycosurias were discussed with the carbohydrates (p. 310). Benedict test or glucose oxidase used for detection.

Protein: Proteinuria occurs in kidney diseases (also urinary tract infections) because of increased glomerular permeability and decreased tubular reabsorption. Detected by heat coagulation or more precise tests (p. 229).

Ketone bodies: Excess amounts may occur in diabetes mellitus, carbohydrate deprivation, high-fat diets. May cause acidosis by depletion of sodium ions and potassium ions required for neutralization; sodium ions and potassium ions spared by ammonia.

Blood: Hematuria—blood in urine; hemoglobinuria—hemoglobin in urine. Detected by red color of urine or benzidine test (blue color).

Bile: Detected in urine in cases of obstructive jaundice (blocking of bile ducts).

REVIEW QUESTIONS

1. What type of wastes must be excreted by the kidney? In what type of medium? Describe the nephron and how it functions in the production of urine.
2. Glomerular urine is referred to as an ultrafiltrate of blood. What does this mean? What volume (average) of glomerular filtrate is produced daily? What is the average daily volume of final urine? What must occur between production of glomerular filtrate and final urine? How is this accomplished?
3. List, in one column, the normal characteristics and substances in the urine. In a parallel column, list changes that may occur in disease states and the metabolic background.
4. Are proteins normal constituents of urine? carbohydrates? How may they be tested for? What test is used to determine the completeness of a 24-hr urine collection?
5. What are clearance tests? How may they be carried out? What is the significance of clearance tests?

REFERENCES

Annual review of physiology, Palo Alto, Calif., Annual Reviews.

De Wardener, H. E.: Kidney: an outline of normal and abnormal structure and function, ed. 4, New York, 1973, Churchill Livingstone, Inc.

Free, A. H., and Free, H. M.: Urinalysis in clinical laboratory practice, Cleveland, Ohio, 1975, CRC Press.

Lippman, R. W.: Urine and the urinary sediment: a practical manual and atlas, ed. 2, Springfield, Ill., 1977, Charles C Thomas, Publisher.

Pitts, R. F.: Physiology of the kidney and body fluids, ed. 3, Chicago, 1974, Year Book Medical Publishers, Inc.

Rouiller, C., and Muller, A., editors: Kidney: morphology, biochemistry, physiology, 4 vols., New York, 1969-1971, Academic Press, Inc.

Smith, H. W.: The kidney: structure and functions in health and disease, New York, 1951, Oxford University Press.

White, W. L., Erickson, M. M., and Stevens, S. C.: Chemistry for the clinical laboratory, ed. 4, St. Louis, 1976, The C. V. Mosby Co.

27 Water, electrolyte, acid-base balance

The subjects of water balance, electrolyte balance, and acid-base balance are so intricately related that it is difficult to separate them for teaching or any other purpose. If, as a single aspect is being discussed, it is always remembered that in vivo most of the processes are occurring at the same time because they are dependent on each other, the true significance of the basic principles, events, and results will be more apparent. Although the importance of these subjects has been appreciated for many years and much work has been performed in these areas, experts are still in disagreement on the interpretations of many observations.

Although many numbers must be used in the following discussions for reasons of accuracy and completeness, these numbers of themselves are not very important. There are always lists at hand from which the actual figures for normal and abnormal situations may be obtained. What is much more important is an understanding of the basic mechanisms underlying the maintenance of a normal status or the reasons for a change that occurs in response to an abnormal condition.

WATER BALANCE

The term *water balance* refers to the equilibrium that is established between the intake and output of water in an individual. Any water taken in or made in the body also comes into equilibrium with the various body water compartments. The three major compartments and their relative percent of total body weight are listed here:

Compartment		*Percent body weight*
Intracellular fluid		50
Extracellular fluid		20
Interstitial fluid	15	
Blood plasma	5	
TOTAL WATER		70

The interstitial fluid is fluid that surrounds the exterior of cells and is included with the blood plasma as extracellular fluid. The blood plasma also functions in the excretion or output of water by way of the lungs and the kidneys.

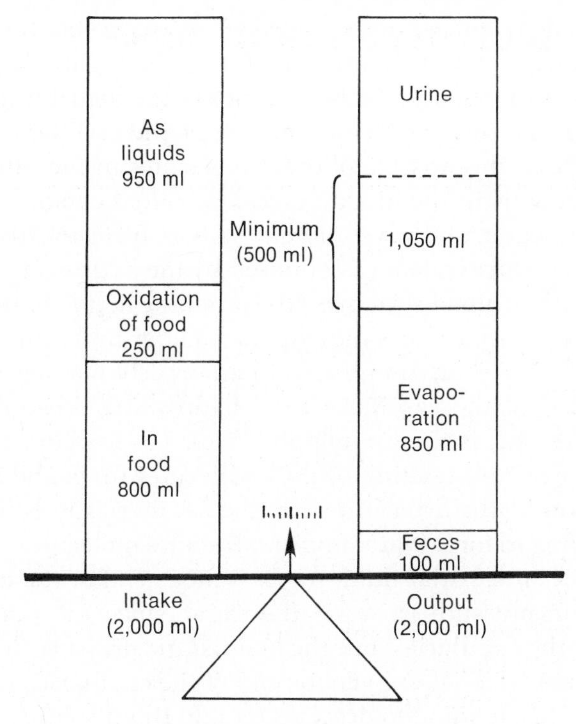

Fig. 27-1. Water balance.

Intake and output of water

Water is made available to the body (1) as drinking water, (2) as water in foods (most of which are mainly water), and (3) as water produced in the body as a result of oxidation of food molecules (Fig. 27-1).

Water is lost from the body (1) in urine, (2) in feces, (3) in sweat, and (4) by evaporation from skin and lungs (insensible perspiration) (Fig. 27-1). There is a minimum amount of water, approximately 500 ml, that must be excreted as urine in order to excrete metabolic waste products such as urea. This makes a total of close to 1500 ml as the minimum total excretion of water and the minimum required intake. Any intake above this figure generally would represent a matter of individual habit and would be balanced normally by an increased excretion of water in the urine.

Should the intake of water exceed the output in significant amounts for a period of time, edema (excess fluid in tissues) would result. On the other hand, should the output be excessive, as in severe cases of vomiting or diarrhea, dehydration would result. Abnormalities of these types are always more serious in younger children—one reason being that the body pool of water is smaller in children.

Water distribution system

In order to distribute water to, or collect water from, the tissues to satisfy their requirements, the body depends on the highly developed transport facilities of the blood. The key feature of the distribution and collection system in the semipermeable nature of the capillary walls (p. 84). With very little in the way of colloidal molecules in the interstitial fluid, the plasma proteins in the blood develop a colloid osmotic pressure that tends to draw water and crystalloidal solutes from the tissues into the blood. However, this tendency is opposed by the hydrostatic (water) pressure built up in the blood system by the pumping action of the heart. Thus there are two opposing forces moving the smaller molecules in and out of the capillaries. This is accomplished in an orderly manner as follows. At the arterial end of the capillaries, the hydrostatic pressure, tending to push molecules out, is approximately 32 mm of mercury (Hg). The colloid osmotic pressure, tending to pull molecules in, is about 22 mm Hg, or the net effect at the arterial end of the capillaries is 10 mm Hg ($32 - 22 = 10$), tending to force water and small solute molecules out of the capillaries into the interstitial fluid. At the venous end of the capillaries the colloid osmotic pressure is about the same, since the plasma proteins remain inside the capillaries, but the hydrostatic pressure drops to 12 mm Hg. Thus the net effect at the venous end of the capillaries is -10 mm Hg ($-22 + 12 = -10$), tending to draw water and small solute molecules into the capillaries from the interstitial fluid. These relationships are illustrated in Fig. 27-2.

Water interchange between tissue cells and interstitial fluid also takes place across a semipermeable membrane—the cell wall. Cell walls are freely permeable to water and small, uncharged molecules but are not freely permeable to proteins or, in some instances, to certain inorganic ions (e.g., sodium and potassium ions).

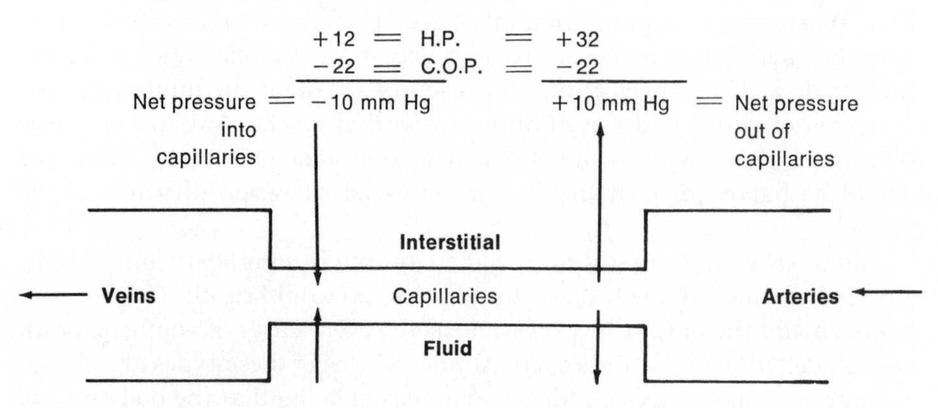

Fig. 27-2. Colloid osmotic pressure (C.O.P.) and hydrostatic pressure (H.P.) relationships in capillaries.

As a result of the equilibria established between plasma, interstitial fluid, and intracellular fluid, it should be noted that any significant disturbances in plasma protein synthesis may have serious effects on the normal distribution of water. This is especially true if the synthesis of albumin decreases because most of the colloid osmotic pressure developed by the plasma proteins is caused by the presence of albumin (p. 375). If the albumin concentration drops low enough, collection of water from the tissues decreases and edema results. A similar situation may occur in heart disorders accompanied by increased blood pressure that cannot be counteracted by the colloid osmotic pressure.

Water uses

In between intake and output of water, the water is used for various purposes in the body. It should be remembered that chemical reactions usually take place in water. Therefore all cells must have some water available. As noted before, water is required to dissolve certain waste products excreted in the urine, and water in the form of perspiration is also used to maintain the body temperature at a constant level. A large volume of water is required daily for the production of digestive juices, as can be seen from the following list:

Digestive juice	*Daily volume (ml)*
Saliva	1500
Gastric juice	2500
Intestinal juice	3000
Bile	500
Pancreatic juice	700
TOTAL DIGESTIVE JUICES	8200

Fortunately, almost all of this volume of water is reabsorbed, only a very small amount being lost in the feces.

ELECTROLYTE BALANCE

Nomenclature and definitions

In the aqueous medium of the body fluids, electrolytes exist in their ionized forms. By convention, positive ions are called cations (travel to the cathode [−] in an electrical field) and negative ions are called anions (travel to the anode [+]). In clinical chemistry, the concentration of the various ions is usually specified in terms of milliequivalents per liter (mEq/liter).

$$1 \text{ mEq} = \frac{1 \text{ Equivalent (in grams)}}{1000}$$

Since $\frac{\text{grams}}{1000}$ = milligrams,

$$1 \text{ mEq} = 1 \text{ Equivalent (in milligrams)}$$

Also, since $1 \text{ Eq} = \frac{\text{Molecular weight}}{\text{Valence}}$,

$$1 \text{ mEq} = \frac{\text{Molecular weight}}{\text{Valence}} \text{ (in milligrams)}$$

Thus, if 1 mEq of sodium ions is 23 mg, a solution containing 2.3 g (2300 mg) of sodium ions per liter has a sodium ion concentration of 100 mEq per liter:

$$\frac{2300 \text{ mg Na}^+/\text{liter}}{23(1 \text{ mEq Na}^+)} = 100 \text{ mEq Na}^+/\text{liter}$$

In the development of osmotic pressure, it is the number of particles that is the important factor. Therefore, a molecule of an electrolyte such as

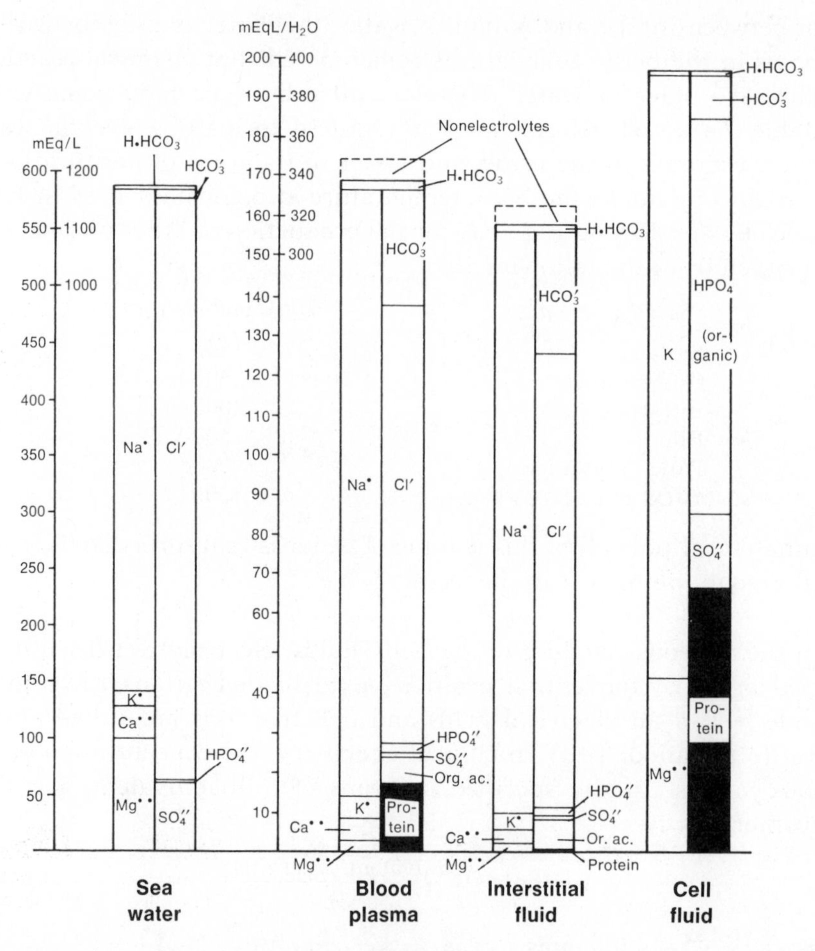

Fig. 27-3. Electrolyte composition of body fluids. (Adapted from Gamble, J. L.: Chemical anatomy, physiology and pathology of extracellular fluids, Cambridge, Mass., Harvard University Press. Copyright 1942, by J. L. Gamble; 1947, 1954, by the President and Fellows of Harvard College.)

sodium chloride (NaCl) contributes two particles per molecule in its ionized form and a molecule such as sodium sulfate (Na_2SO_4) would contribute three particles to the development of the osmotic pressure.

Electrolyte composition of body fluids

The approximate electrolyte composition of the blood plasma, interstitial fluid, and intracellular fluid is presented in bar graph form in Fig. 27-3. There are several important points to be noted. Protein, which is present in plasma and intracellular fluid, is not present (except for a trace) in interstitial fluid. The major cation in intracellular fluid is potassium, whereas the major cation in interstitial fluid is sodium. This would indicate that there is some kind of mechanism for keeping potassium ions inside the cell while keeping sodium ions from entering the cell. Also, while the major anion of the plasma and interstitial fluid is chloride, there is no chloride ion in intracellular fluid.

The electrolyte composition of the various digestive juices and fluids is compared to that of blood plasma in Fig. 27-4. The most striking differences to be noted among these fluids are the very large hydrogen ion concentration found in gastric juice and the much larger concentration of

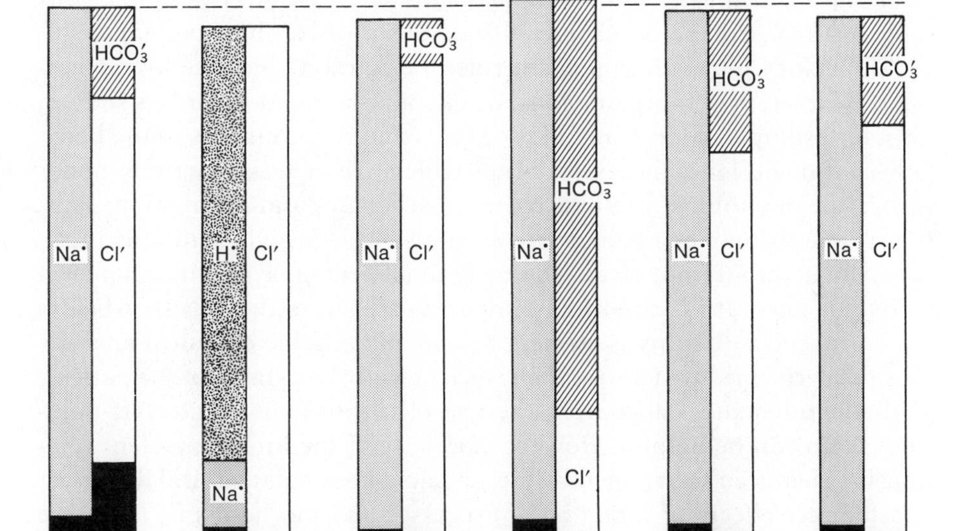

Fig. 27-4. Electrolyte composition of digestive fluids. (Adapted from Gamble, J. L.: Chemical anatomy, physiology, and pathology of extracellular fluids, Cambridge, Mass., Harvard University Press. Copyright 1942, by J. L. Gamble; 1947, 1954, by the President and Fellows of Harvard College.)

bicarbonate ion found in pancreatic juice. These differences become important considerations in the maintenance of water, electrolyte, and acid-base balance—particularly when large volumes of digestive juices are lost in persistent vomiting or diarrhea.

Maintaining electrolyte balance

A key feature in the maintenance of electrolyte balance is the selective permeability of cell membranes. If size were the only criterion, all of the inorganic ions could pass through the cell membrane because they are small enough. Yet, as noted before, potassium ions stay inside cells and sodium ions stay outside. This calls for a process that can counteract the normal tendency of small ions to pass through semipermeable membranes. Various theories have been proposed and are in the process of being tested experimentally. One explanation of the sodium–potassium ion phenomenon invokes the existence of a sodium pump or an active transport system that, with the expenditure of energy, keeps shuttling any sodium ions that stray into the cell back out again and accomplishes the same thing in reverse for potassium ions keeping them in the cell.

Another factor of importance in following the shift of electrolyte ions from one fluid compartment to another is the requirement for maintaining electrical neutrality in both compartments involved. An example of this type of response is found in the chloride-bicarbonate shift (p. 385).

The kidney plays an important role in electrolyte balance by regulating the excretion of inorganic ions, i.e., it can excrete more or less sodium ions, for example, depending on the status and requirements of the body. Other strategic ions, the excretion of which are at least partially under control of the kidney, are hydrogen, bicarbonate, and ammonium ions.

At times it may be necessary to administer electrolyte solutions intravenously in an attempt to aid the body in achieving or maintaining electrolyte balance. It is exceedingly important to stress the fact that before any intravenous therapy is started, thoroughly reliable chemical analysis must be performed to determine the exact electrolyte status of the patient. With the adequate safeguards, the use of intravenous electrolyte solutions often can be helpful. However, because of the interdependent relationships between water, electrolyte, and acid-base balance, and the direct and indirect effects of many body processes and mechanisms, there are many clinical situations in which maintaining the electrolyte balance becomes a most troublesome aspect of patient care.

ACID-BASE BALANCE

As noted before, the blood is in equilibrium with the interstitial fluid, which is in equilibrium with the intracellular fluid. Because these equilibria are attained within relatively short periods, it becomes possible to determine the acid-base status of an individual by making the proper measurements on the blood. It may also be added that it is much simpler to do so with blood than with interstitial fluid or individual cells. The normal pH of blood is about 7.4 with a narrow range of approximately ±0.1 pH unit. This means that the pH of the interstitial fluid must be close to 7.4. Furthermore, this pH must be maintained despite constant production of acids as a result of metabolic reactions.

Carbonic acid (H_2CO_3) is the principal acid produced in cellular oxidations in the amount of 10 to 20 moles per day. Other acids formed in much smaller amounts include pyruvic, lactic, β-hydroxybutyric, phosphoric, and sulfuric acids. Because metabolic reactions produce little in the way of alkaline substances to neutralize these acids, the strategic problem of acid-base balance is to move relatively large amounts of acids from the tissues, through the interstitial fluid, into the blood and to the lungs and kidneys for excretion—all with very little change in the normal pH of the various fluid compartments. To carry out this task, the body relies on buffer systems (p. 117); the lungs, which function in the excretion of carbon dioxide and control of carbonic acid concentration in the blood; and the kidneys, which function in the excretion of hydrogen and bicarbonate ions. In a passive way, an individual in a normal state of water balance also has dilution of the acids produced as a factor operating to keep pH changes at a minimum, whereas under conditions of dehydration, production of acid will have relatively greater effects.

Buffer systems in acid-base balance

As explained previously, buffer systems are capable of neutralizing acids or bases (p. 117). The chief mechanism in the body for keeping pH changes at a minimum involves the use of buffer systems. The major ones found in plasma and red cells are listed here, each consisting of a weak acid (H · A) and a salt of the weak acid (B · A):

$$\text{In plasma:}\ \frac{BHCO_3}{H_2CO_3} \quad \frac{B\cdot \text{protein}}{H\cdot \text{protein}} \quad \frac{B_2\cdot HPO_4}{B\cdot H_2PO_4}$$

$$\text{In red cells:}\ \frac{BHCO_3}{H_2CO_3} \quad \frac{B\cdot Hb}{H\cdot Hb} \quad \frac{B\cdot HbO_2}{H\cdot HbO_2} \quad \frac{B_2\cdot HPO_4}{B\cdot H_2PO_4}$$

Interstitial fluid lacks the protein and hemoglobin buffer systems, and intracellular fluid lacks the hemoglobin system. The systems of greatest quantitative significance are the bicarbonate system in blood plasma and red cells, the plasma protein system in plasma, and the hemoglobin systems in the red cell.

The efficiency and capacity of a particular buffer system depend on the ratio of the concentration of the salt to that of the acid. This can be seen from the very useful Henderson-Hasselbalch equation:

$$pH = pK + \log \frac{(\text{salt})}{(\text{acid})}$$

Because $\frac{(\text{salt})}{(\text{acid})} = 1$ when (salt) = (acid), and log 1 = 0,

$$\text{pK} = \text{pH of a buffer system when (salt)} = \text{(acid)}$$

The pK represents the midpoint of the range of efficient buffer action and is different for each buffer system. This range extends approximately two pH units in either direction from the pK.

Bicarbonate buffer system. The bicarbonate buffer system, in conjunction with the activities of the lungs and kidneys (see later), handles most of the metabolic production of acids (as carbon dioxide). The constituents of this buffer system would neutralize added acid or base as follows:

$$Na^+ + HCO_3^- + H^+ + Cl^- \rightarrow Na^+ + Cl^- + H_2CO_3$$

$$H_2CO_3 + Na^+ + OH^- \rightarrow Na^+ + HCO_3^- + H_2O$$

Note that in the production of carbonic acid and water, which prefer to remain un-ionized, the hydrogen and hydroxyl ions, which could affect the pH, have been removed from solution.

At the pH of the blood, which is 7.4, it can be determined from the Henderson-Hasselbalch equation, and verified by actual analysis, that the ratio of the concentrations of $BHCO_3$ to H_2CO_3 is 20:1 (pK of the bicarbonate buffer system is 6.1):

When pH = 7.4, and pK = 6.1

$$7.4 = 6.1 + \log \frac{(BHCO_3)}{(H_2CO_3)}$$

or

$$\log \frac{(BHCO_3)}{(H_2CO_3)} = 7.4 - 6.1 = 1.3$$

Because log 20 = 1.3,

$$\frac{(BHCO_3)}{(H_2CO_3)} = \frac{20}{1}$$

The actual concentrations found in normal plasma are 27 mEq per liter of $BHCO_3$ and 1.35 mEq per liter of H_2CO_3. Analytic determination of these substances can be important in assessing the acid-base status of an individual.

The lungs in acid-base balance

Slight increases in carbon dioxide or hydrogen ions in the blood cause large increases in pulmonary ventilation (rate and depth of breathing increase). As a result, the increased amount of carbon dioxide is excreted from the body as the following reactions are forced to the right:

$$H^+ + HCO_3^- \rightarrow H_2CO_3 \rightarrow H_2O + CO_2 \uparrow$$

These reactions also remove H^+ from the blood as un-ionized water molecules are formed. Decreases in blood CO_2 or H^+ cause the reverse responses, i.e., breathing rate and depth slow down in an attempt to retain CO_2 and thereby keep the concentration of H^+ at a higher level, as the reactions above are pushed to the left. The object of these responses is to keep the ratio of $(BHCO_3):(H_2CO_3)$ as close as possible to the normal ratio of 20:1. At such a point, the acid-base imbalance is termed *compensated.*

The kidneys in acid-base balance

The kidney, as part of its function in producing urine, is capable of reabsorbing certain ions and excreting others, and in some instances, can substitute one ion for another depending on the situation. On the basis of such mechanisms, the kidneys can produce urine of pH 4.5 to 8.2 at the extremes as conditions require the elimination of excess acid or base from the blood in order to maintain the normal pH of 7.4. Two of the more important aspects of the function of the kidneys in acid-base balance involve the control of HCO_3^- and Na^+ or H^+ excretion.

As the plasma H^+ concentration tends to rise, the glomerular filtrate (which becomes the urine) becomes more acid. This results in more HCO_3^- being converted to H_2CO_3, then to H_2O and CO_2, the CO_2 diffusing across the kidney tubule into the blood. In alkalosis, as the plasma pH rises, the reactions reverse themselves and there is less reabsorption, or more excretion, of HCO_3^-. In addition, as required by varying conditions, the kidney can substitute more or less H^+ for Na^+ to be excreted in the urine. Under normal conditions the urine is slightly acid. As the plasma pH decreases, more H^+ and less Na^+ go out in the urine and as the plasma pH increases, less H^+ and more Na^+ are excreted in the attempt to keep the pH of the blood plasma at the normal level of 7.4.

The mechanisms described represent only a small part of the complicated phenomena of the acid-base balance story with its direct connections with electrolyte and water balance. However, with the details discussed, and remembering that a large number of reactions are occurring at the same time, it should be easy to see how a small imbalance in one phase of the entire situation can cause serious difficulty in the remaining areas.

Abnormalities in acid-base balance

The various abnormalities in acid-base balance are illustrated with a simplified diagram in Fig. 27-5. Although compensatory mechanisms tend to alleviate these conditions, the buffer systems do have finite capacities that may sometimes be exceeded. The object of the compensatory mechanisms is to normalize the $(BHCO_3):(H_2CO_3)$ ratio after it has been changed from the usual 20:1 (see Henderson-Hasselbalch equation, p. 411).

Acidosis. Conditions of acidosis are characterized by a decrease in pH (increase in hydrogen ion concentration) in the blood:

1. *Metabolic acidosis.* This condition results from the metabolic production of an excess of acid substances. This would require more than normal amounts of HCO_3^- to remove excess H^+ ions as un-ionized molecules of H_2CO_3. As more HCO_3^- is used for this purpose, the $(BHCO_3):(H_2CO_3)$ ratio becomes smaller and the pH drops. To compensate for this, the lungs can increase the excretion of CO_2, which leads to a decrease in concentration of H_2CO_3 and the $(BHCO_3):(H_2CO_3)$ ratio returns to 20:1 (Fig. 27-5). Metabolic acidosis may occur in diabetic patients in whom the disorder in carbohydrate metabolism leads to increased production of acidic ketone bodies (p. 310).

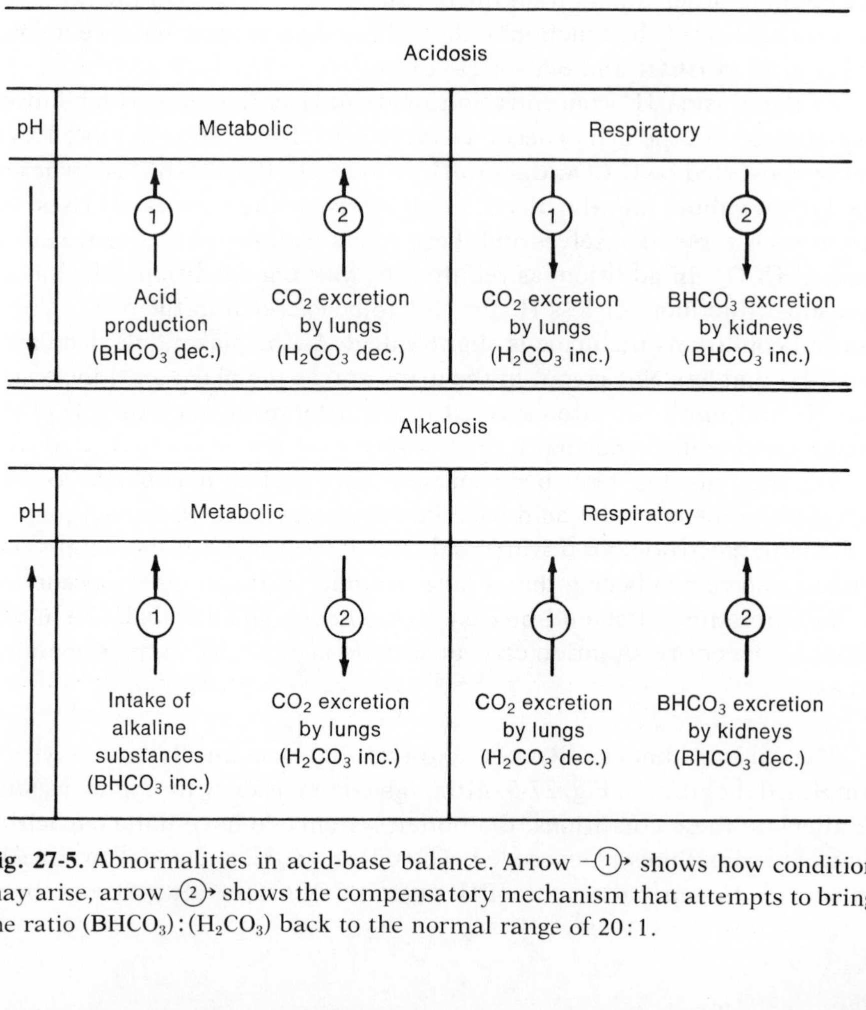

Acidosis

pH	Metabolic		Respiratory	
↓	↑ (1) Acid production ($BHCO_3$ dec.)	↑ (2) CO_2 excretion by lungs (H_2CO_3 dec.)	↓ (1) CO_2 excretion by lungs (H_2CO_3 inc.)	↓ (2) $BHCO_3$ excretion by kidneys ($BHCO_3$ inc.)

Alkalosis

pH	Metabolic		Respiratory	
↑	↑ (1) Intake of alkaline substances ($BHCO_3$ inc.)	↓ (2) CO_2 excretion by lungs (H_2CO_3 inc.)	↑ (1) CO_2 excretion by lungs (H_2CO_3 dec.)	↑ (2) $BHCO_3$ excretion by kidneys ($BHCO_3$ dec.)

Fig. 27-5. Abnormalities in acid-base balance. Arrow –(1)→ shows how condition may arise, arrow –(2)→ shows the compensatory mechanism that attempts to bring the ratio $(BHCO_3):(H_2CO_3)$ back to the normal range of 20:1.

2. *Respiratory acidosis.* This condition results from a lower than normal rate of loss of CO_2 from the lungs that may occur in cases of pulmonary edema or other situations in which gas exchange in the alveoli is decreased. This causes an increase in H_2CO_3 concentration that may be matched by an increase in $BHCO_3$ concentration resulting from decreased excretion of $BHCO_3$ by the kidney (Fig. 27-5).

Alkalosis. Conditions of alkalosis are characterized by an increase in pH (decrease in hydrogen ion concentration) in the blood:

1. *Metabolic alkalosis.* This condition is usually caused by the ingestion of an excess of alkaline substances, such as sodium bicarbonate ($NaHCO_3$). To compensate for the increase in $BHCO_3$ concentration, the concentration of H_2CO_3 can be increased as a result of decreased excretion of CO_2 by the lungs (Fig. 27-5).
2. *Respiratory alkalosis.* This condition can result from increased excretion of CO_2 from the lungs, leading to a decreased concentration of H_2CO_3. To compensate for this, the kidneys can excrete more $BHCO_3$ until the $(BHCO_3):(H_2CO_3)$ ratio once again returns to 20:1 (Fig. 27-5). Respiratory alkalosis may result from an increased breathing rate in cases of high fevers or as a result of psychologic disturbances.

SUMMARY

Water, electrolyte, and acid-base balance are closely related.

WATER BALANCE

Equilibrium between water intake and output. The major body compartments of water are the intracellular and interstitial fluids, and the blood plasma.

Intake and output of water: Water is taken in as drink, in food, or produced by oxidations in the body. Water is lost in urine, feces, sweat, and evaporation from skin and lungs (Fig. 27-1).

Water distribution system: Water is distributed to tissues by way of the blood and depends on differences between hydrostatic and colloid osmotic pressures at arterial and venous ends of capillaries (Fig. 27-2) and properties of semipermeable membranes.

Water uses: Water is required for reactions to take place, dissolves waste products for excretion, helps maintain normal body temperature, and is required for the production of digestive juices and fluids.

ELECTROLYTE BALANCE

Nomenclature and definitions: Positive ions are cations; negative ions are anions. Concentrations are expressed as mEq per liter. Electrolytes contribute to osmotic pressure in proportion to the number of particles (ions) produced on ionization.

Electrolyte composition of body fluids: (Fig. 27-3). Protein is found in intracellular fluid and blood plasma, in only very small amounts in interstitial fluid. Potassium ions found intracellularly, sodium ions interstitially. Chloride is major anion of plasma and interstitial fluid, not found intracellularly. Fig. 27-4 shows electrolyte composition of digestive juices and fluids. Major differences are the high concentration of hydrogen ion in gastric juice and bicarbonate ion in pancreatic juice. These differences are important in effects of persistent vomiting or diarrhea on maintenance of water, electrolyte, and acid-base balance.

Maintaining electrolyte balance: Key feature is selective permeability of cell membranes because all inorganic ions are small enough to pass through. Existence of a sodium pump has been proposed to explain how sodium ions are kept out and potassium ions are kept inside the tissue cells. Shift of electrolytes requires maintenance of electrical neutrality as in chloride-bicarbonate shift. Kidney plays an important role by regulating excretion of such ions as sodium, hydrogen, bicarbonate, ammonium. Before intravenous administration of electrolyte solutions, chemical analyses must be performed to determine exact electrolyte status of the patient.

ACID-BASE BALANCE

Acid-base status is determined by analysis of blood (normal pH is 7.4), which is in rapid equilibrium with other body fluids. Carbonic acid (H_2CO_3) is main acid produced in body. Strategic problem of acid-base balance is to eliminate acids with little or no change in normal pH of body fluids. This is done with the aid of buffer systems, the lungs, and the kidneys.

Buffer systems in acid-base balance: Buffer systems can neutralize added acid or base. Major buffer systems, consisting of weak acid ($H \cdot A$) and its salt ($B \cdot A$), are:

$$\text{In plasma: } \frac{BHCO_3}{H_2CO_3} \quad \frac{B \cdot \text{protein}}{H \cdot \text{protein}} \quad \frac{B_2 \cdot HPO_4}{B \cdot H_2PO_4}$$

$$\text{In red cells: } \frac{BHCO_3}{H_2CO_3} \quad \frac{B \cdot Hb}{H \cdot Hb} \quad \frac{B \cdot HbO_2}{H \cdot HbO_2} \quad \frac{B_2 \cdot HPO_4}{B \cdot H_2PO_4}$$

Similar systems appear in other body fluids. Efficiency and capacity depend on ratio $(B \cdot A):(H \cdot A)$ (see Henderson-Hasselbalch equation, p. 411).

Bicarbonate buffer system: Handles most of the metabolic production of acids (as CO_2). Normal ratio of $(BHCO_3):(H_2CO_3)$ in blood is 20:1 (27 mEq per liter $BHCO_3$: 1.35 mEq per liter H_2CO_3).

The lungs in acid-base balance: Increase or decrease CO_2 excretion and thereby H_2CO_3 or H^+ concentration:

$$H^+ + HCO_3^- \longleftrightarrow H_2CO_3 \longleftrightarrow H_2O + CO_2$$

The kidneys in acid-base balance: Increase or decrease excretion of HCO_3^-, Na^+, and H^+. The kidneys can produce urine in a pH range of 4.5 to 8.2 depending on requirements.

Abnormalities in acid-base balance: When the normal $(BHCO_3):(H_2CO_3)$ ratio is disturbed, the buffer systems, lungs, and kidneys combine to return the ratio to its normal value of 20:1. Fig. 27-5 is a simplified diagram of the situations as they occur in the acid-base abnormalities of metabolic acidosis and alkalosis, respiratory acidosis and alkalosis.

REVIEW QUESTIONS

1. What are the sources of water intake or production in the body? Output? What are some conditions under which intake and output may become unbalanced? Results? What are some of the mechanisms by which the body maintains water balance?
2. What is a cation? Anion? How is electrical neutrality ordinarily maintained inside and outside cells? How is it possible for cells to maintain higher concentrations of potassium ions inside as compared to outside of cells and the reverse for sodium ions?
3. What is a buffer system? Explain how it functions (with equations, if possible). What are some of the more important buffer systems of the body? What is the approximate effective range of a buffer system? Since the pKs (Henderson-Hasselbalch equation) of the usual buffer systems are known, what other information is required to determine the pH of a particular buffer system?
4. Explain the mechanisms by which the kidneys and lungs affect water, electrolyte, and acid-base balance. What effect may excessive vomiting or prolonged diarrhea have on water, electrolyte, and acid-base balance? Why?
5. What is meant by the term *metabolic acidosis?* How do the lungs function to correct this situation? In metabolic alkalosis? How do the kidneys function to correct the situations of respiratory acidosis and respiratory alkalosis? How do these actions relate to the Henderson-Hasselbalch equation?

REFERENCES

Christensen, H. N.: Neutrality control in the living organism, Philadelphia, 1971, W. B. Saunders Co.

Christensen, H. N.: Body fluids and the acid-base balance, Philadelphia, 1964, W. B. Saunders Co.

Christensen, H. N.: pH and dissociation, Philadelphia, 1964, W. B. Saunders Co.

Davenport, H. W.: The ABC of acid-base chemistry: the elements of physiological blood-gas chemistry for medical students and physicians, ed. 6, Chicago, 1974, University of Chicago Press.

Masoro, E. J., and Siegel, P. D.: Acid-base regulation: its physiology, pathophysiology, and the interpretation of blood-gas analysis, Philadelphia, 1978, W. B. Saunders Co.

Quintero, J. A.: Acid-base balance: a manual for clinicians, St. Louis, 1978, Warren H. Green, Inc.

Schwartz, A. B., and Lyons, H., editors: Acid-base balance and electrolyte balance: normal regulation and clinical disorders, New York, 1977, Grune & Stratton, Inc.

White, W. L., Erickson, M. M., and Stevens, S. C.: Chemistry for the clinical laboratory, ed. 4, St. Louis, 1976, The C. V. Mosby Co.

28 Vitamins

Vitamins are recognized as organic compounds, usually of low molecular weight, which are generally required in the diet in rather small amounts for normal growth, maintenance, and reproduction. In the complete absence of a vitamin, particularly any of those with an absolute dietary requirement, clinical conditions known as *deficiency diseases* develop, sometimes with fatal consequences. In contrast to other food substances, vitamins are not used for structural or energy requirements or as raw material for synthesizing other compounds. Although the exact functions of a number of vitamins have not been established yet, all of the B group vitamins are now known to act as coenzymes in various metabolic reactions.

HISTORIC DEVELOPMENT OF VITAMIN THEORY

The first known recognition of a vitamin deficiency disease apparently dates back to 2600 BC when, it is believed, the Chinese were aware of beriberi (vitamin B_1 deficiency). Night blindness (vitamin A deficiency) and its cure, the administration of liver, are included in a famous piece of recorded history, the Ebers papyrus, which is dated as 1600 BC. In spite of such early beginnings, it was not until the last few hundred years that much progress was made in this field—mostly within the last sixty years. The most disheartening aspect of this long struggle for knowledge is the fact that, in addition to overcoming many great and unavoidable technical difficulties, many important discoveries had to be made independently more than once before the results or ideas were generally accepted. Establishing the hypothesis that these and other diseases could result from the dietary deficiency of small amounts of certain substances, rather than from the work of evil spirits or infectious bacteria, was a difficult task.

In the search for causes of beriberi, nothing of importance occurred from 2600 BC until 1882 when Kanehiro Takaki, a Japanese navy physician, found that the condition could be cured by adding vegetable, fish, meat, and barley to the diet. He believed that insufficient protein in the diet was the cause of beriberi. In studies of the 1890s, Christiaan Eijkman, a Dutchman, accidentally discovered that hens fed polished rice would develop beriberi. He also found that they could be cured by adding back to the diet the bran that had been removed, or an extract of the bran. From these results he concluded that the bran contained a curative agent, not a missing factor. A few years later (1901), Gerrit Grijns, a colleague, suggested for the first time that a dietary lack was the cause of beriberi.

Scurvy (vitamin C deficiency) was the most dreaded companion on the early exploratory sea trips. The first recorded outbreak of scurvy occurred during Vasco da Gama's voyage from Spain to India in 1497. Most of the crew died on this trip as on other similar trips of more than a few months' duration. The great success of the British East India Company (maritime traders) was probably mainly because they supplied citrus fruits on their ships—as suggested by Sir James Lancaster in 1601. As a result of the use of limes, oranges, and lemons, British sailors became known as "limeys." This simple expedient was rediscovered in 1720 by Johann Kramer, an Austrian army physician who reported that consumption of orange or lime juice by itself would cure scurvy. In 1753 a complete treatise by James Lind on scurvy, including its simple cure, was published, but scurvy remained a serious problem until more recent times. In 1907, scurvy was accidentally produced in guinea pigs by A. Holst and T. Frölich, Norwegians.

Rickets (vitamin D deficiency) was first described by Daniel Whistler, a British physician, in 1645 in a thesis for the doctorate of medicine. In 1782, the successful use of cod-liver oil to cure two cases of rickets was reported by Robert Darley, an Englishman. This was rediscovered in 1865 by A. Trousseau, a Frenchman, who described the use of cod-liver oil as the perfect cure for rickets. This disease was produced in puppies in 1838 by Jules Guérin, a French surgeon, and, in 1889, cod-liver oil was used to cure rickets in lion cubs at the London Zoo by Sir John Bland-Sutton. Again, in spite of these observations, rickets remained a serious problem until recent years.

With the availability of animals that developed the same deficiency diseases as humans, the rate of progress in nutritional experiments in this field increased. In the early 1900s, famous men (including several Nobel prize winners) in addition to those already mentioned, such as W. Stepp, Sir Frederick G. Hopkins, Thomas B. Osborne, Lafayette B. Mendel, Elmer V. McCollum, Casimir Funk, and others, gradually demonstrated that purified diets of proteins, fats, carbohydrates, and salts, then known to be required for normal growth, were still missing certain factors that were found in natural foods. In two landmark publications (1911, 1912) Casimir Funk, a Polish scientist working in England, proposed the theory that diseases such as beriberi, scurvy, pellagra, and rickets were "caused by a deficiency of some essential substances in the food." As a result of his work on the beriberi factor, which he concluded was an amine, he coined the term *vitamine.* Because work on other factors showed that an amine structure was not common to all of these substances, the final "e" was dropped, leaving the word *vitamin,* as it is now used.

Although the vitamin theory was not readily accepted, research in this field flourished in the 1920s and 1930s with the discovery of more vitamins, determination of their chemical structure, and laboratory synthesis of most of the vitamins. There are few people who have not heard of vitamins and their importance to health, but, on the other hand, deficiency diseases still crop up in the most affluent societies, as well as in undernourished areas of the world.

The purpose of this short, historic excursion is to emphasize several important points: (1) scientific progress generally does not occur in a simple, straight line; (2) new ideas are difficult to establish; (3) the rate of progress in research in the health sciences is faster when the same disease that occurs in humans also occurs in animals that can be used in experimentation; and (4) although the vitamin story may appear to be a closed book in some respects, the contribution of nutrition research to the well-being of humanity is only at the beginning of its potentialities.

GENERAL CONSIDERATIONS

Classification

For classification purposes the vitamins are generally divided on the basis of solubility in water or lipid solvents. The fat-soluble vitamins are A, D, E, and K. The water-soluble vitamins are C and the B group, including B_1, B_2, B_6, B_{12}, niacin, pantothenic acid, biotin, and folic acid. The individual vitamins will be discussed in the order listed.

Vitamin assays

It is sometimes necessary to determine the vitamin status of an individual or whether or not a particular condition will respond favorably to the administration of a single vitamin. A favorable response to a single vitamin was the original assay method for vitamin research. Although this is the ultimate test for effectiveness of a vitamin, it is also time-consuming and costly because in many instances it takes a long period to develop a desired deficiency in an animal.

The search for faster, simpler, less expensive assay methods has led to the development of a number of useful techniques. In some instances it may be possible to isolate a pure sample of the vitamin and identify it by its chemical and physical properties. In other cases, a microorganism may be found that cannot grow in the absence of a particular vitamin. Its rate of growth in the presence of the vitamin can then be used to determine the concentration of the vitamin in the test solution.

The general methods just listed, and other specialized procedures, may be used to determine the concentration of a vitamin in the blood or urine of an individual. This forms the basis of a vitamin saturation determination. If the individual is deficient in a particular vitamin, the tissues may take up more of the vitamin than under normal conditions, and therefore the concentration of the vitamin in the blood or urine will be decreased. Vitamin deficiencies may also be determined on the basis of tissue examinations or metabolic reactions in which substances are missing or piling up as a result of the missing vitamin. Because of the great difference in properties and mode of action of the vitamins, many unique procedures exist in this field.

Vitamin requirements

The dietary requirements for some vitamins are absolute, i.e., the vitamin must be supplied in the food because it cannot be obtained any other way. In other cases, some may be synthesized in the body while the remainder is obtained in the diet. It is also possible, in some instances, to obtain the vitamin as a result of bacterial activity in the small intestine. In the case of vitamin B_{12}, an accessory substance, intrinsic factor (p. 434), is

Table 28-1. Vitamin requirements of adults

Vitamin	Recommended daily requirement*	How met
A	4000-5000 IU	Required in food
D	400 IU†	Synthesized in body
E	8-10 mg αTE‡	Average diet contains about 15 mg
K	70-140 μg	Synthesized by intestinal bacteria
C	50-60 mg	Required in food
B_1	1.0-1.5 mg	Required in food
B_2	1.2-1.7 mg	Required in food
B_6	1.8-2.2 mg	In food
B_{12}	3 μg	In food
Niacin	13-18 mg	Amount required in food depends on tryptophan content
Pantothenic acid	4-7 mg	In food, possibly intestinal synthesis
Biotin	Not known; 100-200 μg§	Synthesized by intestinal bacteria
Folic acid	Not known; 400 μg§	May be synthesized by intestinal bacteria

*Recommended dietary allowances, revised 1980, Food and Nutrition Board, National Academy of Sciences–National Research Council (US).
†400 IU recommended during skeletal growth, thereafter not required in diet.
‡α-tocopheral equivalents, 1 mg d-α-tocopherol = 1 mg αTE.
§Recommended daily intake. Recommended daily requirement is not known.

required in addition to the vitamin itself. The specific information for each vitamin is listed in Table 28-1.

The actual quantitative requirement for each vitamin is usually listed as the recommended dietary allowance (RDA), or recommended intake in instances in which the RDA has not been accurately determined. These amounts for the individual vitamins are included in Table 28-1.

Absorption problems

Only a part of the problem has been met with the intake of a diet containing the necessary amounts of the required vitamins. To be of use in metabolism, the vitamins must be properly absorbed. Therefore any condition that interferes with the processes of digestion and absorption may lead to difficulties with one or more vitamins. For example, obstruction of the bile duct stops the flow of bile into the intestine and interferes with the absorption of fats and also the fat-soluble vitamins. Under these conditions, a deficiency of vitamin K may develop rapidly (p. 428). Drug treatments that decrease the growth of intestinal bacteria or change the balance of growth may also cause some difficulties. Individuals with generalized inflammations of the absorptive surfaces of the small intestine must be watched closely for the possibility of developing vitamin deficiencies. Vitamin B_{12} must have a supply of intrinsic factor available in order to be absorbed (p. 434).

Storage organs

Some vitamins are stored in the body to some extent so that a deficiency does not develop immediately during periods of inadequate intake. The liver is the storage organ for vitamins A, D, B_{12}, and folic acid. The white cells may act as the storage organ for vitamin C, several months of inadequate intake being required for a deficiency to develop. The white cells may also act as the storage organ for vitamin B_2. Storage organs have not been detected for the remaining vitamins.

Increased requirements

At times, there may arise a need for a greater than normal amount of one or more vitamins. It is generally agreed that the requirements for most vitamins go up during pregnancy and lactation. Growing children may require additional vitamins because of metabolic needs and aging adults may require increased amounts because of absorption problems. Additional amounts of vitamins may also be required during periods of convalescence. The absence of one vitamin may increase the requirement for one or more other vitamins.

Excessive intake

Past experience has indicated that very large intakes of vitamins are harmless except for vitamins A and D. The most serious symptom is the deposition of calcium in tissues resulting from extremely large amounts of vitamin D. The symptoms generally disappear when the excessive intake is stopped.

Reports have appeared indicating possible toxicity of excessive quantities of vitamin E, also a fat-soluble vitamin as are A and D. However, humans appear to have tolerance for large amounts of vitamin E.

In recent years, many people have been ingesting huge doses of vitamin C (p. 430), a water-soluble vitamin, apparently without any ill effects. However, several investigators have reported that some of these individuals have turned up with complications resulting from the deposition of crystals of oxalic acid (a breakdown product of vitamin C, ascorbic acid) in their kidneys.

FAT-SOLUBLE VITAMINS

Vitamin A

Deficiency manifestations. The most specific symptom of a vitamin A deficiency is night blindness. Characteristic dermatoses (skin conditions) and *xerophthalmia,* or dry eyes (Fig. 28-1), also develop, and growth of bones and teeth is retarded. As a result of these and other less specific effects, growth, in general, is reduced or stopped.

Chemistry. Vitamin A occurs in many plants in the form of the provitamins α-, β-, and γ-carotene, and cryptoxanthin, which are converted to vitamin A to a varying extent in the body (liver). In the conversion of β-carotene, which has a symmetrical structure, the molecule splits in half with the possible production of two molecules of vitamin A, the structure of which is shown here:

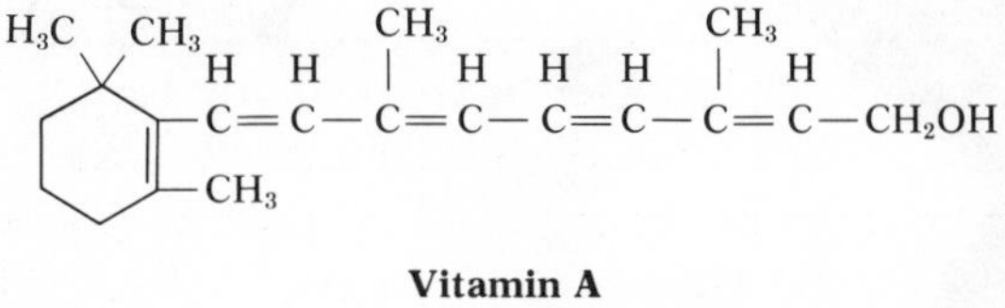

$$H_3C \quad CH_3 \qquad CH_3 \qquad CH_3$$
$$\text{(ring)}-\overset{H}{C}=\overset{H}{C}-\overset{CH_3}{C}=\overset{H}{C}-\overset{H}{C}=\overset{H}{C}-\overset{CH_3}{C}=\overset{H}{C}-CH_2OH \quad (\text{ring})-CH_3$$

Vitamin A

Vitamin A is relatively stable to heat but is readily oxidized. A synthetic product is available for use by individuals allergic to fish products.

Function. Vitamin A in the aldehyde form, in combination with a specific protein opsin, makes up rhodopsin (visual purple) a photosensitive pigment that functions in the visual process in the retina. Light bleaches rhodopsin as vitamin A aldehyde is converted to the alcohol. In the recovery process, the alcohol is reconverted to the aldehyde and rhodopsin is regenerated. In night blindness, the regeneration of rhodopsin is reduced.

Vitamin A also functions in maintenance of normal epithelial tissue and normal growth of bones and teeth but how is not known. Vitamin A became known as an antiinfective vitamin because in its function in main-

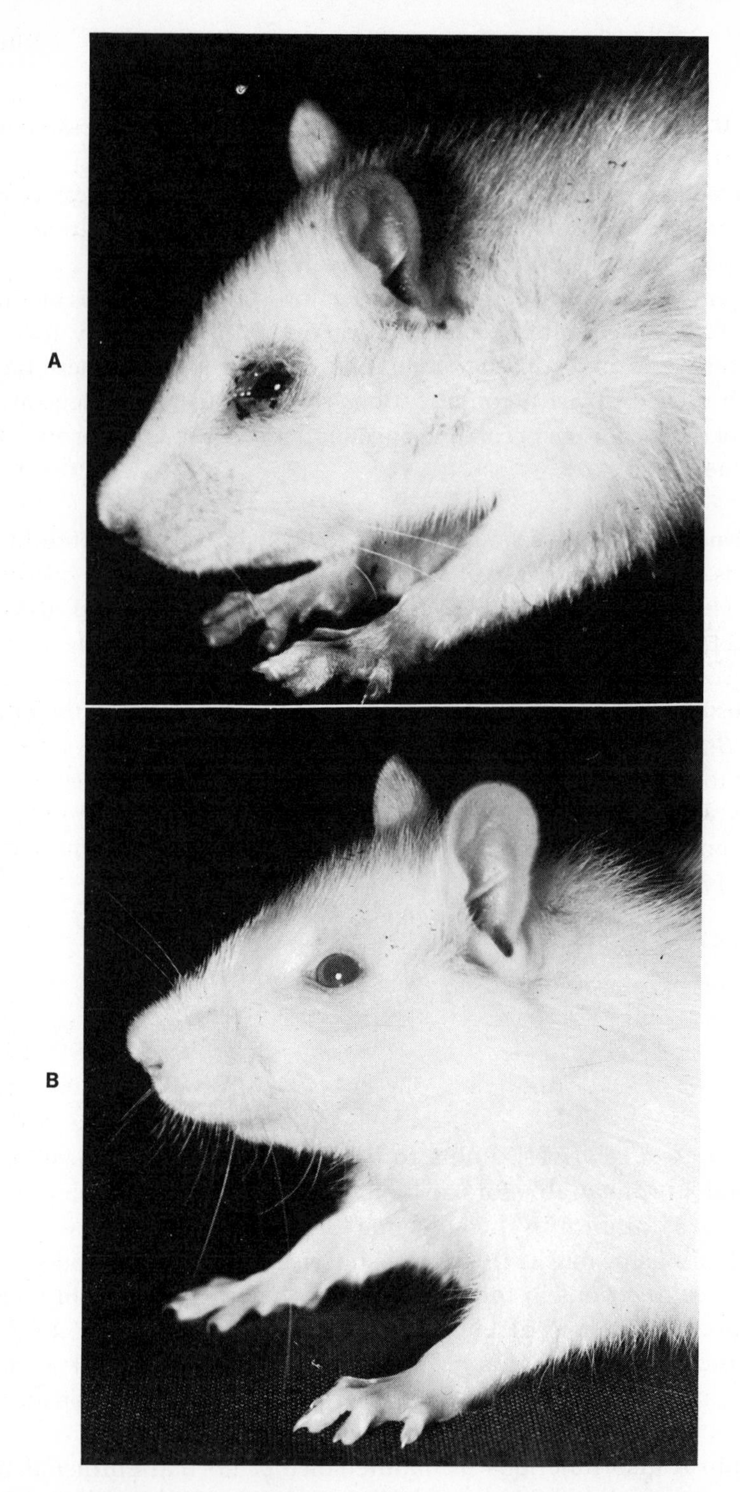

Fig. 28-1. A, Xerophthalmia, dryness of the cornea, and conjunctiva resulting from vitamin A deficiency. **B,** Same rat after 8 days of vitamin A therapy. (From Vitamin manual. Courtesy the Upjohn Co.)

taining normal skin and mucous membranes, the healthy tissues ward off infections better than is possible under deficiency conditions.

Food sources. As carotenoid precursors, vitamin A is found in the yellow pigments of fruits and vegetables. As preformed vitamin A, the richest source is fish-liver oil.

Requirements. The recommended daily allowance (National Academy of Sciences–National Research Council, Table 28-1) for adults is 4000 to 5000 IU of vitamin A. Excessive intake of vitamin A may be harmful (p. 422).

Vitamin D

Deficiency manifestations. In growing children during the period of active skeletal growth, vitamin D deficiency causes *rickets,* which results from defective mineralization of the ends of the growing bones. As a consequence, the ends remain abnormally pliable and eventually assume a bent form, the condition of bowlegs being a characteristic manifestation of this situation (Fig. 28-2). Other deformities characteristic of vitamin D deficiency include "rachitic rosary," "pigeon" or "chicken breast," and "potbelly."

In adults, with fully grown bones, vitamin D deficiency results in defective mineralization of osteoid tissue, a condition known as *osteomalacia,* which leads to the softening of bone structure.

With a vitamin D deficiency, secretion of parathyroid hormone is increased and the level of calcium in the serum is maintained by demineralization of bone (p. 460). A polypeptide hormone, calcitonin, produced in the thyroid, acts to lower the serum calcium level by inhibiting the release of calcium from bone. This appears to be another factor in the control of calcium metabolism.

Chemistry. Two important forms of vitamin D are known. Vitamin D_2, calciferol, is produced by the irradiation with ultraviolet light of ergosterol (provitamin D_2), a sterol found in ergot mold. Vitamin D_3, cholecalciferol, is formed by the irradiation in the skin of 7-dehydrocholesterol (provitamin D_3), formed elsewhere in the body. The structures of the two forms of vitamin D, which differ only slightly in the large, hydrocarbon side chain, are shown here:

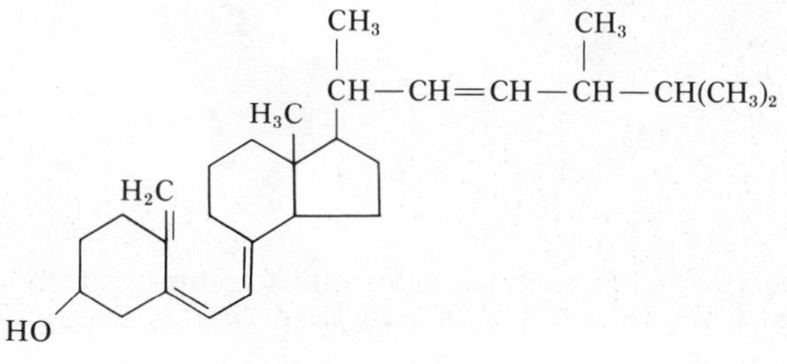

CH_3 CH_3
$CH-CH=CH-CH-CH(CH_3)_2$
H_3C
H_2C
HO

Vitamin D_2
(Calciferol; activated ergosterol)

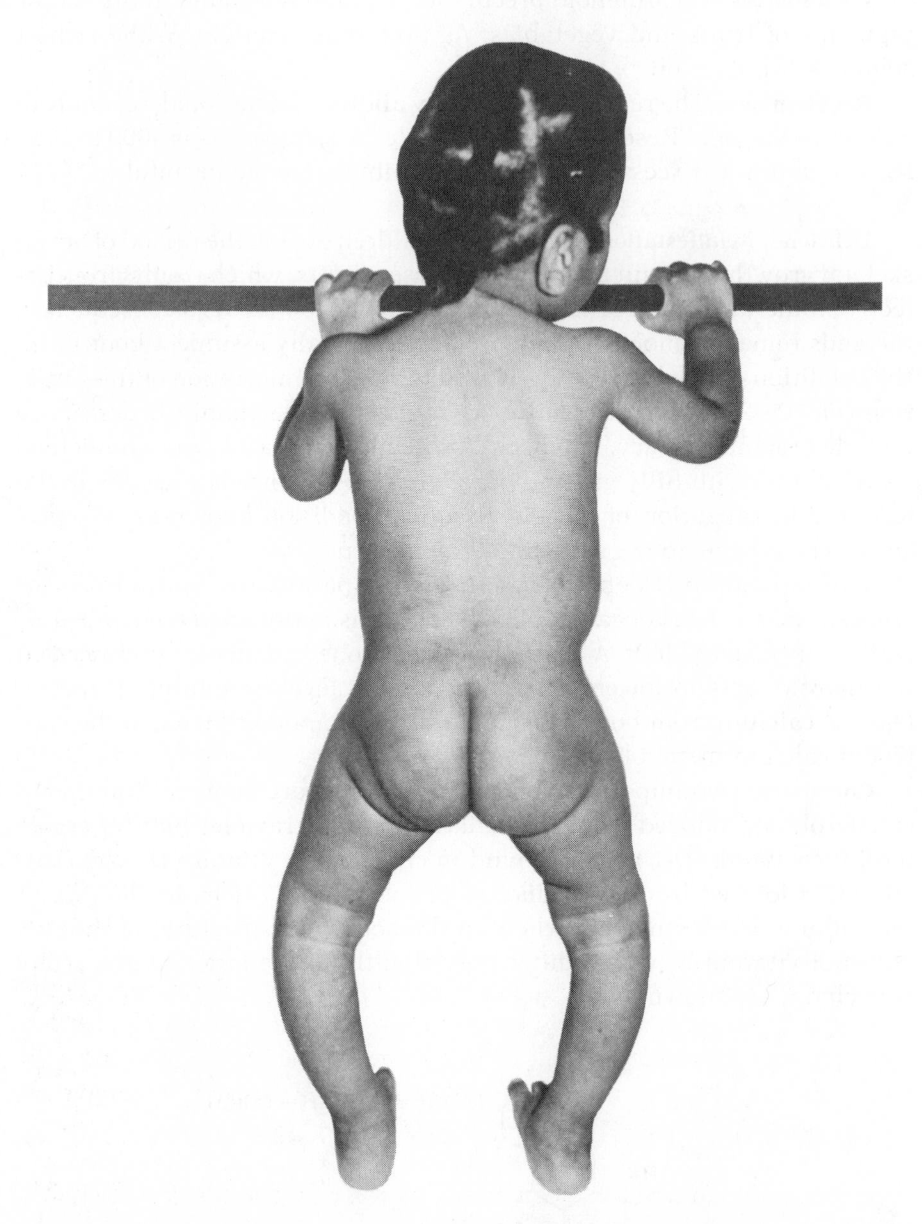

Fig. 28-2. Rickets showing skeletal deformities resulting from vitamin D deficiency. (From Nemir, R. L., New York University, Bellevue Medical Center. Courtesy the Upjohn Co.)

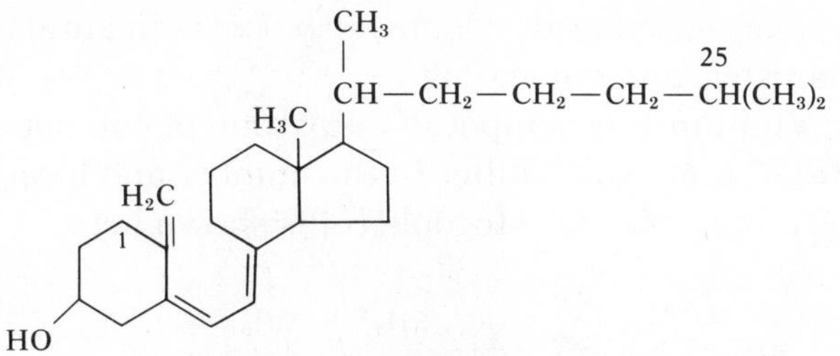

Vitamin D_3
(Cholecalciferol; activated 7-dehydrocholesterol)

Early in the history of vitamin D, it became known as the "sunshine vitamin" because of the lower prevalence of rickets in sunny countries.

Function. Vitamin D aids in the absorption of calcium from the small intestine and subsequently with the formation of normal bone. The specific manner in which these functions are carried out is still the subject of intensive research. Although calcium and phosphate absorption and metabolism appear to be synchronized, the details of exactly how vitamin D metabolites, parathyroid hormone, and calcitonin interact in these processes remain under active investigation.

The active form of vitamin D_3 in animals appears to be an hydroxylated product, 1,25-dihydroxycholecalciferol, which is derived from vitamin D_3 by hydroxylation in the liver to the 25-hydroxy compound and then to the 1,25-dihydroxy compound in the kidney:

$$\text{Cholecalciferol} \xrightarrow{\text{liver}} \text{25-hydroxycholecalciferol} \xrightarrow{\text{kidney}} \text{1,25-dihydroxycholecalciferol}$$

This explains some of the clinical difficulties in calcium metabolism that may occur with liver or kidney diseases.

Food sources. Fish-liver oils are the best natural source of vitamin D. Other sources include milk, butter, and oleomargarine, which are commonly supplemented with added vitamin D.

Requirements. Because of the synthesis of vitamin D that occurs in the body, adults on a normal diet do not require dietary sources. The recommended daily allowance (Table 28-1) for children is 400 IU. The same allowance is recommended for pregnant or lactating females. Excessive intake of vitamin D may be harmful (p. 422).

Vitamin E

Deficiency manifestations. Definite effects of a vitamin E deficiency in humans are not known. In rats, sterility of the male occurs and becomes permanent if the vitamin E deficiency continues. Resorption of the fetus occurs in the female, but normal litters may result if vitamin E is restored to the diet. A type of muscular dystrophy also occurs in animals such as the rat, rabbit, hamster, and guinea pig.

Chemistry. Vitamin E is composed of a group of compounds known as tocopherols (α, β, γ, δ), which differ in the number and location of methyl side chains. The structure of α-tocopherol is shown here:

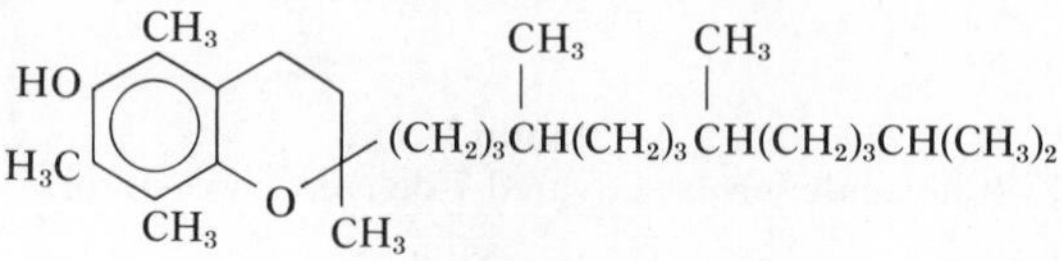

$$\text{HO, } H_3C, CH_3, CH_3 \text{ (chromane ring, O)}, CH_3-(CH_2)_3\overset{CH_3}{\overset{|}{C}H}(CH_2)_3\overset{CH_3}{\overset{|}{C}H}(CH_2)_3CH(CH_3)_2$$

α-Tocopherol

The tocopherols are very susceptible to oxidation and thereby protect other substances from oxidation. Because of this property, the tocopherols are used commercially as inhibitors of oxidation in fats.

Function. There is no definitely established function in humans. In animals, vitamin E functions in maintaining normal reproductive processes and muscle metabolism. Studies have been reported attempting to relate vitamin E to the production of hemoglobin and normal functioning of red cells, but the evidence for its requirement in humans has not yet been generally accepted. However, there have been several reports describing a hemolytic anemia in premature infants that responded completely to treatment with α-tocopherol.

Food sources. Vitamin E is widely distributed in the plant world, particularly in green, leafy vegetables and wheat-germ oil. It is also readily synthesized.

Requirement. The recommended dietary requirement is 8 to 10 mg α-tocopherol equivalents (Table 28-1). Although not as well established clinically as with vitamins A and D, there have been some reports of possibly harmful effects of excessive intake of vitamin E (p. 423).

Vitamin K

Deficiency manifestations. The most important symptom of a vitamin K deficiency is a prolonged clotting time that may lead to internal hemorrhaging and uncontrolled bleeding. Because of its production by intestinal bacteria, a vitamin K deficiency in humans rarely occurs for dietary reasons. It may, however, occur as a result of faulty fat absorption (obstruction of bile duct) or drug treatments that retard the growth of intestinal bacteria (p. 422).

Chemistry. Several compounds with vitamin K activity are known. All are derivatives of 1,5-naphthoquinone (condensed ring structure) and differ only in the composition of the long hydrocarbon side chain. The structure of vitamin K_1 is shown here:

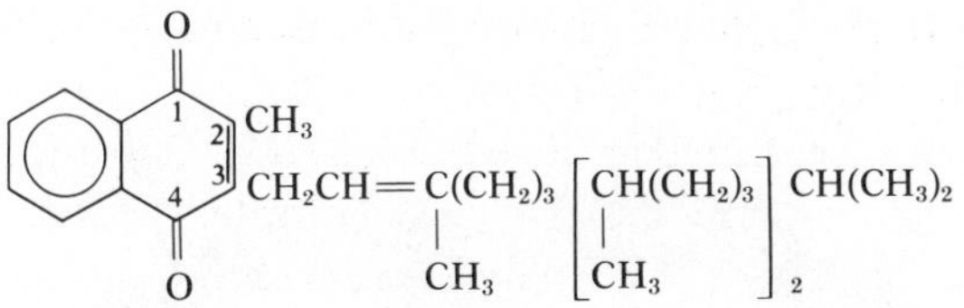

Vitamin K_1

Function. Vitamin K is necessary for the normal production of prothrombin and other factors required for the clotting of blood. A naturally occurring substance, dicumarol, acts as a metabolic antagonist of vitamin K, thereby functioning as an anticoagulant (p. 372). The structure of dicumarol is shown here:

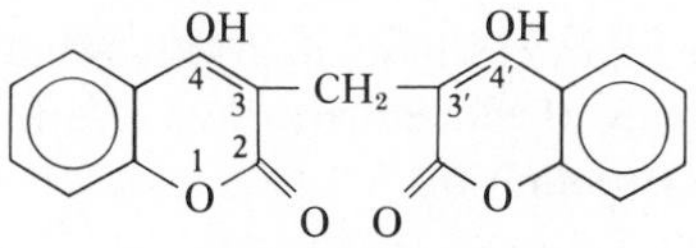

Dicumarol

When split in half, the fragments of dicumarol resemble vitamin K in structure, this being the probable basis of the antagonism.

Recent research results point to the possibility that vitamin K may be involved in the process of bone formation. Vitamin K is required in a reaction in which a second carboxyl group is put on the γ carbon atom of glutamic acid residues (p. 217) already in peptide chains. The resulting amino acid is called γ-carboxyglutamic acid. This increases the capacity of such peptide chains to bind calcium ions, which are necessary for bone formation. A γ-carboxyglutamic acid–containing protein associated with bone formation has been isolated and named *osteocalcin.*

Vitamin K, in the form of coenzyme Q, is also believed to be involved in oxidative phosphorylation as an electron carrier or as a coupling agent in the production of ATP.

Food sources. Vitamin K is found in alfalfa and green, leafy vegetables. A dietary source is usually of little importance because of its synthesis by intestinal bacteria.

Requirement. From clinical results, the requirement for vitamin K appears to be small—in the range of several milligrams. Newborn babies, because of the lack of intestinal bacteria, may develop hemorrhagic difficulties. To avoid this, the mother is generally given additional vitamin K in anticipation of the birth of the baby or it may be given directly to the newborn baby.

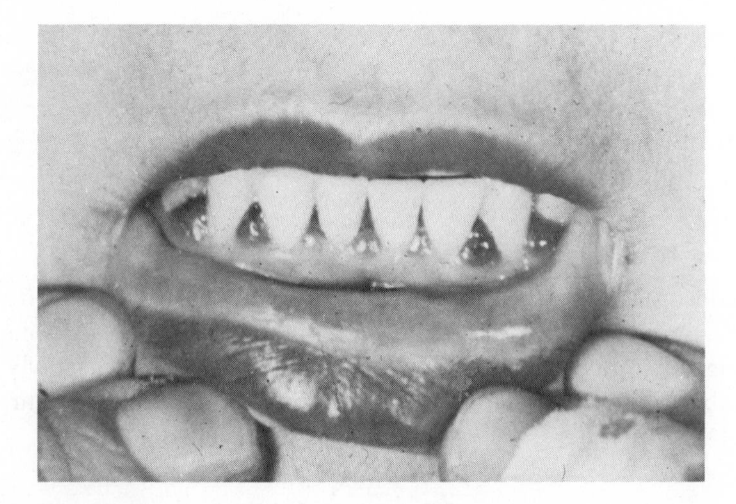

Fig. 28-3. Scurvy with the characteristic gingivitis—one of the symptoms resulting from vitamin C deficiency. (From Darby, W. J., Vanderbilt University, School of Medicine. Courtesy the Upjohn Co.)

WATER-SOLUBLE VITAMINS

Vitamin C (ascorbic acid)

Deficiency manifestations. A vitamin C deficiency in humans over a period of 4 to 6 months results in a condition known as *scurvy*. Among the difficulties encountered in scurvy are internal hemorrhaging, bleeding of the gums, loosening and falling out of teeth, and defective bone and tooth formation (Fig. 28-3). Unless the diet is supplemented with vitamin C, death results.

Chemistry. Ascorbic acid is a strong acid with a structure similar to that of glucose. As in those animals that can do so, ascorbic acid is synthesized from glucose. Human, monkey, and guinea pig lack certain enzymes and therefore must obtain ascorbic acid in the diet. Ascorbic acid readily undergoes a reversible oxidation-reduction reaction with dehydroascorbic acid as shown here:

```
  OC ──┐                       OC ──┐
  |    |                       |    |
 HOC   |                       OC   |
  ||   O                       |    O
 HOC   |        −2H            OC   |
  |    |      ⇌                |    |
  HC ──┘        +2H            HC ──┘
  |                            |
 HOCH                         HOCH
  |                            |
  CH2OH                        CH2OH
```

L-Ascorbic acid ⇌ **Dehydroascorbic acid**

Ascorbic acid is rapidly destroyed by heating in the presence of oxygen and trace metals encountered in cooking utensils.

Function. The specific function of ascorbic acid in metabolic processes is not known. There has been much speculation that its function must be related to its oxidation-reduction properties but there is no conclusive evidence to support these ideas. Ascorbic acid is apparently necessary for the normal formation of collagen fibers and mucopolysaccharides of connective, bone, and tooth tissue and the intercellular "cement substance" of capillaries. Most of the manifestations of scurvy—hemorrhaging and abnormal formation of bones and teeth—are related to this function. A capillary fragility test is sometimes used to detect scurvy in its early stages.

There has long been an interest in the possibility that ascorbic acid in relatively large doses may act as a preventative for the common cold in human beings. Recent studies along these lines have not borne out this contention. In fact, some individuals ingesting these large doses (gram quantities) have developed kidney trouble (p. 423).

Food sources. Citrus fruits are the best natural source of ascorbic acid. It is also synthesized commercially.

Requirement. The recommended daily allowance (Table 28-1) for adults is 50 to 60 mg of vitamin C.

The B group vitamins

One important reason for grouping together the following vitamins in what is known as the B group is that deficiency diseases involving these vitamins are usually multiple in nature. The natural sources of these vitamins are similar and therefore a diet lacking one of them usually is lacking in others of this group. As a result, the symptoms of deficiency are usually not very specific, i.e., they are not definitely correlated with a single, specific missing vitamin.

Vitamin B_1 (thiamine)

Deficiency manifestations. A vitamin B_1 deficiency results in a disease known as beriberi, which is characterized by cardiovascular and neurologic difficulties, sometimes accompanied by edema. Biochemically, a thiamine deficiency manifests itself by increased concentrations of pyruvic and lactate acids in the blood, resulting from decreased levels of thiamine and cocarboxylase, which are involved in the further conversions of pyruvic and lactic acids.

Chemistry. The structure of thiamine contains two heterocyclic rings, a pyrimidine ring (p. 178) and a thiazole ring (p. 179), joined by a single carbon atom, as shown on the top of the next page:

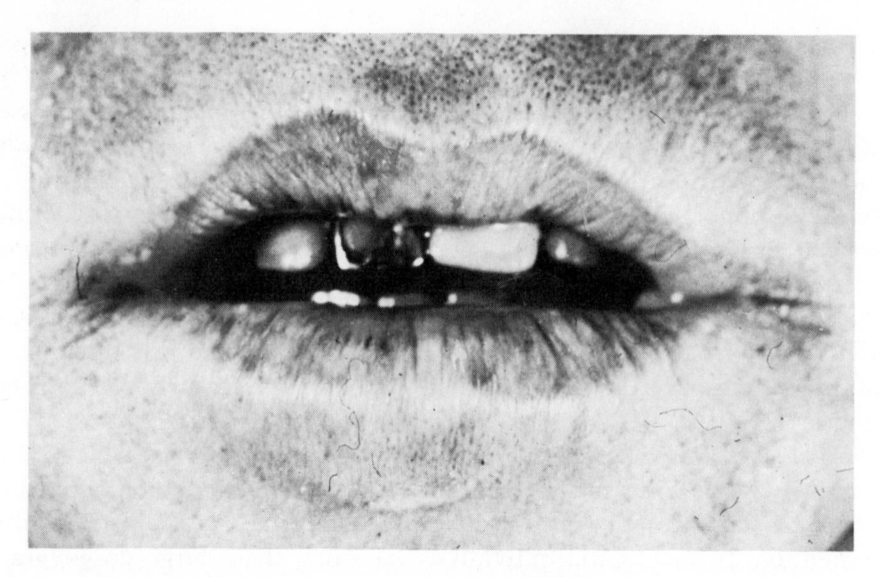

Fig. 28-4. Cheilosis (angular stomatitis) showing characteristic fissures at the sides of the mouth resulting from riboflavin (vitamin B_2) deficiency. (From Horwitt, M. K., St. Louis University, School of Medicine.)

```
              N        NH2    S
            /   \\    /      /   \
  H3C — C         C       HC       C — CH2 — CH2OH
        ||        |       ||       ||
        N         C       N ——— C — CH3
          \     //  \    / +
              C        C     Cl-
              H        H2
```

Thiamine chloride

Because of its high water solubility, most of the vitamin B_1 available in foods may be lost in the water used in cooking. The presence of an amine group in this compound led Funk to use the term "vitamine" (p. 420).

Function. As thiamine pyrophosphate, the active form of vitamin B_1 is a coenzyme (cocarboxylase) in the decarboxylations of the α-keto acids, pyruvic and α-ketoglutaric acids (pp. 302, 303). Increased concentrations of pyruvic acid in blood are looked for in suspected cases of thiamine deficiency.

Food sources. The best food sources of vitamin B_1 are plant products such as beans, nuts, cereal grain germ layers, and certain fruits. Polished cereal grains lose their vitamin B_1 content. Synthetic thiamine is added back to white flour used in baking bread and other products.

Requirement. The recommended daily allowance (Table 28-1) for adults is 1.0 to 1.5 mg of thiamine.

Vitamin B_2 (riboflavin)

Deficiency manifestations. The most specific symptom of a vitamin B_2 deficiency is the occurrence of painful fissures at the angles of the mouth (cheilosis or angular stomatitis, Fig. 28-4). The tongue, nose, and eyes are also involved and a dermatitis around these areas also occurs.

Chemistry. Riboflavin contains a ribitol group (the alcohol corresponding to ribose) attached to a substituted isoalloxazine condensed ring structure as shown here:

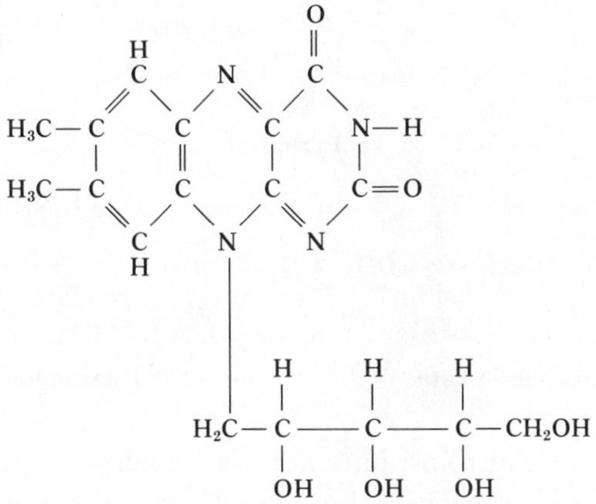

```
                                  O
                                  ‖
             H                    C
             C         N        /   \
          //   \     /   \\   /       \
H3C— C           C          C          N—H
     |           ‖          |          |
H3C— C           C          C          C=O
        \\    /     \     /   \\     /
             C         N          N
             H         |
                       |
                       |      H       H       H
                       |      |       |       |
                      H2C— C —————— C —————— C—CH2OH
                              |       |       |
                             OH      OH      OH
```

Riboflavin

The active forms are phosphorylated derivatives, riboflavin phosphate, or flavin mononucleotide (FMN) and flavin adenine dinucleotide (FAD, p. 346).

Function. As the mono- and dinucleotides, FMN and FAD, riboflavin acts as a coenzyme in oxidation-reduction reactions (p. 324). In these reactions, the FMN or FAD serve as carriers of the hydrogen atoms, which the enzyme removes from a substrate, transferring them to another carrier in the same way as the coenzyme NAD functions (p. 282).

Food sources. Riboflavin is found in all plant and animal tissues. The best sources are liver, milk, and yeast. It is also synthesized commercially.

Requirement. The recommended daily allowance (Table 28-1) for adults is 1.2 to 1.7 mg of riboflavin.

Vitamin B_6 (pyridoxine)

Deficiency manifestations. Specific symptoms of a vitamin B_6 deficiency in humans are not supported by consistent evidence. Among the symptoms generally attributed to such a deficiency are neuritis, dermatitis, anemia, convulsions in infants, and interference in tryptophan metabolism. However, some individuals who have anemia respond to relatively large doses of vitamin B_6 with increased hemoglobin (p. 379) and red cell production.

Chemistry. There are three forms of pyridoxine, a substituted pyridine ring, that apparently exist in equilibrium with each other in animals. The active forms are the phosphorylated derivatives, pyridoxal phosphate and pyridoxamine phosphate. The structures of these related compounds are shown here:

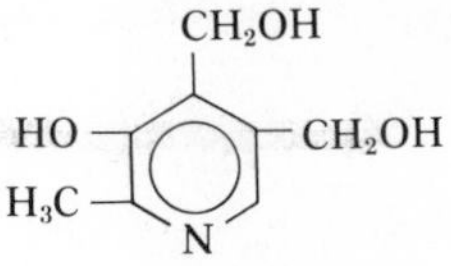

Pyridoxol 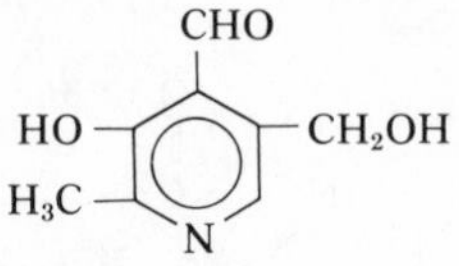 **Pyridoxal** 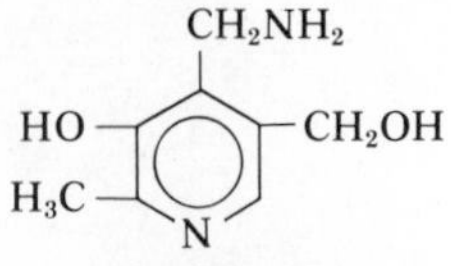 **Pyridoxamine**

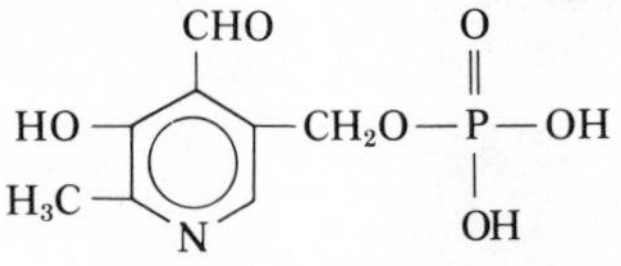

Pyridoxal phosphate 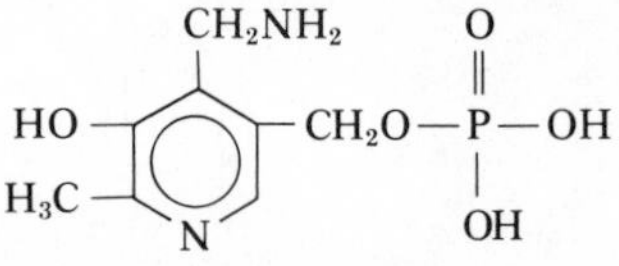 **Pyridoxamine phosphate**

Function. Pyridoxal phosphate acts as a coenzyme in transaminase reactions (p. 360) and other reactions involving amino acids, including the synthesis of hemoglobin (p. 379). It has also been proposed that it acts as a transport agent for moving amino acids across cell membranes.

Food sources. Vitamin B_6 occurs in many plant and animal tissues. Among the best sources are liver, yeast, and wheat germ.

Requirement. The recommended daily allowance for adults is 1.8 to 2.2 mg (Table 28-1).

Vitamin B_{12}

Deficiency manifestations. A long-term deficiency of vitamin B_{12} results in an anemia known as *pernicious anemia.* The developing red cells enlarge (megaloblasts) but do not mature properly, the red cell count decreasing to dangerously low levels (Fig. 28-5). If not reversed in time, degenerative changes occur in the nervous system by way of the spinal cord, leading to sensory disturbances, general weakness, and paralysis. In humans, vitamin B_{12} deficiencies are usually not dietary in origin but result from a lack of intrinsic factor, produced in the stomach, which is required for the absorption of the vitamin (p. 422). Vitamin B_{12} deficiencies may also occur in vegetarians (lack of intake) and patients with total gastrectomies (loss of intrinsic factor) or extensive small intestinal resections (loss of absorptive surface), unless given the vitamin by injection.

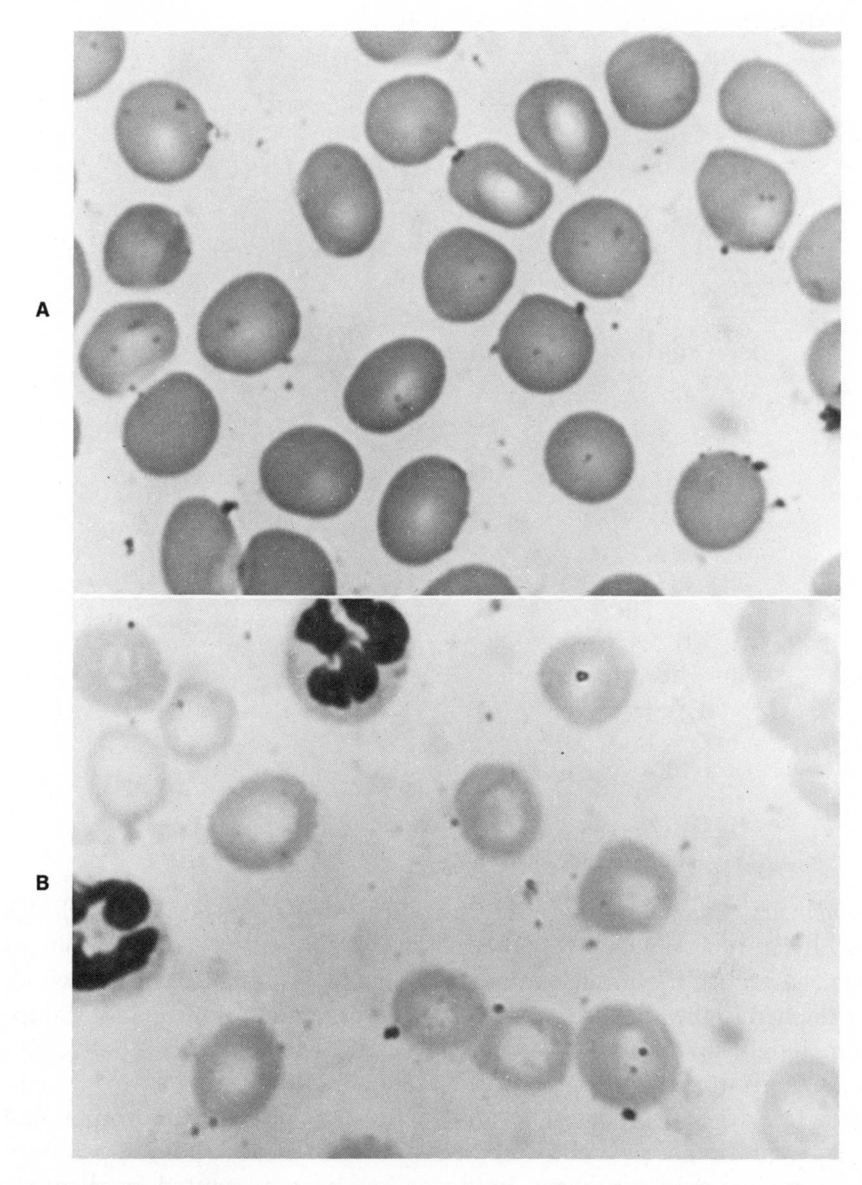

Fig. 28-5. Peripheral human blood smears. (×1000) **A,** Normal. **B,** From a pernicious anemia patient. Note the difference in number and density of the red blood cells. Pernicious anemia results from a vitamin B_{12} deficiency usually arising from a lack of intrinsic factor. (From Haurani, F. I., Cardeza Foundation, Jefferson Medical College, Thomas Jefferson University.)

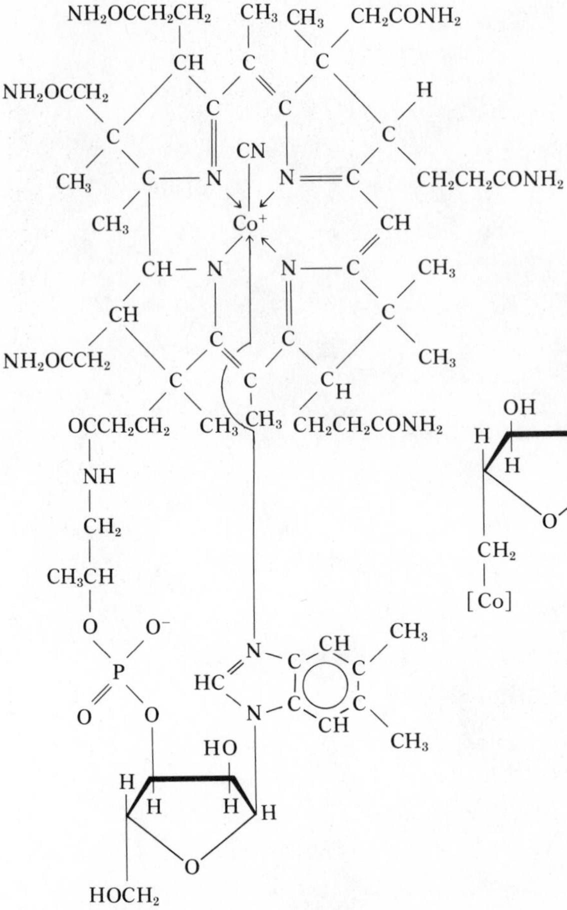

5′-Deoxyadenosine

Vitamin B_{12} (cyanocobalamin)

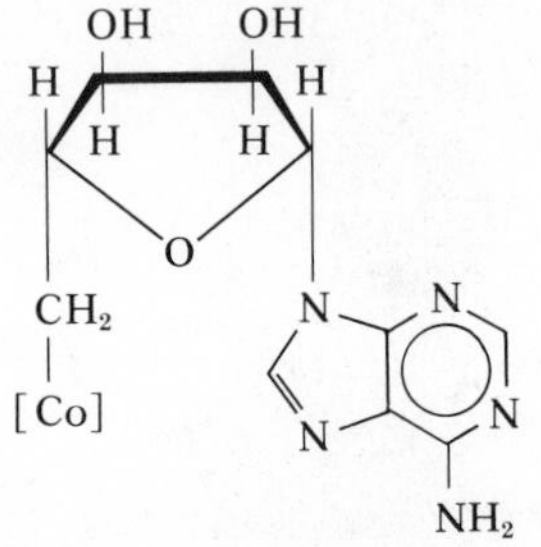

Chemistry. Cyanocobalamin has a complicated structure that includes a porphyrin-type nucleus (p. 377) containing a central cobalt atom to which is attached a cyanide radical and a nucleotide-like structure, which is reattached to the porphyrin portion by way of an aminopropanol grouping. In its coenzyme form, the cyanide radical is replaced by a deoxyadenosine grouping. The structures are shown above.

Function. The specific manner in which vitamin B_{12} acts to promote the normal formation of red cells is not known. It is believed that it acts somehow in nucleic acid synthesis. Evidence has been presented implicating vitamin B_{12} in the conversion of ribonucleotides to deoxyribonucleotides and in methylation reactions. The specific function of vitamin B_{12} in the nervous system is also not known.

Food sources. The best source of vitamin B_{12} is liver, which was used in the first successful therapeutic treatment of pernicious anemia. It is also

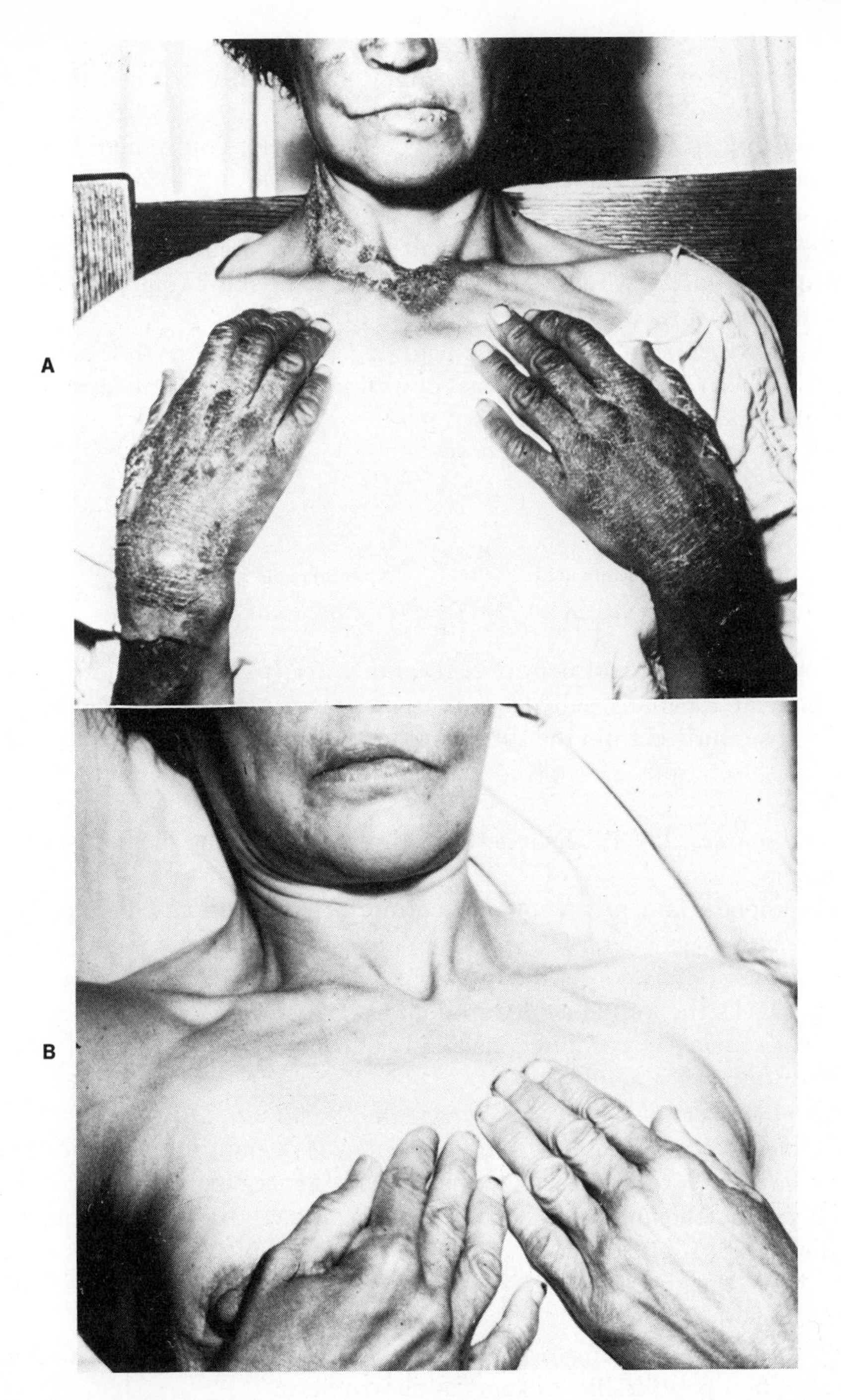

Fig. 28-6. A, Pellagra showing the dermatitis characteristic of nicotinamide deficiency. **B,** Same patient after nicotinamide therapy. (From Jolliffe, N. Courtesy the Upjohn Co.)

found in other meats (kidney) and milk. Commercially it is concentrated from cultures of certain microorganisms.

Requirement. Clinical data indicate that the daily requirement is about 3 μg of vitamin B_{12} (Table 28-1).

Niacin

Deficiency manifestations. A niacin deficiency results in the disease known as pellagra, which literally means "rough skin" (Fig. 28-6). Diarrhea and dementia also occur in this condition.

Chemistry. Niacin, or nicotinic acid, is a substituted pyridine derivative. In its active form it appears as the amide, niacinamide (nicotinic acid amide). These structures are shown here:

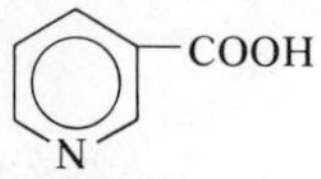

COOH
N

Nicotinic acid

(Niacin)

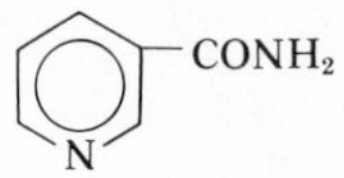

$CONH_2$
N

Nicotinic acid amide

(Niacinamide)

Function. As a constituent of NAD and NADP (p. 245) niacin functions as coenzyme for many dehydrogenases, acting as a carrier for the hydrogen atoms removed from the substrates in the oxidation reactions (p. 282). How this function is involved in the development of pellagra is not known.

Food sources. Liver, wheat germ, and yeast are among the best sources of niacin.

Tryptophan is a precursor of nicotinic acid in humans. Pellagra became a serious problem in the South in the early 1900s when many people were living on diets consisting largely of degerminated cornmeal, a product that lacks the amino acid tryptophan.

Requirement. The recommended daily allowance (Table 28-1) for adults is 13 to 18 mg of niacin.

Pantothenic acid

Deficiency manifestations. No symptoms of a pantothenic acid deficiency have been recognized in humans. Many general symptoms of a vitamin B group deficiency occur in animals, one interesting feature being the graying of hair in the black rat. This caused a flurry of excitement as a possible clue to the graying of hair in humans, but pantothenic acid was found to have no effect.

Chemistry. Pantothenic acid consists of β-alanine combined with pantoic acid in a peptide-like linkage, as shown here:

$$CH_2OHC(CH_3)_2CHOHCONHCH_2CH_2COOH$$

Pantothenic acid

The β-alanine, an unusual type of amino acid in animal tissue, is formed by the decarboxylation of aspartic acid.

Function. As a constituent of coenzyme A, pantothenic acid functions in many important reactions involving the transfer of acyl groups, such as acetyl, butyryl, and succinyl groups.

Food sources. Pantothenic acid is found in all plant and animal tissues. Among the best sources are liver, wheat bran, yeast, and peas.

Requirement. The daily requirement for humans is not known. An intake of 4 to 7 mg of pantothenic acid has been recommended.

Biotin

Deficiency manifestations. Unless provoked by the administration of a vitamin antagonist, biotin deficiency does not occur in humans. A substance in raw egg white acts as a biotin antagonist. When administered in large amounts to humans on a low biotin diet, raw egg whites caused the development of dermatitis, anorexia, nausea, anemia, and hypercholesterolemia.

Chemistry. Biotin is a sulfur-containing heterocyclic compound. In its active form, biocytin, it is combined with lysine. The structures of these compounds are shown here:

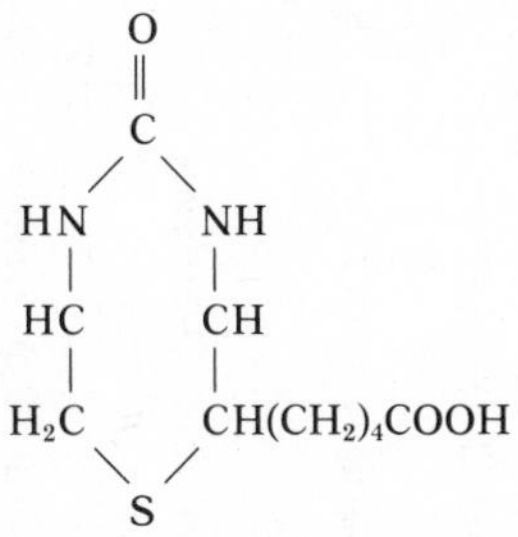

```
        O
        ‖
        C
       / \
     HN   NH
     |    |
     HC   CH
     |    |
    H2C   CH(CH2)4COOH
      \  /
       S
```

Biotin

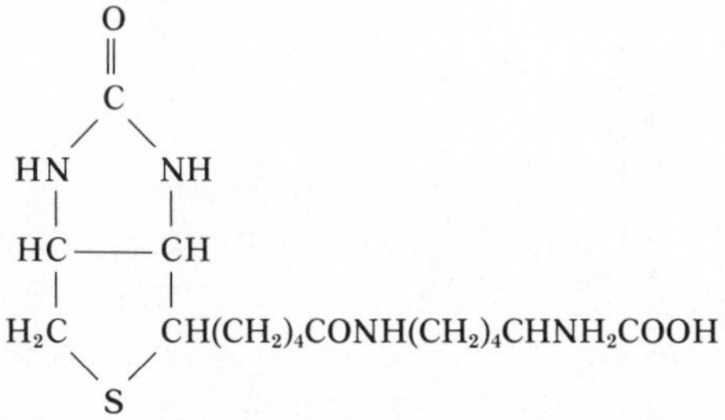

```
        O
        ‖
        C
       / \
     HN   NH
     |    |
     HC───CH
     |    |
    H2C   CH(CH2)4CONH(CH2)4CHNH2COOH
      \  /
       S
```

Biocytin

(ε-N-biotinyl-L-lysine)

Function. Biotin acts as coenzyme in reactions involving the transfer of carbon dioxide, as in the conversion of acetyl-CoA to form malonyl-CoA (fatty acid synthesis, p. 318).

Table 28-2. The vitamins and their properties

Vitamin	Symptoms of deficiency	Chemical nature	Function	Food sources
A	Night blindness, keratinization of epithelial cells	Fat-soluble, unsaturated aralkyl alcohol	Vision (rhodopsin), maintenance of epithelial cells	Yellow pigment of fruits, vegetables, also fish-liver oils
D	Rickets (weak, deformed bones)	Fat-soluble sterols	Regulates calcium and phosphorus metabolism	Fish-liver oils, supplemented milk, butter, oleomargarine
E	Sterility, muscular dystrophy in animals (not humans)	Fat-soluble, O-heterocyclic phenols	?	Green, leafy vegetables, wheat germ oil
K	Prolonged clotting time, hemorrhages	Fat-soluble quinones	Synthesis of factors in blood coagulation (prothrombin, etc.)	Green, leafy vegetables
C (ascorbic acid)	Scurvy (hemorrhages, painful joints)	Water-soluble acid, sugar-like structure	Formation and maintenance of intercellular connective and cement substances	Citrus fruits
B_1 (thiamine)	Beriberi (polyneuritis, heart failure)	Water-soluble, N,S-heterocyclic base	Forms part of coenzyme involved in decarboxylation and oxidation (thiamine pyrophosphate)	Beans, nuts, fruits, cereal germ layers
B_2 (riboflavin)	Sore lips, tongue and skin lesions, cheilosis	Water-soluble, N-heterocyclic, ribitol-containing compound	Forms part of coenzymes involved in hydrogen transport during biologic oxidations (FMN and FAD)	Liver, milk, yeast

B_6 (pyridoxine)	Nervous system disorders? Anemias	Water-soluble, N-heterocyclic alcohols, aldehydes, amines	Forms part of coenzyme involved in transfer of amino groups in amino acid metabolism (transamination)	Liver, yeast, wheat germ
B_{12} (cyanocobalamin)	Macrocytic anemia, specifically pernicious anemia (usually due to lack of intrinsic factor)	Water-soluble complex, porphyrin-like structure containing cobalt	Involved in amino acid and nucleic acid metabolism, red cell maturation, methylation reactions	Liver, kidney, milk
Niacin (nicotinic acid)	Pellagra (dermatitis, diarrhea, dementia)	Water-soluble, N-heterocyclic acid	Forms part of coenzymes involved in hydrogen transport during biologic oxidations (NAD, NADP)	Liver, wheat germ, yeast
Pantothenic acid	?	Water-soluble, peptide-like acid	Forms part of coenzyme involved in metabolism of acetic acid and fatty acids (CoA)	Liver, wheat bran, yeast, peas
Biotin	Dermatitis?	Water-soluble, S,N-heterocyclic acid	Involved in transfer of CO_2 (fatty acid synthesis)	Egg yolk, liver, kidney, yeast
Folic acid	Macrocytic anemias	Water-soluble, N-heterocyclic base plus amino acid	Forms part of coenzymes involved in 1-carbon unit transfers; also acts in red cell maturation	Green, leafy vegetables, liver, yeast

Food sources. Biotin occurs in most plant and animal tissues. Among the best sources are egg yolk, liver, kidney, and yeast.

Requirement. Because of its synthesis by intestinal bacteria, the human requirement for biotin is not known. The recommended intake is 100 to 200 μg (Table 28-1). From the experiments on humans, injections of less than 0.5 mg per day rapidly relieved the symptoms of deficiency.

Folic acid

Deficiency manifestations. As in the case of vitamin B_{12}, a deficiency of folic acid results in an anemia characterized by the appearance of large, immature red cells (megaloblasts) in the blood. Deficiencies of folic acid may occur in gastrointestinal disorders that interfere with absorption of the vitamin or during pregnancy, which increases the requirement for folic acid.

Chemistry. The term *folic acid,* although applied to pteroylglutamic acid, also refers to a series of compounds that are interconvertible in the body. The coenzyme form is tetrahydrofolic acid. These structures are shown here:

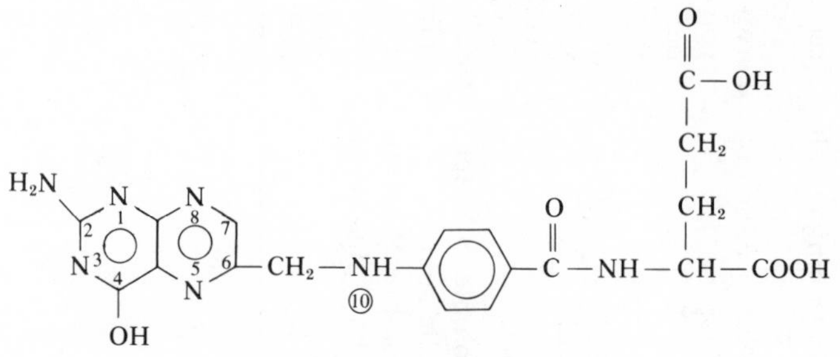

Pteroylmonoglutamic (folic) acid

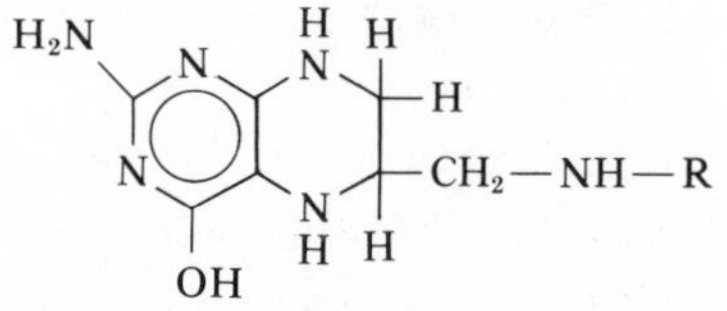

5,6,7,8-Tetrahydrofolic acid (FH_4)

Function. Tetrahydrofolic acid acts in reactions in which 1-carbon units are transferred, as in the biosynthesis of purines (p. 337). It is also involved in methylation reactions with vitamin B_{12}.

Food sources. Among the best sources are green, leafy vegetables, liver, and yeast.

Requirement. The recommended daily intake of folic acid (folacin) is 400 μg (Table 28-1).

A summary of the properties and functions of the vitamins is presented in Table 28-2.

SUMMARY

SUMMARY

Vitamins are low–molecular weight organic compounds, required in small amounts for normal growth, maintenance, and reproduction. In the absence of vitamins, deficiency diseases develop. Many vitamins act as coenzymes.

HISTORIC DEVELOPMENT OF VITAMIN THEORY

Dates back to 2600 BC when beriberi was known in China. Cures for night blindness, beriberi, scurvy, and rickets were discovered many times before being completely accepted. By the late nineteenth century, deficiency diseases were being discovered in animals and nutritional experiments were indicating that certain diseases were caused by the lack of important factors in the diet. In 1911, 1912, Funk proposed the theory that diseases such as beriberi, scurvy, pellagra, and rickets were "caused by a deficiency of some essential substances in food." He named these substances "vitamines," the "e" being dropped later. Deficiency diseases are still important problems in undernourished areas of the world and still occur in the most affluent countries.

GENERAL CONSIDERATIONS

Classification:

Fat-soluble vitamins: A, D, E, and K.

Water-soluble vitamins: C and B group. B group includes B_1, B_2, B_6, B_{12}, niacin, pantothenic acid, biotin, folic acid.

Vitamin assays: Animal, microorganism, chemical, and physical assays used. Saturation tests used to determine vitamin status of individuals. Many special tests are also used.

Vitamin requirements: Table 28-1 lists the daily requirements.

Absorption problems: Fat-soluble vitamins are absorbed better in presence of bile. Drug treatments may decrease bacterial synthesis of vitamins in intestines. Absorptive surfaces in intestines must be functioning normally. Intrinsic factor is required for the absorption of vitamin B_{12}.

Storage organs: Liver is storage organ for A, D, B_{12}, and folic acid. White cells may be storage tissue for C and B_2. No others recognized.

Increased requirements: Greater amounts needed during pregnancy, lactation, childhood, aging, convalescence. If one vitamin is missing, greater amounts of others may be needed.

Excessive intake: Harmful effects noted with fat-soluble vitamins A and D and possibly E. Harmful effects also reported for water-soluble vitamin C.

SUMMARY Table 28-2 is a summary of vitamins, including symptoms of deficiency, chemical nature, biologic functions, and food sources of the vitamins. Requirements are listed in Table 28-1.

REVIEW QUESTIONS

1. What are vitamins? How are most of them normally obtained? Do humans benefit from the production of vitamins by intestinal bacteria?
2. In what way do the B vitamins function? Are they water or fat-soluble? List the deranged metabolic states that occur as a result of the various vitamin B deficiencies and the best sources of these vitamins. Do vitamin B deficiencies generally occur on the basis of a single B vitamin or a multiple set of B vitamins? Explain.
3. What is the significance of bile to vitamin nutrition? What is intrinsic factor, where is it made, and how does it function in vitamin nutrition? What may result from its absence? How may drug treatments and intestinal inflammations affect vitamin nutrition?
4. Name some fat-soluble vitamins, associated deficiency states, functions, and best sources.
5. What is the range of daily adult requirements for the various vitamins? Are any vitamins stored in the body? If so, which and where? What are the usual conditions under which greater amounts of vitamins may be required? Are any vitamins toxic in excessively large doses? Are vitamin deficiencies ever encountered in affluent regions with large food supplies?

REFERENCES

Annual review of biochemistry, Palo Alto, Calif., Annual Reviews.

Annual review of physiology, Palo Alto, Calif., Annual Reviews.

Bailey, H.: Vitamin pioneers, New York, 1970, B. J. Publishing Group.

Goodwin, T. W.: The biosynthesis of vitamins and related compounds, New York, 1964, Academic Press, Inc.

Robinson, F. A.: Vitamin co-factors of enzyme systems, Elmsford, N.Y., 1966, Pergamon Press.

Sebrell, W. H., Harris, R. S., György, P., and Pearson, W. N., editors: The vitamins: chemistry, physiology, pathology, methods, Vols. 1-3, 5-7, New York, 1967-1972, Academic Press, Inc.

Vitale, J. J.: Vitamins, Kalamazoo, Mich. 1976, The Upjohn Co.

Vitamins and hormones (annual publication), New York, Academic Press, Inc.

Wagner, A. F., and Folkers, K.: Vitamins and coenzymes, Huntington, N.Y., 1975 (reprint of 1964 edition), R. E. Krieger Pub. Co., Inc.

Hormones

29

The classic concept of a hormone is that of a substance produced in one part of the body and carried elsewhere by the blood to a "target organ" or tissue that is stimulated into activity by the hormone. The term *hormone* was used for the first time by Bayliss and Starling in 1902 with regard to secretin, which is secreted by the duodenal mucosa and acts by stimulating the flow of pancreatic juice and bile (p. 267). The tissue or organ by which the hormone is synthesized is referred to as an endocrine gland, endocrine tissue, or a ductless gland, because the hormone is secreted directly into the bloodstream. The locations in the body of the major endocrine glands are shown in Fig. 29-1.

Much of the early knowledge of hormones came from observations of cases involving diseases of the endocrine glands or tissues. In some instances, production of a particular hormone was stopped and in others, an excessive amount of hormone was produced. At present the structures of most hormones are known and many are prepared commercially. Intensive research efforts on the functions and mechanism of action of hormones are continuing to supply further details about these strategic substances.

GENERAL CONSIDERATIONS

Hormone activity

Hormones represent an important element in the fine control system by which the body modulates the many cellular reactions that keep it functioning normally. A crucial factor in the ability of hormones to function in this manner is the very high degree of specificity shown by various cells; i.e., for a hormone to be able to affect cell reactions selectively, it must be recognized only by cells where its control activities are required. This would imply some sort of "receptor" on the surface of the target cell not present on other cells. Such receptors have been described. Cells with receptors for a particular hormone respond to relatively small concentrations of that hormone but cells without such specific receptors will not respond to massive concentrations of that same hormone.

Another key aspect in the fine control exerted by hormones is their relatively high turnover rate. Thus if only a small metabolic change is

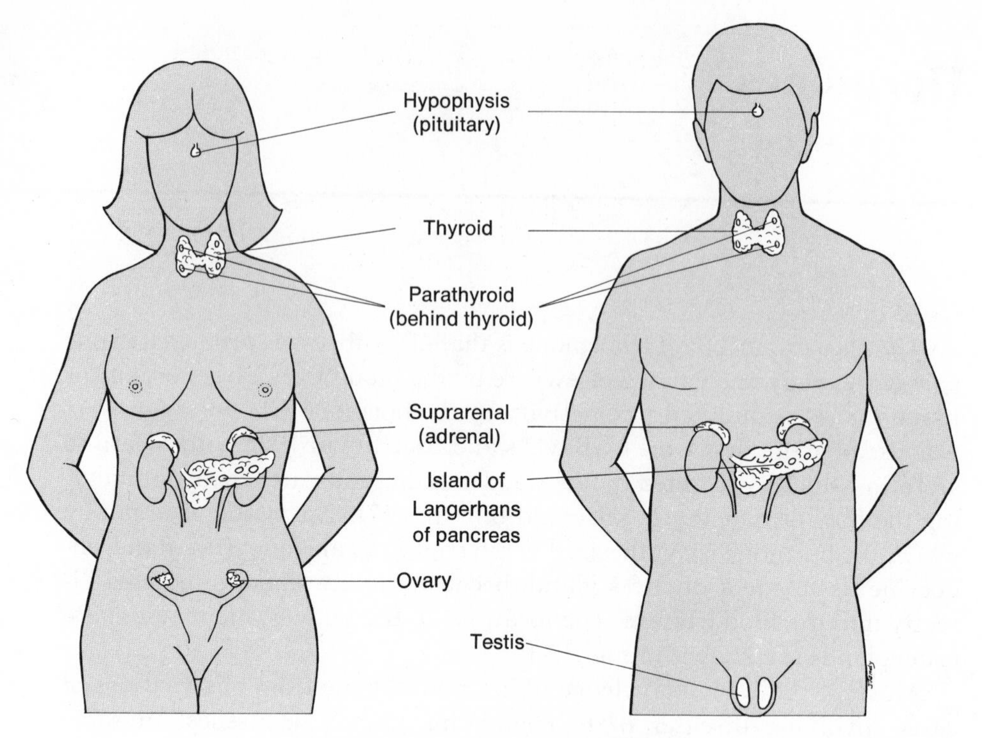

Fig. 29-1. Endocrine glands.

required, a small amount of hormone can act and the molecules are broken down rapidly or inactivated. If a continuing metabolic change is required, continued synthesis of a particular hormone would be required to sustain such activity. Also, as will be noted in subsequent discussions, some hormones may exhibit multiple actions and some may elicit different responses from different tissues in order to accomplish the net result required.

Many of the polypeptide hormones have been shown to be secreted in an inactive prohormone form that is subsequently activated by splitting off a portion of the polypeptide. This is probably part of a mechanism to make certain that the hormone will act in the right place at a time when it is required. This is similar to the phenomenon observed with the digestive enzymes (preenzymes, p. 255).

Mechanism of hormone action

Attempts in the past to explain how hormones act did not meet with great success until recently. At present, the second messenger concept of Sutherland, who won the Nobel prize for his efforts, appears to have gained the general acceptance required of a thoroughly documented, well-established mechanism of action.

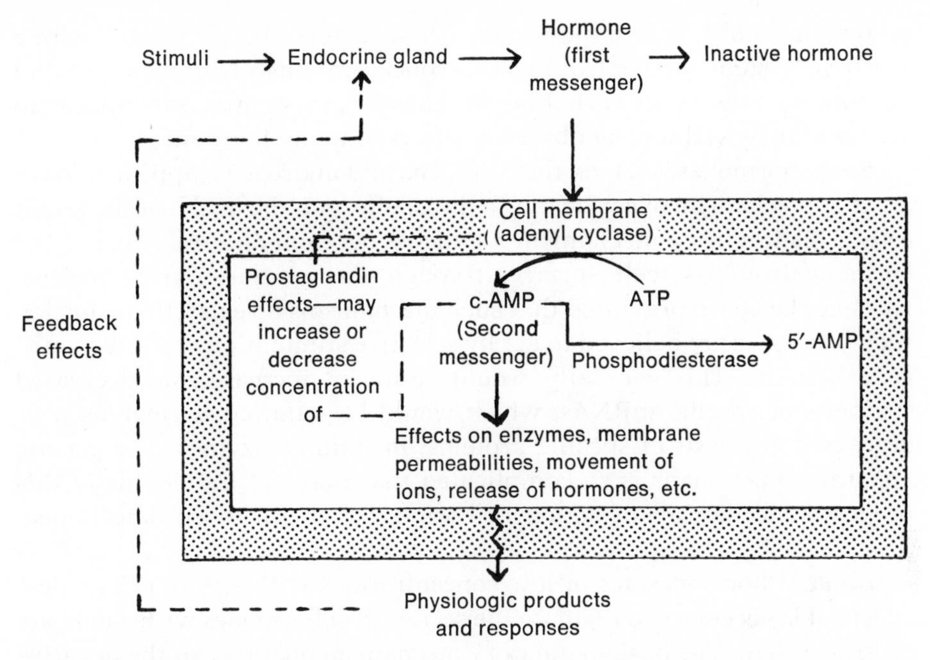

Fig. 29-2. Diagrammatic representation of the mechanism of action of hormones at the cellular level, incorporating the second messenger concept of Sutherland and the modulating effects of the prostaglandins.

In Sutherland's hypothesis, a hormone secreted as the result of stimulation of an endocrine gland is known as the first messenger. At the cellular level, the first messenger influences the activity of adenyl cyclase, a membrane-bound enzyme, responsible for the conversion of ATP to c-AMP, cyclic adenosine 3′,5′-monophosphate, which is termed the second messenger. c-AMP then exerts its influence on enzymatic reactions of many types, membrane permeabilities, movement of ions, and release of hormones. These activities result in the production of a large variety of physiologic products and responses. In Fig. 29-2 is presented a diagrammatic representation of this proposed mechanism of action of hormones at the cellular level.

It should be noted that there are four modulating effects shown in Fig. 29-2. The concentration of hormone, or first messenger, may be adjusted by variations in the rates of processes that inactivate it. The cellular concentration of c-AMP, or second messenger, may be adjusted by the rate at which the phosphodiesterase converts it to 5′-AMP and by the effects of the various prostaglandins on the activity of adenyl cyclase. Another modulating factor involves the feedback effects of the physiologic products on the secretion activity of the endocrine gland that produces the original hormone, or first messenger. These factors provide the cell with a sensitive system for controlling the concentrations of first and second messengers to meet its varied needs. Thus far, these proposals appear to account fairly well for the observed effects of many hormones.

Some hormones, such as the estrogens and androgens, appear to have another mechanism of action, i.e., a direct effect on the nuclei of the target cells. This implies a mechanism for the recognition and transport of the hormone from the cell surface, through the cytosol to the nucleus. Evidence has been presented that such mechanisms do exist. In the nuclei, such hormones are believed to act by selective stimulation or inhibition of gene activity. This generally would result in increased or decreased synthesis of specific mRNAs, which would be followed by increased or decreased synthesis of specific proteins, including enzymes. The genetic apparatus of humans is so complicated that many of the details of this proposed mechanism of action of hormones remain to be determined.

Regulation of hormone secretion

Because hormones act in low concentrations with strong physiologic effects, it is necessary to regulate the secretion of hormones with a delicate control system. The basic regulatory mechanism operates on the negative feedback principle. In the case of insulin, its concentration in the blood rises and falls as the blood sugar concentration rises and falls. The concentration of parathyroid hormone is controlled in the same way, rising and falling as the blood calcium concentration falls and rises.

The secretion of some other hormones are stimulated by the action of what are known as tropic hormones, e.g., the production of thyroid hormone, thyroxine, by the thyroid gland is stimulated by the thyrotropic hormone (thyrotropin) produced in the adenohypophysis (anterior pituitary). As the blood concentration of thyroxine increases, the production of thyrotropic hormone is decreased, which causes a decrease in production of thyroid hormone. This is another example of negative feedback control of hormone secretion. Some aspects of these relationships are summarized in Fig. 29-3.

Hormone-releasing and release-inhibiting factors. An important aspect of the regulation of hormone secretion, especially the tropic hormones produced by the anterior pituitary or adenohypophysis, involves the nervous system. The secretion of these tropic hormones is stimulated by neurohumoral substances formed in the hypothalamus (in functional nervous system units called nuclei) and then released into the blood (hypophyseal portal system) and carried to the adenohypophysis. (Thus communication between the hypothalamus and adenohypophysis is by way of nerve cells and then portal blood system, whereas communication

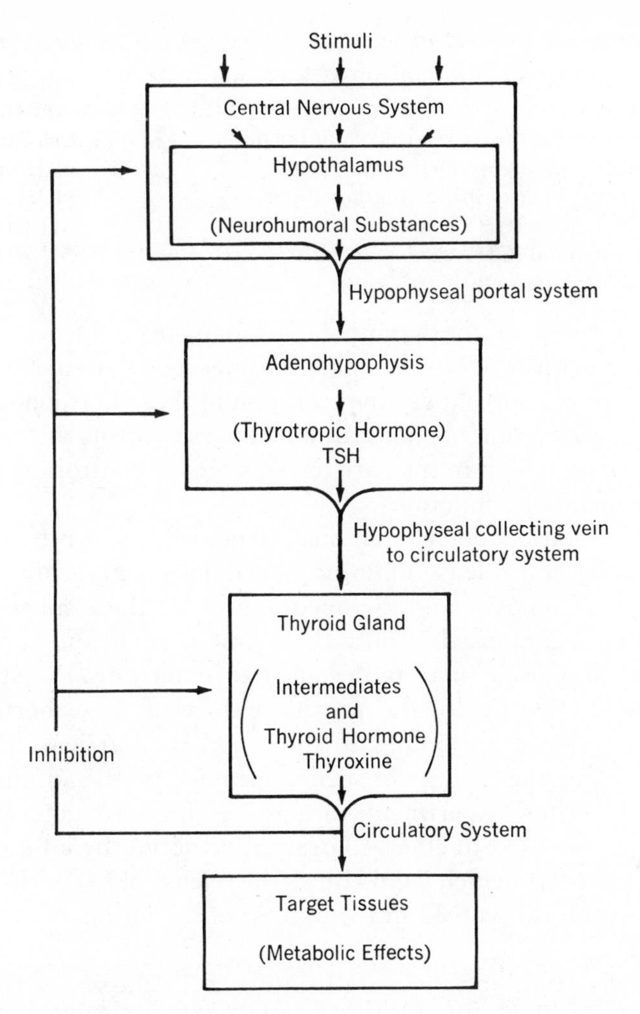

Fig. 29-3. Neuroendocrine and feedback inhibition control of hormone secretion as illustrated with thyroid hormone.

between the hypothalamus and neurohypophysis or posterior pituitary is by way of nerve cells only.) These neurohumoral substances, peptide in nature and fitting the definition of hormones, are now known as releasing factors and some have been recognized as release-inhibiting factors. Included here are the names and abbreviations indicating the functions of those factors that have been characterized to date:

Hypothalamic factors	*Abbreviation*
Thyrotropin (TSH)-releasing hormone	TRH or TRF
Corticotropin (ACTH)-releasing hormone	CRH or CRF

Continued.

Hypothalamic factors—cont'd	*Abbreviation—cont'd*
Follicle-stimulating hormone (FSH)–releasing hormone	FSH-RH or FSH-RF
Luteinizing hormone (LH)–releasing hormone	LH-RH or LH-RF
Prolactin (P)-releasing hormone	PRH or PRF
Prolactin (P) release–inhibiting hormone	PRIH or PIF
Growth hormone (GH)–releasing hormone	GH-RH or GH-RF
Growth hormone (GH) release–inhibiting hormone (somatostatin)	GH-RIH or GIF

Since the activity of the hypothalamus may be influenced by stimuli reaching the central nervous system, the nervous system thereby exerts some measure of control over the secretion of these hormones (Fig. 29-3). Thus, a delicate system of checks and balances controls the production of these hormones, which in turn are responsible for controlling many other significant metabolic functions.

With the isolation, determination of structure, and synthesis of some of these releasing and release-inhibiting hormones, significant possibilities now exist for diagnostic and therapeutic uses for these substances. TRH, for example, is composed of only three amino acids and can be readily synthesized and used in tests designed to determine the status of the thyroid gland. Because of the peptide nature of these hormones, it is possible to produce antibodies to them and it is at least theoretically possible to block the activity of such a hormone by the administration of its antibody. Thus experiments are in progress to determine if this approach can be used to effect contraceptive action by administering an antibody to LH-RH, which would block the release of LH or the luteinizing hormone, which is involved in the process of ovulation.

Chemical classification of hormones

Chemically the hormones fall into three classes of compounds: steroids, polypeptides, and amino acid derivatives.

The basic structure of the steroid hormones is the cyclopentanoperhydrophenanthrene nucleus, which, in simple terms, consists of a cyclopentane ring (D) fused, by one side, to a saturated phenanthrene structure (rings A-B-C), as shown here:

A B C D

Because substituents and side chains have to be properly located and identified, a standard numbering system has been adopted, as shown here:

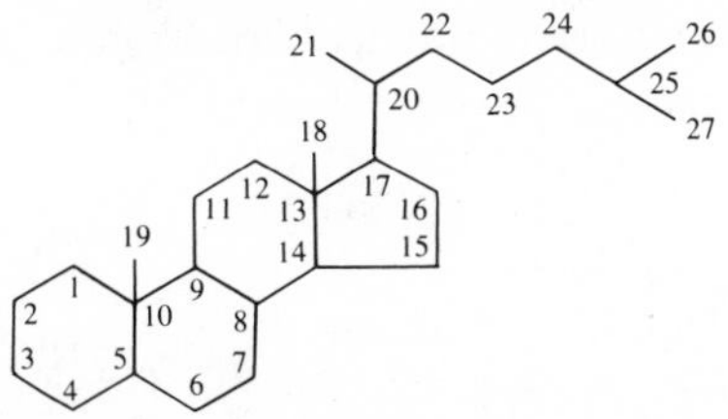

The possibilities for isomerism of various types are many, although only a few have been found in nature. The system or systems of nomenclature are very complicated and constantly changing as newer problems in identification are uncovered.

In some ways, the polypeptide and amino acid–derived hormones present simpler problems. These hormones are usually known by simple, trivial names. Complete details of the structures of some are known.

Synthesis of hormones

The steroid hormones are synthesized from the fundamental acetate unit by way of cholesterol (p. 209), as shown in the simplified scheme that follows:

$$\text{Acetate} \rightarrow \text{Cholesterol} \rightarrow \text{Progesterone}$$
$$\swarrow \qquad \searrow$$
$$\text{Estrogen} \leftarrow \underset{\text{(androgens)}}{\text{Testosterone}} \qquad \underset{\text{corticoids}}{\text{Adrenal}}$$

Although many complicated reactions are involved, complete details of the biosynthesis of many of the steroid hormones are known.

The polypeptide hormones, some in the form of glycoproteins, are presumably made by the usual procedures of peptide synthesis (p. 354). The hormones derived from amino acids require special reactions to convert the amino acid to the necessary hormone structure. The details of many of these reactions are also known.

ANDROGENS—MALE SEX HORMONES

Androgens are the steroid hormones responsible for the development of the secondary sex characteristics in the male, such as the typical distribution of body hair and deepening of the voice at puberty, and the normal growth and function of the seminal vesicles and prostate gland. These hormones are synthesized in the testis (Leydig or interstitial cells), adrenal cortex, and ovary. The production of androgens is stimulated by the gonadotropic hormone (luteinizing hormone, p. 462) from the adenohypophysis (anterior pituitary).

Chemically, the androgens all have nineteen carbon atoms, seventeen in the ring structure and methyl groups at C-10 and C-13. The structures of testosterone, the most active natural androgen, and androsterone, are shown here:

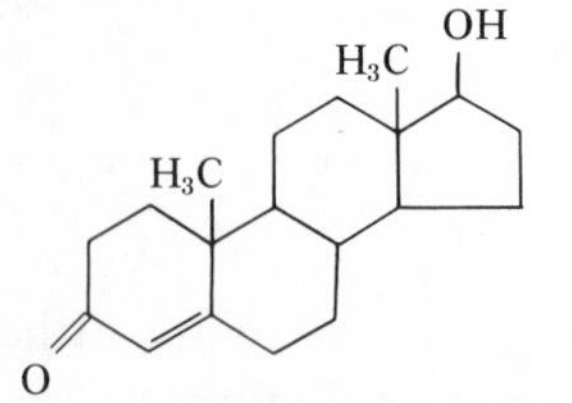

Testosterone 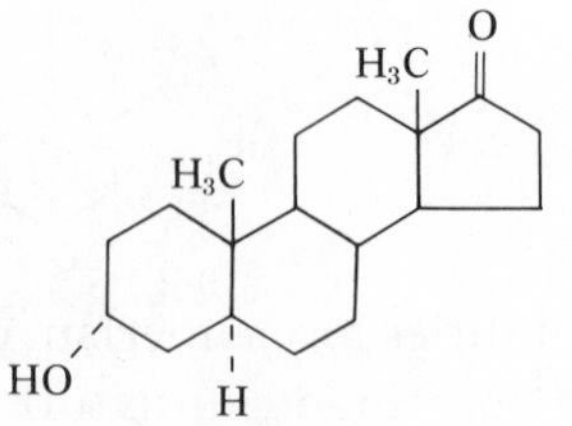 **Androsterone**

In the testis, testosterone is the major end product of steroid synthesis, whereas in the ovary, testosterone is used as a precursor of estrogen.

The androgens produce a general protein anabolic effect. Observations of a prompt increase in prostatic mRNA after administration of testosterone has led some investigators to believe that androgens carry out their functions by the stimulation of the synthesis of mRNA.

ESTROGENS—FEMALE SEX HORMONES

Estrogens are the steroid hormones responsible for the development of the secondary sex characteristics in the female, such as the typical distribution of body hair and growth of the mammary glands at puberty, and the normal growth of the female genital organs. Synthesis of estrogens, from androgenic steroid intermediates, takes place in the ovary (maturing follicles and corpus luteum), adrenal cortex, placenta, and testis. The gonadotropic hormones (follicle-stimulating hormone, p. 461; luteinizing hormone, p. 462; and prolactin, p. 462) stimulate the production of estrogens.

The estrogens differ from the androgens by having one less methyl group (no methyl group at C-10) and therefore have eighteen carbon atoms. Another difference is the presence of three double bonds in ring A of the estrogens, giving the hydroxyl group at C-3 phenolic properties. The structures of estrone and estradiol, both naturally occurring estrogens, are shown here:

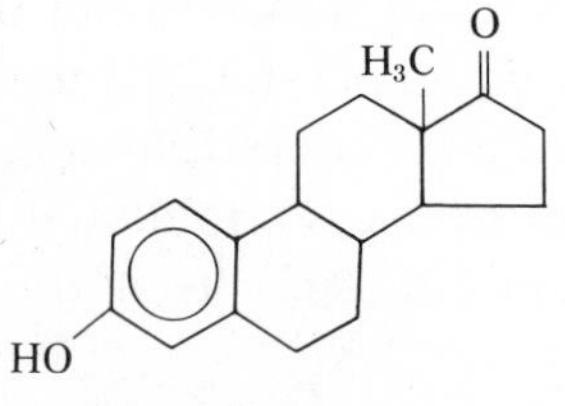

Estrone 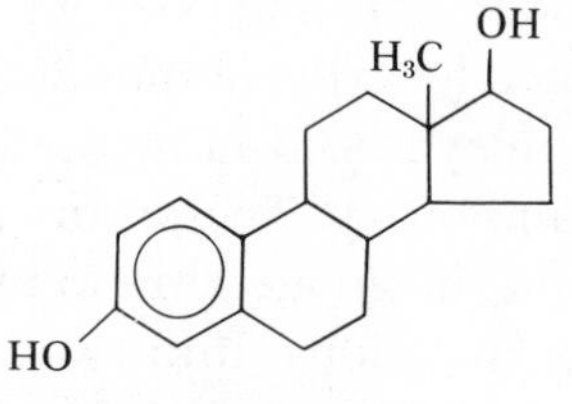 **Estradiol**

A commercially synthesized compound, stilbestrol, which exhibits strong estrogenic activity, has been widely used as a therapeutic agent but has also been cited for its carcinogenic activity. The structure of this compound is shown here:

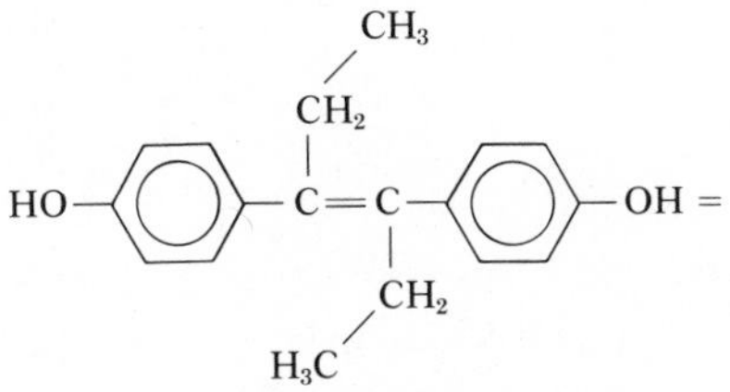

Stilbestrol 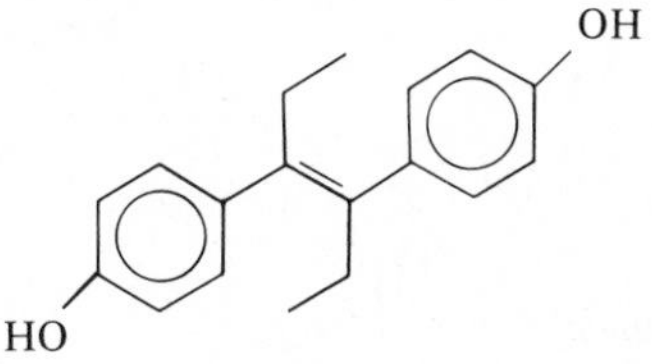 Steroid form

In the nonpregnant female during the reproductive years, the production of estrogens occurs in a cyclic pattern in response to variations in secretion of the gonadotropic hormones. The various events that take place during the usual 4-wk menstrual cycle are correlated in Fig. 29-4. If conception takes place, the corpus luteum does not degenerate and it continues to produce a high level of estrogens and progesterone until the placenta takes over the major synthesis after about 3 mo.

The cyclic changes in estrogen secretion and production decrease at menopause and continue to decrease thereafter.

Metabolically the estrogens, as the androgens, have a protein anabolic effect with indications that this is on the same basis of stimulation of the synthesis of mRNA.

Compounds such as androsterone and estrone are known as 17-ketosteroids because of the keto group on C-17. Analysis for 17-ketosteroids in urine is frequently called for in various clinical disorders involving the testes or adrenal cortex.

PROGESTERONE

As noted before, progesterone is synthesized in the ovary by the corpus luteum. It is also made by the placenta, adrenal cortex, and testis. The appearance of progesterone on ovulation is usually detected by the presence of its excretory derivative, pregnanediol, in the urine. The structures of these two compounds are shown here:

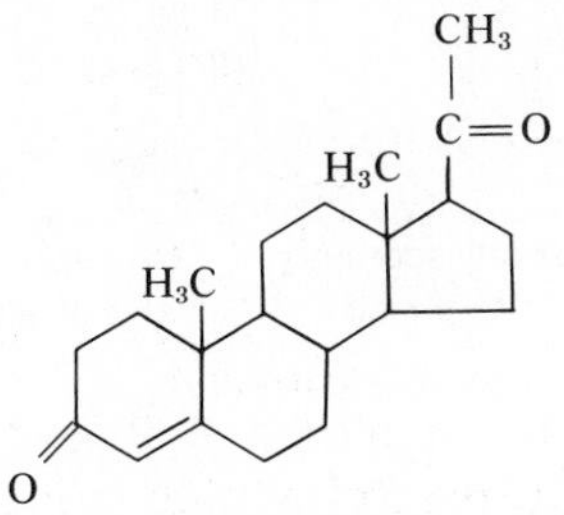

Progesterone 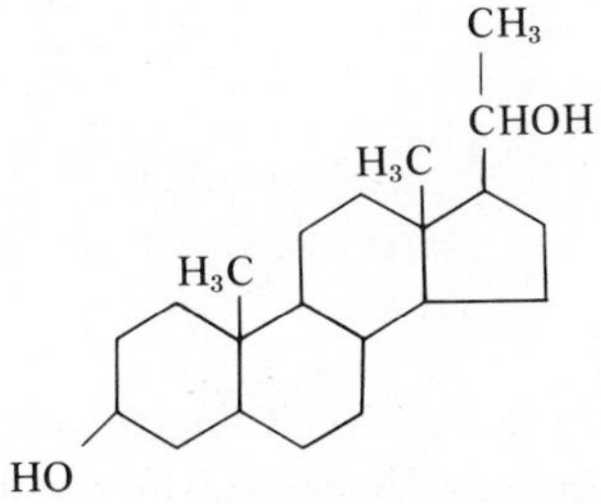 Pregnanediol

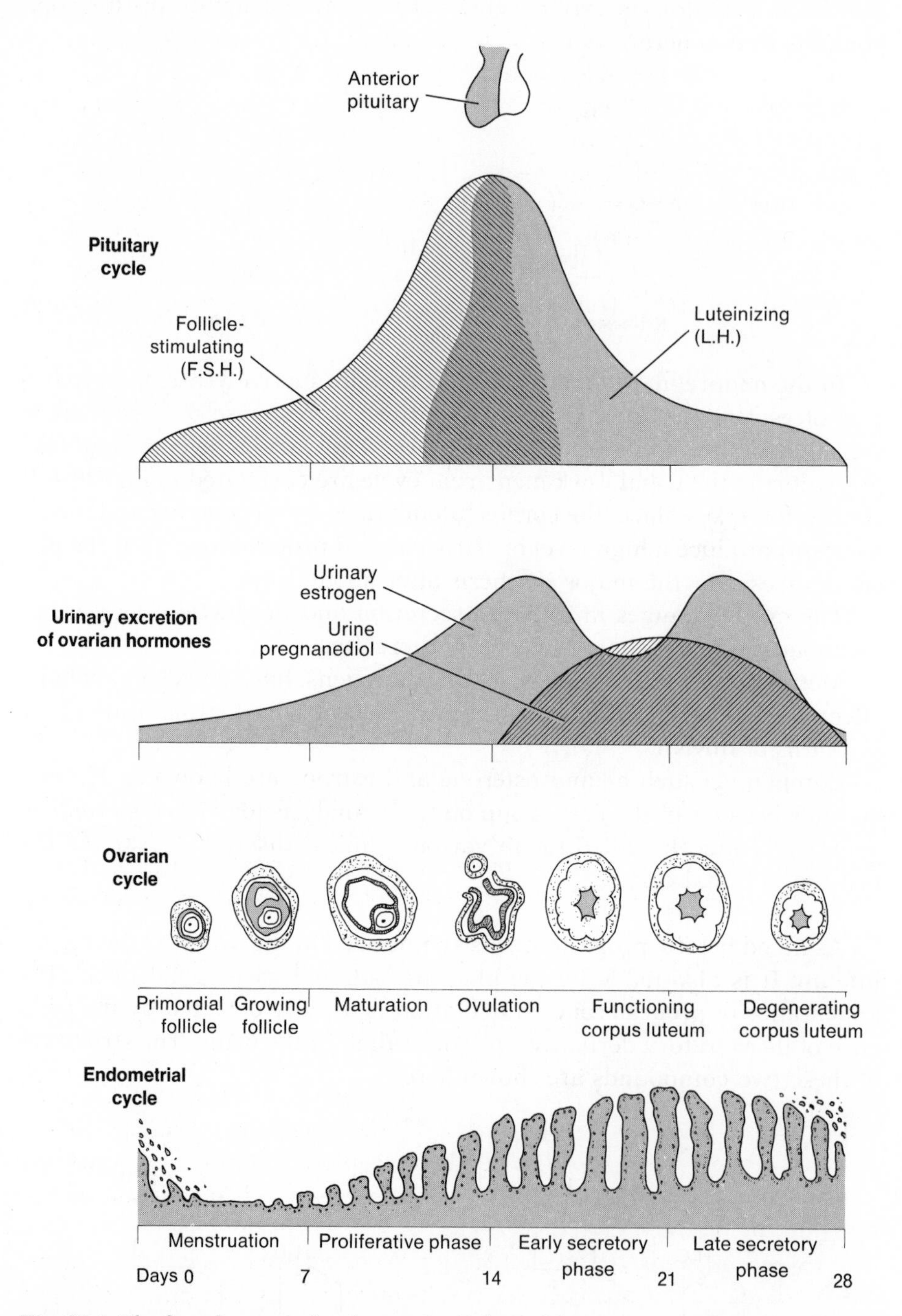

Fig. 29-4. The female sex endocrine cycle. (Adapted from Dr. A. E. Rakoff, Jefferson Medical College, Thomas Jefferson University.)

Progesterone is synthesized by the corpus luteum during early pregnancy, after which it is made mainly by the placenta. It can serve as a precursor of androgens, estrogens, and adrenal corticoids.

Metabolically, progesterone causes changes in the endometrium in the second half of the menstrual cycle (Fig. 29-4) that are designed to receive and support the growth of the fertilized ovum. If fertilization does not occur, the secretion of progesterone by the corpus luteum decreases after about 2 weeks and menstrual bleeding starts.

ADRENAL MEDULLA HORMONES

The adrenal glands (suprarenal glands) are made up of two distinct and independent portions, the medulla (inner part) and cortex (outer layer). First, the hormones of the adrenal medulla will be discussed.

The hormones epinephrine (adrenaline) and norepinephrine (noradrenaline) are synthesized in the adrenal medulla from the amino acids tyrosine or phenylalanine. Epinephrine differs from norepinephrine in having a methyl group attached to the amine nitrogen, as shown here:

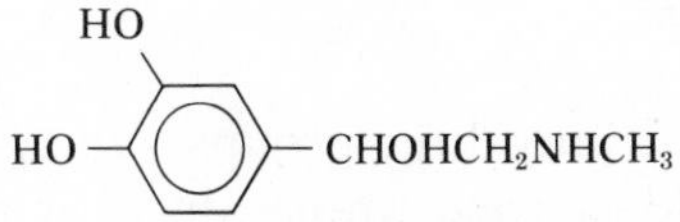

$$(HO)_2C_6H_3-CHOHCH_2NHCH_3$$

Epinephrine

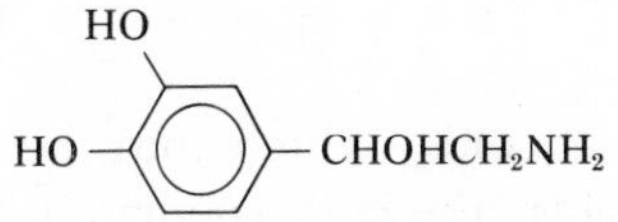

$$(HO)_2C_6H_3-CHOHCH_2NH_2$$

Norepinephrine

Evidence indicates that these hormones are also made in ganglia and in nerve cells of the sympathetic nervous system.

These hormones are secreted in response to various emotional states (anger, fear, pain), hypoglycemia, and hypotension (low blood pressure). Some stimuli produce a preferential effect, e.g., hypoglycemia causes the release of epinephrine mainly, whereas hypotension stimulates release of more norepinephrine.

The major metabolic effect of epinephrine is its hyperglycemic effect, i.e., it causes a rapid rise in blood sugar concentration as a result of an increased rate of glycogenolysis in the liver, followed by a rise in blood lactic acid concentration caused by an increased rate of glycolysis in the muscles. The hyperglycemic effect of norepinephrine is much smaller than that of epinephrine. In adipose tissue, these hormones stimulate the release of free fatty acids to the blood. Oxygen consumption is increased, possibly as a result of increased oxidation of fatty acids or a general increase in metabolic activity.

Because of their hyperglycemic and hypertensive effects, epinephrine and norepinephrine are used in conditions such as insulin shock (low blood glucose), hypotensive shock, and heart stoppage (injection directly into the heart muscle).

ADRENAL CORTEX HORMONES: ANDROGENS, ESTROGENS, PROGESTERONE

These steroid hormones have already been discussed (pp. 451-453).

ADRENAL CORTEX HORMONES: CORTICOIDS

These steroid hormones are called corticoids because of the custom of naming these compounds according to their relationship to corticosterone, the structure of which is shown here:

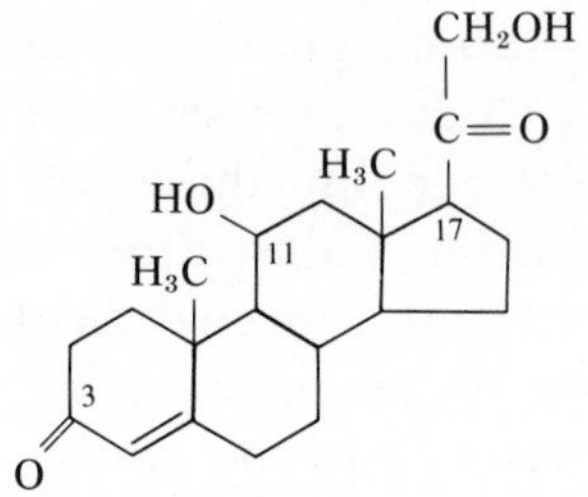

Corticosterone

The corticoids, which may be subdivided into three groups—the glucocorticoids, the electrocorticoids, and aldosterone (mineralocorticoid)—provide excellent examples of how small changes in structure result in great differences in function. Structural features common to all the natural corticoids are: (1) a double bond at C-4-5, (2) a ketone group at C-3, and (3) a ketone group at C-20.

Glucocorticoids

These hormones, which include an oxygen atom at C-11 (as hydroxyl or ketone group) in their structures, have their greatest effect on carbohydrate and protein metabolism, with a minor effect on electrolyte and water metabolism. Examples of this group are corticosterone (see before), cortisol, and cortisone (see here). Cortisol is the major glucocorticoid in humans.

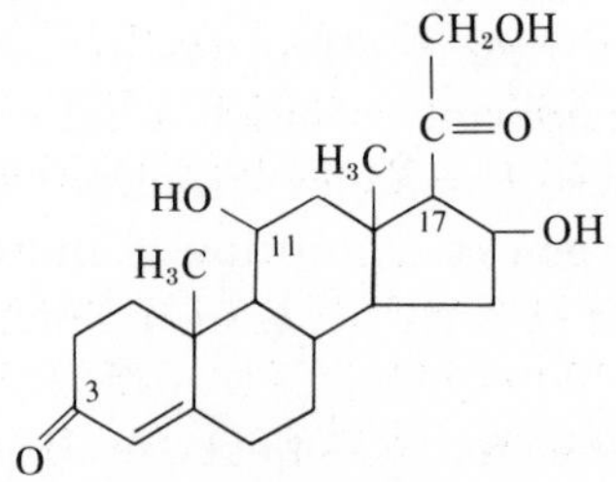

Cortisol

(17-hydroxycorticosterone)

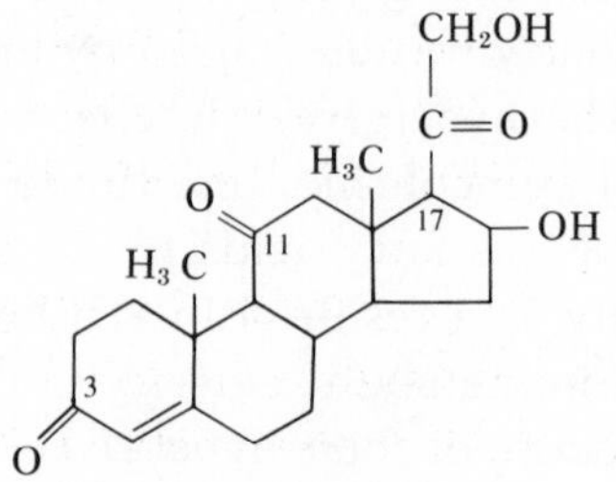

Cortisone

(11-dehydro-17-hydroxy-corticosterone)

The glucocorticoids stimulate the breakdown of muscle protein, increasing the concentration of available free amino acids, which then can be used in gluconeogenesis in the liver.

Electrocorticoids

Corticoids without an oxygen atom at C-11 have very little activity in carbohydrate and protein metabolism but are active in electrolyte and water metabolism, and are therefore known as electrocorticoids. Important examples of this group are 11-deoxycorticosterone and 11-deoxycortisol, shown here:

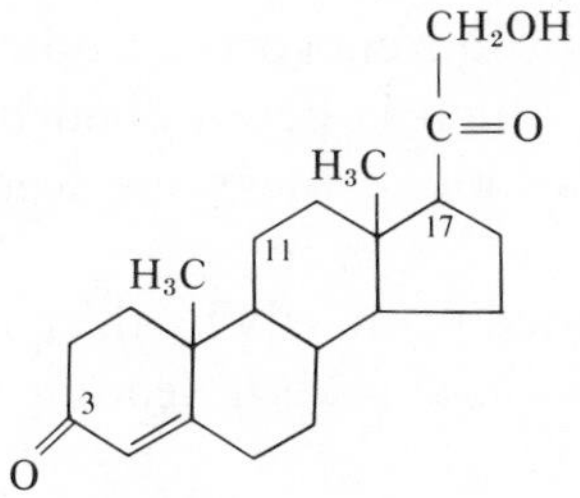

CH_2OH — $C{=}O$ — H_3C — 17 — H_3C — 11 — 3 — O

11-Deoxycorticosterone

CH_2OH — $C{=}O$ — OH — H_3C — 17 — H_3C — 11 — 3 — O

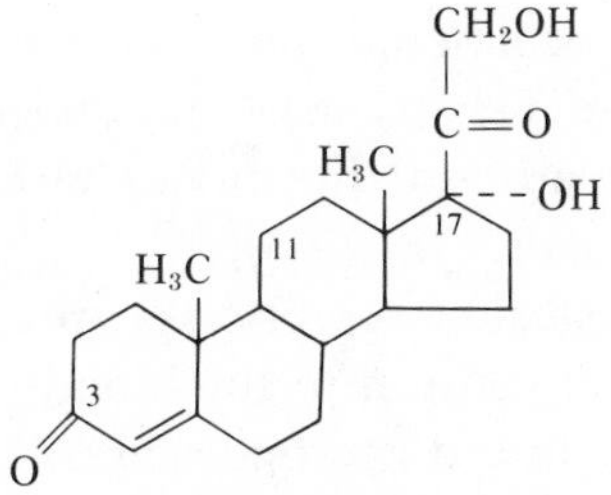

11-Deoxycortisol
(11-deoxy-17-hydroxy-corticosterone)

Mineralocorticoid

Aldosterone, with an aldehyde group instead of a methyl group at C-18, is the most active mineralocorticoid known. Its structure is shown here:

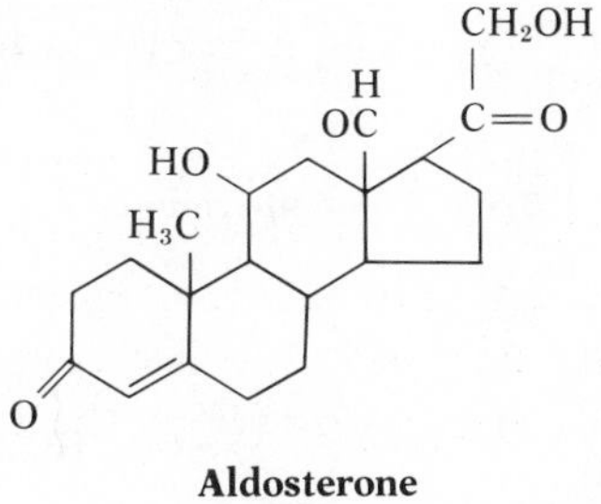

CH_2OH — H — OC — $C{=}O$ — HO — H_3C — O

Aldosterone

Aldosterone exerts its major effect on sodium and potassium metabolism and is the main mineralocorticoid in humans. It may also have effects in carbohydrate, protein, and lipid metabolism, depending on dosage.

Functions of the corticoids

The major effects of the glucocorticoids are to increase the blood glucose concentration and to stimulate the formation of liver glycogen. This is accomplished by decreasing the uptake and use of glucose by the body, and increasing the rate of gluconeogenesis.

The most important function of the mineralocorticoid, aldosterone, is to promote the retention of sodium in the body by stimulating its reabsorption by the tubular cells of the kidney. As a result, potassium excretion and water retention are increased.

Cortisol was found to be very effective in treatment of the inflammation of the joints that occurs in rheumatoid arthritis and rheumatic fever. Because of some adverse side effects, research on synthetic derivatives produced new compounds such as prednisone, prednisolone, and dexamethasone, which have greater anti-inflammatory activity but decreased side effects. One very serious side effect of prolonged corticoid administration is the development or exacerbation of gastrointestinal ulcers.

Regulation of corticoid secretion

Glucocorticoid secretion is stimulated by adrenocorticotropic hormone (ACTH, p. 461), the concentration of which is determined by the level of adrenal hormones circulating in the blood (negative feedback mechanism, Fig. 29-3).

Aldosterone secretion is apparently regulated primarily by the concentration of sodium in the blood, a decrease in blood sodium leading to an increase in aldosterone secretion.

THYROID HORMONES

The thyroid gland produces thyroglobulin, an iodine-containing glycoprotein, which hydrolyzes to give a series of iodinated tyrosines and thyronines. Among the more active forms are a triiodo- and tetraiodothyronine (thyroxine), the structures of which are shown here:

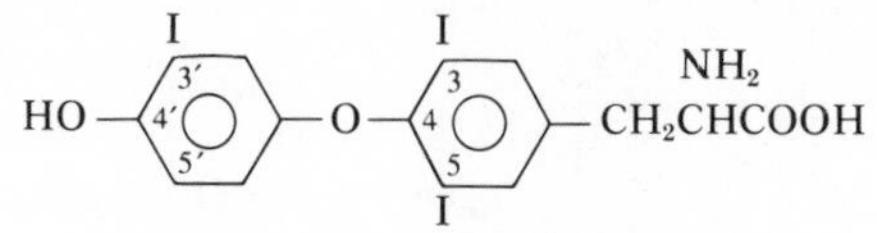

3,5,3′-Triiodothyronine

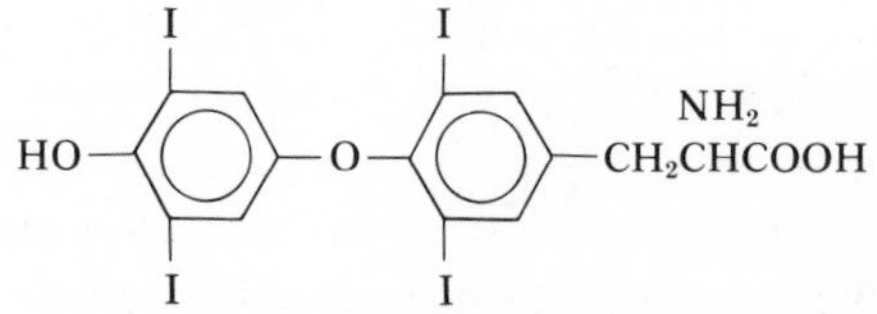

Thyroxine

(tetraiodothyronine)

In general, the activity of triiodothyronine is about four to ten times the activity of thyroxine.

Thyroid hormones act to increase the general rate of metabolism in the body. This phenomenon is referred to as an increase in the *basal metabolic rate* (BMR). Among the metabolic responses attributed to the thyroid hormones are: (1) increase in protein anabolism; (2) increase in hepatic glycogenolysis, gluconeogenesis, and utilization of glucose in the tissues; and (3) increase in oxidation of fatty acids. Some of these effects may be direct, occurring as a result of the influence of thyroid hormones on the effects of other hormones, such as insulin and epinephrine.

Hypothyroidism in infancy causes a condition known as *cretinism*, in which mental and physical growth are severely retarded. If detected soon enough and thyroxine administration is started and continued, the child may develop normally. This effect may also involve growth hormone.

Hypothyroidism in the adult results in *myxedema*, with symptoms of decreased metabolic rate, leading to decreased mental and physical activity. These symptoms are relieved by administration of an appropriate dosage of thyroxine.

In certain geographic areas where the iodine content of the water and food (lack of fresh salt-water fish) is low, hypothyroidism may occur as a result of the lack of a raw material required in the synthesis of thyroid hormone. To circumvent this possibility, most table salt is iodized to ensure an adequate supply of iodine in the diet. Most of the absorbed iodide is very effectively concentrated in the thyroid gland.

Enlargement of the thyroid gland, *goiter*, usually accompanies hypothyroidism, as the tissue attempts to supply the required amount of thyroid hormone.

Hyperthyroidism, caused by excessive development of the thyroid tissue, results in an increase in the metabolic rate, physical and mental activity, and irritability. One of the usual symptoms is a bulging of the eyes, *exophthalmos*. The enlargement of the thyroid gland in this condition is therefore known as *exophthalmic goiter*. Treatment consists of surgical removal of part of the gland, or partial destruction by the administration of radioactive iodine (^{131}I, p. 32). The iodine concentrates in the thyroid tissue, where its radioactivity can effectively destroy some of the thyroid gland, with relatively little effect on the rest of the body. Certain antithyroid chemical agents, which are goitrogenic because of their action in inhibiting the synthesis of thyroid hormone, are also used in treatment of hyperthyroidism. Among these are compounds such as thiouracil and sulfonamide derivatives, structures of which are shown here:

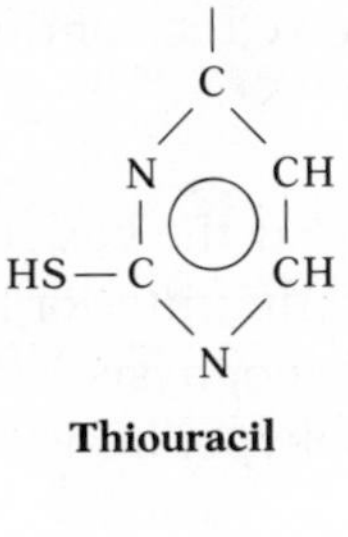

Thiouracil

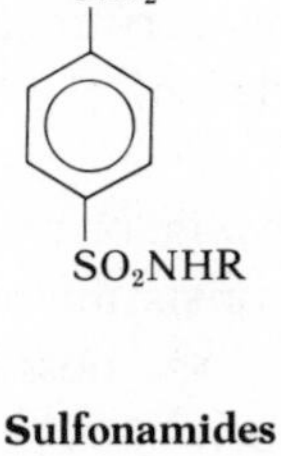

Sulfonamides

Among the procedures used in diagnosing thyroid conditions are determination of: (1) the rate of concentration of ^{131}I, (2) the concentration of protein-bound iodine in the plasma, and (3) the BMR.

Calcitonin

The thyroid gland produces another hormone, calcitonin, secretion of which is apparently stimulated by increasing serum calcium concentration. There appears to be some disagreement in the literature as to whether or not the parathyroid glands also secrete some calcitonin. This hormone is a polypeptide (molecular weight, 3600) that acts to lower the serum calcium concentration by inhibiting the mobilization of calcium from bone. Calcitonin is one of the important factors involved in the control of calcium metabolism in the body, the other major factors being parathyroid hormone and vitamin D metabolites.

PARATHYROID HORMONE

As the designation implies, the parathyroid glands are several small glands attached to the thyroid gland. The hormone produced by these glands is a straight-chain polypeptide (molecular weight, 9500), the chief function of which is to increase the level of calcium ions in the blood serum whenever the level might tend to fall below normal.

The parathyroid hormone carries out its function in increasing the serum calcium by: (1) increasing the intestinal absorption of calcium (adequate amounts of vitamin D required), (2) increasing the rate of mobilization of calcium from bone, (3) increasing the tubular reabsorption of calcium in the kidney, and (4) decreasing the tubular reabsorption of phosphate. This action of the parathyroid hormone is opposed by the action of calcitonin, which inhibits the mobilization of calcium from bone. The secretion of parathyroid hormone appears to depend only on the serum calcium concentration, being stimulated when serum calcium is low and depressed when serum calcium is high. Together with the action of vitamin D metabolites and calcitonin, a fine control of serum calcium concentration can be maintained normally.

On the basis of the effects described, continued withdrawal of minerals from bone in hyperparathyroidism results in decalcification of the bones and possibly formation of calcium phosphate stones in the kidney. Hypoparathyroidism may result in tetany (uncontrollable muscle twitching resulting from low serum calcium), which if allowed to continue may result in convulsive death.

ADENOHYPOPHYSEAL (ANTERIOR LOBE) HORMONES

The hypophysis, or pituitary gland, located at the base of the brain, is a small gland consisting of two major portions, the anterior lobe, or adenohypophysis, and the posterior lobe, or neurohypophysis.

All of the hormones produced by the adenohypophysis are proteins or polypeptides. Most of these, the tropic hormones, control the functional activity of other endocrine glands. At least one, the growth hormone (somatotropin), has a direct effect on metabolism.

Thyrotropic hormone (TSH, thyroid-stimulating hormone)

This hormone, although not yet isolated in pure form, appears to be a glycoprotein (molecular weight approximately 30,000) containing glucose and galactose. Its function is to regulate the growth of the thyroid gland and the secretion of thyroid hormone (Fig. 29-3). The secretion of thyrotropic hormone is controlled mainly by two mechanisms: (1) the concentration of thyroid hormone in the circulating blood (negative feedback); and (2) the production of neurohumoral factors in the hypothalamus by way of stimulation of the central nervous system (Fig. 29-3). In effect, the administration of either thyrotropic hormone or thyrotropin-releasing factor (TRF) would produce the same result as administration of thyroxine.

Adrenocorticotropic hormone (ACTH)

ACTH is a straight-chain polypeptide (molecular weight, 3500) containing thirty-nine amino acid residues, only twenty-four of which are apparently necessary for full activity. Some active compounds have been prepared synthetically. In addition to its major function in stimulating the secretion of hormones by the adrenal cortex, ACTH appears to influence other processes, e.g., it increases the movement of fatty acids from adipose tissue to the blood. The secretion of ACTH is controlled in a manner similar to the control of TSH secretion (Fig. 29-3). In addition, there appears to be a neurohypophyseal factor, corticotropin-releasing factor (CRF), involved in the control mechanisms.

Gonadotropic hormones

Three hormones that influence the development and function of the gonads are produced by the adenohypophysis: (1) follicle-stimulating hormone (FSH); (2) luteinizing hormone (LH, or interstitial cell–stimulating hormone, ICSH); and (3) prolactin (lactogenic hormone, LTH). In the female, they affect growth of the ova and secretions of the ovary; in the male, they affect spermatogenesis and production of androgens. Secretion of the gonadotropic hormones is controlled mainly by the neurohumoral factor mechanism, which is responsive to stimuli from the central nervous system (Fig. 29-3). In the female, these hormones are produced in cycles (Fig. 29-4).

Follicle-stimulating hormone

FSH is a water-soluble glycoprotein (molecular weight, 34,000). As the name implies, FSH stimulates the growth and maturation of ovarian follicles in the female and spermatogenesis in the male. In the presence of LH, FSH stimulates the secretion of estrogen by the ovary.

Luteinizing hormone

LH is a water-soluble glycoprotein, which has been isolated in pure form (molecular weight, 26,000). In conjunction with FSH, in the female, LH influences ovulation of mature follicles and secretion of estrogen by the ovary. LH is also involved in the formation of the corpus luteum and, with LTH, influences the production of estrogen and progesterone by the corpus luteum. In the male, LH is concerned with the development of the interstitial cells (Leydig cells) of the testis and secretion of androgen (testosterone).

Prolactin

This lactogenic hormone (LTH) is a protein that has been isolated in pure form (molecular weight, 24,000). It is responsible for causing lactation in women after childbirth and after prior stimulation of the breast by estrogen and progesterone. The function of LTH in the male is not known.

Growth hormone (somatotropin)

Human growth hormone is made up of a single polypeptide chain (molecular weight, 21,500) and has been prepared in crystalline form. It is interesting to note that human and monkey growth hormones are active in all species tested, but the beef growth hormone, which could be a good source for therapeutic use, is not effective in humans or monkeys. This indicates species specificity, which is a troublesome problem many times in life science research.

Growth hormone affects many metabolic processes. It has a protein anabolic effect, a hyperglycemic, anti-insulin effect, and it stimulates the mobilization of fatty acids from adipose tissue and subsequently, increased fatty acid oxidation and ketogenesis.

Insufficient growth hormone in childhood results in *dwarfism,* but mental development is not retarded. Excessive secretion of growth hormone in childhood causes *gigantism* and, in adults, causes *acromegaly.*

NEUROHYPOPHYSEAL (POSTERIOR LOBE) HORMONES

Two hormones, vasopressin and oxytocin, have been isolated from the posterior lobe of the pituitary. Both hormones are octapeptides with similarities in structure, and both have been synthesized in the laboratory, the first peptide hormones to be prepared in a test tube. Because of their similar structures, each has a degree of activity for processes for which the other is mainly responsible. These hormones are stored in nerve cell endings of the neurohypophysis and are released in the blood after appropriate stimulation.

Vasopressin

Administration of vasopressin (Pitressin) causes an increase in blood pressure known as the pressor effect and an increased concentration of solids in a decreased urine volume known as the antidiuretic effect. The importance of the pressor effect in humans is not certain, but the antidiuretic effect is definitely significant and causes increased reabsorption of water by the cells of the distal tubules of the kidney as the urine is being formed (p. 392). In the absence of vasopressin the urine cannot be concentrated and, as a result, large volumes of urine of low specific gravity (diuresis) are excreted. The secretion of vasopressin is responsive to stimuli from the nervous system, changes in osmotic pressure of the blood, changes in volume of the water compartments of the body, and various drugs, such as morphine and ether.

Oxytocin

Oxytocin (Pitocin) causes contraction of the uterus and ejection of milk from the lactating breast, and it stimulates the contraction of the intestine, gallbladder, ureter, and urinary bladder. Suckling increases the secretion of oxytocin.

PANCREATIC HORMONES

The two pancreatic hormones are synthesized by cells in the islands of Langerhans, insulin by the beta cells and glucagon by the alpha cells. In general terms, insulin has a hypoglycemic effect (lowers blood sugar concentration), whereas glucagon has a hyperglycemic effect (raises blood sugar concentration).

Insulin

Insulin is a protein hormone (molecular weight, 5700), the structure of which has been completely determined. Two research groups, one in the United States and the other in China, were the first to claim some measure of success in the complete synthesis of insulin in the laboratory. Insulin consists of two straight polypeptide chains, one of twenty-one amino acids, the other of thirty amino acids, bridged in two places by the amino acid, cystine. Because of its protein nature, insulin is destroyed by proteolytic enzymes in the gastrointestinal tract and must be administered by injection for therapeutic purposes.

Insulin acts to decrease blood sugar by stimulating liver and muscle glycogenesis, decreasing gluconeogenesis, and increasing glucose utilization by way of oxidation, lipogenesis, and protein synthesis. The secretion of insulin appears to be regulated directly only by the blood sugar level, rising as the blood sugar increases and decreasing as the blood sugar becomes lower. Effects of other substances on insulin secretion are indirect by way of their influence on the blood sugar concentration.

Diabetes mellitus, in which insulin is lacking or is present in insufficient amount, is an irreversible condition. This means that an individual who has diabetes will require treatment of some kind at all times. Since insulin must be given by injection because of its protein nature, a search for oral agents to regulate carbohydrate metabolism has been in progress for a long time. Some success has been achieved in this direction with sulfonylurea derivatives, such as Orinase. These agents seem to work better if there is at least a minimum production of insulin by the patient, which indicates that stimulation of the release of insulin from the pancreas may be involved. Long-acting forms of insulin, such as protamine zinc insulin, have also been developed to decrease the frequency of injections. Research is continuing on the use of very small portable computerized pumps that will be able to sense the glucose concentration of the blood, calculate the amount of insulin required, and introduce this amount of insulin into the bloodstream.

Glucagon

Glucagon is a small, straight-chain polypeptide (molecular weight, 3485) made up of twenty-nine amino acids. This hormone acts to raise the blood sugar level by increasing the rate of glycogenolysis in the liver. Glucagon functions in this way by accelerating the reactivation of the enzyme phosphorylase, which catalyzes the breakdown of glycogen. The action of glucagon is rapid, the blood sugar rising and falling within an hour of administration. In comparison, it should be noted that epinephrine increases both hepatic and muscle glycogenolysis. Glucagon also increases mobilization of fatty acids from adipose tissue and ketogenesis in support of its hyperglycemic activity.

GASTROINTESTINAL HORMONES

Secretin

This hormone is a relatively small polypeptide of twenty-seven amino acid residues that has been isolated in crystalline form. It is synthesized by the mucosal glands of the duodenum and jejunum and is secreted in response to various substances normally found in the digestive tract, such as acid (from the stomach), polypeptides, and fatty acids. It increases the flow of pancreatic juice and bile. It is interesting to note that the pancreatic juice produced on stimulation with secretin is of greater volume and bicarbonate content but lower enzyme concentration in comparison to juice flowing in response to vagus stimulation.

Cholecystokinin-pancreozymin

Cholecystokinin-pancreozymin, at one time considered to be two different hormones, has now been determined to be a single substance—a peptide of thirty-three amino acid residues.

It is secreted by the mucosal glands of the upper regions of the intestines (as is secretin), especially in response to fat and fatty acids. It stimulates the flow of bile by causing the contraction and emptying of the gallbladder and also stimulates the increased output of enzymes by the pancreas.

Gastrin

Gastrin is made in the pyloric region of the stomach in response to mechanical distention or vagus stimulation. It stimulates the production of hydrochloric acid in the fundic region of the stomach. Because of this action and its similarity to the effect of histamine, gastrin and histamine were at one time considered to be the same substance. Human gastrin has been isolated in two forms, I and II. On the basis of structural studies, these substances have been identified as amides of heptadecapeptides (peptides containing seventeen amino acid residues) and synthetic human gastrin I has been prepared.

ERYTHROPOIETIN

Erythropoietin, also known as hemopoietin, or erythrocyte-stimulating factor (ESF), is a glycoprotein the secretion of which appears to be stimulated by severe anemias or anoxia (lack of oxygen). Although generally regarded as a hormone secreted by the kidney, erythropoietin is also found in bilaterally nephrectomized animals, indicating at least an auxiliary source of the hormone. Erythropoietin acts to stimulate the formation and release of red cells from bone marrow.

SUMMARY

A hormone is a substance produced in one part of the body and carried elsewhere by the blood to a tissue or tissues that are stimulated into activity.

HORMONE ACTIVITY

Hormones have very specific receptors on the surface of target cells and have a relatively high turnover rate, i.e., short duration of action. Some hormones have multiple actions and some cause different responses in different tissues. Many polypeptide hormones are secreted in prohormone form and are then activated by splitting off a small peptide from the prohormone.

Mechanism of hormone action: According to the second-messenger concept, stimulation of an endocrine gland causes release of a hormone, the first messenger, which at the cellular level influences the activity of membrane-bound adenyl cyclase, which causes conversion of ATP to c-AMP, the second messenger. c-AMP in turn influences many enzyme reactions, membrane permeabilities, ion movements, release of hormones, etc., involved in the production of many physiologic products and responses. Modulating effects, including those of the prostaglandins, provide a delicately sensitive system of control of the concentrations and activities of the first and second messengers (Fig. 29-2).

Some hormones, such as estrogens and androgens, are transported through the cell membrane and the cytosol to the nucleus, where they selectively stimulate or inhibit gene activity, resulting in increased or decreased synthesis of specific mRNAs responsible for the synthesis of specific proteins.

REGULATION OF HORMONE SECRETION By way of nervous system stimulation, tropic hormones, tropic hormone release and release-inhibiting hormones, negative feedback mechanisms (Fig. 29-3).

CHEMICAL CLASSIFICATION OF HORMONES Hormones fall into three classes of compounds: steroids, polypeptides, and amino acid derivatives.

SYNTHESIS OF HORMONES Steroid hormones are synthesized from the acetate unit by way of cholesterol. Polypeptide hormones are made in the same way as other proteins.

Table 29-1. Summary chart of the hormones

Hormones	Site of formation	Chemical structure*	Actions
Androgens (male sex hormones)			
Testosterone (and other androgens)	Interstitial cells of testis	Steroid	Secondary sex characteristics in males, protein anabolic effect
Estrogens (female sex hormones)			
Estrone, estradiol	Ovarian follicle	Steroid	Secondary sex characteristics in females, protein anabolic effect
Progesterone			
Progesterone	Corpus luteum	Steroid	Preparation of uterus for implantation of ovum
Adrenal medulla hormones			
Epinephrine Norepinephrine	Adrenal medulla, nerves	Phenolic amines	Stimulate glycogenolysis, hypertensive effect
Adrenal cortex hormones			
Electrocorticoids (11-deoxy) 11-deoxycortisol	Adrenal cortex	Steroids	Electrolyte, water balance
Glucocorticoids (11-oxy) Cortisone, cortisol	Adrenal cortex	Steroids	Gluconeogenesis from amino acids, antiinsulin effects on glucose metabolism
Mineralocorticoid (aldosterone)	Adrenal cortex	Steroid	Retention of sodium ions, excretion of potassium ions

*In instances where the hormones are composed of amino acids in peptide linkage, the chemical structure is indicated as *peptide* regardless of size. The actual size of each of these hormones is indicated in the text, as well as the designations such as peptide, polypeptide, and protein, which are commonly used to indicate the relative size of these substances.

Table 29-1. Summary chart of the hormones—cont'd

Hormones	Site of formation	Chemical structure	Actions
Thyroid hormones			
Thyroid hormone	Thyroid gland	Amino acid (iodinated): thyroxine or derivatives	Regulates rate of metabolism
Calcitonin	Thyroid gland	Peptide	Lowers serum calcium
Parathyroid hormone			
Parathyroid hormone	Parathyroid glands	Peptide	Regulates blood calcium level
Adenohypophyseal (anterior lobe) hormones			
Thyrotropic hormone (TSH)	Anterior pituitary	Peptide	Stimulates development and secretion of thyroid gland
Adrenocorticotropic hormone (ACTH)	Anterior pituitary	Peptide	Stimulates growth and secretion of adrenal cortex
Follicle-stimulating hormone (FSH)	Anterior pituitary	Peptide	Stimulates growth of follicles and production of estrogen, formation of spermatozoa in male
Luteinizing hormone (LH)	Anterior pituitary	Peptide	Formation of corpus luteum and production of progesterone, production of androgens by interstitial cells in male
Prolactin (lactogenic hormone, LTH)	Anterior pituitary	Peptide	Initiates lactation
Growth hormone	Anterior pituitary	Peptide	Growth (also affects fat and carbohydrate metabolism)
Neurohypophyseal (posterior lobe) hormones			
Vasopressin, (Pitressin, antidiuretic hormone, ADH)	Posterior pituitary	Peptide	Increases blood pressure, stimulates reabsorption of water in kidney tubule
Oxytocin (Pitocin)	Posterior pituitary	Peptide	Contracts uterus

Continued.

Table 29-1. Summary chart of the hormones—cont'd

Hormones	Site of formation	Chemical structure	Actions
Pancreatic hormones			
Insulin	Islet cells of pancreas, beta cells	Peptide	Facilitates carbohydrate catabolism
Glucagon	Islet cells of pancreas, alpha cells	Peptide	Raises blood sugar by hepatic glycogenolysis
Gastrointestinal hormones			
Secretin	Duodenum (in presence of acid food)	Peptide	Stimulates flow of pancreatic juice and bile to a much smaller extent
Cholecystokinin-pancreozymin	Duodenum (fat or acid stimulus)	Peptide	Contraction of gallbladder; stimulates secretion of pancreatic enzymes
Gastrin	Pyloric mucosa of stomach	Peptide	Stimulates secretion of gastric juice (HCl)
Erythropoietin			
Erythropoietin (ESF)	Kidney	Peptide	Stimulates red cell formation

REVIEW QUESTIONS

1. What is a hormone? Discuss the key concepts of the "second messenger" proposal with respect to the mechanism of action of hormones.
2. How do controls on the secretion of hormones tie in with the control of metabolic processes? Explain with examples. What may be the reason for some controls overlapping?
3. What types of chemical structures are found among the hormones? How are these structures synthesized?
4. For the hormones which are emphasized in your class, list the site of formation, chemical type, action, and possible associated metabolic disorders.
5. What is the major difference in structure between testosterone and estrone? between the corticoids? How do these changes affect the functions of these substances?

REFERENCES

Annual review of biochemistry, Palo Alto, Calif., Annual Reviews.

Annual review of physiology, Palo Alto, Calif., Annual Reviews.

Antoniades, H. N., editor: Hormones in human blood: detection and assay, Cambridge, Mass., 1976, Harvard University Press.

Butt, W. R., editor: Hormone chemistry, vol. 1, ed. 2, New York, 1975, Halsted Press.

Frieden, E. H.: Chemical endocrinology, New York, 1976, Academic Press, Inc.

Gray, C. H., and Bacharach, A. L., editors: Hormones in blood, ed. 2, 2 vols. New York, 1967, Academic Press, Inc.

Heftmann, E., editor: Modern methods of steroid analysis, New York, 1973, Academic Press, Inc.

Litwack, G., editor: Biochemical actions of hormones, 6 vols., New York, 1970-1979, Academic Press, Inc.

Pasqualini, J. R., editor: Receptors and mechanism of actions of steroid hormones, I, II, New York, 1976, 1977, Marcel Dekker, Inc.

Recent progress in hormone research, New York, Academic Press, Inc.

Rickenberg, H. V., editor: Biochemistry of mode and action of hormones, Baltimore, Md., 1978, University Park Press.

Sutherland, E. W.: On the biological role of cyclic AMP, J.A.M.A. **214:**1281, 1970.

Vitamins and hormones (annual publication), New York, Academic Press, Inc.

White, W. L., Erickson, M. M., Stevens, S. C.: Chemistry for the clinical laboratory, ed. 4, St. Louis, 1976, The C. V. Mosby Co.

30 Biochemical basis of nutrition

Nutrition information, policies, and possible intervention or treatment plans designed to combat a malnutrition problem of one sort or another should be based on proved information available from studies of intermediary metabolism. It would then follow naturally that most of the subjects discussed in the previous chapters of Unit Four must be considered as being very closely and inseparably related to the field of nutrition. This is not to imply that a good knowledge of intermediary metabolism will supply all the answers to nutritional problems, but rather that such knowledge should always be used to its fullest extent in attempting to solve such problems. In addition, a good knowledge of intermediary metabolism will serve to indicate the direction of additional studies required to fill in gaps in current nutrition information.

In this chapter the discussions of nutritionally related abnormalities and examples of some disease states and conditions with important nutritional aspects emphasize the very close relationship of nutrition to intermediary metabolism. Some general considerations of nutrition not covered elsewhere in the text are also included in this chapter. This coverage is not meant to be exhaustive but rather to provide broad indication of proper concerns of nutritionists.

NUTRITIONALLY RELATED ABNORMALITIES

Diabetes mellitus

The basic metabolic defect in diabetes mellitus is the lack of production of sufficient insulin required for the normal utilization of glucose (p. 310). It is well known that diabetes can usually be controlled by injections of insulin. Why is it that insulin cannot be taken by mouth? A simple consideration of the chemical nature of insulin (p. 463) and what happens to it if taken by mouth supplies the answer. Insulin is a small polypeptide that would be broken down to amino acids if taken by mouth. Therefore, it must be supplied by injection in order to bypass the intestinal tract. This has led to an important area of research, namely that of discovering adequate oral agents (p. 463).

In some individuals who have diabetes, the condition can be controlled by the proper manipulation of their diets, one of the most important factors being the control of the intake of glucose in one form or another. However, the tendency toward ketosis and acidosis in diabetes might be aggravated by decreasing the intake of carbohydrates too drastically. Therefore, careful dietary adjustments on an individual basis are required for this type of intervention to be successful. The acidosis that may result in untreated diabetes is a possible life-threatening situation that brings into play the metabolic control systems involved in acid-base balance (p. 414).

Retinopathy is another serious consequence in many diabetic persons. The molecular mechanisms or the biochemistry responsible for the condition are not well enough understood at present to provide a proper basis for devising a means of treatment.

Immune system

Many studies have shown that the ability of the body to combat infections—bacterial or viral—is diminished if the individual is malnourished. This is reflected in many statistics on incidence of disease, mortality, and life expectancy in populations known to have relatively poor nutrition. Much of these effects is undoubtedly due to poor protein synthesis resulting from lack of sufficient essential and nonessential amino acids in such populations, particularly since the enzymes required to keep cellular metabolism operating normally are proteins. It is unfortunate that some grains, which many people rely on to a very large extent for their food supply, lack essential amino acids such as lysine. This has led to research efforts to develop strains of grains with more complete amino acid composition or higher concentrations of the essential amino acids.

Mental retardation

One of the most publicized conditions that can lead rapidly to mental retardation in newborn infants if not treated correctly is phenylketonuria (PKU). The nutritional problem involved is serious and difficult to overcome. The metabolic difficulty is a genetic defect in the formation of the enzyme phenylalanine hydroxylase, which is required for the normal conversion of the essential amino acid phenylalanine to the amino acid tyrosine. As a result, the concentration of phenylalanine and some of its normally minor metabolic products increase to dangerous levels. The exact molecular mechanism that causes mental retardation is still under investigation.

As soon as the condition and its metabolic significance were recognized, a rational plan of treatment was instituted. All infants can be tested easily very soon after birth and those with PKU are placed on special diets low in phenylalanine for at least the first few years of life. Because phenylalanine is an essential amino acid, it cannot be removed completely from the diet. For some years it appeared as if it were possible eventually to take these children off the special diet. Recently, however, disturbing reports have begun to appear indicating possible continuing brain damage if these children return to normal diets.

Malnutrition

It is beyond the scope of this text to discuss to any great extent the serious difficulties related to malnutrition throughout the world. However, it is important to point out some significant problems that appear in rather unexpected areas.

In recent studies, the nutritional status of surgical patients and of general medical patients in the wards of a municipal teaching hospital was surveyed. The results showed that 50% of the surgical patients and 44% of the general medical patients were suffering from protein-calorie malnutrition. It appears to be difficult for most people living in a relatively affluent society to believe that malnutrition is an immediate problem. Nevertheless, it is obvious from the discussion above on the metabolic effects of malnutrition on the response of the immune system that the patients in a hospital are at a disadvantage in their ability to recover unless their malnutrition is recognized and alleviated.

Another survey of patients on special diets showed that the patients were actually eating diets that were quite different from those prescribed because of the unrestricted access of these patients to other sources of food. As a result, all sort of erroneous conclusions might be drawn about the effectiveness of the prescribed diets. Even when a patient has no access to other food, generally no record is kept as to actual intake other than a routine watch on weight.

An unseen and unexpected problem is that of long-range effects of malnutrition. In some instances, the serious effects may not show up until the second or third generation. An example of this is vitamin B_{12} malnutrition. Certain strict vegetarians, known as vegans, who do not eat any fish, meat, or dairy products are subject to pernicious anemia. However, although the symptoms may not always appear with the mother, there have been reports of serious anemia problems in infants born to such mothers because if they consume a diet of vegetables only, they are taking in an inadequate supply of vitamin B_{12} which is required for the proper production of hemoglobin (p. 380).

Fad diets

The current obsession with weight-losing diets has generally created more problems for the dieters than they have solved. Metabolically the body seems to function best when the composition of the diet is approximately 60% carbohydrate, 30% lipids, 10% protein. Many proposed diets involve a decrease in the relative or absolute amount of carbohydrate, to the point where ketosis can become a serious health problem. The individual who seriously wants to lose some excess weight and is normal in all other respects, can do so simply by eating slightly less each day of what would be considered a balanced diet on the basis of the best current information. The benefit of such an approach is that the metabolic stress to the body is minimal whereas crash diets almost invariably are accompanied by the dangers of ketosis.

In all of these examples, there is a significant relationship between nutrition and the metabolic problem of the patient. However, in the understandable rush to solve the more acute aspects of the patient's difficulties, this relationship is often overlooked for extended periods of time and sometimes forgotten altogether.

GENERAL CONSIDERATIONS

The relationship between nutrition and biochemistry is a very close one. In a very real sense, biochemistry describes what happens to nutrients from the moment of ingestion, through energy-producing and anabolic pathways, until breakdown products are excreted as waste materials. The ultimate aim of nutritional studies is to make it possible to devise specific diets for any specific conditions an individual may encounter in a lifetime, which would allow the body to meet all of its requirements of the moment. An adequate diet may be defined as one under which growth, maintenance, and reproduction occur in a normal manner. That requirements change as the individual grows is a well-known fact and is reflected in the dietary recommendations of the Food and Nutrition Board of the National Academy of Sciences–National Research Council (Tables 30-1, 30-2). The challenge then becomes (1) determination of the value to the individual of each dietary constituent in calories, minerals, vitamins, and other essential constituents, and (2) determination of the changes in requirements with changes in conditions of the individual.

Nutritional experiments are beset with all sorts of difficulties. Most such experiments are performed on animals other than the human. This means that the results obtained must be extrapolated to human requirements and an estimate based on the best available information is made. Another major difficulty stems from the fact that many animal experiments start with a standard diet and evaluate the differences resulting from changes in one ingredient at a time. Should more than one major ingredient be changed at the same time, the results of such changes on the status of the animal are not necessarily determined by taking into account the results of single changes. It is known that the protein requirement depends not only on the carbohydrate content of the diet, but also on the total calories (protein-sparing effect, p. 354). As a result of such problems, most dietary recommendations include a wide margin of safety to take care of species and individual differences.

Although nutrition is often considered simply on the basis of "three square meals a day," research has indicated that it is far more important than simply filling the stomach. Observations have been made that point to the probability that the adequacy of nutrition has a definite relationship to longevity, especially during the early periods of life. Evidence has also accumulated that indicates a high correlation between malnutrition and mental retardation, particularly when the malnutrition occurs at very early ages. Studies have also indicated the importance of good nutrition to development and social behavior.

Table 30-1. Food and Nutrition Board, National Academy of Sciences–National Research Council Recommended Daily Dietary Allowances.* Revised 1980

	Age (years)	Weight		Height		Energy needs (kcal)	Protein (g)	Fat-soluble vitamins		
		kg	lbs	cm	in			Vitamin A (μg RE)†	Vitamin D (μg)‡	Vitamin E (mg α TE)§
Infants	0.0-0.5	6	13	60	24	kg × 115	kg × 2.2	420	10	3
	0.5-1.0	9	20	71	28	kg × 105	kg × 2.0	400	10	4
Children	1-3	13	29	90	35	1300	23	400	10	5
	4-6	20	44	112	44	1700	30	500	10	6
	7-10	28	62	132	52	2400	34	700	10	7
Males	11-14	45	99	157	62	2700	45	1000	10	8
	15-18	66	145	176	69	2800	56	1000	10	10
	19-22	70	154	177	70	2900	56	1000	7.5	10
	23-50	70	154	178	70	2700	56	1000	5	10
	51+	70	154	178	70	2400	56	1000	5	10
Females	11-14	46	101	157	62	2200	46	800	10	8
	15-18	55	120	163	64	2100	46	800	10	8
	19-22	55	120	163	64	2100	44	800	7.5	8
	23-50	55	120	163	64	2000	44	800	5	8
	51+	55	120	163	64	1800	44	800	5	8
Pregnant						+300	+30	+200	+5	+2
Lactating						+500	+20	+400	+5	+3

*The allowances are intended to provide for individual variations among most normal persons as they live foods in order to provide other nutrients for which human requirements have been less well defined. (From vised 1980, Washington, D.C., 1980, U.S. Government Printing Office.)

†Retinol equivalents. 1 Retinol equivalent = 1 μg retinol or 6 μg β-carotene.

‡As cholecalciferol. 10 μg cholecalciferol = 400 IU vitamin D.

§α-Tocopherol equivalents. 1 mg d-α-tocopherol = 1 α TE.

‖1 NE (niacin equivalent) is equal to 1 mg of niacin or 60 mg of dietary tryptophan.

¶The folacin allowances refer to dietary sources as determined by *Lactobacillus casei* assay after treatment

#The RDA for vitamin B_{12} in infants is based on average concentration of the vitamin in human milk. The diatrics) and consideration of other factors such as intestinal absorption.

**The increased requirement during pregnancy cannot be met by the iron content of habitual American recommended. Iron needs during lactation are not substantially different from those of nonpregnant to replenish stores depleted by pregnancy.

Nutrition is rapidly becoming a more urgent worldwide problem as the world population daily increases faster than the rate of total food production. Many warnings have been voiced in recent years of impending widespread famines in the near future.

Some nutritional aspects of biochemistry have already been discussed with the metabolism of carbohydrates (p. 291), lipids (p. 316), proteins (p. 352), and nucleic acids (p. 336). Dietary requirements of vitamins are included in Chapter 28. Certain other considerations and the metabolism of some strategic inorganic substances will be discussed next.

Water-soluble vitamins							Minerals					
Vitamin C (mg)	Thiamin (mg)	Riboflavin (mg)	Niacin (mg NE)‖	Vitamin B_6 (mg)	Folacin¶ (μg)	Vitamin B_{12} (μg)	Calcium (mg)	Phosphorus (mg)	Magnesium (mg)	Iron (mg)	Zinc (mg)	Iodine (μg)
35	0.3	0.4	6	0.3	30	0.5#	360	240	50	10	3	40
35	0.5	0.6	8	0.6	45	1.5	540	360	70	15	5	50
45	0.7	0.8	9	0.9	100	2.0	800	800	150	15	10	70
45	0.9	1.0	11	1.3	200	2.5	800	800	200	10	10	90
45	1.2	1.4	16	1.6	300	3.0	800	800	250	10	10	120
50	1.4	1.6	18	1.8	400	3.0	1200	1200	350	18	15	150
60	1.4	1.7	18	2.0	400	3.0	1200	1200	400	18	15	150
60	1.5	1.7	19	2.2	400	3.0	800	800	350	10	15	150
60	1.4	1.6	18	2.2	400	3.0	800	800	350	10	15	150
60	1.2	1.4	16	2.2	400	3.0	800	800	350	10	15	150
50	1.1	1.3	15	1.8	400	3.0	1200	1200	300	18	15	150
60	1.1	1.3	14	2.0	400	3.0	1200	1200	300	18	15	150
60	1.1	1.3	14	2.0	400	3.0	800	800	300	18	15	150
60	1.0	1.2	13	2.0	400	3.0	800	800	300	18	15	150
60	1.0	1.2	13	2.0	400	3.0	800	800	300	10	15	150
+20	+0.4	+0.3	+2	+0.6	+400	+1.0	+400	+400	+150	**	+ 5	+25
+40	+0.5	+0.5	+5	+0.5	+100	+1.0	+400	+400	+150	**	+10	+50

in the United States under usual environmental stresses. Diets should be based on a variety of common
National Academy of Sciences–National Research Council: Recommended daily dietary allowances, re-

with enzymes ("conjugases") to make polyglutamyl forms of the vitamin available to the test organism.
allowances after weaning are based on energy intake (as recommended by the American Academy of Pe-

diets nor by the existing iron stores of many women; therefore the use of 30-60 mg of supplemental iron is
women, but continued supplementation of the mother for 2-3 mo after parturition is advisable in order

Caloric requirements

The total caloric requirement of an individual is based on the expenditure of energy for three major purposes: (1) basal metabolism, (2) specific dynamic action of food, and (3) muscular activity. Inadequacies in dietary calories must be supplied by stored components or tissue breakdown. The approximate calories supplied by the major food groups are listed here:

Food	*kcal/g*
Carbohydrate	4
Lipid	9
Protein	4

Specific recommendations of total caloric intakes for various individuals are listed in Table 30-1.

Basal metabolism. By basal metabolism is meant the expenditure of energy that takes place at complete rest (no voluntary muscular activity) and in the postabsorptive state (about 16 hours after the last food). Under such conditions, the energy must come from stored substances and is dissipated almost entirely in the form of heat. Each individual has a characteristic basal metabolism, or *basal metabolic rate* (BMR), which classically was determined in an indirect manner on the basis of the amount of oxygen used and the amount of carbon dioxide produced. Tables of average BMRs based on measurements from large numbers of individuals are available. The BMR is related to the surface area of the body rather than the weight and may be expressed as (1) kilocalories (kcal, Cal) per square meter of body surface per hour, (2) liters of oxygen consumed per square meter per hour; or (3) percent above or below the appropriate normal average BMR.

The BMR may be influenced by age, sex, body size, temperature, barometric pressure, states such as pregnancy and lactation, level of anxiety, and drug treatments. BMR measurements were used at one time in the diagnosis of abnormal thyroid conditions. About 40% of the BMR results from the action of thyroid hormone in normal concentration. The BMR of athyroid individuals is therefore about 40% below the normal BMR. The administration of 1 mg of thyroxine raises the BMR about 3% in such individuals. On this basis, the BMR was used to evaluate the extent of hypo- or hyperactivity of the thyroid gland in patients. A range of ±10% of the normal BMR is generally considered normal.

More modern methods that measure the status of thyroid function more directly are generally used at present. Iodine is a prominent part of the thyroid hormone structure (p. 458). When thyroxine appears in the blood it is bound by plasma protein. This protein can be precipitated and its iodine content determined (protein-bound iodine, PBI). The value obtained is then compared to established ranges including normal, low (hypothyroid), and high (hyperthyroid) values. Another direct method involves the uptake of radioactive iodine (p. 32), the rate and amount of uptake being compared to established ranges.

Specific dynamic action of food (SDA). An increase in heat production is observed in individuals under otherwise basal conditions after ingestion of food. This phenomenon is termed the *specific dynamic action* (SDA) of food and is greatest for protein. Therefore, in order to meet an individual's caloric requirement, after the caloric content of a recommended diet is calculated, up to 15% or 20% additional calories must be provided in order to account for the SDA.

Muscular activity. This category of caloric requirement is the most variable. Increases in hourly caloric requirements may range from 35 kcal for sitting quietly to 1000 kcal for walking rapidly, and higher for more strenuous activities.

Carbohydrate in the diet

At the present time, it is believed that all necessary carbohydrates can be synthesized in the body. Therefore, carbohydrates are not essential ingredients in the diet. However, for best functioning of metabolic processes, it appears that a good diet should contain approximately 60% carbohydrates.

As explained in the metabolism of carbohydrates, the body is geared to use carbohydrates as the preferred fuel for energy purposes. In situations of dietary carbohydrate restriction or starvation, increased fatty acid oxidation and ketogenesis take place, with possibly serious effects. Also, carbohydrates exhibit a protein-sparing effect. An additional benefit from carbohydrates in the diet is the roughage supplied by indigestible substances such as cellulose from plant products. Roughage apparently is of help in keeping the intestinal contents moving through at a suitable rate.

Lipid in the diet

Except for the essential fatty acids (linoleic, linolenic, arachidonic), it is believed that all other necessary lipid components can be synthesized in the body. This means that there must be at least enough lipid in the diet to supply these essential components. It is recommended that the usual diet contain about 30% lipids. At such levels, lipids serve mainly as a fuel for caloric requirements.

In recent years, much interest and controversy have centered around the relationship of dietary lipids, especially cholesterol, to the development and progress of atherosclerosis. As yet, there is no definitive evidence on which to base conclusions about the significance of including or omitting specific lipids from the diet. It is known, however, that plasma cholesterol levels may be lowered by restricting lipid intake or by increasing the ratio of unsaturated to saturated fatty acids in the diet.

Protein in the diet

The main purpose of proteins in the diet is to supply the essential and, in optimum situations, also the nonessential amino acids required for the synthesis of enzymes, tissue proteins, plasma proteins, and hormones, to list a few of the important protein constituents of the body. Protein is used as a fuel to a significant extent only in emergency circumstances. The recommended intakes of protein in various situations are listed in Table 30-1.

Important factors that must be kept in mind in determining the amount of protein to be included in a diet are the need for maintaining or increasing the nitrogen balance (depending on the growth stage of the individual or requirements of convalescence) and the biologic value of the available dietary proteins. The usual recommendation for adults is approximately 1 g protein per kg body weight, the figure being increased for children, convalescents, and pregnant women.

Important considerations about the biologic value of food proteins have already been discussed (pp. 352, 353). A deficiency of dietary protein for any length of time must be considered as a very serious situation. Unfortunately, the world supply of foodstuffs is most lacking in the main strategic group, the proteins. The typical potbellied appearance of undernourished children is caused mainly by insufficient intake of protein, usually accompanied by an inadequate total caloric intake.

Inorganic constituents in the diet

Calcium. In addition to its more obvious function in the formation of bones and teeth, calcium also functions in (1) blood coagulation (p. 372), (2) formation of milk, (3) muscle contraction and excitability, (4) transmis-

Table 30-2. Estimated safe and adequate daily dietary intakes of additional selected vitamins and minerals*

		Vitamins			Trace elements†	
	Age (years)	Vitamin K (μg)	Biotin (μg)	Pantothenic acid (mg)	Copper (mg)	Manganese (mg)
Infants	0-0.5	12	35	2	0.5-0.7	0.5-0.7
	0.5-1	10-20	50	3	0.7-1.0	0.7-1.0
Children and adolescents	1-3	15-30	65	3	1.0-1.5	1.0-1.5
	4-6	20-40	85	3-4	1.5-2.0	1.5-2.0
	7-10	30-60	120	4-5	2.0-2.5	2.0-3.0
	11+	50-100	100-200	4-7	2.0-3.0	2.5-5.0
Adults		70-140	100-200	4-7	2.0-3.0	2.5-5.0

*Because there is less information on which to base allowances, these figures are not given in the main table
Board, National Academy of Sciences–National Research Council: Recommended dietary allowances, re
†Since the toxic levels for many trace elements may be only several times usual intakes, the upper levels for

sion of nerve impulses, (5) activation of enzymes, and (6) changes in membrane permeability.

Approximately 99% of the body calcium is found in the bones. Bone calcium is in equilibrium with serum calcium, the concentration of which is under the influence of vitamin D (p. 427), parathyroid hormone (p. 460), and calcitonin (p. 460). One difficulty in obtaining an adequate intake of calcium is the fact that milk is the only good source of calcium. This leads to serious problems at times, for children or adults who cannot tolerate milk. The recommended daily intake of calcium is 0.8 g for adults and children.

Iron. Iron is important as a constituent of hemoglobin (p. 379), required for the transport of oxygen to the tissues, and of the cytochromes (p. 284), required for cellular oxidations. Iron is also found in myoglobin (muscle protein) and other enzymes such as catalase and peroxidase. Iron is transported in the blood in the form of a β_1-globulin known as transferrin.

An interesting feature of iron metabolism is the fact that there is no provision for the excretion of iron. Iron resulting from the breakdown of hemoglobin or other prosthetic groups reenters the body pool of iron and is reutilized as necessary. Inevitably some iron is lost by accidental or functional bleeding and by the constant replacement of skin tissue, which contains a small amount of iron. As a result, there is a dietary requirement for replacing such losses. It appears that the absorption of iron depends on the pool of iron available in the body, i.e., more iron is absorbed when the body pool is low. Exactly how this is accomplished is still under investigation.

Trace elements†				Electrolytes		
Fluoride (mg)	**Chromium (mg)**	**Selenium (mg)**	**Molybdenum (mg)**	**Sodium (mg)**	**Potassium (mg)**	**Chloride (mg)**
0.1-0.5	0.01-0.04	0.01-0.04	0.03-0.06	115-350	350-925	275-700
0.2-1.0	0.02-0.06	0.02-0.06	0.04-0.08	250-750	425-1275	400-1200
0.5-1.5	0.02-0.08	0.02-0.08	0.05-0.1	325-975	550-1650	500-1500
1.0-2.5	0.03-0.12	0.03-0.12	0.06-0.15	450-1350	775-2325	700-2100
1.5-2.5	0.05-0.2	0.05-0.2	0.1-0.3	600-1800	1000-3000	925-2775
1.5-2.5	0.05-0.2	0.05-0.2	0.15-0.5	900-2700	1525-4575	1400-4200
1.5-4.0	0.05-0.2	0.05-0.2	0.15-0.5	1100-3300	1875-5625	1700-5100

of the RDA and are provided here in the form of ranges of recommended intakes. (From Food and Nutrition vised 1980, Washington, D.C., 1980, U.S. Government Printing Office.)
the trace elements given in this table should not be habitually exceeded.

Deficiency of iron may lead to anemia and excess iron (from transfusions or injections) may be deposited in liver and other tissues, leading to serious conditions (hemosiderosis, hemochromatosis) in the presence of large excesses. The recommended daily intake of iron is about 10 to 18 mg.

Phosphorus. The importance of phosphorus in the capture, storage, and transmission of energy (p. 286) has already been stressed. Phosphorus is also a constituent of many strategic substances in metabolic reactions. Some of these are the sugar-phosphate esters in carbohydrate metabolism (p. 295), the nucleic acids (p. 240), the nucleotides NAD and NADP (p. 245), phospholipids (p. 204), and the phosphate buffer system (p. 411). In addition to these functions, phosphorus is closely involved with calcium in the formation of bones, teeth, and milk. The daily requirement for phosphorus is 0.8 to 2 g and is easily obtained in a normal, well-balanced diet because of the wide distribution of phosphorus in foods.

Sodium, potassium, chloride. The functions of sodium, potassium, and chloride ions in water balance (osmotic pressure), electrolyte, and acid-base balance were discussed in Chapter 27. The sodium and chloride ions are also important in digestive fluids, as sodium bicarbonate in bile and pancreatic juice, and as hydrochloric acid in gastric juice.

It should also be recalled that sodium is the major cation and chloride the major anion of extracellular fluid (interstitial fluid and blood plasma), and potassium is the major cation of intracellular fluid. The concentrations of sodium and potassium in the plasma are influenced by the mineralocorticoid hormone, aldosterone (p. 457). Recommended daily intakes of sodium, potassium, and chloride are listed in Table 30-2.

Iodine. The importance of iodine in the normal production of thyroid hormone has been discussed (p. 458). The small, daily requirement of about 0.15 mg is sometimes difficult to obtain in a normal diet in geographic areas far from sea water. For this reason, an iodide salt is added to table salt, which is then known as iodized salt. The thyroid gland has a remarkable capacity for trapping and storing iodine that reaches it by way of the blood.

Copper. Copper is a constituent of certain oxidative enzymes and appears to function in this way in the synthesis of hemoglobin (p. 380). It appears in the α_2-globulin fraction of the plasma proteins as ceruloplasmin, a blue-green protein. This is apparently the transport form of copper. The recommended daily intake is 2 to 3 mg of copper.

Magnesium. Magnesium ion acts as an activator of several important enzymes such as phosphorylase, phosphoglucomutase, and enolase, which function in carbohydrate metabolism (p. 297). It also inhibits the inactivation of ATP and affects neuromuscular irritability. The recommended daily intake for magnesium is 300 to 400 mg.

Other trace elements. Metals such as zinc, manganese, and molybdenum are activators or possibly constituents of enzymes. Cobalt is an important constituent of vitamin B_{12} (p. 436). Fluorine is a strategic constituent of tooth enamel and is found helpful in preventing dental caries when present in drinking water at an appropriate concentration (1 to 1.5 parts per million).

Among other trace elements, the functions of which are still being defined by increasing research efforts, are selenium, chromium, tin, vanadium, and lithium. Because in many instances it is yet to be established that the trace elements are actually required in human nutrition, it is very difficult to recommend any definite daily allowances. Also, because many of the trace metals are toxic in more than trace amounts, it would be wise not to overdo the ingestion of such substances until more specific information is available. Recommended intakes for some of these trace elements are listed in Table 30-2.

SUMMARY

Nutrition information should be based on proved data from studies of intermediary metabolism, the biochemical basis for nutrition.

NUTRITIONALLY RELATED ABNORMALITIES

Diabetes mellitus: Diabetes results from lack of the hormone insulin, which must be supplied by injection, because if taken by mouth, it would be broken down to constituent amino acids. Tendency toward ketosis in individuals who have diabetes might be aggravated by drastic cuts in intake of carbohydrates. Adjustments in diet must be made on an individual basis. Acid-base balance mechanisms attempt to alleviate acidosis in untreated diabetic persons. The molecular mechanism of diabetic retinopathy has not yet been established.

Immune system: Effectiveness of the immune system is diminished in malnutrition, probably by way of decreased protein synthesis.

Mental retardation: Phenylketonuria (PKU) is a well-known example of condition leading to mental retardation if not treated rapidly. Because of a genetic defect, a missing enzyme allows the accumulation of phenylalanine and some normally minor products to rise to dangerous levels. Treatment involves early detection and decrease of phenylalanine dietary intake. Since phenylalanine is an essential amino acid, it cannot be removed entirely from the diet.

Malnutrition: Malnutrition exists even among patients in hospitals. This lowers their ability to recover from their medical conditions. Some long-range effects of malnutrition may not show up until the second or third generation, e.g., vitamin B_{12} malnutrition in vegans.

Fad diets: Many fad and crash diets lead to ketosis because of a decrease in intake of carbohydrates. An otherwise normal individual can lose excess weight safely by eating slightly less each day of a completely balanced diet.

GENERAL CONSIDERATIONS

An adequate diet is one under which growth, maintenance, and reproduction occur normally. Dietary requirements change with the stage of growth and health status of the individual. Results of nutritional experiments on animals must be interpreted carefully when extrapolated to human needs. Observations have indicated that nutrition has a direct bearing on longevity, mental development, and social behavior.

Caloric requirements: Sum of calories expended in (1) basal metabolism, (2) specific dynamic action of food, and (3) muscular activity.

Basal metabolism: The basal metabolic rate (BMR) is the expenditure of energy when at complete rest and about 16 hr after last food. The caloric yields of the major food groups are: carbohydrate, 4 kcal/g; lipid, 9 kcal/g; protein, 4 kcal/g. Total caloric recommendations are found in Table 30-1. BMR determinations have been found very useful in the diagnosis of thyroid conditions because thyroxine activity affects approximately 40% of the BMR.

Specific dynamic action of food (SDA): SDA is the phenomenon of heat production after the ingestion of food under otherwise basal conditions.

Muscular activity: This is the most variable part of the total caloric requirement and depends on the activity of the individual.

Carbohydrate in the diet: Not essential but metabolism functions optimally with approximately 60% of the diet in the form of carbohydrates. Indigestible cellulose adds roughage to intestinal contents.

Lipid in the diet: Necessary at least to supply the essential fatty acids. Recommended at the level of 20% to 35% of the diet. No clear connection yet between blood cholesterol level, dietary supply, and atherosclerosis.

Protein in the diet: Necessary to supply the essential amino acids, and for optimum protein synthesis, also a supply of nonessential amino acids. The biologic value of food proteins must be taken into account in determining an adequate diet.

Inorganic constituents in the diet:

Calcium: Required for bone and tooth formation, blood coagulation, milk formation, muscle contraction and excitability, nerve impulses, activation of enzymes, changes in membrane permeability. Blood calcium concentration is influenced by vitamin D, parathyroid hormone, and calcitonin.

Iron: Important constituent of hemoglobin (oxygen transport), cytochromes (cellular oxidations), myoglobin (muscle protein), and other enzymes. The transport form of iron is transferrin, a β_1-globulin. There is no provision for excretion of iron and absorption depends on the body pool. Deficiency leads to anemia.

Phosphorus: Required in the capture, storage, and transfer of energy; as a constituent of sugar-phosphate esters, nucleic acids, nucleotides, phospholipids, and a buffer system.

Sodium, potassium, chloride: Important in water, electrolyte, acid-base balance, digestive juices. Plasma concentrations of sodium and potassium are influenced by the hormone, aldosterone.

Iodine: Required for the formation of thyroxine.

Copper: Required as an enzyme constituent for the formation of hemoglobin. Transport form of copper is ceruloplasmin, an α_2-globulin.

Magnesium: Acts as enzyme activator, inhibits inactivation of ATP, and influences neuromuscular irritability.

Other trace elements: Zinc, manganese, and molybdenum are found as activators or enzyme constituents. Cobalt is a constituent of vitamin B_{12}. Fluorine is important in preventing dental caries. Other trace elements possibly required in human nutrition are selenium, chromium, tin, vanadium, and lithium.

REVIEW QUESTIONS

1. How can the nutritional difficulties that individuals who have diabetes are subject to be explained on a metabolic basis? Phenylketonuria? Mental retardation? Other conditions?
2. A sailor marooned on an uncharted island finds a storehouse of vitamins and carbohydrates as his only food supplies and an adequate water supply. Can he survive on such a diet? Suppose his food supply consisted of vitamins, proteins, and water. Could he survive on such a diet? Or, vitamins, lipids, and water? Could he survive? Explain your answers on a metabolic basis.
3. What is the relationship of the state of thyroxine metabolism to the basal metabolic rate? To protein-bound iodine (PBI)?
4. List some trace elements and inorganic constituents in the diet and indicate their uses in metabolic processes.
5. Which of the major food classes provides the highest caloric value per unit weight? What are the recommended percentages of the major food classes in a normal diet? How may large variations in these percentages affect the metabolism of the body?

REFERENCES

Advances in food research, New York, Academic Press, Inc.

Burton, B. T.: Human nutrition, ed. 3, New York, 1976, McGraw-Hill Book Co.

Feldman, E. B., editor: Nutrition and cardiovascular disease, New York, 1976, Appleton-Century-Crofts.

Kallen, D. J., editor: Nutrition, development and social behavior, Washington, D.C., 1973, Superintendent of Documents, U.S. Government Printing Office.

Merritt, D. H., editor: Infant nutrition, New York, 1976, Academic Press, Inc.

Moss, N. H., and Mayer, J., editors: Food and

nutrition in health and disease, Ann. N.Y. Acad. Sci. **300:**1, 1977.

Olson, R. E., editor: Protein-calorie malnutrition, New York, 1975, Academic Press, Inc.

Prasad, A. S.: Trace elements and iron in human metabolism, New York, 1978, Plenum Press.

Schneider, H. A., Anderson, C. E., and Coursin, D. B., editors: Nutritional support of medical practice, New York, 1977, Harper & Row, Publishers.

Wurtman, R. J., and Wurtman, J. J., editors: Nutrition and the brain, 2 vols., New York, 1977, Raven Press.

Appendix

UNITS OF MEASUREMENT AND COMMON EQUIVALENTS

Length:

meter (m) = 1,650,763.73 wavelengths of the orange-red radiation from krypton-86
kilometer (km) = 1000 meters
centimeter (cm) = 0.01 meter
millimeter (mm) = 0.001 meter = 0.1 centimeter
1 inch = 2.54 centimeters
1 meter = 39.5 inches = 1.1 yards

Weight:

kilogram (kg) = weight of a standard platinum-iridium cylinder
kilogram = 1000 grams
gram (g) = 0.001 kilogram
milligram (mg) = 0.001 gram = 0.000001 kilogram
1 kilogram = 2.2 pounds
1 ounce = 30 grams
1 gram = 15 grains

Volume:

liter (l) = volume of a cubic decimeter
liter = 1000 cubic centimeters
cubic centimeter (cc) = 0.001 liter
milliliter (ml) = 0.001 liter
1 cc = 1 ml
1 quart = 950 cc
1 fluid ounce = 30 cc
1 teaspoon = 4-5 cc
1 tablespoon = 15 cc = ½ fluid ounce

Heat:

calorie (cal) = heat required to raise the temperature of 1 g of water from 15° to 16° C
1 Calorie (Cal) = 1 kilocalorie (kcal) = 1000 calories (cal)

SELECTED NORMAL CLINICAL VALUES

Blood:	
Erythrocytes	
Men	4.5-6 million per cu mm
Women	4.3-5.5 million per cu mm
Leukocytes	5000-10,000 per cu mm
Platelets	200,000-500,000 per cu mm
Hematocrit	
Men	40%-54%
Women	37%-47%
Hemoglobin	
Men	14-18 g per 100 ml
Women	12-16 g per 100 ml
Blood chemistry:	
Albumin (serum)	4.0-5.2 g per 100 ml
Globulins (serum)	1.3-2.7 g per 100 ml
Bilirubin (total, serum)	0.1-0.8 mg per 100 ml
Cholesterol (total, serum)	120-260 mg per 100 ml
Glucose (fasting, blood)	70-110 mg per 100 ml
pH (serum)	7.35-7.45
Urea nitrogen (serum)	10-25 mg per 100 ml
Uric acid (serum)	3.0-6.0 mg per 100 ml
Urine:	
Specific gravity	1.010-1.030
pH	6 (4.7-8.0)
Volume	1200 (600-2500) ml per 24 hr
Creatinine	1-1.6 g per 24 hr
17-Ketosteroids	
Men	8-20 mg per 24 hr
Women	6-15 mg per 24 hr
Total solids	55-70 g per 24 hr

Enzymes of diagnostic significance

Organ or tissue of origin or analysis	Enzyme assays*	Conditions for which assayed
Liver and biliary tract	GPT, GOT, AP, GLDH	Hepatitis, jaundice, cholestasis, cirrhosis, metastasis
Heart	CPK, GOT, LDH, GPT	Myocardial infarction, inflammatory heart disease, heart failure
Pancreas	α-Amylase	Pancreatitis
Tumors	LDH, acid phosphatase, GOT, GPT, GLDH	Tumors, prostatic carcinoma, liver metastases
Blood	LDH, transaminases	Pernicious anemia, jaundice
Muscle	CPK, ALD	Muscular dystrophy, dermatomyositis

*ALD, aldolase; *AP,* alkaline phosphatase; *CPK,* creatine phosphokinase; *GLDH,* glutamate dehydrogenase; *GOT,* glutamate-oxaloacetate transaminase; *GPT,* glutamate-pyruvate transaminase; *LDH,* lactate dehydrogenase.

Table of relative atomic weights 1975* (based on the atomic mass of the isotope ^{12}C = 12)

Name	Symbol	Atomic number	Atomic weight
Actinium	Ac	89	227.0278
Aluminum	Al	13	26.98154
Americium	Am	95	(243)
Antimony	Sb	51	121.75
Argon	Ar	18	39.948
Arsenic	As	33	74.9216
Astatine	At	85	(210)
Barium	Ba	56	137.33
Berkelium	Bk	97	(247)
Beryllium	Be	4	9.01218
Bismuth	Bi	83	208.9804
Boron	B	5	10.81
Bromine	Br	35	79.904
Cadmium	Cd	48	112.41
Calcium	Ca	20	40.08
Californium	Cf	98	(251)
Carbon	C	6	12.011
Cerium	Ce	58	140.12
Cesium	Cs	55	132.9054
Chlorine	Cl	17	35.453
Chromium	Cr	24	51.996

Continued.

*As adapted from a compilation by the International Union of Pure and Applied Chemistry. A value in parentheses is the atomic mass number of the isotope of that element of longest known half-life. For further details see Pure and Applied Chemistry **47:**75, 1976.

Table of relative atomic weights 1975 (based on the atomic mass of the isotope ^{12}C = 12)—cont'd

Name	Symbol	Atomic number	Atomic weight
Cobalt	Co	27	58.9332
Copper	Cu	29	63.546
Curium	Cm	96	(247)
Dysprosium	Dy	66	162.50
Einsteinium	Es	99	(254)
Erbium	Er	68	167.26
Europium	Eu	63	151.96
Fermium	Fm	100	(257)
Fluorine	F	9	18.998403
Francium	Fr	87	(223)
Gadolinium	Gd	64	157.25
Gallium	Ga	31	69.72
Germanium	Ge	32	72.59
Gold	Au	79	196.9665
Hafnium	Hf	72	178.49
Helium	He	2	4.00260
Holmium	Ho	67	164.9304
Hydrogen	H	1	1.0079
Indium	In	49	114.82
Iodine	I	53	126.9045
Iridium	Ir	77	192.22
Iron	Fe	26	55.847
Krypton	Kr	36	83.80
Lanthanum	La	57	138.9055
Lawrencium	Lw	103	(260)
Lead	Pb	82	207.2
Lithium	Li	3	6.941
Lutetium	Lu	71	174.97
Magnesium	Mg	12	24.305
Manganese	Mn	25	54.9380
Mendelevium	Md	101	(258)
Mercury	Hg	80	200.59
Molybdenum	Mo	42	95.94
Neodymium	Nd	60	144.24
Neon	Ne	10	20.179
Neptunium	Np	93	237.0482
Nickel	Ni	28	58.70
Niobium	Nb	41	92.9064
Nitrogen	N	7	14.0067
Nobelium	No	102	(259)
Osmium	Os	76	190.2

Table of relative atomic weights 1975 (based on the atomic mass of the isotope ^{12}C = 12)—cont'd

Name	Symbol	Atomic number	Atomic weight
Oxygen	O	8	15.9994
Palladium	Pd	46	106.4
Phosphorus	P	15	30.97376
Platinum	Pt	78	195.09
Plutonium	Pu	94	(244)
Polonium	Po	84	(209)
Potassium	K	19	39.0983
Praseodymium	Pr	59	140.9077
Promethium	Pm	61	(145)
Protactinium	Pa	91	231.0359
Radium	Ra	88	226.0254
Radon	Rn	86	(222)
Rhenium	Re	75	186.207
Rhodium	Rh	45	102.9055
Rubidium	Rb	37	85.4678
Ruthenium	Ru	44	101.07
Samarium	Sm	62	150.4
Scandium	Sc	21	44.9559
Selenium	Se	34	78.96
Silicon	Si	14	28.0855
Silver	Ag	47	107.868
Sodium	Na	11	22.98977
Strontium	Sr	38	87.62
Sulfur	S	16	32.06
Tantalum	Ta	73	180.9479
Technetium	Tc	43	(97)
Tellurium	Te	52	127.60
Terbium	Tb	65	158.9254
Thallium	Tl	81	204.37
Thorium	Th	90	232.0381
Thulium	Tm	69	168.9342
Tin	Sn	50	118.69
Titanium	Ti	22	47.90
Tungsten	W	74	183.85
Uranium	U	92	238.029
Vanadium	V	23	50.9414
Xenon	Xe	54	131.30
Ytterbium	Yb	70	173.04
Yttrium	Y	39	88.9059
Zinc	Zn	30	65.38
Zirconium	Zr	40	91.22

Periodic table

KEY: 1 ·· Atomic number; H ·· Symbol of element; 1.008 ·· Atomic weight

Period	IA	IIA	IIIB	IVB	VB	VIB	VIIB	VIIIB			IB	IIB	IIIA	IVA	VA	VIA	VIIA	O Inert gases
1	1 H 1.008																1 H 1.008	2 He 4.003
2	3 Li 6.941	4 Be 9.012											5 B 10.81	6 C 12.011	7 N 14.007	8 O 15.999	9 F 18.998	10 Ne 20.179
3	11 Na 22.990	12 Mg 24.305	Transition elements										13 Al 26.982	14 Si 28.086	15 P 30.974	16 S 32.06	17 Cl 35.453	18 Ar 39.948
4	19 K 39.098	20 Ca 40.08	21 Sc 44.956	22 Ti 47.90	23 V 50.941	24 Cr 51.996	25 Mn 54.938	26 Fe 55.847	27 Co 58.933	28 Ni 58.70	29 Cu 63.546	30 Zn 65.38	31 Ga 69.72	32 Ge 72.59	33 As 74.922	34 Se 78.96	35 Br 79.904	36 Kr 83.80
5	37 Rb 85.468	38 Sr 87.62	39 Y 88.906	40 Zr 91.22	41 Nb 92.906	42 Mo 95.94	43 Tc 97	44 Ru 101.07	45 Rh 102.906	46 Pd 106.4	47 Ag 107.868	48 Cd 112.41	49 In 114.82	50 Sn 118.69	51 Sb 121.75	52 Te 127.60	53 I 126.905	54 Xe 131.30
6	55 Cs 132.905	56 Ba 137.33	57-71 La-Lu Rare Earths	72 Hf 178.49	73 Ta 180.948	74 W 183.85	75 Re 186.207	76 Os 190.2	77 Ir 192.22	78 Pt 195.09	79 Au 196.967	80 Hg 200.59	81 Tl 204.37	82 Pb 207.2	83 Bi 208.980	84 Po 209	85 At 210	86 Rn 222
7	87 Fr 223	88 Ra 226.025	89-103 Ac-Lw Actinides															
Rare earths			57 La 138.906	58 Ce 140.12	59 Pr 140.908	60 Nd 144.24	61 Pm 145	62 Sm 150.4	63 Eu 151.96	64 Gd 157.25	65 Tb 158.925	66 Dy 162.50	67 Ho 164.930	68 Er 167.26	69 Tm 168.934	70 Yb 173.04	71 Lu 174.97	
Actinides			89 Ac 227.028	90 Th 232.038	91 Pa 231.036	92 U 238.029	93 Np 237.048	94 Pu 244	95 Am 243	96 Cm 247	97 Bk 247	98 Cf 251	99 Es 254	100 Fm 257	101 Md 258	102 No 259	103 Lw 260	

SUPPLEMENTARY READINGS

Science, weekly journal published by the American Association for the Advancement of Science.
The Sciences, published monthly by the New York Academy of Sciences.
Scientific American, published monthly.

Index

t indicates table; n indicates footnote

C

E

H

M

T

U